Dokumentation
zum Forschungs- und Entwicklungsprogramm
des Bundesministeriums für Forschung und Technologie
(BMFT)

TRIBOLOGIE

Reibung · Verschleiß · Schmierung

Band 1

Abrasivverschleiß · Schwingungsverschleiß
Oberflächenbehandlung · Bearbeitungsverfahren

Herausgeber:
Projektträgerschaft „Metallurgie, Werkstoffentwicklung, Rückgewinnung“ des BMFT bei der DFVLR, Köln
Wissenschaftliche Leitung und Redaktion:
W. Bunk, J. Hansen, M. Geyer

Springer-Verlag Berlin Heidelberg New York 1981

Professor Dr. rer. nat. Wolfgang Bunk
Direktor des Instituts für Werkstoff-Forschung
der Deutschen Forschungs- und Versuchsanstalt für Luft- und Raumfahrt e.V., Köln-Porz
Lehrstuhl für angewandte Metallkunde, RWTH Aachen

Dr. rer. nat. Jörn Hansen
Wissenschaftlicher Mitarbeiter am Institut für Werkstoff-Forschung
der Deutschen Forschungs- und Versuchsanstalt für Luft- und Raumfahrt e.V., Köln-Porz

Dipl.-Ing. Manfred Geyer
Wissenschaftlicher Mitarbeiter am Institut für Werkstoff-Forschung
der Deutschen Forschungs- und Versuchsanstalt für Luft- und Raumfahrt e.V., Köln-Porz

Mit 359 Abbildungen und 66 Tabellen

CIP-Kurztitelaufnahme der Deutschen Bibliothek
Tribologie: Reibung, Verschleiß, Schmierung; Dokumentation zu dem Forschungs- und Entwicklungsprogramm des Bundesministeriums für Forschung und Technologie (BMFT) / Hrsg.: Projektträgerschaft „Metallurgie, Werkstoffentwicklung, Rückgewinnung" des BMFT bei der DFVLR, Köln. Wissenschaftliche Leitung und Redaktion: W. Bunk ... – Berlin; Heidelberg; New York: Springer
NE: Bunk, Wolfgang (Red.); Deutsche Forschungs- und Versuchsanstalt für Luft- und Raumfahrt ⟨Köln⟩ / Projektträgerschaft Metallurgie, Werkstoffentwicklung, Rückgewinnung
Band 1: Abrasivverschleiß, Schwingungsverschleiß, Oberflächenbehandlung, Bearbeitungsverfahren. – 1981.

ISBN 978-3-540-10800-9 ISBN 978-3-642-52220-8 (eBook)
DOI 10.1007/978-3-642-52220-8

Buch- und Offsetdruckerei Richard Thierbach, Bachstraße 24, 4330 Mülheim (Ruhr)
2362/3020 – 5 4 3 2 1 0

V O R W O R T

Reibung und Verschleiß verursachen an Maschinen und technischen Anlagen in der Industrie erhebliche Verluste durch Minderung der Lebensdauer und des Wirkungsgrades. Weitgehend unbekannt ist der erschreckend hohe Betrag, zu dem sich hierdurch die Kosten für Energieaufwand, Instandhaltung, für Ersatzteile und Ausfallzeiten in unserer deutschen Volkswirtschaft aufaddieren. Verschärft wird die Situation dadurch, daß die internationale Konkurrenz der deutschen Industrie Maschinenanlagen mit größeren Leistungsdichten, erhöhten Bauteilauslastungen und verbesserten Wirkungsgraden abverlangt. Mit dieser Entwicklung wachsen aber auch die durch Reibung und Verschleiß bedingten Schwierigkeiten erheblich an. Diese Erkenntnisse und die Einsparmöglichkeiten an Rohstoffen und Energie erfordern große Anstrengungen, um mit den Problemen der Reibung und des Verschleißes fertig zu werden. Hier kann das systematische Zusammenwirken mehrerer Wissensgebiete in dem interdisziplinären Fach "Tribologie" einen beachtlichen Beitrag liefern.

Nach Offenlegung der Situation in der Bundesrepublik Deutschland in dem BMFT-Forschungsbericht "Tribologie" entschloß sich das Bundesministerium für Forschung und Technologie (BMFT) im Rahmen seines Rohstofforschungs-Programms zu einer schwerpunktmäßigen Förderung der Tribologie mit einer Laufzeit von zunächst 6 Jahren und einem Fördervolumen in Höhe von ca. 37 Mio DM im Rahmen der Projektträgerschaft des Instituts für Werkstoff-Forschung der Deutschen Forschungs- und Versuchsanstalt für Luft- und Raumfahrt e. V. (DFVLR), Köln. Die Förderung begann 1978, heute werden ca. 100 Vorhaben gefördert.

Ein Bericht der Deutschen Gesellschaft für Mineralölwissenschaft und Kohlechemie e.V. (DGMK) von 1977 gibt an, daß rund 50 % der Verluste in Folge Reibung und Verschleiß durch intensive Nutzung der bekannten tribologischen Erkenntnisse einzusparen sind. Im Kampf gegen Reibungs- und Verschleißverluste müssen der Forschung und Entwicklung in der Tribologie die Kommunikation, Information und Innovation konsequent folgen. So ist neben der Forschungs- und Entwicklungsarbeit die Wissensverbreitung eine notwendige Ergänzung des Förderprogramms Tribologie. Diesem Ziel soll die mit dem vorliegenden ersten Band begonnene Buchreihe dienen, in der über die im Tribologie-Förderprogramm erarbeiteten Ergebnisse berichtet wird.

Köln, im Juni 1981

W. Bunk, J. Hansen, M. Geyer

P R O J E K T T R Ä G E R S C H A F T

Das Forschungs- und Entwicklungsprogramm "Tribologie" ist Teil der Projektträgerschaft "Metallurgie-Werkstoffentwicklung-Rückgewinnung". Diese wird vom Institut für Werkstoff-Forschung (Prof. Bunk) der DFVLR im Auftrage des BMFT durchgeführt. Die administrative Betreuung liegt beim Bereich für Projektträgerschaften der DFVLR.

Zur technisch-wissenschaftlichen Beratung des Projektträgers für den Förderbereich "Tribologie" dient ein Gutachterkreis. Dieser wird gebildet von den Herren

Oberingenieur H. Böer
Dr.-Ing. E. Gülker
Dr.-Ing. G. Heinke
Ing.(grad.) W.H. Kara
Dr.-Ing. O. Lang
Prof. Dr.-Ing. H. Peeken
Prof. Dr. E. Pollmann
Dr.rer.nat. E. Roemer
Dr.-Ing. K.-H. Treptow ab August 1980
Prof. Dr.-Ing. H. Uetz
(Dipl.-Ing. P. Greis bis März 1980)

INHALTSVERZEICHNIS

Begründung und Durchführung der BMFT-Förderung "Tribologie"

von

Dr. Jörn Hansen

Institut für Werkstoff-Forschung
der Deutschen Forschungs- und
Versuchsanstalt für Luft- und
Raumfahrt (DFVLR), Köln

Zusammenfassung

Verschiedene Studien zeigten auf, wie hoch die volkswirtschaftlichen Verluste durch Reibung und Verschleiß an Maschinen und technischen Anlagen sind und welchen Beitrag zur Senkung dieser Verluste die Tribologie zu leisten imstande ist. Hierdurch wurde das Bundesministerium für Forschung und Technologie (BMFT) bewogen, das Gebiet der Tribologie durch ein Forschungs- und Entwicklungsprogramm zu fördern. Im Jahre 1977 wurde eine Förderung über ca. 6 Jahre mit einem Fördervolumen von ca. 37 Mio DM konzipiert.

Die Projektträgerschaft wurde dem Institut für Werkstoff-Forschung der DFVLR übertragen. 1978 begann die Förderung, heute werden ca. 100 F+E-Vorhaben gefördert. Über die Ergebnisse der ersten 67 Vorhaben wurde in einer ersten Präsentation berichtet, eine zweite Präsentation wird voraussichtlich 1984 folgen. Die Schlußberichte zu fast allen Vorhaben werden in dieser Buchreihe veröffentlicht.

Reibungs- und Verschleißverluste, Tribologie

Durch Reibung und Verschleiß von Maschinen und technischen Anlagen werden in der Industrie erhebliche volkswirtschaftliche Verluste verursacht: Neben der Lebensdauerverringerung und Verminderung des Nutzungsgrads bewirken sie beachtliche Instandhaltungs- und Ersatzteilkosten sowie Verluste durch verschleißbedingte Ausfallzeiten. Die Situation verschärft sich durch den Trend der Industrie - aus Rationalisierungsgründen und aus Gründen gegenseitiger Konkurrenz - zu immer leistungsfähigeren Maschinen und technischen Anlagen. Mit der Steigerung der Baugröße, der Belastung und der Arbeitsgeschwindigkeiten der Einheiten nehmen auch die Probleme mit Reibung und Verschleiß stark zu.

Die Grenzen, solche Reibungs- und Verschleißprobleme mit einfachen Mitteln zu lösen, sind rasch erreicht und es wird eine komplexere Behandlung im Sinne der Tribologie nötig: Durch systematisches Zusammenwirken mehrerer Wissensgebiete wie z.B. Kinematik, Hydrodynamik, Thermodynamik, Werkstofftechnik, Verfahrenstechnik, Schmierstoffchemie u.a. in dem interdisziplinären Fachgebiet Tribologie lassen sich Reibungs- und Verschleißverluste teilweise beträchtlich mindern. Hierin liegt die große Bedeutung dieses Fachgebietes für die Industrie.

Wie erschreckend hoch die Verluste durch Reibung und Verschleiß in ganzen Volkswirtschaften sind und welche Einsparmöglichkeiten man durch die Tribologie sieht, zeigen Studien aus Großbritannien, den USA und aus der Bundesrepublik: In dem 1966 in Großbritannien erschienenen sogenannten Jost-Report [1)] wird angegeben, daß durch gezielte Maßnahmen gegen Reibungs- und Verschleißverluste in Großbritannien eine Summe in Höhe von etwa 2 % des Bruttosozialproduktes eingespart werden könnten. Ähnlich hohe Verluste sowie Einsparmöglichkeiten - speziell auf dem Energie-Sektor - hat eine amerikanische Studie der ASME [2)] im Jahre 1977 abgeschätzt: In den Vereinigten Staaten sollen sich 11 % des nationalen Energieverbrauches durch tribologische Maßnahmen einsparen lassen. Hierfür sollte ein Forschungs- und Entwicklungsaufwand in Höhe von 70 Mio DM ausreichen. Eine neuere Studie von H.Peter Jost und J. Schofield [3)] berichtet über ähnliche Untersuchungen in Großbritannien.

Auf Anregung der Gesellschaft für Tribologie hat das BMFT zur Vorklärung der Situation in der Bundesrepublik Deutschland eine Studie "Tribologie" [4)] erstellen lassen. Sie wurde 1975 bis 1976 unter Federführung des Forschungskuratoriums Maschinenbau e.V. erstellt. Darin werden die Verluste durch Reibung und Verschleiß in unserem Lande mit mehr als 10 Mrd DM pro Jahr (für 1975) angegeben, das sind etwa 1 % des Bruttosozialproduktes. Einen noch größeren Betrag gibt ein Bericht der Deutschen Gesellschaft für Mineralölwissenschaft und Kohlechemie (DGMK) [5)] von 1977 an mit 12,5 Mrd DM pro Jahr (1974), der anhand der Zahlen aus dem Jost-Report abgeschätzt wurde.

Programm der Förderung

Nachdem durch die BMFT-Studie die Situation in der Bundesrepublik Deutschland bekannt gemacht wurde, entschloß sich das BMFT zu einer schwerpunktmäßigen Förderung der Tribologie im Rohstofforschungs-Programm [6)] mit einer Laufzeit von zunächst 6 Jahren und einem Fördervolumen in Höhe von ca. 37 Mio DM. Mit Berücksichtigung der Industrie-Beteiligung konnten und können damit Forschungs- und Entwicklungsarbeiten auf dem Gebiet der Tribologie für ca. 60 Mio DM durchgeführt werden. Die Projektträgerschaft wurde der DFVLR,

Institut für Werkstoff-Forschung, übertragen und hier in die Projektträgerschaft "Metallurgie-Werkstoffentwicklung-Rückgewinnung" eingegliedert.

Zur fachlichen Beratung des Projektträgers wurde ein Gutachterkreis gebildet. Hierfür stellten sich die Autoren der zehn Unterabschnitte der BMFT-Tribologie-Studie zur Verfügung. Erste Aufgabe des Projektträgers war es, zusammen mit den Gutachtern ein Rahmenprogramm für die Förderung zu erstellen. Dieses Programm wurde 1977 in den BMFT-Mitteilungen veröffentlicht [7)]: "Es sollen praxisbezogene Forschungs- und Entwicklungsvorhaben gefördert werden, die bei Maschinen und technischen Anlagen eine Lebensdauerverlängerung, Einsparung von Rohstoffen und Energie, Produktionsausfallminderung, Verbesserung des Umweltschutze sowie Schwingungs- und Geräuschminderung zum Ziele haben."

Neben den allgemeinen Bewilligungskriterien des BMFT werden folgende Kriterien für die Beurteilung von Anträgen zur Tribologie-Förderung aufgestellt: interdisziplinärer tribologischer Charakter, exemplarische Bedeutung, volkswirtschaftliche Wichtigkeit sowie Industrie-Relevanz der geplanten Arbeiten.

Für die Abwicklung der Förderung wird eine Abstimmung mit den anderen laufenden Aktivitäten für wichtig gehalten. Entsprechend wurden gute Kontakte mit verschiedenen Gesellschaften und Verbänden hergestellt und gepflegt. Von diesen erfolgte eine Reihe wertvoller Impulse für die Projektträgerschaft.

Wegen der besonders in der Tribologie notwendigen und noch oft fehlenden Kommunikation und Information wurden schon zu Anfang die Durchführung von Präsentationen und die Veröffentlichung der erarbeiteten Ergebnisse in Form von Berichten geplant, um nicht nur Wissen erarbeiten zu lassen, sondern dieses auch einem möglichst weiten Kreis von Interessenten zur Verfügung zu stellen.

Durchführung der Förderung

Anfang 1978 begann die Förderung, bis heute wurden und werden 100 Vorhaben gefördert und nach dem derzeitigen Stand werden am Ende der Förderung Ergebnisse von ca. 140 Vorhaben vorliegen [8)]. Von den bisher geförderten Vorhaben wurden bzw. werden 41 Vorhaben in der Industrie (hiervon 33 in der Groß- und 8 in der Mittel- und

Kleinindustrie), 43 an Hochschulen und 16 an anderen Forschungsinstituten bearbeitet. Wegen des im Rahmenprogramm geforderten Praxisbezuges sind alle Vorhaben der Hochschulen und anderer Forschungsinstitute eng an die Industrie gekoppelt, teils durch direkte Zusammenarbeit mit Industrie-Firmen, teils durch Mitarbeit in industriellen Arbeitskreisen.

Um eine möglichst effektive Arbeit zu erreichen, wurden thematisch ähnliche Vorhaben zu 13 Koordinationskreisen zusammengefaßt. Jeder Koordinationskreis tagt etwa zweimal im Jahr. Bei diesen Sitzungen wird über Ziel und Stand der Vorhaben berichtet und diskutiert:

		Anzahl der Vorhaben
1.	Abrasivverschleiß	7
2.	Schwingungsverschleiß	3
3.	Gleitlager-Ausführungsformen	10
4.	E H D	8
5.	Mischreibung	6
6.	Meß- und Prüfverfahren	9
7.	Gleitlager-Werkstoffe	7
8.	Oberflächenbehandlung	12
9.	Schmierstoffe für Verbrennungsmotoren	4
10.	Bearbeitungsverfahren	11
11.	Konstruktive Gestaltung	8
12.	Betriebsverhalten von Reibungssystemen	8
13.	Dieselmotoren (Lebensdauerverlängerung)	2
	ohne Zuordnung	5

Veröffentlichung der Ergebnisse

Die Tribologie ist eine noch sehr junge Wissenschaft, deren Bedeutung in weiten Kreisen noch nicht erkannt ist. Dies zeigt deutlich die 1980 veröffentlichte und 1981 ergänzte Repräsentativ-Umfrage der Fa. Omnikote [9)10)], die in der Industrie durchgeführt wurde. Sie gibt an, daß 64 % der Befragten die volkswirtschaftliche Bedeutung der Tribologie nicht kennen und daß 83 % meinen, daß der Tribologie nicht der ihr gebührende Rang eingeräumt sei. Das Ergebnis signalisiert, daß noch eine erhebliche Arbeit auf den Gebieten Motivation, Information, Kommunikation sowie Aus- und

Weiterbildung zu leisten ist. Besonders muß die große Zahl an Konstrukteuren, Herstellern und Betreibern von Maschinen und technischen Anlagen angesprochen werden, die eben nicht tribologisch geschult sind und die immer wieder vor tribologischen Problemen stehen, deren Lösung ein tribologisches Systemdenken mit guten Kenntnissen in verschiedenen Disziplinen wie Werkstoff- und Schmierstoffkunde, Meßtechnik, Konstruktionslehre und anderen verlangt.

Auch in der BMFT-Förderung muß dieser Situation Rechnung getragen werden. Der Forschung und Entwicklung muß konsequent die Information als notwendige Ergänzung des Förderprogramms folgen. Diesem Ziel sollen insbesondere zwei Aktivitäten dienen:

- Präsentationen, bei denen die erarbeiteten Ergebnisse der Öffentlichkeit vorgestellt und diskutiert werden. Die erste Präsentation fand vom 12. bis 14. Mai 1981 in Koblenz statt, bei der über die Ergebnisse von 67 Vorhaben berichtet wurde.
  Eine zweite Präsentation wird voraussichtlich 1984 folgen, die im Umfang mit der ersten Veranstaltung vergleichbar sein wird.
- Die Berichtsreihe "Tribologie", in der die Schlußberichte fast aller Vorhaben veröffentlicht werden.
  Der vorliegende erste Band enthält Berichte zu 13 Vorhaben. Die nächsten beiden Bände sind für Anfang 1982 geplant, weitere werden nach Anfall der Berichte herausgegeben.

Zur besseren Übersichtlichkeit der Arbeiten entwickelte die Bundesanstalt für Materialprüfung Berlin eine Tribologie-Deskriptorenliste, die an alle Projektleiter zum Ausfüllen verteilt wurde. Die Deskriptoren zu den in diesem Band beschriebenen Vorhaben sind am Ende dieses Buches aufgeführt. Eine Auswertung aller Deskriptorenlisten hat Professor Czichos vorgenommen. Die Ergebnisse beschreibt er in dem übernächsten Beitrag.

Literatur

1) H.Peter Jost, "Lubrication (Tribology) - A Report of the Present Position and Industry's Needs", Department of Education and Science, HM Stationary Office (1966), London

2) ASME, "Strategy for Energy Conservation through Tribology", ASME Publications Department (1978), N.Y.

3) H.Peter Jost, J. Schofield "Energy Saving through Tribology: A techno-economic Study", James Clayton Lecture, Proc. Instn. Mech. Engrs. Vol 195 (1981)

4) BMFT, "TRIBOLOGIE Reibung - Verschleiß - Schmierung", BMFT-Forschungsbericht T 76-38 (1976), Bonn

5) DGMK, "TRIBOLOGIE Eine Gliederung und Erfassung des Fachgebietes", FB 218 (1977), Hamburg

6) BMFT, Rahmenprogramm Rohstofforschung 1976-1979 (1976), Bonn

7) BMFT, BMFT-Mitteilungen 11 1977 (1977), Bonn

8) W. Bunk, J. Hansen, M. Geyer, "Tribologie-Übersicht", DFVLR (1981), Köln

9) Fa. Omnikote GmbH, OKS Report '80 (1980), München

10) Fa. Omnikote GmbH, OKS Euro Report '80 (1981), München

THE ECONOMIC IMPORTANCE OF TRIBOLOGY IN THE CONSERVATION OF ENERGY

by

Dr. H. Peter Jost, C.B.E.

given at
The First Presentation of the
BMFT FÖRDERUNG "TRIBOLOGIE"
on 12th May, 1981
at Koblenz, Federal German Republic

INTRODUCTION

The danger of a shortage of energy, at least in the form it is traditionally used, is now recognized by all industrial nations. Their Governments have therefore embarked on strategies to effect savings of wasted* (or rejected) energy.

However, whilst much attention has been paid by Governments to the saving of wasted heat in space heating through insufficient insulation and to thermal cycles of machinery, insufficient thought seems to have been given to the direct and indirect loss of energy occasioned by wear and by friction, and to the savings of materials. Then in 1977, an American Government financed report suggested that $ 16 1/4 billion per annum (at 1976 values) could be saved by a "Strategy for Energy Conservation through Tribology". Updated to 1981 American values and taking into account the increased world price of oil and American inflation, this figure of saving would now exceed $ 40 billion** a year.

This magnitude of energy savings through tribology alone seemed to be very large. Therefore, in 1980, Dr. Schofield and myself, started our own investigation, the results of which were presented in the 1981 James Clayton Lecture to the Institution of Mechanical Engineers; some of the findings are included in this paper.

---

*"Wasted" or "rejected" is defined as that portion of input energy which does not appear as useful output in the form of work, electricity, thermal processing and the like. This non-utilization of energy results both from constraints imposed by Carnot cycle efficiencies, and from losses such as heat transfer, friction, wear and leakage.

** a U.S. billion = 10^9

ENERGY SAVINGS THROUGH TRIBOLOGY

As is now well-known, the concept of tribology - and indeed the name "tribology" - was established in 1966 and was first acclaimed on 9th March 1966 when it appeared in a British Ministry of Education & Science Report, which has since become universally known as the "Jost" Report. Its main findings were that the subject of interacting surfaces in relative motion, i. e. in the main friction and wear, being multi-disciplinary, and embracing engineering, physics, chemistry, metallurgy, etc., had previously not been recognized as such and as a result great economic losses were suffered; also that a large proportion of these losses were of an avoidable nature. The report estimated that by the application of mainly known principles the science and technology of tribology in Great Britain could save in the region of £ 515 million per annum (at 1965 values). Of this, £ 28 million were by a reduction in energy consumption due to lower friction (Fig. 1). Largely as the result of the "Jost" Report, tribology was recognized worldwide as a source of practical economic savings, and professional activities in tribology have spread throughout the world. In the main promoted in the German Federal Republic by the Gesellschaft für Tribologie, they have directly led to this presentation of Government sponsored work in tribology.

In the original "Jost" Report, it was estimated that a reduction of energy of some 25 % of the wasted energy, approximately 8 % of the total generated energy, was possible through tribology. In those days coal, fuel oil and electricity were relatively cheap.

Today, largely as the result of the oil crisis, energy prices have risen and the £ 28 (approx. DM 132*) million saving in the U.K. in 1965 would now be £ 175 (approx. DM 825) million per annum. Such savings would be only primary and direct energy savings, e.g. savings through a reduction of friction. As we shall shortly see, this is probably the smallest section of energy savings obtainable through the application of tribology.

*The mid-exchange rate of 4.72 DM per £ (1.5.81) is used throughout this paper.

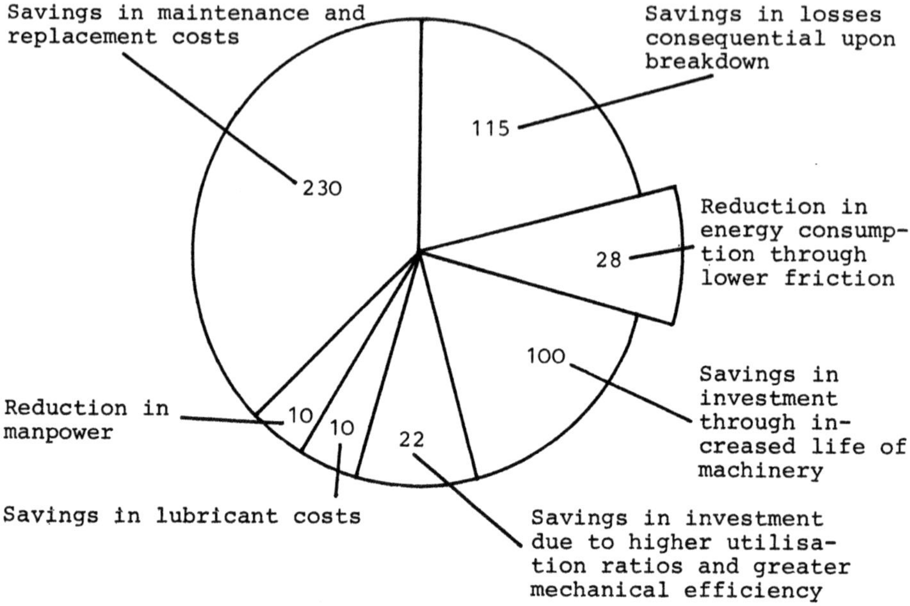

Fig. 1 The figures in the chart represent (in £M) the breakdown of the total saving of £515 M p.a. (1965 values) (2)

COMPONENTS OF ENERGY SAVING THROUGH TRIBOLOGY

However, before we can establish where and how energy can be saved through tribology, a structure for the investigation has to be established and the components of this structure were identified to be

(a) Direct and Indirect Savings,

(b) Savings in R. & D. Related Groups (Fig. 2)

(c) Sectors of Application

DIRECT SAVINGS

Direct savings come in three modes:

Mode 1 - Primary Savings of energy through tribology are mainly

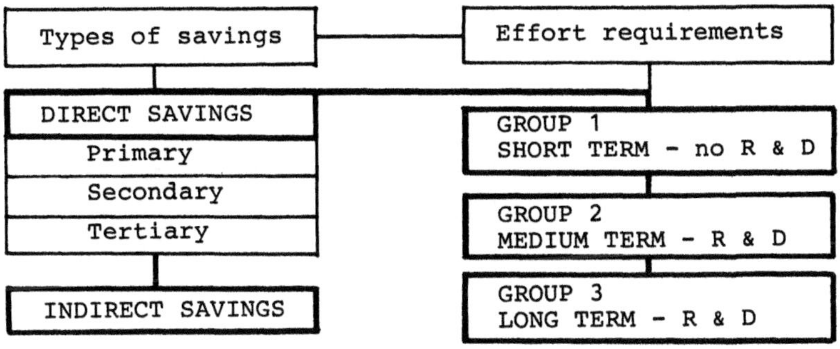

Fig. 2 Structure of investigation of assessment of energy savings through tribology

those resulting from a reduction in friction of whatever type, e.g. fluid, mixed, solid, etc.

Mode 2 - Secondary Savings of energy through tribology are, in the main, those which obviate the need to replace, e.g. to manufacture the worn parts. This group of savings is of particular importance, since in many cases the purpose of interacting surfaces in relative motion is to maximise friction, as in the case of railway and automotive tyres, brakes, clutches, belt drives, shoe soles and others. The problem of minimising wear when high friction is required, is often more difficult and complex than that pertaining when the tribological aim is to minimise friction, as in bearings.

The energy required for manufacturing parts to replace those worn, and which can be saved if wear is avoided, can be substantial.

Mode 3 - Tertiary Savings of energy through tribology are mainly those obtained by not having to make the materials from which the worn replacement parts are made.

The extent of such savings depends largely on the energy content of the materials involved. Some typical examples are shown in Fig. 3.

	APPROX. ENERGY CONTENT OF MATERIALS GJ/TONNE	APPROX. COST OF ENGINEERING MATERIALS Price per Tonne (Jan. 1980) £	DM
Aluminium	300	910 - 1,110	3.200
Plastics	100	400 - 4600	1.500 - 4.400
Copper	100rising to 500	1,024 - 1360	3.950
Zinc	70	470 - 791	1.375
Mild Steel	50	200 - 220	850 - 950
Glass	20	680	650 - 800
Cement	8	24	70 - 110
Brick	4		
Timber	2	196 - 340	400 DM/m^3
Gravel	01		9,5 - 12
		(October 1980)	
Oil	44	100 - 115	314
Coal	29	37	205
(Energy costs UK £2 US$ 4.4) per GJ in 1980			

Fig. 3 Approximate Energy Content and cost of some Engineering Materials

In total terms, the order of magnitude can be significant; for example, in Europe every 1/4 million vehicles that are taken off the roads involve about 250 thousand tonnes of scrap, ferrous and non-ferrous material at about 100 GJ of energy per tonne, that is to say 25 million GJ of energy lost. Since before machining and forming, etc., more metal is required, than that contained in the finished, i. e. scrapped product, the energy content in these materials probably lies between 37 1/2 to 50 GJ.

In other words, primary and secondary losses through tribological causes are only a proportion, and not always the major proportion of energy losses. Thus any savings in primary losses, e.g. by a reduction of metal to metal friction, can often lead to larger energy savings of a secondary and, even more so, of a tertiary nature.

The distinct identities of these three modes of savings, and that of indirect savings, are considered to be of importance for a realistic assessment of potential savings obtainable through them. Regrettably, there is an almost complete absence in the world of statistics on secondary and tertiary savings that can be made through tribology and also of equivalent indirect energy savings obtainable through tribology. The distinct existence of these three modes was not overlooked in the United Kingdom study, nor should it be in future tribological studies relating to energy. However, when other (non-energy) savings through tribology are investigated, it may often be expedient and practicable to combine secondary and tertiary savings.

INDIRECT SAVINGS

Indirect savings through tribology are those which are consequential upon direct savings of whatever mode. They include the savings of capital investment, not only in money, but also in energy terms. For instance, if by better tribological bearing design, the rejected friction energy of 500/650 MW turbo-alternator sets, could be reduced by 2 MW per set, this would result in indirect investment savings of £ 1/2 (DM 2.4) million per MW in the case of fossil stations, or £ 3/4 (DM 3.5) million per MW in the case of nuclear stations. Since the making of generating equipment requires substantial amounts of energy, such indirect savings in monetary terms are invariably accompanied by substantial savings in energy.

GROUPING OF ENERGY SAVINGS RELATED TO R. & D. IN TRIBOLOGY

Group 1

Savings capable of being made by the use of existing knowledge and technology, i. e. involving motivation, education and training, and communication, but requiring no Research and Development. The majority of the savings would be expected to show economic visibility within three to five years and can therefore be described as SHORT TERM.

Group 2

Savings requiring R. & D. effort - in addition to education and training and communication - for periods generally not exceeding four years, the majority of savings would be expected to show economic visibility in periods around five to seven years and can be described as MEDIUM TERM.

Group 3

Savings requiring longer term and sustained R. & D. effort - the outcome of which would, to a degree, be speculative. Savings could take up to twenty years to show economic visibility, but could be large, can be described as LONG TERM. In the current paper this group of long term savings is omitted from consideration. However, strategic and, to a certain degree, speculative R. & D. efforts are of importance and should not be overlooked in overall considerations of the subject.

AREAS OF ENERGY SAVING ASPECTS OF TRIBOLOGY

The sources of energy in most industrial countries are coal, petroleum, natural gas and traditionally electricity, generated by nuclear and hydro means. The relationship between these sources varies from country to country and even within a particular country over a period of years. This changing pattern in both sources and uses of energy affects the rate at which savings can be made. For the purpose of this paper, savings, both actual and potential, are quoted in million tonnes of coal equivalent, irrespective of the energy source. Where expressed in £ sterling, this is converted at the rate of £ 2 per GJ. (Note: This is known to be low. In the Federal Republic values of 7.5 - 18 DM per MJ apply for coal and electricity, excluding light oil, gas and hydrogen or £ 1.6 to £ 3.8 million per MJ.) A typical source and usage diagram of energy for the United Kingdom is shown in Fig. 3, to which has been added the main customers of energy in 1980. Like the United Kingdom economy, those of many other countries embrace a wide spectrum of industries. However, in most industrial countries three major sectors seem to dominate the energy users field. They are (i) Transport, (ii) Industrial and (iii) Domestic. In the case of the United Kingdom the energy usage in these three sections amounted to 87 % in 1980 and in the Federal German Republic 82,4%(Figs. 4 and 5). It would therefore be expedient to look at these three main headings from the point of view of potential energy savings.

Transport	23 %	subdivided into	Road	18 %
			Air	3 %
			Rail and Water	2 %
Domestic	27 %			
Industrial	37 %	subdivided into	Iron and Steel	8 %
			Other Industries	29 %
Public Administration	6 %			
Agriculture	1 %			
Miscellaneous	6 %			

Fig. 5 Principal Energy Users in the United Kingdom (1980)

TRANSPORT SECTOR

The first of these is the Transport Sector. It is a principal direct consumer of energy in liquid form and, in industrial countries, could be as high as 15 % to 20 % (17.96 % in the F.G.R.) of the total energy consumption of a country. Therefore, if only a few percent of this use could be saved through tribology, some very vast sums could be saved.

Looking at the areas of potential savings through tribology in the Transport Sector, it becomes clear that there are four to which attention should be drawn. Firstly, the power unit, i. e. the engine, secondly, the transmission, thirdly, the oil, in particular the viscometrics of oils, and fourthly, interaction between tribology and fuel efficiency.

Fig. 4 United Kingdom energy flows in 1977 (therms × 10⁹) (**15**) and final users in 1980 (percentages) (**16**)

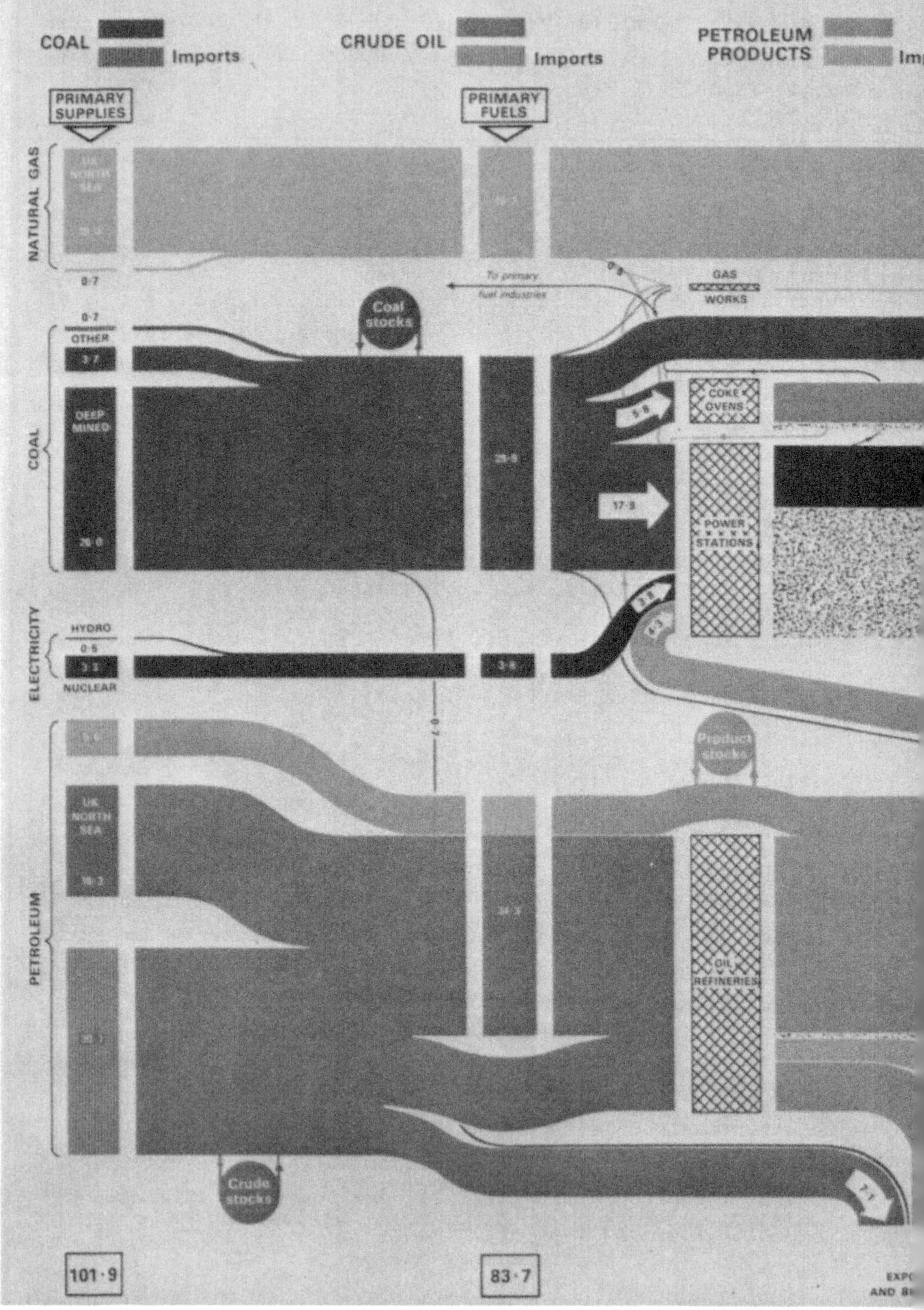

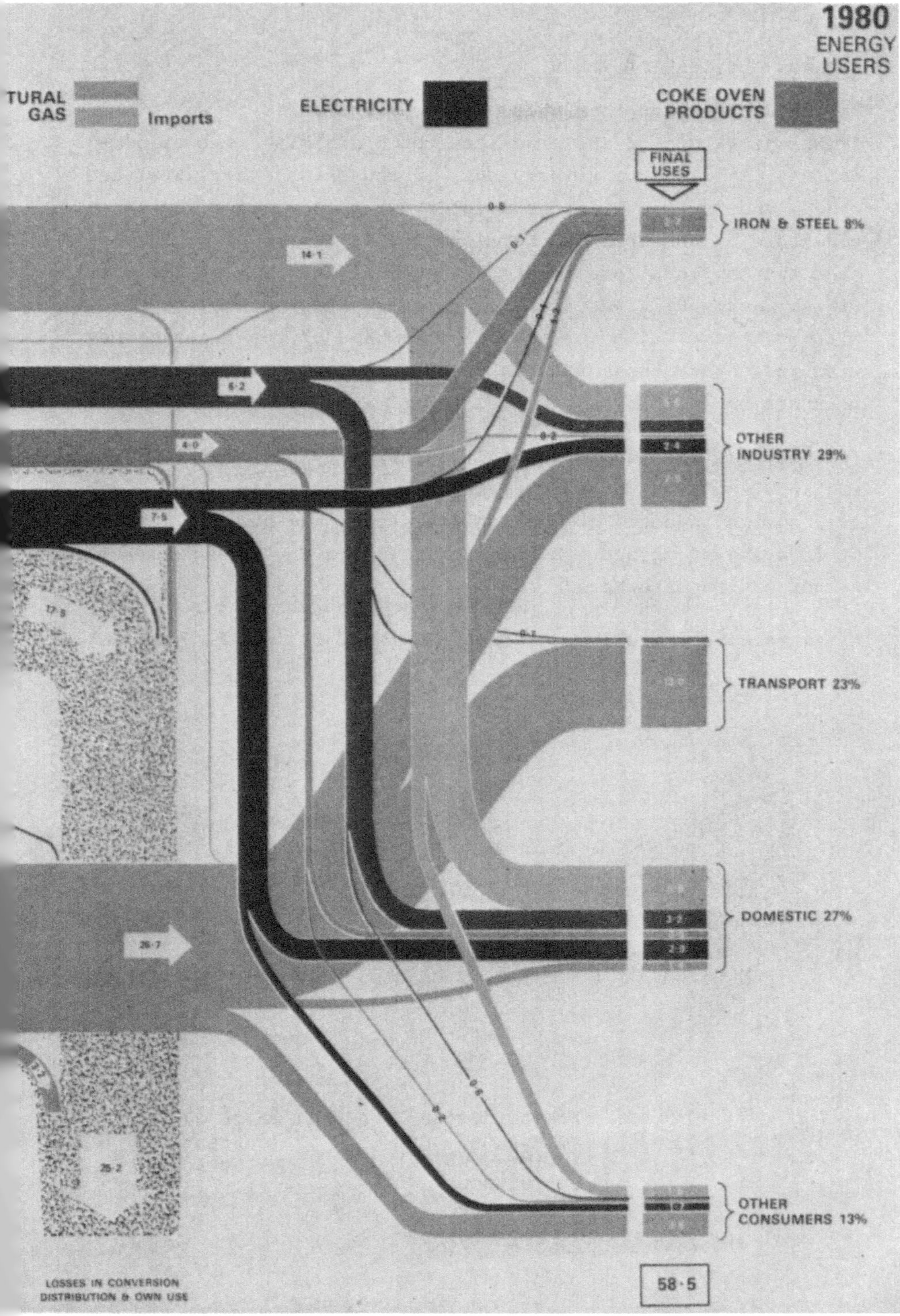

1980
ENERGY
USERS
TURAL
GAS
Imports
ELECTRICITY
COKE OVEN
PRODUCTS
FINAL
USES
IRON & STEEL 8%
OTHER
INDUSTRY 29%
TRANSPORT 23%
DOMESTIC 27%
OTHER
CONSUMERS 13%
LOSSES IN CONVERSION
DISTRIBUTION & OWN USE
58·5

Group 1

Savings obtainable without R. & D.

Our study revealed that there are many instances in which correct tribological design and practices could minimise wear of components, and lead to primary but, in the main, substantial secondary and tertiary savings of energy - this without any expenditure in Research and Development. It is interesting to note that two American researchers estimated that 2.6 % of the total U.S.A. Energy Bill was spent on repairs, maintenance and replacement parts of motor vehicles. Members of the audience may wish to convert these figures into equivalent ones appertaining to their own countries.

Tribological losses are not only incurred by not applying existing knowledge and technology, but not infrequently such losses are "design induced" or "design assisted", whether for reasons of lowering of production costs or "eliminating" the need for relubrication or others.

An example is the universal coupling (Fig. 6), which, even in

Original design

Second design retaining boss (painted white) but without grease nipple

Third design (current) without boss

Fig. 6 Universal coupling cruciform

involving energy, materials, machining and forming of components and maintenance and material usage, could lead to savings with an estimated energy content in the U.K. of 2.2 million to 4.2 million tonnes of coal equivalent per annum.

Group 2

Savings obtainable from medium term R. & D. Projects

From the available evidence these fall into three major headings.

- Piston Rings and Cylinders
- Transmissions
- Lubricants and Viscometrics

In the American study, improved thermal cycles feature predominantly, but since they are not considered primarily tribological problems, they are only referred to briefly.

Piston Rings and Cylinders

The results of many investigations show that friction and leakage account for 20 % to 50 % of all mechanical engine losses in motor vehicles. To reduce these significant losses could involve R. & D. work on piston rings, materials, coatings, arrangements, design of rings, rates of wear, analytical models of lubrication regimes.

Piston rings/cylinder wall losses, for example, represent 12 % of brake horsepower delivered by a typical automotive engine; this can be due to friction, blow-by hysteresis and other relatively well understood phenomena at the piston ring/cylinder wall interface, such as ring geometry, their number, tension, methods of mounting, etc. After discussion with leading experts in the field, including thermodynamicists, a reduction of losses from 12 % to 9 %, equivalent in the U.K. to savings of £ 33 million per annum, is regarded as being applicable.

In the area of transmissions, one of the limitations of the traditional piston engine is that, in order to meet load requirements, it must run at speeds incompatible with optimum engine efficiency. A transmission whose output would range from

its simplest form, takes considerable energy to produce and also contains high energy materials. Unlike its lubricated predecessor, which lasted for the life of a car, the current version of the coupling with its one charge of special lubricant, which however cannot be replenished, has a limited life, say, of two to three years. Because of this unnecessary replacement requirement, it is a waster of energy and material.

From an energy point of view, modern motor vehicles display many similar examples, including those of wear that takes place because of reduction in materials hardness specifications, etc. It is a fact that top rocker shafts, clutches, brakes, do not often require replacing for reasons other than wear.

Motor vehicles are made from many materials, including steel, iron, copper, aluminium, rubber, plastic, glass and contain alloying elements such as nickel, chromium, cobalt, manganese, tin, etc. Most of these materials must be mined and converted from the raw ore to the primary product; thereafter all have to be finally finished and fabricated. They are moved to the car factories where they are assembled. Each step takes energy.

After a comparatively short life, in many cases a thousand hours of running, the car needs replacement items, such as clutch, brakes, tyres, camshafts, etc. All these take energy. The energy content of most manufactured items is generally between 5 % and 20 % of product cost. The total product cost for the motor vehicle industry in the U.K. in 1980 was about £11.8 million, taking the two percentages this would give an energy cost between £ 0.59 to £ 2.3 million.

Engine and transmission life is seldom below one thousand hours and not often above ten thousand hours, say between 80,000 and 800,000 kilometres, or a life span of between 5 and 17 years. Yet existing tribological technology can increase minimum life to between 2,000 and 4,000 hours without significantly affecting cost.

Therefore in this sector, measures to deal with such avoidable, possibly even design- induced wear and tribological failures,

idling or zero to maximum speed, at the same time permitting the engine to run at a speed approaching optimum, would considerably raise engine, i. e. fuel efficiency.

Such system is known as CVT (continuously variable transmission). It has been investigated for some years. It would make an engine run always near its maximum efficiency and permit the use of brake energy storage systems. The American investigation paid particular attention to this problem and claimed that the development and widespread adaptation of a CVT system would yield 25 % to 30 % improvement in automotive fuel economy. The additional gains could be obtained by coupling the CVT with a flywheel regenerative braking system, in which case 50 % to 80 % gain in fuel economy could be obtained.

The British study showed that such magnitude of savings was not practically possible, at least not without considerable long term R. & D. efforts in tribology and - even more so - in other areas of engineering. Nevertheless the study revealed that considerable savings could be made by tribological improvements in the area of transmissions and that several promising projects are being worked on, including multi-stepped gearboxes and lock-up clutches, a modified CVT and other innovations. There is reasonable probability that such developments will yield the desired results over a period of 7 years, and that these could account for a 10 % to 15 % horsepower gain.

Lubricants and Viscometrics

Reduction in fuel consumption is obtainable by the lowering of internal friction of lubricating and traction fluids, by using low viscosity lubricants for both engine and transmission, by the development of low friction and antiwear additive, by identification of those specific areas within the engine where frictional and viscous energy losses are occurring.

Oils used in new types of transmission systems would have to cater for the required elastohydrodynamic demand and dynamics of the fluid films. Existing fluids would have to be modified to meet the extreme pressures and temperatures of traction contact, so as to be usable over long periods without causing erosion or wear of essential parts and without chemical degradation.

In the United Kingdom, the estimate of savings accruing under this heading is in the order of 2 % to 2 1/2 % of fuel used for automotive purposes. American estimates of savings are much higher but are not believed to be reasonably obtainable in practice. This view is shared by Professor Bartz who suggests that practical savings obtainable through improvement in the sector of motor and transmission lubricants and viscometrics could be between 1 % and 5 %, depending on the size of car and its usage.

Improved thermal cycles

Whilst these were the main savings of the original American report, they do not primarily fall under the heading of tribology. Improved thermal cycles have been the aim of thermodynamicists for many years. Considering that the efficiency of the petrol engine is 20 % to 22 %, and that of the diesel engine 33 % to 37 %, there appears to be scope for improvement, expecially with modern development in chemical-thermal treatments, oxides, ceramics, carbides and intermetallic coatings, advanced lubricants, etc.

In the traditional engine about one third of the fuel energy is lost through the transfer of heat into the cooling water via the cylinder walls. Improved thermal cycles would raise thermodynamic efficiency. In the ultimate, if the cooling system were dispensed with and the cylinders insulated, then the energy produced could be removed by turbo compounding, and the efficiency of the piston engine significantly improved. Cylinder wall temperatures of 540 ^{o}C would create major problems, admittedly of a tribological nature, and these would have to be resolved. However, in spite of the reported running of adiabatic engines for a period of 100 hours in the U.S.A., we consider that the development of such engines is largely outside the field of tribology and not likely to be accomplished during the next twenty years.

Summing up, the totality of the savings under the heading of

Transportation could be in the region of 8.6 % of total automotive fuel consumption, nearly all of which in the F.G.R. originates from imported oil or, in the case of the United Kingdom, between £ 332 million and £ 500 million per annum (1980). The equivalent savings appertaining to the Federal Republic could be DM 3630 millions p.a. (1980). In practical terms, these savings could make the hundred miles per gallon or 2,7 litre per hundred km small family car a realistic probability and, together with some engine thermal improvements, an absolute certainty.

THE INDUSTRIAL SECTOR

In the United Kingdom, the manufacturing industries, including the metallurgical industries, use 37 % of total U.K. energy production, a matter of 131.7 million tonnes of coal equivalent. This figure will vary from country to country, but it is bound to be significant in every industrial country.

The difference between energy savings through tribology in the Transportation Sector and that in the Industrial Sector is that in the former such savings can generally be made in concentrated areas. This is not so in industries where, with the exception of power generating utilities and turbines, applications cover a vast range, the generic nature of tribology being much in evidence.

On the industrial side, savings fall under two headings. Firstly, the savings through improved performance which can be divided into (a) power generation and turbines, (b) manufacturing and process industries, including metal forming, and secondly, savings mainly secondary and tertiary, in industries where substantial maintenance-caused energy losses can be alleviated by the application of tribological principles and practices.

Power Generation and Turbines

The prime mover of electrical energy generation is the steam turbine (Fig. 7), the trend in the new power stations being towards larger turbines of 500 MW and 650 MW capacity and over. In the Federal Republic there are now 41 Power Stations with turbo-alternators of 500 MW and over. Losses through tribological causes of 0.5 MW per bearing have been measured; to generate these could take about 21 MW per turbo-alternator set. Through tribology a saving of 15 % to 20 % of this loss is considered to be possible. If the sets were run for an average of 6,500 hours per annum, the estimate in saving of energy arising from this could be 0.2 % of the total nationally generated energy, or in the F.G.R. approaching 60,000 GWH (1979).

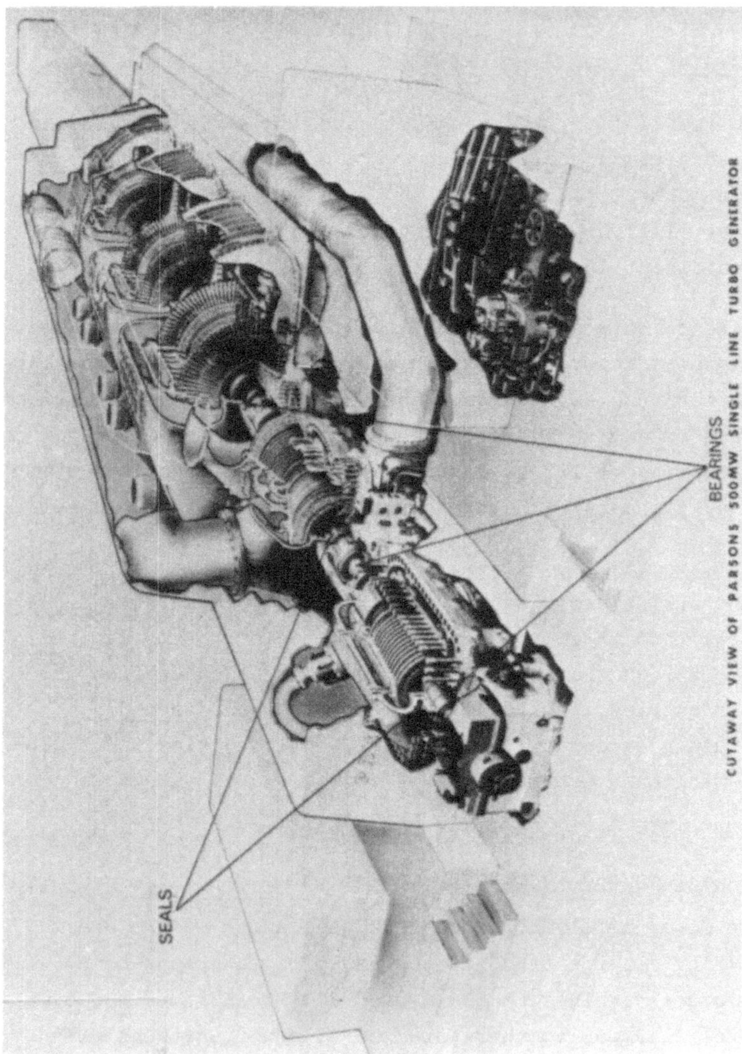

Fig. 7

The reduction of inertia and turbulence effects and of shear losses in the control of the effects of turbulent flow and the mechanism and condition of turbulent bearing operation are suitable subjects for R. & D. efforts. In addition, improved bearing materials would permit designs of shorter bearings.

Seal Losses

These include leakage past the blade tips through the labyrinth sealing and past the turbine stages. Estimates differ from 2 % to 4 %, the age of the plant being a significant factor. On the basis that about half of the energy leakage, i. e. 1 % to 2 %, could be recovered, with which experts agree, the U.K. saving of national energy consumption of 459,000 tonnes of coal equivalent or £ 17 million per annum would result.

Similar savings in gas path losses in turbo-machinery, not being generating plant, could double this figure of savings. Since the major energy loss is the leakage of working fluids through shaft misalignments, due to centrifugal forces, thermal and mechanical distortion, cumulative manufacturing tolerances, etc., it is believed reasonable to develop through R. & D., machines that can obtain such savings.

Rotary Pump Seals

Total losses in 1974, in respect of rotary pumps of units generating 60,000 MW, or approximately 4 1/2 % of total output, were reported in Germany. Of this 5 % to 10 % was due to leakage and friction connected with seals, i. e. approximately 0.2 % to 0.5 % of the total output was lost in this manner. Energy required in respect of these losses was in the order of 1.35×10^6 MWh.

The conservative German BMFT Report suggests that 10 % of this could be saved. The United Kingdom estimate is that 5 % would be a more realistic estimate of the potential savings of avoidable energy losses through this cause.

Savings through improved performance - Manufacturing & Process Industries

Some 45,000 KWh of electrical energy is used in the United Kingdom for driving machines, about one-third of which was estimated to be lost by avoidable friction. There are many examples showing how substantial reduction of energy through lower friction in driving machinery in the manufacturing industry can be obtained by the application of correct tribological principles and practices; for instance, in the case of a 5 Stand Cold Mill, the design change from hydrodynamically lubricated sleeve bearings to oil mist lubricated rolling bearings led to a reduction of absorbed friction energy from 15 % to 5 % and to a total power saving of the Mill of 1127 horse power (840.7 KW) or 1.15 times 10^{10} KJ per annum - this on the basis of a 15 shift week and 8o % plant utilization.

Most of the savings through lower driving friction are Group 1 savings, i. e. savings by the use of existing knowledge and technology and require no R. & D. effort. Therefore a reduction of such tribologically caused losses in the manufacturing industries could lead to enormous savings of 8 % of the energy used to drive machines. New types of bearings, such as magnetic bearings, would increase the energy savings, but would fall under Group 3, Long Term Savings.

Coatings, powder, metallurgy and similar processes

There are considerable gaps of knowledge in the area of mechanism of lubrication by solid coatings, such as polymers, waxes, stearates, phosphates, graphite. Metal cutting removes excess material which - from an energy and materials point of view - should not have been there in the first place.

In some cases 90 % of a forging has to be removed during finish machining. If parts could be formed more closely to the final shape, significant secondary and tertiary energy savings would result. For these reasons, consideration should be given to close forming, casting, and metallurgy, etc., from a tribological point of view - neglected subjects.

Cutting lubricants, tool wear and metal cutting

Tribological factors often limit the utilization of the machine tool. Any reduction in tool wear, whether by improved materials, new coatings, improved lubricant formulations and their supply to the tool point, will lead to energy savings, the sum total of which can be considerable.

Tried and tested mathematical models, developed for machine lubricants, may well find an application in metal forming fluids. Furthermore, there are sophisticated models of rolling and drawing lubricants which could prove useful in high speed, high load conditions to predict film and tractions. In the area of cutting metals, the American investigation concluded that an average of 1.25 % to 2.5 % of the energy used could be saved through tribology. In the United Kingdom we consider this estimate as being very conservative.

Metal forming and processing

This area also involves a large number of operations; it is a good example of the generic nature of tribology and of its complexity, for it embraces hydrodynamics and elastohydrodynamics, phase changes in materials, molecular phenomena, etc.

Processes, such as hot forming, provide areas of considerable potential savings through tribology; for instance, a reduction in frictional loss of energy of up to 35 % has been reported during the use of high viscosity rolling oils (Cold Mills) and a horsepower reduction of 15 % when roll lubrication is used in hot strip mills. In addition, up to 9 % fuel savings for a slab re-heating furnace are reported to be possible through tribologically more efficient rolling.

Cichelli and Poplawaki estimated that in hot and cold rolling mills alone 3.6 million barrels of oil equivalent energy could be saved through tribology in the U.S. steel industry per annum, equivalent to a saving of 828,000 tonnes of coal in respect of a steel production of 97 million tonnes (1978).

The changeover from hot forming to warm forming can further

significantly improve energy savings.

Altogether, there are very many other examples in the areas of metal forming and processing, demonstrating how the application of tribological principles and practices can lead to appreciable savings of energy. However, much work and education is required to achieve these savings.

Secondary and Tertiary energy savings in the Metallurgical and other industries in the area of Maintenance

Metallurgical Industries - Tribological losses through wear in the metallurgical industries are considerable, and are generally taken to account for 45 % to 50 % of all maintenance costs, with high energy and materials content; a figure of 60 % for the iron and steel industry has been reported in the DDR. The total maintenance cost in the United Kingdom in the metallurgical industry, which in 1979 had an output of about 21 million tonnes reported to be in the order of £ 750 millions, about 50 % of which was in direct materials.

Costs through tribological causes - mainly wear - are particularly severe at the "heavy end" of the industry, namely ore preparation, iron and coke making, casting and primary rolling, where considerable savings can be made by the application of existing tribological knowledge.

Other Industries - In the mining industry, tribological causes of wear are also very heavy, decreasing rapidly after the stone, clay, glass, paper and dried products, and chemical industries have been taken into consideration. Indeed the tribological maintenance energy content of most of the lighter industries is very small. For the whole of the remaining industry it is estimated to be 40 % of that of the metallurgical industries (in monetary units).

The result of a thorough investigation showed that in energy alone, about £ 25 (DM 120) million could be saved in the United Kingdom through tribological reduction in maintenance.

THE DOMESTIC SECTOR

In this area the overwhelming primary energy losses (rejection) are incurred in the thermal field, notably through insufficient insulation. This has led to investigations into potential savings through tribology omitting to consider the domestic energy consumption scenario. As far as primary energy savings are concerned, this is largely correct. However, when considering secondary and tertiary energy savings, the domestic sector was found to be one of considerable importance, especially as most of such earnings fall within Group 1, i. e. savings obtainable by the application of known technologies and knowledge.

The design of consumer durables of the short cycle design concept, as practiced on an ever increasing scale, takes no account of the rising costs and shortage of energy and raw materials, or - in general - of the tribological technology that can extend the life of many products.

A typical example of short life consumer components can be found in many popular makes of front loading washing machines and automatic spin dryers, e. g. the programmer and the motor module of the same machine (Fig. 8). Failures are fairly common and generally require the replacement of both of these two items; labour costs are high as replacement is difficult. The tribological cause of the trouble is metal transfer between the contacts. It generally starts at the programmer and, in turn, introduces full mains voltage to the motor module, no fuse being fitted to the programmer. The design is such that it is impossible to dismantle the programmer, in order to replace individual contacts. This means that standard very cheap electronic components, having failed through built-in tribological weaknesses, cannot be changed easily, even if this were desired, which apparently it is not since no wiring diagram is issued.

The trouble-free life of many washing machines and similar machines has dropped since the end of the last war down to 5 years or even less. In a world of rapidly escalating energy costs, the continuation of such built-in tribological failure

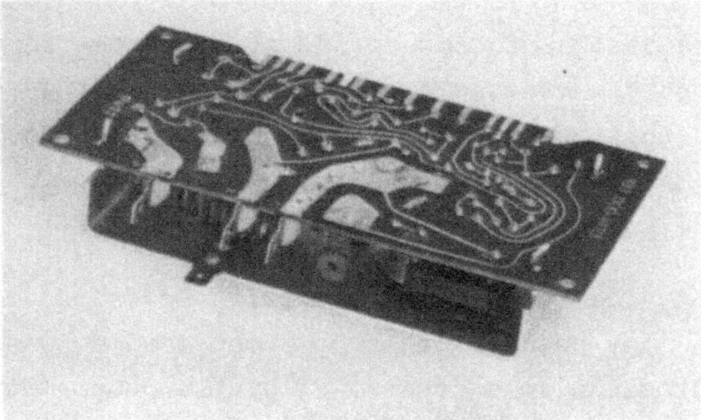

Fig. 8 Electrical Units from an Automatic Washing Machine requiring replacement due to tribological failure

liabilities, together with a practical impossibility of remedying these failures, except by whole unit replacement, is -from an energy point of view - an economic and social malpractice. Such malpractice is not limited to washing machines, of which 1.3 million were sold in the United Kingdom in 1979.

On the basis of a gross energy content of only 10 %, and of improved design - through the application of existing tribological knowledge, preferably helped by a new type of international standard - being able to reduce wear and its consequential effects by only 8 % to 12 %, secondary and tertiary savings in energy, as well as indirect savings of the order of £ 80 (DM 377) millions to £ 120 (DM 560) millions per annum, would be the direct result in the United Kingdom.

SUMMARY OF SAVINGS

In the Summary of our investigation (see Fig. 9) after allowing 20 % for overlap - not all savings are additive - we came to a conservative estimate of energy savings through tribology in the United Kingdom of £ 468 - £ 700 millions per annum. This, incidentally, for a research and development cost of only £ 12.3 millions spread over 5 years.

In coal equivalent terms this amounts to 13 to 19 million tonnes of coal equivalent per annum or, very roughly between 3 1/2 % and 5 % of total energy consumption in the United Kingdom. Equivalent figures may well pertain in the Federal Republic.

SUMMARY OF ESTIMATED SAVINGS

Primary, Secondary and Tertiary Energy Savings through Tribology covering 87 % of U.K. energy users (not including Public Administration 6 %, Agriculture 1 %, and Miscellaneous 6 %).

	£	DM
(a) Transport Sector		
Group 1 Saving	79-140 M p.a.	373 - 661 Mio p.a.*)
Group 2 Saving	253-361 M p.a.	1.194 - 1.704 Mio p.a.
(b) Industrial Sector		
Group 1 & 2 Saving	173-253 M p.a.	817 - 1.194 Mio p.a.
(c) Domestic sector		
Group 1 Saving	80-120 M p.a.	378 - 566 Mio p.a.
	585-874 M p.a.	2.761 - 4.125 Mio p.a.
Less: Overlap factor of 20 % (since not all savings are additive, an overlap factor has to be applied, which the US Report found to be of the order of 20 %)	117-174 M p.a.	552 - 821 Mio p.a.
Total Estimated Savings (after overlap adjustment)	468-700 M p.a.	2.209 - 3.304 Mio p.a.

*) 1£ = 4,73 DM

Fig. 9

It may be of interest to know that the original estimated savings through tribology in America were 10.9 % of U.S. consumption. After the revision deleting long-term and speculative research, potential energy savings were reduced to 5.1 % of total consumption. This however dealt almost wholly with primary savings; secondary, tertiary and indirect savings were covered to an extremely small extent only. Therefore the American estimates of energy savings obtainable through tribology are somewhat higher than the ones established for the United Kingdom.

FINDINGS & CONCLUSIONS

In the foregoing paragraphs an outline has been given of how tribology can affect energy savings, and some examples illustrated the means of obtaining such savings.

(1) The findings of the U.K. investigation has led to the conclusion that the application of tribological principles and practices can affect national energy savings of considerable magnitude, and this is true for all industrial countries. The spectrum of possible and significant savings of energy through tribology extends from the simple prevention and minimization of"design-induced" or "design-assisted" wear, to the application of the results of advanced technology, obtained by sophisticated R. & D. effort.

(2) There is a lack of awareness of the role tribology can play in the area of energy conservation. Contrary to the awareness of thermal factors, neither manufacturers nor users of equipment, nor Government, nor the Professional Bodies have fully identified the part which tribology can play in energy saving; yet significant and considerable savings in energy are obtainable, not only from reduced friction, but - more so - from reduced wear and from the consequences arising from such wear.

(3) The benefits of direct savings of energy through reduced friction (primary savings) are often readily identifiable,

but they are frequently only a small portion of the total potential energy savings that can be made through tribology. Energy savings by not having to make spares (secondary savings), and in particular savings of materials used for such spares (tertiary savings), and other advantages, e.g. more efficient and more compact machines, derived from tribological considerations, will in magnitude often heavily outweigh primary savings.

(4) In the area of the generic technology of tribology, the difficulty in distinguishing between savings of energy and those of materials is pronounced, especially in the case of wear.

(5) It is very important that not only governments but also technical bodies and educational bodies should ensure that tribology is included in all public educational activities dealing with energy in order to produce tribology expertise at craft, technician and technologist levels. Furthermore, government sponsored research and development efforts should not be limited to conventional tribological subjects but should be extended to those dealing with direct and indirect savings of energy.

To conclude:

Tribology has already shown that it can play an important part in the economy of countries, and that it can assist in producing better products for a small outlay and without harming anyone.

Now we can see that, in addition, it can be a significant source for the saving of energy and materials. This is probably the reason for the sharp upsurge in the world-wide interest in tribology during the last few years. However, by definition tribology is a generic technology, i. e. a multi-disciplinary technology which is very thinly spread over a wide variety of industries, so much in fact that no industry alone might receive the benefit which the cost of research and development and of education might involve.

It is therefore up to the Governments to take the measures to ensure that all steps are taken to save the materials and energy, which can in practice be saved by tribology.

As far as the Federal German Republic is concerned, in addition to the programme, the progress of which will be discussed during the next 2 1/2 days, it might not be inappropriate to commission an Institute, a University or other organization to undertake a study of the energy, including the materials, that can reasonably be expected to be saved by the application of the principles and practices of tribology. In such work the structure of investigation, laid down in the 1981 James Clayton Memorial Lecture and outlined in this paper may be found to be of assistance.

Energy and materials, in the form we are using them now, will not last forever. To conserve them through tribology is a task both noble and worthwhile, and one which I commend to all German tribologists.

R E F E R E N C E S

1. 'Strategy for energy conservation through tribology', ASME, New York, Nov. 1977.

2. Jost, H. Peter and Schofield J. 'Energy saving through tribology: a Techno-Economic Study' Proceedings, Institution of Mechanical Engineers, London, U.K., Volume 195, No. 16, 1981.

3. 'Lubrication (tribology) education and research (Jost Report), Department of Education and Science, HMSO, London 1966.

4. 'The introduction of a new technology', Department of Trade and Industry, HMSO 1973.

5. Research Report T76-36 'Tribologie' (Code BMFT-FB-(76-36). Bundesministerium für Forschung und Technologie (Federal Ministry for Research and Technology), West Germany, 1976.

6. Barnes, R.S. 'The Energy involved in producing engineering materials', Proc. Instn. Mech. Engrs. 1976 190 153-162.

7. Ashby, M.J. and Jones, R.H.D. 'Engineering Materials', Pergamon Press. International Series on Materials Science and Technology, Volume 34, 1980.

8. 'Digest of United Kingdom energy statistics', Department of Energy, HMSO 1979.

9. 'Digest of United Kingdom energy statistics', Department of Energy, HMSO 1980.

10. 'Energy statistics year book supplement', OECD Publication IEA, 1977.

11. Hirst and Herendeen, 'Total energy demand for automobiles' Society of Automotive Engineers, Publication N. 730065, U.S.A.

12. Smithson, D.J. 'Energy savings in the U.K. iron and steel industry', Energy Word 1979.

13. Bartz, W.J. 'Tribologie und Schmierungstechnik', VDI Nachrichten No. 5, 1971, West Germany.

14. Jost, H. Peter, 'Economic impact of tribology', Proceedings of the Mechanical Failures Prevention Symposium, 8-10 May 1974, U.S. Department of Commerce, U.S. Bureau of Standards.

15. Holland, J. 'Tribologie, Billion losses through friction and wear', Fachberichte Hüttenpraxis Metallverarbeitung, 1979, 732-733, West Germany (in German).

16. Moore, D.F. 'Tribology and energy savings', Tribology in Industry Conference and Exhibition, University of Leeds, September, 1980.

17. Cichelli, A.E. and Poplawaki, J.V. 'Tribology and energy considerations in the rolling of steel' Bethlehem Steel Corporation, ASLE Annual Meeting 1980.

18. 'Energy consumption in manufacturing', Conference Board, Ballinger Publishing Co., Cambridge, Mass., 1974.

19. 'Evaluation of energy use', The Watt Committee on Energy, November 1977.

20. Rolide, S.M., Wilcock, D.F. and Chang, H.S. 'Energy conservation through fluid film lubrication technology', Frontiers in research and design, ASME, New York, 1979.

21. Wahl, W. 'Verschleissminderung im Hüttenwerk', Schmierungstechnik 1978 9.6, GDR, based on Eisen und Stahl, 1976 96 No.19.

22. Stubbs, P.W.R. 'The development of a Perbury traction transmission for motor car applications', Century 2 International Power Transmission and Gear Conference, San Francisco, August 1980, Trans. ASME, J. Mech. Des. 1980.

23. Morton, P.G. 'Measurements of the dynamic characteristics of a large sleeve bearing', Trans. ASME Paper 70-LUB-14/1971.

24. Bartz, W.J. 'Gedanken zur Kraftstoffeinsparung durch reibungssenkende Motoren- und Getriebeöle', Technical Academy, Esslingen, West Germany, 1980.

25. Pinkus, O. and Wilcocks, D.F. 'The role of tribology in energy conservation', ASLE Paper 78-AM-3A-1. 1978.

26. McGeeham, J.A. 'Literature review of the effects of piston and ring friction and lubricating oil viscosity on fuel economy', SAE Paper 78-0673. 1978.

27. German Working Group, Energy Balances, Düsseldorf, W. Germany

28. V.D.E.W., Association of German Electricity Undertakings, Statistics, 1979.

29. Transport in Figures, 1980, German Federal Ministry for Transport.

30. Facts & figures of the Transport Economics, 1980. Association of the German Automotive Industry.

Entwicklung eines Tribologie-Thesaurus und Auswertung des Tribologie-Förderprogramms mittels Deskriptoren

von

Prof. Dr. H. Czichos

Bundesanstalt für Materialprüfung (BAM)

Berlin-Dahlem

Förderkennzeichen: 01 ZT 227

Laufzeit: 1.4.1978 - 31.3.1982

Zusammenfassung

Das interdisziplinäre Gebiet der Tribologie ist durch eine Vielzahl technischer Problemstellungen, tribologischer Prozesse und Einflußgrößen gekennzeichnet. Dementsprechend sind auch bei der Darstellung, Erschließung und Anwendung der Ergebnisse tribologischer Forschungs- und Entwicklungsarbeiten zahlreiche Parameter zu berücksichtigen. Unter Verwendung der im Vorhaben "Systematik tribologischer Prüfungen" durchgeführten Analyse tribologischer Meß- und Prüfkenngrößen wurde bei Berücksichtigung der Systematik von DIN 50 320 sowie der Klassifikation der BAM-Dokumentation Tribologie ein systematischer Thesaurus für die Tribologie entwickelt. Mit Hilfe der Deskriptoren des Thesaurus wurde eine Auswertung von 95 Vorhaben des bisherigen Tribologie-Förderprogramms vorgenommen, wobei die folgenden hauptsächlichen Aspekte betrachtet wurden: Anwendungsgebiete, Beanspruchungsfälle, Stoffpaarungen und Werkstoffe, Flüssigkeiten und Schmierstoffeigenschaften, Reibungsarten und Schmierungszustände, Verschleißarten und Verschleißmechanismen, Untersuchungsverfahren und Meßgrößen, Interessenhintergrund. Die Auswertung zeigt, daß mit dem Tribologie-Förderprogramm neben volkswirtschaftlich wichtigen Beiträgen zur Minderung reibungs- und verschleißbedingter Material- und Energieverluste auch Impulse zur Sicherheitstechnik, zum Technologie-Transfer und zur technischen Innovation gegeben werden.

Gliederung

1. Einleitung

Das Forschungsprogramm Tribologie basiert bekanntlich auf einer von Fachleuten aus Industrie und Wissenschaft im Jahre 1976 erstellten Studie /1/, die eindrucksvoll die große volkswirtschaftliche Bedeutung der Tribologie anzeigt und eine Anzahl von Forschungs- und Entwicklungsvorhaben von hoher Priorität vorschlägt. Durch deren Verwirklichung sollen vorhandene Wissenslücken geschlossen und die durch Reibung und Verschleiß verursachten volkswirtschaftlichen Verluste sowie die Kosten für Instandhaltung und Produktionsausfälle vermindert werden. Eine weitere wesentliche Aufgabe des Forschungsprogramms besteht darin, "vorhandenes Wissen aufzubereiten und der allgemeinen Anwendung zugänglich zu machen".

Der Aufbereitung und Zugänglichkeit von erzielten Forschungsergebnissen sowie ihrer Umsetzung in die Anwendungspraxis kommt heute ganz allgemein eine ständig wachsende Bedeutung zu /2/. Neben Anstrengungen zur Schließung "technologischer Lücken" werden verstärkt Notwendigkeiten von "Produktinnovationen" /3/ und des dafür erforderlichen "Technologie-Transfer" /4/ diskutiert. Hierbei wurde bereits vor Jahren in einer in den USA publizierten Arbeit nachdrücklich auf den Nutzen einer systematischen Literaturauswertung hingewiesen /5/. So verursachte z.B. allein in einem einzigen speziellen Fall die Nichtbeachtung bereits veröffentlichter Forschungsergebnisse unnötige Kosten von 1,5 Millionen Dollar.

Nachdem im Forschungsprogramm Tribologie etwa 100 Vorhaben in der Förderung sind und die ersten Ergebnisse von Einzelvorhaben vorliegen, wurde vom Projektträger des Förderprogramms angeregt, im Rahmen des in der Bundesanstalt für Materialprüfung (BAM) bearbeiteten Forschungsvorhabens "Systemanalytische Methoden zur Planung, Systematik und Auswertung tribologischer Messungen" (Förderkennzeichen 01 ZT 227) eine Auswertung des gesamten Förderprogramms Tribologie vorzunehmen. Im folgenden wird ein Thesaurus in Form eines "Deskriptorensystems zur Kennzeichnung und Erschließung tribologischer Forschungs- und Entwicklungsarbeiten vorgestellt und über die Ergebnisse der Auswertung des bisherigen Tribologie-Förderprogramms berichtet.

2. Systematik tribologischer Parameter und Entwicklung eines Thesaurus

Auf dem Gebiet der Tribologie, das als interdisziplinäres Wissensgebiet durch eine Vielzahl von Anwendungsbereichen und technisch-wissenschaftlichen Problemstellungen gekennzeichnet ist, muß beim Darstellen und Anwenden von Forschungsergebnissen eine große Anzahl von Einflußgrößen und Parametern berücksichtigt werden /6/. Zur Beschreibung komplexer Sachverhalte und umfassenden Klassifizierung von Einflußgrößen und Parametern werden heute in verschiedenen interdisziplinären Wissensgebieten mit Vorteil die in neuerer Zeit entwickelten systemtechnischen Methoden eingesetzt /7/. Auch in der Tribologie ist in zunehmendem Maße auf die Bedeutung von Systembetrachtungen für die Bearbeitung und Ergebnisdarstellung der komplexen Reibungs-, Schmierungs- und Verschleißfragen hingewiesen worden /8/ bis /17/.

Für das Verschleißgebiet ist die Systemmethodik in DIN 50320 "Verschleiß: Begriffe, Systemanalyse von Verschleißvorgängen, Gliederung des Verschleißgebietes" dargestellt. Ergänzend zu der Systematik von DIN 50320 wurde für den Aufbau des Deskriptorensystems die Gliederung der BAM-Dokumentation Tribologie herangezogen. Im Rahmen dieser Dokumentation wird seit 1966 die gesamte Tribologie-Weltliteratur mit jährlich etwa 6000 bis 8000 Titeln in Form einer fortlaufenden Titelbibliographie von der BAM publiziert. Die Sachgebietseinteilung der BAM-Dokumentation Tribologie mit ihren 10 Hauptabschnitten und mehr als 100 Unterabschnitten hat sich bei der Erfassung, Klassifizierung und Erschließung der Tribologie-Literatur bewährt /18/.

Unter Berücksichtigung aller dieser Gesichtspunkte wurde für die Darstellung und Auswertung des Tribologie-Forschungsprogramms ein Deskriptorensystem erstellt, dessen Gliederung in Bild 1 wiedergegeben ist. Bei einer Aufteilung in 7 Hauptgruppen wurden insgesamt etwa 300 Deskriptoren zusammengestellt, mit denen alle wichtigen Aspekte tribologischer Forschungsarbeiten beschreibbar erscheinen. Das Deskriptorensystem kann somit als ein systematischer "Thesaurus" für das Gebiet der Tribologie angesehen werden. Der erste Entwurf des Deskriptorensystems wurde zunächst dem Gutachterkreis des Tribologie-Forschungsprogramms zur Stellungnahme und ggf. Modifikation oder Ergänzung zugeleitet. Nach Berücksichtigung einiger Änderungsvorschläge wurde dann eine 4-seitige Fragebogenliste erstellt, die die Deskriptoren zusammen mit einer numerischen Notation enthält, womit auch eine Auswertung mittels EDV möglich wird.

3. Auswertung des Tribologie-Förderprogramms mittels Deskriptoren

3.0 Allgemeines und Überblick

Die Fragebogenlisten mit den Deskriptoren wurden Anfang Februar 1981 an die Projektleiter der einzelnen Forschungsvorhaben des Tribologie-Förderprogramms gesandt. Ende März 1981 lagen die Deskriptorennennungen zu insgesamt 95 Vorhaben vor, die für die folgende Auswertung herangezogen wurden und von denen 40 in der Industrie, 40 in Hochschulinstituten und 15 in außeruniversitären Forschungsinstituten bearbeitet werden. Von den in den Fragebogen aufgeführten 302 Deskriptoren wurden 270 von den Forschungs-Sachbearbeitern zur Beantwortung benutzt und weitere 20 ergänzend vorgeschlagen. Insgesamt erfolgten mehr als 3800 Deskriptorennennungen, d.h. im Durchschnitt wurden je Vorhaben etwa 40 unterschiedliche Deskriptoren benannt.

Durch die Zuordnung der Einzelvorhaben zu Deskriptoren ist eine einfache "Erschließung" des Tribologie-Förderprogramms möglich, die darin besteht, daß einem Verzeichnis der Deskriptoren sofort die zu jedem fachlichen Deskriptor bearbeiteten Vorhaben entnommen werden können. Durch die Kombination von Deskriptoren ist außerdem eine spezifische Selektion der Vorhaben möglich. Darüberhinaus wurde für jeden Deskriptor die Häufigkeit seiner Nennung bestimmt. Hieraus kann sowohl auf die

0 Technische Funktion des Tribosystems

1 Anwendungsgebiet
- 11 System, Maschine, Gerät, Bauteil, Produkt
- 12 Fertigungsverfahren, Werkzeuge

2 Beanspruchungsfall
- 21 Kontakt- und Bewegungsart
- 22 Betriebsparameter

3 Systemstruktur
- 31 Stoffliche Eigenschaften und Stoffparameter
  - 311 Stoffpaarung Grund-/Gegenkörper
  - 312 Stoffart von Grundkörper, Gegenkörper, Zwischenstoff, Umgebungsmedium
  - 313 Stoffparameter
- 32 Formparameter
  - 321 Konstruktive Gestaltung
  - 322 Oberflächenrauheit, Welligkeit

4 Tribologische Problemstellung
- 41 Oberflächen und Kontakte
- 42 Reibung und Energieverlust
- 43 Verschleiß und Materialabtrag
- 44 Schmierung und Schmierstoffeigenschaften

5 Untersuchungsverfahren
- 51 Charakter der Untersuchung
- 52 Art des Untersuchungsverfahrens

6 Interessenhintergrund

Bild 1: Aufbau des Deskriptorensystems

Anwendungsgebiet	Deskriptornennungen	
	Anzahl	Anteil [%]
Antriebstechnik (111)	127	28,2
Kraft- u. Arbeitsmaschinen (112)	105	23,3
Grundstoffindustrien (116)	86	19,1
Fertigungsverfahren, Werkzeuge (12)	54	12,0
Fahrzeuge, Verkehr (114)	53	11,8
Feinwerktechnik (113)	18	4,0
Elektrotechnik (115)	7	1,5
Σ	450	100,0

Tribologische Problemstellung	Deskriptornennungen	
	Anzahl	Anteil [%]
Verschleiß u. Materialabtrag (43)	369	34,4
Schmierung u. Schmierstoffeigenschaften (44)	352	32,9
Reibung u. Energieverlust (42)	217	20,3
Oberflächen u. Kontakte (41)	133	12,4
Σ	1071	100,0

Untersuchungsverfahren	Deskriptornennungen	
	Anzahl	Anteil [%]
Praxisnahe Laborprüfung (511.2)	60	46,9
Modellprüfung (511.1)	35	27,3
Feldversuch, prakt. Einsatz (511.3)	23	18,0
Nicht-tribologische Werkstoff- u. Schmierstoffuntersuchung (512)	10	7,8
Σ	128	100,0

Bild 2: Verteilung der Deskriptorennennungen in den Bereichen "Anwendungsgebiet", "Tribologische Problemstellung", Untersuchungsverfahren"

schwerpunktmäßig in dem Tribologie-Forschungsprogramm bearbeiteten fachlichen Aspekte als auch auf "Bearbeitungslücken" geschlossen werden.

Für eine erste allgemeine Übersicht über das Forschungsprogramm Tribologie sollen zunächst die folgenden generellen Fragen betrachtet werden:

- Welche technischen Anwendungsgebiete wurden bisher hauptsächlich bearbeitet?
- Wo lag schwerpunktmäßig die tribologische Fragestellung?
- Welche Untersuchungsverfahren wurden bevorzugt eingesetzt?

Die Antworten zu diesen Fragen können Bild 2 entnommen werden, das eine Aufschlüsselung der Anzahl der Nennung fachlicher Detailaspekte in den mit ihrer Kennziffer-Notation aufgeführten Deskriptorengruppen enthält. Danach wurden bisher am häufigsten tribologische Probleme der Antriebstechnik bearbeitet, wobei Verschleißprobleme im Vordergrund standen, die hauptsächlich mit Methoden der praxisnahen Laborprüfung untersucht wurden. Die Übersichts-Darstellung von Bild 2 läßt bereits erkennen, daß in den 95 zur Auswertung herangezogenen Forschungsvorhaben ein weites Spektrum von Anwendungsgebieten, tribologischen Problemstellungen und eingesetzten Untersuchungsverfahren behandelt wird.

3.1 Anwendungsgebiete

Für die technische Praxis ist die Frage, welche Anwendungsgebiete im Tribologie-Forschungsprogramm bisher schwerpunktmäßig bearbeitet wurden, von besonderem Interesse. In dem Deskriptorensystem sind 7 hauptsächliche Anwendungsgebiete mit insgesamt 62 Unterbereichen aufgeführt. Die Verteilung der im Rahmen des Tribologie-Programms bearbeiteten Anwendungsgebiete ist zusammen mit einem Vergleich der Verteilung publizierter Arbeiten aus den einzelnen Anwendungsgebieten in der tribologischen Weltliteratur im oberen Teil von Bild 3 zusammengestellt. Die Auswertung der tribologischen Weltliteratur zu den Anwendungsgebieten bezieht sich auf den Zeitraum

Anwendungsgebiet	Anteil [%]	
	Tribol.-Vh	Weltliteratur
Antriebstechnik (111)	28,2	23,8
Kraft- u. Arbeitsmaschinen (112)	23,3	16,0
Grundstoffindustrien (116)	19,1	6,5
Fertigungsverfahren, Werkzeuge (12)	12,0	25,7
Fahrzeuge, Verkehr (114)	11,8	25,2
Feinwerktechnik (113)	4,0	1,3
Elektrotechnik (115)	1,5	1,5
Σ	100,0	100,0

Systeme der Antriebstechnik	Anteil [%]	
	Tribol.-Vh	Weltliteratur
Gleitlager (111.11)	34,2	41,1
Zahnradgetriebe (111.21)	19,0	13,5
Wälzlager (111.12)	12,6	23,2
Reibgetriebe (111.22)	10,1	0,4
Kurvengetriebe (111.23)	8,9	0,4
Kupplungen, Bremsen (111.3), (111.4)	6,3	7,3
Dichtungen (111.5)	6,3	12,3
Spitzenlager (111.13)	1,3	0,1
Dämpfungselemente (111.6)	1,3	0,7
Magnetlager (111.14)	–	1,0
Σ	100,0	100,0

Bild 3: Verteilung der bearbeiteten Anwendungsgebiete im Tribologie-Förderprogramm und in der internationalen Tribologie-Literatur

Beanspruchungsfall	
Kontaktform (211)	konform (39,8 % der Vh) kontraform (20,4 %)
Bewegungsart (212)	Gleiten (75,3 %) Wälzen (28,0 %) Strömen (11,8 %) Normal-, Haftkontakt (10,8 %) Stoßen/Prallen (8,6 %)
Bewegungsablauf (221)	kontinuierlich (41,9 %) osz., schwingend (22,6 %) intermitt., rep. (17,2 %)
Normalkraft F_N [N]	$5 \cdot 10^{-2}$ bis $1,5 \cdot 10^{7}$
Flächenpressung p [N/mm²]	10^{-1} bis $3,2 \cdot 10^{3}$
Geschwindigkeit v [m/s]	10^{-4} bis 65
Temperatur T [°C]	–70 bis 1000

Bild 4: Verteilung der Deskriptorennennungen auf dem Bereich "Beanspruchungsfall"

1977/78 und umfaßt etwa 9000 Literaturstellen. Aus einem Vergleich der Verteilung der Arbeiten in dem Tribologie-Förderprogramm und in der tribologischen Weltliteratur werden z.B. die folgenden Unterschiede deutlich:

- Verglichen mit den in der internationalen Literatur ausgewiesenen Aktivitäten wurden bisher im Tribologie-Programm mit höherem Anteil Problemstellungen aus den Bereichen "Antriebstechnik", "Kraft- u. Arbeitsmaschinen", "Grundstoffindustrien" untersucht.
- In der internationalen Literatur sind dagegen im stärkeren Umfang Arbeiten aus den Bereichen "Fertigungsverfahren und "Werkzeuge" sowie "Fahrzeuge, Verkehr" vertreten, wobei in diesen Bereichen der prozentuale Anteil doppelt so hoch wie im Tribologie-Förderprogramm ist.

Neben den Verteilungen in den Hauptanwendungsgebieten sind im unteren Teil von Bild 3 auch die Verteilungen in dem besonders wichtig erscheinenden Bereich der Antriebstechnik zusammengestellt. Auch hier werden bemerkenswerte Unterschiede deutlich:

- Im Tribologie-Förderprogramm dominieren bisher vergleichsweise die Arbeiten aus den Bereichen "Zahnradgetriebe", "Reibgetriebe", "Kurvengetriebe".
- In der internationalen Literatur überwiegen dagegen Veröffentlichungen zu den Themenkreisen "Gleitlager", "Wälzlager" und "Dichtungen".

3.2 Beanspruchungsfälle

Die Ergebnisse der Analyse der Beanspruchungssituationen bei den im Tribologie-Forschungsprogramm bisher geförderten Vorhaben gibt Bild 4. Danach war in 39,8 % aller Vorhaben eine konforme Kontaktgeometrie gegeben, bei 75,3 % aller Vorhaben war die Bewegungsart "Gleiten" und in 41,9 % aller Vorhaben lag ein kontinuierlicher Bewegungsablauf vor.

Zur numerischen Kennzeichnung der Beanspruchungssituationen waren in dem Deskriptoren-Fragebogen zahlenmäßige Angaben zum Beanspruchungskollektiv erbeten worden. Diese Angaben wurden zu 3/4 aller Vorhaben (allerdings z.T. unvollständig) abgegeben. Die Grenzdaten des Beanspruchungskollektivs sind ebenfalls in Bild 4 zusammengestellt. Man erkennt, daß in der Tribologie eine sehr große Variationsbreite der Beanspruchungsdaten vorliegt. Die in Bild 4 zusammengefaßten Beanspruchungsangaben können für die Auslegung von labormäßigen Reibungs- und Verschleißprüfapparaturen von Bedeutung sein.

3.3 Stoffpaarungen, Werkstoffe, Stoff- und Formparameter

Neben den Beanspruchungskenngrößen sind die in tribologischen Systemen verwendeten Werkstoffe von besonderem Interesse.
In Bild 5 wird eine Übersicht über die untersuchten Werkstoffpaarungen gegeben. Bemerkenswert ist die Dominanz metallischer Werkstoffe, die sich besonders darin ausdrückt, daß in 71 % aller Vorhaben die Werkstoffpaarung Metall/Metall vorlag und nur in jeweils 1 bis 2 % der Vorhaben das tribologische Verhalten nichtmetallischer Werkstoffpaarungen behandelt wurde.
Die große Breite der insgesamt untersuchten Werkstoffe geht aus Bild 6 hervor, wobei hervorzuheben ist, daß in einer sehr großen Anzahl von Vorhaben (84,9 %) Stahl- und Eisenwerkstoffe verwendet wurden. Während in der Technik ganz allgemein ein Trend zur Verwendung von Polymerwerkstoffen oder nicht metallischen Verbundwerkstoffen festzustellen ist, sind Arbeiten über das tribologische Verhalten nichtmetallischer Werkstoffe im Tribologie-Forschungsprogramm noch ziemlich unterrepräsentiert. Diese Feststellung wird auch durch einen Vergleich der Anteile der in dem Tribologie-Forschungsprogramm untersuchten Werkstoffe mit denen der in der internationalen Literatur hauptsächlich behandelten Werkstoffgruppen bestätigt, vgl. Bild 7. Dabei ist festzustellen, daß im internationalen Rahmen bei etwa 25 % der einschlägigen tribologischen Forschungsarbeiten Organische Stoffe und Polymere untersucht wurden, während im Tribologie-Programm der Anteil entsprechender Arbeiten bisher nur 12 % beträgt.

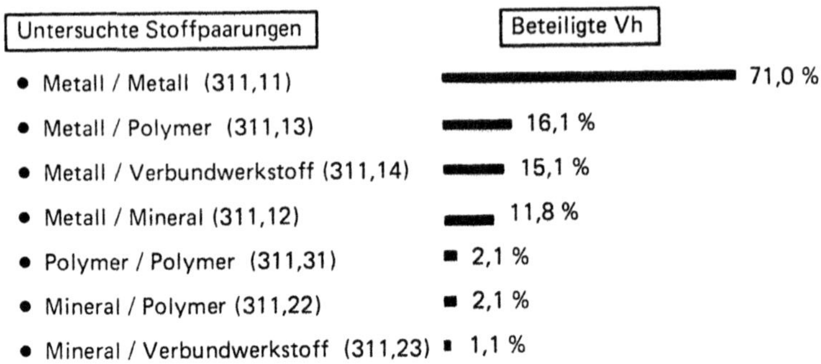

Bild 5: Verteilung der Deskriptorennennungen auf dem Bereich "Stoffpaarungen"

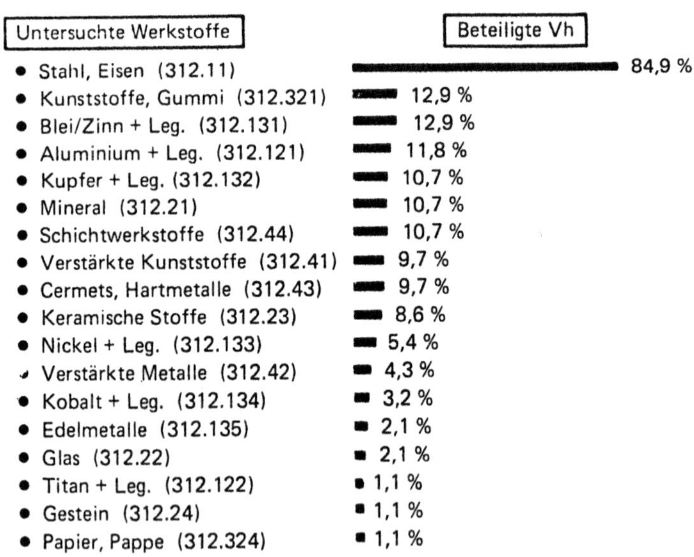

Bild 6: Verteilung der Deskriptorennennungen auf dem Bereich "Werkstoffe"

Werkstoffgruppe	Anteil [%]	
	Tribol.-Vh	Weltliteratur
Metalle u. Halbmetalle	61	47
Verbundwerkstoffe	16	16
Organische Stoffe, Polymere	12	25
Minerale, keramische Stoffe	11	13

Bild 7: Verteilung der untersuchten Werkstoffgruppen im Tribologie-Förderprogramm und in der internationalen Tribologie-Literatur

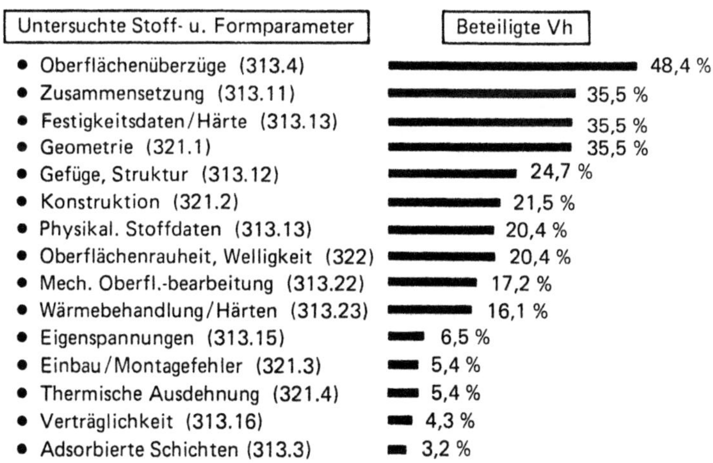

Bild 8: Verteilung der Deskriptorennennungen auf dem Bereich "Werkstoff-Einflußgrößen"

Die im Rahmen des Tribologie-Förderprogramms behandelten Werkstoff-Einflußgrößen sind in Bild 8 zusammengestellt. Neben den für etwa 1/3 aller Vorhaben wichtigen Einflußgrößen "Chemische Zusammensetzung", "Festigkeitsdaten/Härte", "Geometrie" wurden bei etwa der Hälfte aller Vorhaben Oberflächenüberzüge verwendet oder untersucht. Obwohl dieser Anteil relativ hoch erscheint, ist er auch im Vergleich mit der internationalen Entwicklung gerechtfertigt, da Reibungs- und Verschleißprozesse ja primär in Oberflächenbereichen von Werkstoffen ablaufen und eine Verbesserung der tribologischen Eigenschaften in vielen Fällen durch eine gezielte Oberflächenbehandlung zu erreichen ist.

3.4 Flüssigkeiten und Schmierstoffeigenschaften

Bei den in den Vorhaben des Tribologie-Programms verwendeten oder untersuchten flüssigen Stoffen dominieren nach Bild 9 eindeutig die als Schmierstoffe überwiegend eingesetzten Mineralöle, gefolgt von den Additiven und den synthetischen Ölen.

Dagegen sind die weiteren in Bild 9 aufgeführten flüssigen Stoffe in erheblich geringerem Maße eingesetzt oder untersucht worden. Ähnlich wie bei den Werkstoffen sollte auch hier der allgemeine Hinweis gegeben werden, daß eine gewisse Intensivierung der Arbeiten über synthetische Stoffe notwendig erscheint, um den sich abzeichnenden Ressourcen-Engpässen begegnen zu können.

Eine Aufstellung der untersuchten Schmierstoffeigenschaften gibt Bild 10. Die Zusammenstellung bestätigt die bekannte Tatsache, daß für die Tribologie eine Anzahl von Schmierstoffeigenschaften von Interesse sind, wobei nach Bild 10 bei dem Tribologie-Förderprogramm bisher die rheologischen Eigenschaften der Schmierstoffe im Vordergrund stehen.

3.5 Reibungsarten, Schmierungszustände, Verschleißarten und Verschleißmechanismen

Im Bereich der Tribologie nehmen die Begriffe "Reibung", "Schmierung", "Verschleiß" einen zentralen Raum ein. In welchem

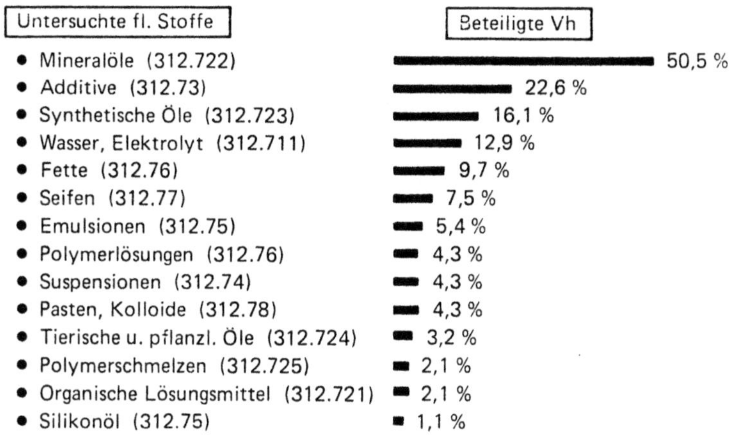

Bild 9: Verteilung der Deskriptorennennungen auf dem Bereich "Flüssige Stoffe"

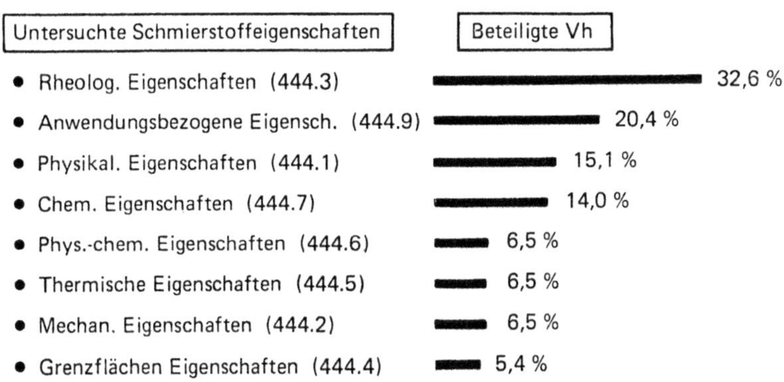

Bild 10: Verteilung der Deskriptorennennungen auf dem Bereich "Schmierstoffeigenschaften"

Umfang die verschiedenen Detailaspekte hierzu im Rahmen des Tribologie-Forschungsprogramms untersucht wurden, geht aus den Bildern 11 und 12 hervor.

Bei der Reibung wurde hauptsächlich die Gleitreibung und in erheblich geringerem Maße die Wälz- oder Rollreibung untersucht. Die betreffenden tribologischen Systeme arbeiteten bei etwa 40 % der Forschungsvorhaben im Bereich der Mischreibung sowie mit Anteilen von 18 bis 26 % im Bereich der Hydrodynamik, der Grenzreibung, der EHD und im ungeschmierten Zustand der Festkörperreibung.

Bei den Verschleißarten, siehe Bild 12, dominierte - in Analogie zur Reibung - bei mehr als 50 % aller Vorhaben der Gleitverschleiß. Mit geringeren Anteilen wurden auch die anderen hauptsächlichen Verschleißarten nach DIN 50 320 untersucht, mit Ausnahme des Strahlverschleißes. Als hauptsächlicher Verschleißmechanismus wurde mit einem Anteil von etwa 42 % aller Vorhaben am häufigsten die Abrasion genannt. Die anderen grundlegenden Verschleißmechanismen - Adhäsion, Oberflächenzerrüttung, tribochemische Reaktionen sowie der Vorgang der Oberflächenverformung - sind mit Anteilen von etwa 19 bis 25 % in vergleichbarem Umfang vertreten. Hieraus kann gefolgert werden, daß häufig eine Überlagerung der verschiedenen Verschleißmechanismen vorliegt.

3.6 Untersuchungsverfahren und Meßgrößen

Wie in allen technischen Disziplinen ist auch in der Tribologie die Meß- und Prüftechnik von erheblicher Bedeutung, da nur durch den sinnvollen Einsatz experimenteller Techniken gesicherte Erkenntnisse über das Verhalten tribologischer Systeme bestimmbar sind. Bereits in Bild 2 ist dargestellt, daß im Rahmen des Tribologie-Förderprogramms am häufigsten praxisnahe Laborprüfungen eingesetzt wurden. Die Art der verwendeten Untersuchungsverfahren geht aus Bild 13 hervor. Es liegt in der Natur der Sache, daß die mechanisch-physikalische Prüfung im Vordergrund steht und daß außerdem optischen Verfahren - wie z.B. bei der Untersuchung von Verschleißoberflächen mit der Licht- oder

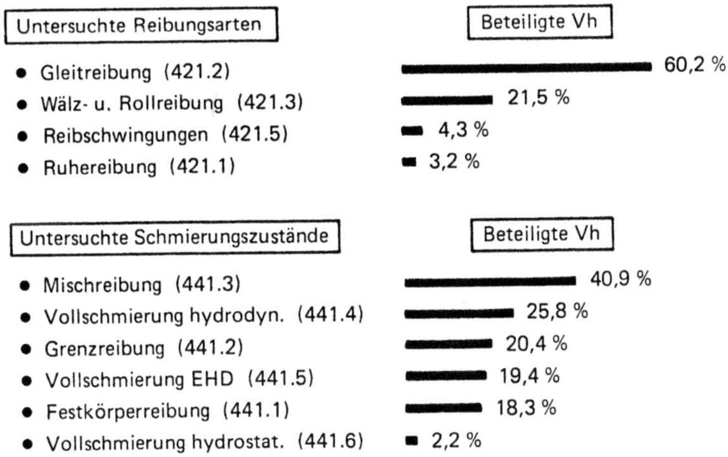

Bild 11: Verteilung der Deskriptorennennungen auf den Bereichen "Reibungsarten" und "Schmierstoffzustände"

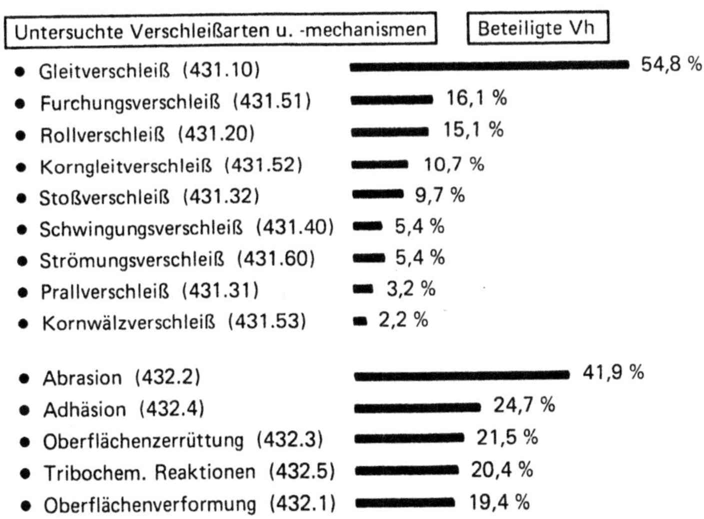

Bild 12: Verteilung der Deskriptorennennungen auf den Bereichen "Verschleißarten" und "Verschleißmechanismen"

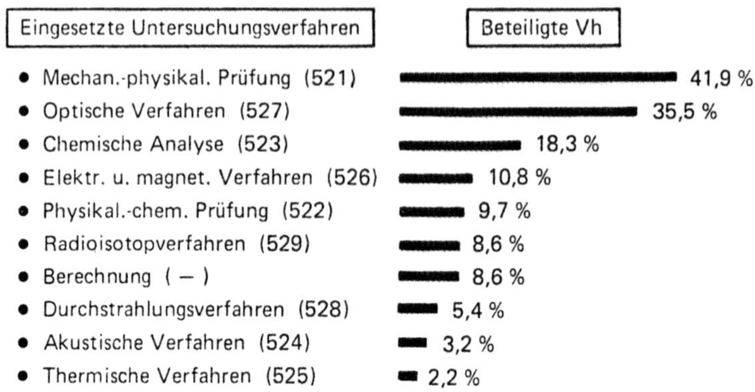

Bild 13: Verteilung der Deskriptorennennungen auf dem Bereich "Untersuchungsverfahren"

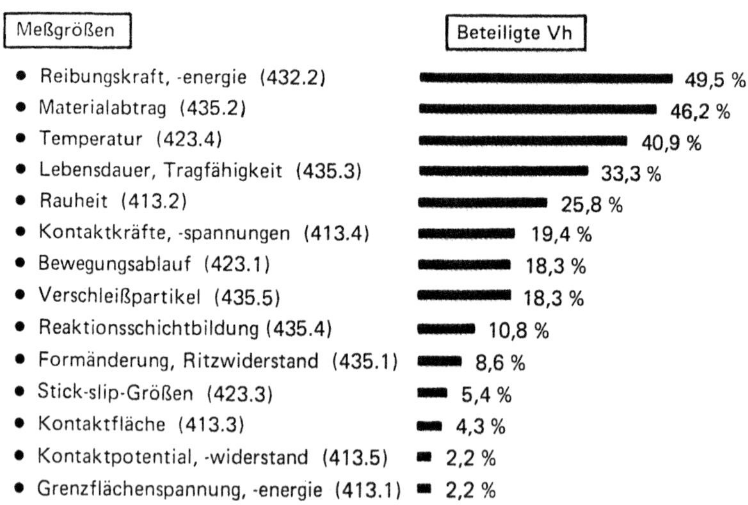

Bild 14: Verteilung der Deskriptorennennungen auf dem Bereich "Meßgrößen"

Rasterelektronenmikroskopie - eine wichtige Rolle zukommt.

Eine Aufstellung der bei den Vorhaben des Tribologie-Programms erfaßten Meßgrößen gibt Bild 14. Erwartungsgemäß sind die hauptsächlichen Meßgrößen die Reibungskraft oder -energie, der Materialabtrag und die Temperatur. Bemerkenswert ist hier, daß im Durchschnitt aller Vorhaben etwa 3 Meßgrößen pro Vorhaben bestimmt wurden. Dies bestätigt die häufig erhobene Forderung, daß Fortschritte bei der experimentellen Bearbeitung tribologischer Forschungs- und Entwicklungsprojekte nur bei gleichzeitiger Erfassung mehrerer Meßgrößen möglich sind.

3.7 Interessenhintergrund

Der Interessenhintergrund der Vorhaben des Tribologie-Forschungsprogramms kann Bild 15 entnommen werden. Die hierzu am häufigsten genannten Deskriptoren lassen sich nach der Häufigkeit ihrer Nennung in zwei Gruppen einteilen:

(a) Bei 38 bis 45 % aller Vorhaben wurden genannt: Werkstoffe und Schmierstoffe, Zuverlässigkeit und Sicherheit, Übertragbarkeit sowie Wirtschaftlichkeit und Kosten.

(b) Bei 32 bis 34 % aller Vorhaben wurden genannt: Neuentwicklungen, Konstruktion, Berechnung, Betriebseigenschaften, Vergleich, Volkswirtschaftliche Bedeutung.

Neben den offensichtlichen tribotechnischen Interessen an Werkstoffen und Schmierstoffen, Konstruktion und Berechnung sowie den Betriebseigenschaften, Vergleich und Übertragbarkeit experimenteller Ergebnisse sollten die folgenden drei Aspekte besonders hervorgehoben werden:

- Zuverlässigkeit und Sicherheit
- Wirtschaftlichkeitsbetrachtungen und Kostenfragen
- Interesse an Neuentwicklungen.

Aus den genannten Interessenlagen wird deutlich, daß mit dem Tribologie-Förderprogramm nicht nur spezifisch tribologische Problemstellungen behandelt werden, sondern auch in erheblichem Umfang Beiträge zur Sicherheitstechnik, zur Wirtschaftlichkeit und - mit einer Beteiligung von immerhin 1/3 aller

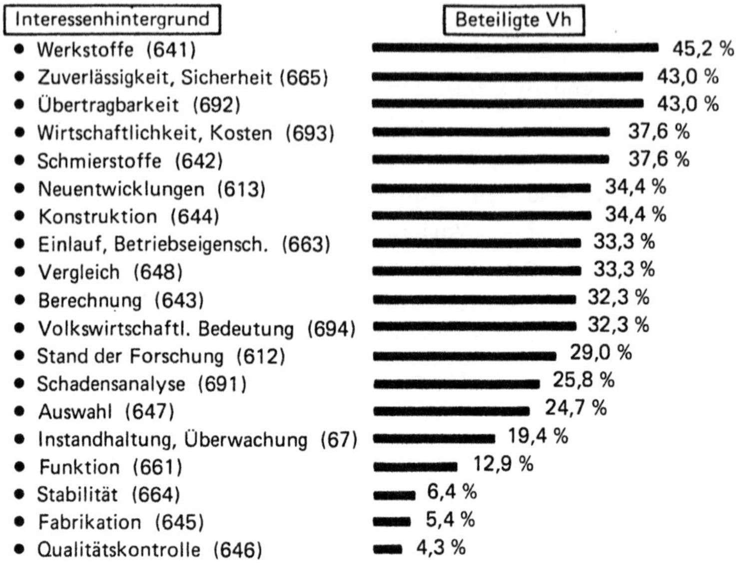

Bild 15: Verteilung der Deskriptorennennungen auf dem Bereich "Interessenhintergrund"

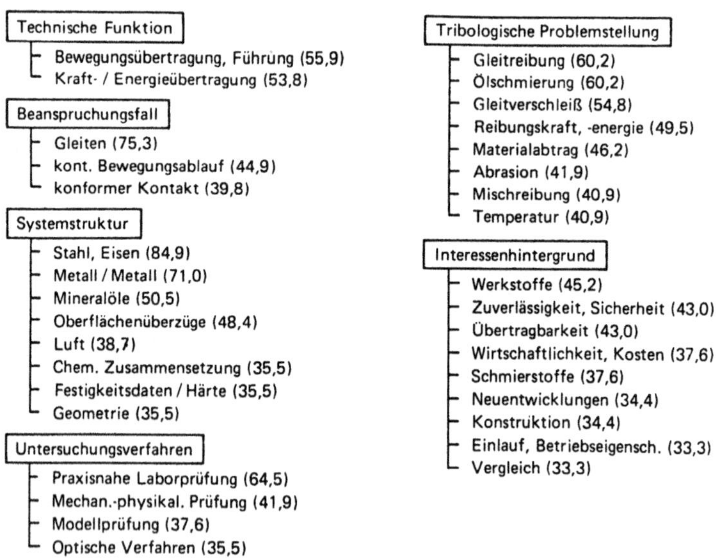

Bild 16: Zusammenfassende Darstellung der häufigsten Deskriptorennennungen zu den verschiedenen Systemkenngrößen

bisherigen Vorhaben - zur technischen Innovation geleistet werden.

4. Zusammenfassung und Ausblick

Mit Hilfe des auf der Basis von Systembetrachtungen entwickelten Thesaurus konnte eine umfassende Darstellung, Erschließung und Auswertung der bisherigen Vorhaben des Tribologie-Förderprogramms mittels Deskriptoren vorgenommen werden. In Bild 16 sind in einer zusammenfassenden Darstellung für die hauptsächlichen Parametergruppen nochmals diejenigen Deskriptoren aufgeführt, die bei mehr als 1/3 aller Vorhaben genannt wurden. (Die Zahlen in Klammern bezeichnen den prozentualen Anteil der Vorhaben,in denen die genannten Themen behandelt wurden.)

Aus der zusammenfassenden Darstellung von Bild 16 geht wiederum die interdisziplinäre Vielschichtigkeit und Komplexheit tribologischer Problemstellungen hervor. Das Bild bestätigt die in den letzten Jahren sowohl von der industriellen Praxis als auch von der wissenschaftlichen Forschung betonte Erkenntnis, daß weitere Fortschritte in der Tribologie nur bei einer sinnvollen Kombination von Forschungsergebnissen zu den einzelnen Systemaspekten möglich sind. In diesem Zusammenhang sei erwähnt, daß in einer kürzlich erschienenen Arbeit über die Notwendigkeiten und Erfordernisse des Technologie-Transfer darauf hingewiesen wurde, daß für die Lösung eines industriellen Problems heute meist die Ergebnisse von mehr als 50 (!) Forschungsprojekten kombiniert werden müssen /4/.

Die in den vorhergehenden Abschnitten dargestellte Auswertung macht deutlich, daß die Ergebnisse der bisherigen tribologischen Forschungsarbeiten verschiedene technische Anwendungsgebiete betreffen. Im Vergleich mit der internationalen Literatur werden jedoch unterschiedliche Anwendungs-Schwerpunkte erkennbar, so daß zu prüfen wäre, ob für einzelne technische Bereiche eine Verstärkung der Forschungsaktivitäten erforderlich ist. Eine deutliche Lücke im bisherigen Forschungsprogramm scheint darin zu bestehen, daß das tribologische Verhalten synthetischer Stoffe - wie z.B. von Polymerwerkstoffen, Verbundwerkstoffen oder auch von synthetischen Schmierstoffen - bisher nur in relativ geringem Umfang untersucht wurde. Da Anregungen zur Schließung eventueller "Forschungslücken" oder zur Initiierung zukunftsträchtig erscheinender Forschungs- und Entwicklungsarbeiten jedoch eine sehr sorgfältige Abwägung und Gewichtung zahlreicher Beurteilungskriterien erfordern, soll hier auf weitere Hinweise verzichtet werden.

Abschließend sei betont, daß das Tribologie-Förderprogramm neben volkswirtschaftlich wichtigen Beiträgen zur Erhaltung von Stoffen und Energie durch Minderung reibungs- und verschleißbedingter Material- und Energieverluste auch wesentliche Impulse zur Sicherheitstechnik, zum Technologie-Transfer und zur technischen Innovation gibt, wie die Auswertung des Interessenhintergrundes der Forschungs- und Entwicklungsvorhaben zeigt.

Dank

Mein herzlicher Dank gilt Herrn Dipl.-Phys. H. Tischer von der Dokumentationsstelle "Rheologie und Tribologie" der BAM für die Vorarbeiten zur Erstellung des Deskriptorensystems und die Abschätzung der Häufigkeitsverteilungen tribologischer Literatur zu bestimmten Forschungsaspekten an Hand der BAM-Dokumentation Tribologie. Außerdem danke ich Herrn Dr. K.-H. Habig, Leiter des Laboratoriums "Verschleißschutz; Tribometrie u. Tribophysik" der BAM, für hilfreiche fachliche Diskussionen sowie den Herren Dr. J. Hansen und Dipl.-Ing. M. Geyer von der DFVLR für die gute Zusammenarbeit bei der Fragebogenerhebung.

Literatur

/1/ H. Böer, G. Graue, P. Greis, E. Gülker, H. Heinke, W.-H. Kara, O. Lang, H. Peeken, E. Pollmann, E. Roemer, H. Uetz: Tribologie - Reibung, Verschleiß, Schmierung. Forschungsbericht T 76-38, Bundesministerium für Forschung und Technologie, Juli 1976, 280 S.

/2/ E. David: Die Technologie im Meinungsumschwung. Umschau, 76 (1976) 551.

/3/ F. Kramer: System und Realisierung der Produktinnovation an Beispielen eines mittleren Industrieunternehmens. Konstruktion 31 (1979) 225.

/4/ M.J. Neale: Technology transfer - the application in industry of the results of research. Proc.Instn.Mech.Engrs., 191 (1977) 333.

/5/ NN: The worth of literature search. Materials Protection, 6 (1967) 22.

/6/ H. Czichos: Tribology - a systems approach to the science and technology of friction, lubrication and wear. Elsevier Publ.Co, Amsterdam, 1978.

/7/ G. Ropohl: Systemtechnik - Grundlagen und Anwendungen. Hanser Verlag, München, 1975.

/8/ H. Wahl: Allgemeine Verschleißfragen. Schriftreihe Verschleißfragen 1941, Heft A1 Verschleißtechnik. Die Technik (1948) 193.

/9/ G. Fleischer: Systembetrachtungen zur Tribologie. Wiss. Z. TH Magdeburg, 14 (1970) 415.

/10/ G. Salomon: Application of systems thinking to tribology. ASLE Trans., 17 (1974) 295.

/11/ H. Czichos: Systemanalyse und Physik tribologischer Vorgänge, Teil 1: Grundlagen, Teil 2: Anwendungen. Schmiertechnik u. Tribologie, 22 (1975) 126 und 23 (1976) 6.

/12/ J. Mølgaard: The dry wear of metals as a process in an open system. Wear, 32 (1975) 353.

/13/ G. Heinke: Verschleiß - eine Systemeigenschaft; Auswirkungen auf die Verschleißprüfung. Z. Werkstofftechnik, 6 (1975) 164.

/14/ H. Uetz und J. Föhl: Tribologie - Stand der Erkenntnisse und Nutzen. VDI-Berichte Nr. 333, 1979, 1.

/15/ H. Krause und J. Scholten: Verschleiß-Grundlagen und systematische Behandlung, Teil 1 und Teil 2. VDI-Z. 121 (1979) 799 und 1221.

/16/ G. Stähli: Verschleiß - eine Systemeigenschaft. Material und Technik (1980) 183.

/17/H. Czichos: Systemtechnische Analyse und Beschreibung von Verschleißvorgängen. Z.f. Metallkde, 71 (1980) 421.

/18/ K. Kirschke: Documentation of international scientific and technical literature in the field of wear, friction and lubrication. Lubrication Engineering, 31 (1975) 362.

Untersuchungen zur Minderung der im wesentlichen durch Abrasiv-Beanspruchung bedingten Verschleißschäden im Großanlagenbau

von
Dr.-Ing. Hans Grewe
Dipl.-Ing. Hans-Friedrich Wilms

Fried. Krupp GmbH
Krupp Forschungsinstitut
Essen

Förderkennzeichen: 01 ZT 518 S - ZK/NT/NTS 1009
Laufzeit: 1.8.1978 - 31.12.1980

Zusammenfassung

Dieses Vorhaben wurde mit der Zielsetzung bearbeitet, möglichst viele Verschleißprobleme aus dem Unternehmen zusammenzutragen, zu analysieren, zu vergleichen, zu systematisieren und zu bewerten.

Zunächst wurden im Rahmen von zwei Befragungsaktionen im Bereich des Krupp-Konzerns insgesamt 110 Verschleißprobleme benannt, die vornehmlich durch Abrasiv-Verschleiß verursacht sind. Der letztlich entwickelte Fragebogen wurde auf der tribologischen Seite in Anlehnung an DIN 50.320 und zur Berücksichtigung wirtschaftlicher Aspekte in Anlehnung an die Methoden der Nutzwert-Kosten-Analyse gestaltet. Die Auswertung dieser Verschleiß-Befragung ließ viele Rückschlüsse auf den allgemeinen Kenntnisstand bezüglich der untersuchten tribologischen Probleme zu. Aufgrund der beschriebenen Wichtung bzw. Bewertung wurden aus den 110 benannten Problemen zunächst 8 Verschleißkomplexe für die weitere tribologische Analyse ausgewählt.

Von diesen 8 Verschleißkomplexen wurden vier Themen als besonders vordringlich eingestuft:

- Verschleiß an Auskleidungen von Schurren, Bunkern, Gutübergabestellen sowie an Austragungsarmen von Bunkerentleerungsmaschinen.

- Verschleiß an Schlagpanzerlamellen im Hochofengichthals.

- Verschleiß an Zähnen und Eimern von Großraumbaggern.

- Verschleiß an Schake und Bolzen von Raupen- und Eimerketten sowie an Raupen- und Eimerketten-Gleitführungen.

Diesen vier Themen galten eine ausführliche Daten- und Informationsbeschaffung sowie eine intensive Literaturauswertung.

Bei der Auswertung der Fragebögen, bei der nachfolgenden Daten- und Informationsbeschaffung, insbesondere aber bei der Literaturauswertung zeigte sich, daß bei der Weiterführung des Forschungsvorhabens über die Minderung des Abrasiv-Verschleisses im Großanlagenbau neben den genannten vier Verschleiß-Themen gleichwertig und parallel zwei grundsätzliche Fragen zu behandeln sein werden:

- Werkstofffragen vor konstruktiven und verfahrenstechnischen Fragen. Aus diesem Grunde wurde u.a. ein hochverschleißbeständiges Verbund-Hartmetall entwickelt, welches in verschiedenen Anwendungsfällen erfolgreich als Alternative zu Manganhartstahl eingesetzt worden ist.

- Fragen zur Klärung und Beschreibung des Mechanismus von Furchungsverschleiß. Aus diesem Grunde wurden Abriebuntersuchungen beim Mahlen von Titancarbid, Schmelzkorund und Quarz-Sand in technischen Rollmühlen durchgeführt. Aus diesen Abriebuntersuchungen konnten Rückschlüsse auf Mechanismus und Kinetik des Abrasiv-Verschleisses im Tribosystem Rollmühle hergeleitet werden.

Ziel der weiteren Arbeiten wird es sein, die an den genannten ausgewählten Bauteilen des Großanlagenbaus auftretenden - vornehmlich mineralisch bedingten - Verschleißschäden systematisch zu analysieren, die zugrunde liegenden Verschleißmechanismen zu erkennen und auf dieser Basis Verbesserungsmöglichkeiten für die Einzelprobleme vorzuschlagen.

Die notwendigen praktischen Versuche sollen vornehmlich im großtechnischen Rahmen (Betriebsversuche) durchgeführt werden.

G L I E D E R U N G

1. Einführung und Zielsetzung

Die in letzter Zeit immer notwendiger werdenden Bemühungen um eine Sicherung von Energie und Rohstoffen finden steigendes Verständnis in einer zunehmend breiter werdenden und zunehmend interessierten Öffentlichkeit. Obwohl die Warnungen vor dem exponentiell wachsenden Trend von Energie- und Rohstoffverbrauch [1,2] bereits Ende der 60er Jahre zu vernehmen waren, können erst seit einigen Jahren verstärkte und erfolgversprechende Aktivitäten festgestellt werden, die geeignet sind, den angedeuteten Trend abzubremsen bzw. eine Stabilisierung oder eine Minderung des Verbrauchs von Energie und Rohstoffen herbeizuführen.

Dabei wird u.a. auch die Verringerung von Verschleiß als eine der wesentlichen Möglichkeiten zur Einsparung von Rohstoffen herausgestellt. Dieses unterstreicht vor allem auch die Tribologie-Studie des BMFT [3]. Die Menschheit lebt mit dem Phänomen Verschleiß in fast allen Bereichen von Technik und Zivilisation, und zwar überall dort, wo zwei einander sich berührende Stoffbereiche eine i.a. tangential gerichtete Relativbewegung zueinander ausführen.

Für die verantwortlichen Techniker und für die entsprechend verantwortliche Industrie kommt zu der oben skizzierten, allgemeinen Problematik hinzu, daß das tribologische Grundlagenwissen noch unzureichend und lückenhaft ist und daß auch die konsequente und systematische Anwendung bereits gewonnener Kenntnisse auf praktische Belange nicht befriedigt.

Das Gebiet der Tribologie ist besonders unübersichtlich in der Beurteilung seiner Einflußgrößen und wurde daher lange Zeit vernachlässigt. Nicht nur die Tribologie allgemein, sondern auch deren Teilgebiet Verschleiß haben so starken interdisziplinären Charakter, daß fachliche Detailkenntnisse über viele klassische Fachrichtungen verstreut und nur aus einer Vielzahl von Veröffentlichungen unter den verschiedensten Aspekten greifbar sind. Um Verschleißprobleme mit vertretbarem Aufwand für praxisgerechte Anwendungen zu lösen, wird es somit einer weiter übergeordneten Bearbeitungsweise als bisher bedürfen.

Ein bedeutsamer Ansatz zur Verbesserung und Vereinheitlichung des allgemeinen Kenntnisstandes auf dem Verschleißgebiet,der bei der Erfassung und Analyse von Verschleißkomplexen hilfreich ist, liegt seit November 1976 mit dem Entwurf und seit Dezember 1979 mit dem Weiß-Druck der Norm DIN 50.320 vor [4]. In dieser Norm wird der Begriff Verschleiß, wie folgt, definiert:

> "Verschleiß ist der fortschreitende Materialverlust aus der Oberfläche eines festen Körpers, hervorgerufen durch mechanische Ursachen, d.h. Kontakt und Relativbewegung eines festen, flüssigen oder gasförmigen Gegenkörpers."

a) Die Beanspruchung der Oberfläche eines festen Körpers durch Kontakt und Relativbewegung eines festen, flüssigen oder gasförmigen Gegenkörpers wird als tribologische Beanspruchung bezeichnet.

b) Verschleiß äußert sich im Auftreten von losgelösten kleinen Teilchen (Verschleißpartikel) sowie in Stoff- und Formänderungen der tribologisch beanspruchten Oberflächenschicht.

c) In der Technik ist Verschleiß normalerweise unerwünscht, d.h. wertmindernd. In Ausnahmefällen, wie z.B. bei Einlaufvorgängen, können Verschleißvorgänge jedoch auch technisch erwünscht sein. Bearbeitungsvorgänge als wertebildende technologische Vorgänge gelten in bezug auf das herzustellende Werkstück nicht als Verschleiß, obwohl im Grenzflächenbereich zwischen Werkstück und Werkzeug tribologische Prozesse wie beim Verschleiß ablaufen.

Dieser Definition entsprechend und vor allem im Sinne von DIN 50.320 hat die Verschleißforschung das Ziel, den durch Verschleiß bedingten volkswirtschaftlichen Verlust möglichst exakt festzustellen und durch systematische tribologische Untersuchungen letztlich zu mindern. Daß trotz aufwendiger Versuchsarbeit im Einzelfall oft nur geringe Erfolge in der Reduzierung des Verschleißes sichtbar werden, kann durch die häufig empirische und wenig systematische Vorgehensweise erklärt werden, die in vielen Fällen von den finanziellen und zeitlichen Notwendigkeiten des praktischen Arbeitsalltags widersinnigerweise diktiert wird. Andererseits können auch spektakuläre Verbesserungen erzielt werden, die mit Genugtuung zur Kenntnis genommen werden, für die aber oft keine ausreichend fundierte Erklärung vorgelegt werden kann, so daß die Übertragung auf anscheinend ähnlich gelagerte Fälle nicht gelingt. Erschwert wird die Übersicht über das Verschleißgebiet noch dadurch, daß Untersuchungsergebnisse über eine große Palette technischer Publikationen verstreut sind und daß damit Doppelarbeit sowie Ergebnisverlust für die Praxis fast zwangsläufig sind. Auch umfangreiche tribologische Dokumentationen, wie sie z.B. von der Bundesanstalt für Materialprüfung (BAM) herausgegeben werden, helfen dem Praktiker im Einzelfall nicht sofort weiter, ohne daß mit dieser Feststellung der große Wert dieser Dokumentation bezweifelt werden soll.

Die Bewältigung der hier nur angedeuteten Probleme bei der praxisgerechten Behandlung von Verschleißfragen ist in jüngster Zeit - wie bereits angedeutet auch aus rohstoff- und energie-politischen Gründen - ein weltweites Anliegen geworden. Damit ist zu hoffen, daß in absehbarer Zeit für die Lösung und Beherrschung der Verschleißprobleme ein ähnlicher Steilanstieg erreicht werden kann wie für die Problemkreise "Festigkeit" und "Korrosion", so wie dieser Anstieg bereits von Hans WAHL [5,6] um das Jahr 1950 in einem nach wie vor aktuellen bzw. aktualisierten Diagramm erwünscht wurde, Bild 1.

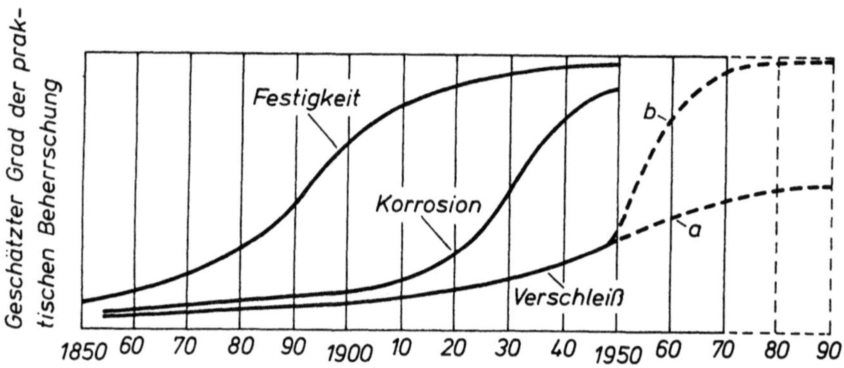

Bild 1: Beherrschung von Festigkeits-, Korrosions- und Verschleißfragen (schematisch nach Hans Wahl)

a) Wahrscheinlicher weiterer Verlauf bei Bearbeitung der Verschleißprobleme in bisheriger Art.

b) Wahrscheinlicher weiterer Verlauf bei intensiver und rationeller Bearbeitung der Verschleißprobleme.

Aus dieser Zielsetzung heraus will der voliegende Bericht versuchen, einen Beitrag zur ordnenden und systematisierenden Betrachtungsweise von Verschleißproblemen zu liefern. Die Ausarbeitung hat es sich zum Ziel gesetzt, nach der tribologischen Analyse einer Vielfalt von Verschleißfällen, die vornehmlich aus dem Großanlagenbau und dem Bereich des Abrasiv-Verschleißes kommen sollen, Kriterien und Vorgehensweisen zu entwickeln, die dem Praktiker die Behandlung von Verschleißfragen erleichtern und die nacheiner eingehenden Bestandsaufnahme des Einzelfalls Lösungsansätze für diesen Einzelfall nahelegen. Dabei erfolgt die Behandlung der Thematik im allgemeinen in Anlehnung an DIN 50.320.

In verschiedenen Schritten werden zunächst 110 Verschleißfälle aus den einzelnen Bereichen des Krupp-Konzerns zusammengetragen, tribologisch methodisch behandelt und nach interessierenden Gesichtspunkten auf weitere wissenschaftlich-analytische Bearbeitung aus tribologischer Sicht überprüft. Diese weiterführende Untersuchung soll dann im Rahmen von Teil II dieses Vorhabens an ca. 4 Verschleißfällen - vornehmlich auf der Basis von Betriebsversuchen - erfolgen.

2. Einige Gedanken zum Thema: Problemstellungen in der Verschleißforschung

Die nachfolgend noch näher zu erläuternden Arbeitsergebnisse gaben Anlaß, der Frage nachzugehen, warum ein Verschleißgeschehen überhaupt in den Blickpunkt rückt und warum es technisch Verantwortliche dazu anregt, sich seiner anzunehmen und mit dem Ziel der Verschleißminderung daran zu arbeiten. Nach den grundsätzlichen Zielen und Arbeitsmethoden vieler Fachleute [5] ließen und lassen sich auch heute noch etwa die folgenden Trends zur Bearbeitung von Verschleißproblemen zusammenfassen:

a) Werkstoffentwicklung mit dem Ziel, noch härtere und verschleißbeständigere Werkstoffe zu suchen und zu entwickeln;

b) Verschleißminderung über die konstruktive (Um-) Gestaltung von betrieblichen Einrichtungen (Schienen, Mahlgefäße, Mahlkugeln, Krümmer, Maschinenelemente mit Werkzeugcharakter);

c) gezielte Grundlagenuntersuchungen zur Aufklärung von spezifischen Verschleißmechanismen und Ausarbeitung von Lösungsvorschlägen aus den beobachteten Gesetzmäßigkeiten;

d) allgemeine Untersuchungen zur Aufhellung größerer tribologischer Zusammenhänge mit dem Ziel der Erarbeitung einer analytischen Systematik, einer planmäßigen Auswertung und der Weitergabe der gesicherten Erkenntnisse bzw. Gesetzmäßigkeiten.

Aus heutiger Sicht lassen sich drei u.E. typische Fälle konzipieren und auch schematisch darstellen, die Verschleißforschung auslösen:

a) Das tatsächliche Verschleißgeschehen weicht mit verstärktem Verschleiß vom erwarteten Verlauf ab, wobei unterschiedliche Möglichkeiten der Abweichung denkbar sind, Bild 2, Teilbilder a u. b.

b) Das tatsächliche Verschleißgeschehen weicht mit vermindertem Verschleiß vom erwarteten Verlauf ab, Bild 2, Teilbild c.

c) Das Verschleißgeschehen entspricht dem erwarteten Verlauf, es soll aber bewußt zu einem verminderten Verschleiß hin verschoben werden, Bild 2, Teilbild d.

Bild 2

Diagramme zur Darstellung möglicher, eine Verschleiß-
forschung auslösender Problemstellungen
[schematische Verschleiß-Zeit-(V, t)-Diagramme]

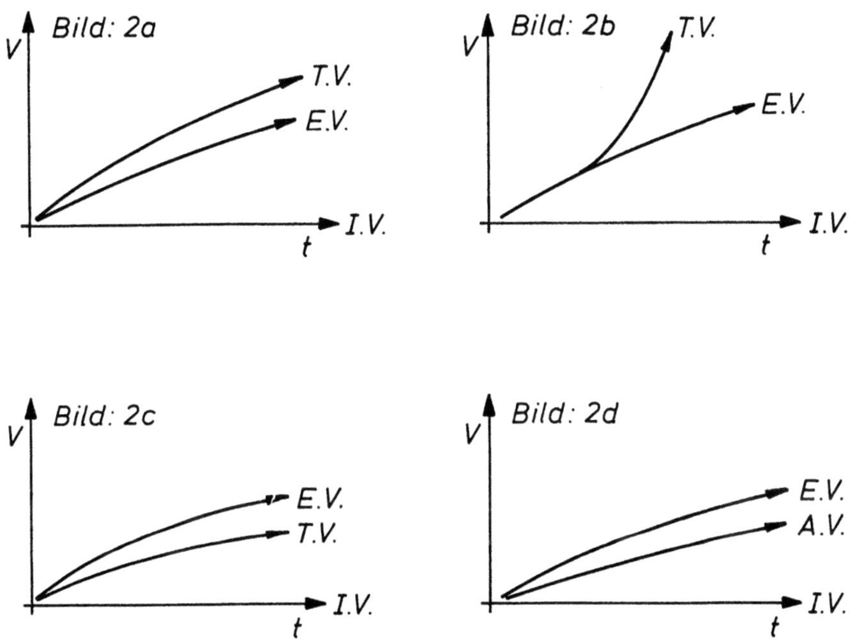

V ≙Verschleißbetrag; t ≙ Zeit
I.V.≙Idealer verschleißfreier Verlauf
E.V.≙Erwarteter verschleißbehafteter Verlauf (Soll-Kurve)
T.V.≙Tatsächlicher Verschleißverlauf (Ist-Kurve)
A.V.≙Angestrebter Verschleißverlauf

Die Probleme, wie sie in Bild 2 skizziert sind, werden von der Praxis vorgegeben durch nicht kalkulierte oder nicht kalkulierbare Ereignisse, die die festgestellte Abweichung des tatsächlichen Verlaufs vom erwarteten Verlauf bedingen. Aufgabe der Verschleißforschung ist es in allen Fällen, die Ursachen für die Abweichungen zu ermitteln. Dabei ist für den Verlauf in Bild 2, Teilbild c, zu prüfen, warum ein besseres Ergebnis erzielt wurde sowie warum und wie der tatsächliche Verlauf in der Zukunft reproduzierbar zum erwarteten Verlauf werden kann.

Erst wenn die Verschleißforschung nach Typ des Bildes 2, Teilbild d, mit gezielt vorgegebener Aufgabenstellung stärker in das Bewußtsein eindringt, wird sich der erhöhte Kenntnisstand in der Verschleißforschung dem Kurvenzug b in Bild 1 angleichen. Es soll somit vornehmlich ein Anliegen dieses Vorhabens sein, dem in der Praxis stehenden Ingenieur Hilfen für ein systematisches, nachvollziehbares und mit anderen Verschleißfällen vergleichbares Arbeiten an seinem speziellen Einzelfall zu geben [7-23].

3. Sammlung und systematische Behandlung von Verschleißfällen

3.1. Befragungen und Auswertung

Zur Sammlung von Verschleißfällen wurde zunächst im Krupp-Konzern eine orientierende Befragung durchgeführt, auf der aufbauend ein detaillierter Fragebogen entwickelt wurde. Damit wurde eine 2. Befragung vorgenommen, deren Ergebnisse eine systematische Auswertung und Bearbeitung der genannten Verschleißfälle ermöglichten.Zur Unterstützung und Beratung der Arbeiten fanden konzerninterne Arbeitskreisbesprechungen statt.

Aufgrund der Fragebogenauswertung wurde eine Anzahl von Verschleißfällen ausgewählt und tribologisch-analytisch näher untersucht. Es wurden für den jeweiligen Verschleißablauf Daten und Informationen, möglichst durch Befragung der entsprechenden Fachkollegen und durch Besichtigung "vor Ort", gesammelt und aufgearbeitet. Wie noch näher auszuführen sein wird, wurden u.E. vertretbar vergleichbare Verschleißfälle gleichzeitig bzw. gleichartig behandelt, Für diese Fälle wurde eine entsprechende umfangreiche Literaturauswertung angeschlossen. Die Ergebnisse des Bewertungsverfahrens wurden kritisch beurteilt, nicht zuletzt aufgrund von Folgerungen aus dem Literaturstudium. Aus zwei zunächst orientierenden Versuchsreihen ergaben sich Erkenntnisse für weitere Werkstoffentwicklungen und für eine bessere grundsätzliche Beurteilung des Verschleißverhaltens von Mahlkörpern in technischen Rollmühlen. Dabei wurden Titancarbid, Schmelzkorund (Aluminiumoxid) und Quarzsand (SiO_2) als Hartstoffe eingesetzt, die allseits von technischer Bedeutung sind.

3.1.1. Fragebogenaktion

Die Grundlage systematischer Arbeit ist eine möglichst umfangreiche und detaillierte Bestandsaufnahme aller relevanten Größen, wobei die Schwierigkeit oft darin besteht, diese Größen bereits bei der Planung eines Fragebogens zu erkennen und in ihrer Bedeutung für die Auswertung abzuschätzen.

Ziel der detaillierten, 2. Fragebogen-Runde war es, möglichst viele Einzelprobleme zusammenzutragen, wobei jedes Problem durch weiter spezifizierte Fragestellungen aufbereitet werden sollte.

Der 2. Fragebogen (detaillierte Form) liegt diesem Bericht als Tabelle 1 bei.

Tabelle 1/1

TRIBOLOGIE-FRAGEBOGEN

(2. Fassung, detaillierte Form, gekürzt)

zur Ermittlung von Verschleißproblemen
bei den Krupp-Konzernunternehmen

Konzernunternehmen	Bearbeiter	Telefon

1. Beschreibung des Verschleißsystems

1.1 Benennung bestehender, erwarteter oder vermuteter Verschleißprobleme

1.2 Kann das Verschleißsystem durch eine Skizze, eine Zeichnung oder ein Foto erläutert werden? (Prospekte?) (bitte ankreuzen)

	Skizze liegt bei	Zeichnung liegt bei	Foto liegt bei	nein
a)				
b)				
c)				
.				
.				
.				

1.3 Welche technische Funktion hat das Tribosystem und welche Elemente eines Tribosystems (Grundkörper, Gegenkörper, Zwischenstoff, Umgebungsmedium) sind am Verschleißvorgang beteiligt? (vgl. auch Tabelle 2)

1.3.1 Technische Funktion des Tribosystems

1.3.2 Elemente des Tribosystems

	Grundkörper	Gegenkörper	Zwischenstoff	Umgebungsmedium
a)				
b)				
c)				

1.3.3 Material und Materialzustand der Elemente (Werkstoff, Werkstoff-Nr., chemische Zusammensetzung, Gefügezustand, Festigkeit, Härte, Viskosität, Aggregatzustand usw.)

	Grundkörper	Gegenkörper	Zwischenstoff	Umgebungsmedium
a)				
b)				
c)				

1.4 Lassen sich Angaben über die technisch-physikalischen Beanspruchungsparameter des Tribosystems machen? (vgl. Tabelle 4)

1.4.1 Beanspruchungsart (z.B. Dauerbetrieb, x % im Jahr, Zweischichtbetrieb oder intermittierender Betrieb, x Vorgänge à y min pro Stunde, Einschichtbetrieb) und Normalkraftbelastung F_N (oder Druck p)

Beanspruchungsart	Normalkraft F_N (oder Druck p)	Qualitativer Verlauf $F_N = f(t)$
a)		
b)		
c)		

1.4.2 Geschwindigkeit v und Temperatur T

Geschwindigk. v	Qual. Verlauf v = f(t)	Temp. T	Qual. Verlauf T = f(t)
a)			
b)			
c)			

1.5 Lassen sich Angaben über das Umgebungsmedium machen?

Druck	Temperatur	Feuchtigkeit
a)		
b)		
c)		

2. Beschreibung des Verschleißvorgangs

2.1 Welche tribologische Beanspruchung bzw. Verschleißart nach Tabelle 9 kann für die aufgeführten Verschleißprobleme angenommen werden?

2.2 Treten neben oder bei der Verschleißbeanspruchung weitere Arten der Oberflächenzerstörung am Grundkörper auf z.B. durch Korrosion, Verzunderung, Sublimation oder andere? (bitte ankreuzen)

Korrosion im Betrieb	Korrosion im Stillstand	Verzunderung im Betrieb	Verzunderung im Stillstand	Sublimation im Betrieb	Sublimation im Stillstand
a)					
b)					
c)					

2.3 Wie macht sich der Verschleiß bemerkbar? Welche Verschleißerscheinungsformen liegen vor? Kann ein Foto den Verschleiß verdeutlichen?

Verschleißerscheinungsform	Foto (ja/nein)
a)	
b)	
c)	

2.4 Welche mittlere Standzeit (Standmenge, Standweg) gilt für den Grundkörper?
Wann gilt der Grundkörper als verschlissen (Ausfallkriterien)? Kann der Grundkörper wieder aufgearbeitet werden z.B. durch Auftragsschweißen? Wenn ja, wie?

mittlere Standzeit (Standmenge) (Standweg)	Ausfallkriterien	Aufarbeitung mögl? (ja,wie?/nein)
a)		
b)		
c)		

2.5 Können Meßmethoden und Meßgrößen für den Verschleiß angegeben werden?
Liegen Meßwerte oder Diagramme vor? Wenn ja, welche?
(z.B. Verschleißbetrag pro Zeit, Verschleißgeschwindigk., verschleißbedingter prozentualer Gewichtsverlust zum Standzeitende)

a)
b)
c)

2.6 Liegen Kenntnisse oder Vermutungen über den Verschleißmechanismus und die Ursachen des Verschleißes vor? Auch spekulative, wissenschaftlich noch nicht begründbare Überlegungen sollten angeführt werden.

a)
b)
c)

3. Wirtschaftliche Aspekte des Verschleißfalls

3.1 Welche Kostenangaben können zu dem Verschleißproblem gemacht werden?

3.1.1 Neukosten des Grundkörpers (=Bauteil) und der Baugruppe (z.B. Lüfterrad = Bauteil, Gebläse = Baugruppe, Entstaubungsanlage = Anlage)

Tabelle 1/4

Neukosten des Grundkörpers	Bezeichnung der Baugruppe	Neukosten der Baugruppe
a)		
b)		
c)		

3.1.2 Kosten bei Eintritt des Verschleißfalles

Reparatur-u.Einfahrkosten an der Baugruppe (ohne Neukosten des Grundkörpers)	kalkulierte Stillstandzeit	kalkul. Folgekosten während der Stillstandzeit
a)		
b)		
c)		

3.2 Wieviele gleichartige Bauteile (≙ Grundkörper) werden schätzungsweise pro Jahr im Krupp-Konzern hergestellt bzw. wieviele gleichartige Bauteile befinden sich im Krupp-Konzern im Einsatz bei vergleichbarer tribologischer Beanspruchung? (bitte ankreuzen)

	1	bis 10	bis 100	bis 1000	über 1000
a)					
b)					
c)					

4. Verbesserungsversuche und Änderungsvorschläge

4.1 Sind bereits Versuche unternommen worden, den Verschleiß zu mindern? Wenn ja, welche und mit welchem Erfolg?

a)
b)
c)

4.2 Bestehen Vorstellungen, wie das Verschleißproblem gelöst werden kann (angestrebte Lösung)?

a)
b)
c)

4.3 Zusätzliche Gedanken und Hinweise zu den einzelnen Verschleißproblemen (z.B. spezielle Literatur)

a)
b)
c)

5. DIN 50.320, Entwurf November 1976,
Tabelle 1, S. 4
Beispiele für die technische Funktion tribologischer Systeme.

6. DIN 50.320, Entwurf November 1976,
Seite 6, Punkt 2.3.1
Elemente eines Tribosystems

7. DIN 50.320, Entwurf November 1976,
Seite 5, Beanspruchungskollektiv
Punkt 2.2.1: Bewegungsform und zeitlicher Bewegungsablauf
Punkt 2.2.2: Technisch-physikalische Beanspruchungsparameter.

8. DIN 50.320, Entwurf November 1976,
Seite 13, Tabelle 4
Gliederung des Verschleißgebietes nach der Art der tribologischen Beanspruchung.

9. DIN 50.320, Entwurf November 1976,
Seite 11, Punkt 2.4.1
Verschleißerscheinungsformen

10. DIN 50.320, Entwurf November 1976,
Seiten 9 + 10, Punkte 2.3.3.3.1
2.3.3.3.2
2.3.3.3.3
Verschleißmechanismen eines Tribosystems

Der Fragebogen wurde in 4 Fragenkomplexe gegliedert, und zwar

1. Beschreibung des Verschleißsystems
2. Beschreibung des Verschleißvorgangs
3. Wirtschaftliche Aspekte des Verschleißfalls
4. Verbesserungsversuche und Änderungsvorschläge

Zur Verdeutlichung einiger Fragen wurden dem Fragebogen Tabellen aus dem DIN-Entwurf 50.320 [24] beigegeben, um zum einen den im DIN-Entwurf zusammengefaßten Kenntnis- und Erfahrungsstand zu nutzen und um gleichzeitig Sprache und Begriffe der erwarteten Antworten zu vereinheitlichen.

Wie ersichtlich ist, wurden die Ausführungen des DIN-Entwurfes unwesentlich erweitert, speziell jedoch um die Fragen nach dem wirtschaftlichen Aspekt (3. Fragen-Komplex) und um die Fragen nach Verbesserungsversuchen und potentiellen Lösungsmöglichkeiten (4. Fragen-Komplex).
Zur Hilfestellung bei der Bearbeitung bzw. bei der Beantwortung der Fragen wurde der Anhang des DIN-Entwurfes 50.320 [24] beigefügt, Tabelle 2.

Insgesamt ist der Fragebogen als eine Art Zwischenstation auf dem Wege zu einem noch weiter detaillierten Fragenraster, wie es am Ende dieses Vorhabens angestrebt wird, zu sehen. Er bietet bei ausreichender Beantwortung einmal die Möglichkeit, einzelne Verschleißfälle im Sinne einer Auswahl gegeneinander abzuwägen und erlaubt andererseits, ein bestimmtes Problem näher zu charakterisieren und für erste Lösungsansätze aufzubereiten. So konnte aus einzelnen Konzernunternehmen beispielsweise berichtet werden, daß bereits für einfacher gelagerte Verschleißfälle Verbesserungslösungen erarbeitet wurden, zu denen die exakte Beantwortung der Fragen angeregt hatte.

Tabelle 2

Anhang

Anlage II

Formblatt zu DIN 50 320	Beschreibung und Systemanalyse eines Verschleißvorganges .	Datum : Blatt-Nr :	Bearbeiter :
Allgemeine Beschreibung des Verschleißvorganges :			

I Technische Funktion des Tribosystems

Bewegungs-übertragung	Bewegungs-hemmung	Energie-übertragung	Informations-übertragung	Material-transport	Material-bearbeitung	Material-umformung	Sonstige Funktion

II Beanspruchungskollektiv

Bewegungs-form: Gleiten ☐ Rollen ☐ Bohren ☐ Prallen ☐ | Überlagerungen :

Bewegungs-ablauf : | Beanspruchungs-dauer t_B[ ] : | Bemerkungen :

Belastung[1] F_N[N]	F_N bei t_0	F_N bei t_B	Geschwindigkeit[2] v [m/s]	v bei t_0	v bei t_B	Temperatur[3] T [°C]	T bei t_0	T bei t_B

F_N — Zeit t[ ] | v — Zeit t[ ] | T — Zeit t[ ]

III Struktur des Tribosystems

Elemente	Grundkörper (1)	Gegenkörper (2)	Elemente	Zwischenstoff (3)	Umgebgsmed. (4)
Eigenschaften[4] Bezeichnung Werkstoff			Eigenschaften[4] Bezeichnung Stoff		
Volumen-eigenschaften: Abmessungen			Aggregatzustand		
Chemische Zusammen-setzung, Gefüge			Chem. Zusammensetzung, chem. Struktur		
Physikalische Stoffdaten[5]			Physikalische Stoffdaten[5]		
Festigkeitsdaten, Härte			Viskositätsdaten		
Oberflächen-eigenschaften: Oberflächenrauheitsdaten			Sonstige Eigenschaften		
Phys.-chem. Oberfl.-daten[6]					

Tribokontaktfläche A_0			Reibungszustand nach DIN 50 281 :	Reibungsdiagramm
Eingriffsverhältnis ε [%]	ε(1)	ε(2)		Reibgs-zahl — Zeit t[ ] bzw. Weg s[ ]
Verschleißmechanismen und Wechselwirkungen :			Bemerkungen :	

IV Verschleißkenngrößen

	Grundkörper (1)	Gegenkörper (2)	Verschleißdiagramm
Verschleißerscheinungsform (Beschreibung, Bild)			Verschleiß-betrag — Zeit t[ ] bzw. Weg s[ ]
Gesamt-Verschleißbetrag nach DIN 50 321 :			
Mittlere Verschleiß-geschwindigkeit $\frac{\Delta l}{\Delta t}$ $\left[\frac{\mu m}{h}\right]$			
Mittleres Verschleiß-Weg-Verhältnis $\frac{\Delta l}{\Delta s}$ $\left[\frac{\mu m}{km}\right]$			Bemerkungen :

1) Aus Belastung F_N und Tribokontaktfläche A_0 ergibt sich die nominelle Flächenpressung $p_0 = F_N/A_0$
2) v ist die Relativgeschwindigkeit zwischen Grundkörper (1) und Gegenkörper (2)
3) T ist die makroskopische Durchschnittstemperatur des Tribosystems
4) Gegebenenfalls sind auch verschleißbedingte Eigenschaftsänderungen zu kennzeichnen
5) Tribologisch relevante Größen sind z.B. Dichte, Ausdehnungskoeffizient, Wärmeleitfähigkeit, Volumen, Druck
6) Eigenschaften der obersten Atomlagen von (1) und (2), z.B. Dicke und Zusammensetzung von Oxidschichten

Dieses Formblatt kann bei Erscheinen der Norm durch Beuth Verlag GmbH, Burggrafenstraße 4–7, 1000 Berlin 30 und Kamekestraße 2–8, 5000 Köln 1 bezogen werden.

Als Ergebnis der Umfrage wurden aus 12 Konzernunternehmen 110 Verschleißprobleme genannt, die in Tabelle 3 in der Reihenfolge des Fragebogen-Rücklaufs aufgelistet sind.

In welchem Maße die Fragen des Fragebogens bei den 110 Verschleißfällen beantwortet wurden, geht aus den Bildern 3 und 4 hervor. Die erhaltenen Antworten wurden in 3 Kategorien gegliedert

a) gute Antwort im Sinne der Fragestellung
b) allgemeine Antwort
c) keine Antwort

Die Darstellungen zeigen deutlich, wo Probleme bei der Beschreibung von Verschleißfällen auftreten. Sind im allgemeinen die Elemente des jeweiligen Tribosystems noch gut zu definieren - obwohl mancher Beantworter durch unpräzise Benennung eines Falles auch hier Schwierigkeiten hatte -, so treten bei den technisch-physikalischen Beanspruchungsparametern erste große Lücken auf. Die kritische Betrachtung der Antworten zeigte dann auch, daß diese Daten oft weder beim Konstrukteur noch beim Betreiber bekannt sind. Damit fehlt aber eine wichtige Grundlage für die spätere Analyse der Verschleißmechanismen - ein nicht zu übersehender Hinweis auf vorprogrammierte Mißerfolge.

Beim Fragenkomplex 2 fällt auf, daß kaum geeignete Fotos zur Dokumentation des Verschleißbildes vorliegen, ein Mangel, der allerdings bei der Bearbeitung eines Einzelfalles beseitigt werden kann. Bedeutungsvoller ist, daß zur Frage der Messung des Verschleißes (Meßgröße, Meßmethode, Meßwerte) nur wenige geeignete Antworten vorgelegt werden konnten.

Tabelle 3/1

Auflistung der benannten Verschleißfälle aus den Krupp-Konzernunternehmen

1. Verschleiß von Seekühlwasser-Rohrleitungen bei Seewassertemperaturen von 45°C
2. Verschleiß von Trockenreibkupplungen in Schiffsmotoranlagen
3. Oberflächenverschleiß von Mühlenwalzen
4. Abtragung der Innenfläche von Zentrifugentrommeln
5. Verschleiß der Schweißnähte an Wabenheizkammern
6. Äußerer Rohrverschleiß bei "langen" Heizkammern
7. Erosionsverschleiß an Innenflächen von Extraktionsanlagen
8. Spurkranzverschleiß der Laufräder von Raupenketten
9. Verschleiß zwischen Schake und Bolzen bei Raupenketten und Eimerketten
10. Verschleiß an Eimerkettengleitführung
11. Verschleiß der Eimerzähne bei kontinuierlich arbeitenden Gewinnungsgeräten
12. Verschleiß im Spitzenbereich der Austragsarme von Bunker-Entleerungsmaschinen

Tabelle 3/2

13. Verschleiß an Formzeugen und Preßstempeln von Strangpressen

14. Unzureichende Schmierung zwischen Walzgut und Walzen beim Kaltwalzen dünner Bänder

15. Nicht optimale Schmierung zwischen Schleifstein und den zu schleifenden Arbeitswalzen für ein Kaltwalzblech

16. Kavitationsverschleiß an Pumpenlaufrädern (Korrosion)

17. Kavitationsverschleiß an Ejektoren (Korrosion)

18. Erosionsverschleiß an Pumpenlaufrädern

19. Erosionsverschleiß an Pumpengehäusen durch Sand im Seewasser (Korrosion)

20. Erosionsverschleiß in Rohrleitungen durch Sand im Seewasser (Korrosion)

21. Tropfenschlagverschleiß an den Zwischenwänden der Verdampferanlage (Korrosion)

22. Verschleiß an Kastenplatten von Kalksandstein-Formwerkzeugen

23. Verschleiß an Kastenplatten von Betonstein-Formwerkzeugen

24. Mechanische oder erosive Abnutzung der Schneckenstege bei der Kaltgummiextrusion

25. Mechanische oder erosive Abnutzung der Zylinderbüchse in kaltgefütterten Einschnecken-Extrudern

26. Verschleiß an Flockierwalzen für Saatpartikel

27. Verschleiß der Schneckengänge von Seiherschneckenpressen

28. Verschleiß der Seiherstäbe von Seiherschneckenpressen

29. Verschleiß von Kette und Schleißblech am Redler

30. Verschleiß an Nußknacker-Prallmühlen

31. Verschleiß von Hydrozyklonen zum Trennen von Schalen und Nüssen

32. Verschleiß an Kugel- und Rollendrehverbindungen von Baggern und Umschlaggeräten

33. Verschleiß an Eimerketten von Eimerkettenbaggern

34. Verschleiß an Eimern und Zähnen von Baggern und Umschlagsgeräten

35. Verschleiß an angetriebenen Laufrädern von langsam fahrenden Schienenfahrzeugen

36. Verschleiß an Tragrollenmänteln von Förderband-Tragrollen

37. Verschleiß an Auskleidungen von Schurren, Bunkern, Gutübergabestellen usw.

38. Verschleiß an Kolbenstangen von Hydraulikzylindern

39. Verschleiß an Brechwerkzeugen von Kreisel- bzw. Kegelbrechern

40. Verschleiß der Exzenterbuchse der Exzenter-Lagerung von Kreisel- bzw. Kegelbrechern

41. Verschleiß der Laufbuchse der Exzenter-Lagerung von Kreisel- bzw. Kegelbrechern

42. Verschleiß des Kegelradpaares in Kreisel- bzw. Kegelbrechern

43. Verschleiß von dünnen Kunststoff-Fahrbahnbelägen unter Fahrzeugverkehr

44. Verschleiß von Brückenlagern mit Teflon-Gleitschichten

45. Führungen (Kühlrohre, Führungsrohre, Windungsleger-Rohre) von Draht-Walzwerken

46. Abrollsegmente für Hochumform-Walzwerke Krupp/Platzer

47. Stranggieß-Kokillen für NE-Metall-Stranggießanlagen

48. Verschleiß an Meißeln in Schlagwerkzeugen

49. Verschleiß beim hydraul. Transport von Flugasche in mehrstufigen Kreiselpumpen (KT-Verg.)

50. Verschleiß beim hydraul. Transport von Flugasche in Krümmern der Rohrleitung (KT-Verg.)

51. Verschleiß beim pneumatischen Transport von Kohlenstaub in der Mischbirne (KT-Verg.)

52. Verschleiß im Verblasungsrohr des KT-Vergasers

53. Verschleiß an den Mahlwerkzeugen der Kohlemahlanlage (KT-Verg.)

54. Verschleiß am Gebläse für pneum. Transport (Kohlemahlanlage der KT-Verg.)

55. Verschleiß der Heißbrecher des gesinterten Gutes (Sinteranlage)

56. Verschleiß an Heißsieben für Sichtung des Gutes (Sinteranlage)

57. Verschleiß an Gebläseschaufeln von Entstaubungsanlagen (rohrgasseitig)

58. Verschleiß an Regel- und Drosselklappen in Abgasleitungen von Entstaubungsanlagen

59. Verschleiß an Mehrschalengreifern für Mineralien und Schrott

Tabelle 3/6

60. Verschleiß an Sinterwasserpumpen

61. Verschleiß an Kalibrier- und Einzugs-
rollen der Drahtdrehanlage

62 Verschleiß im Richtpolierwerk
(Kunststofflineale mit hohem Abrieb)

63. Verschleiß in der hinteren Hohlwellen-
führung der neuen Drahtdrehanlage

64. Verschleiß an Rollgangsrollen

65. Verschleiß an Führungsrollen in
Rollenführungen

66. Verschleiß an Drahtwalzen

67. Verschleiß an Rollgangsrollen

68. Verschleiß an Treibrollen

69. Verschleiß an Leisten für Rechenbett

70. Verschleiß auf Frei- und Spanfläche
bei CVD-beschichteten Wendeschneidplatten

71. Verschleiß an Schrämm-Meißeln beim
Kohleschrämmen

72. Verschleiß an Baggerzähnen

73. Verschleiß an Hartmetall-Werkzeugen
beim Tiefziehen

74. Verschleiß an Hartmetall-Werkzeugen beim Schneiden von Schwarzblechen

75. Verschleiß an Hartmetall-Werkzeugen bei Steinpressen

76. Verschleiß an Hartmetall-Strangpreßwerkzeugen

77. Verschleiß an HM-Warmwalzen für Drahtstraßen

78. Verschleiß an HM-Gleitringen für Kreiselpumpen

79. Verschleiß an HM-Plungern für Hochdruckverdichter

80. Verschleiß der Laufbuchse durch Kolbenringe

81. Verschleiß der Ventile

82. Verschleiß der Lagerlaufschichten

83. Verschleiß an den Kettengliedern einer Kette (Laufpolster)

84. Verschleiß an Triebrädern

85. Verschleiß des Laufrollenbelages

86. Verschleiß an Bremsscheiben

87. Verschleiß des Kettenendverbinders

88. Verschleiß des Mittelführungszahnes

89. Verschleiß an Rohrweichen

90. Verschleiß an 2-Wege-Schiebern in Förderleitungen

91. Verschleiß an Materialeinlaßventilen für Druckgefäßförderer

92. Verschleiß des Entspannungs- und Entlüftungsventils am Druckgefäßförderer

93. Verschleiß an Planetenkühler-Einläufen

94. Verschleiß an Planetenkühler-Einbauten (Hubelemente)

95. Verschleiß an Stützkonsolen und Schleppplatten von Recupolkühlern

96. Verschleiß an Unterlegblechen unter losen Laufringen an Drehöfen

97. Verschleiß am Lüfterrad eines Turbo-Windsichters

98. Verschleiß am Sichtergehäuse eines Turbo-Windsichters

99. Hammerverschleiß am Hammerbrecher bei Kalkstein

100. Verschleiß der Roststäbe am Hammerbrecher bei Kalkstein

101. Hammerverschleiß am Hammerbrecher bei Klinker

102. Verschleiß der Prallwände am Hammerbrecher

103. Verschleiß an den Mahlwerkzeugen einer Walzenschüsselmühle

104. Verschleiß an den Innenteilen einer Walzenschüsselmühle

105. Verschleiß an den Wälzlagern der Mahlrollen

106. Verschleiß an Laufring und Laufrolle des Drehtrichters für Hochofen-Gichtverschlüsse

107. Verschleiß an der kleinen Glocke für Hochofen-Gichtverschlüsse

108. Verschleiß am Flockensitzring für Hochofen-Gichtverschlüsse

109. Verschleiß an Schlagpanzer-Lamellen im Hochofen-Gichthals

110. Verschleiß der Rollen und der Laufflächen an den Laufringen eines Drehkonverters

Bild 3

Beantwortungsgrad zum Tribologie-Fragebogen
Fragenkomplex 1

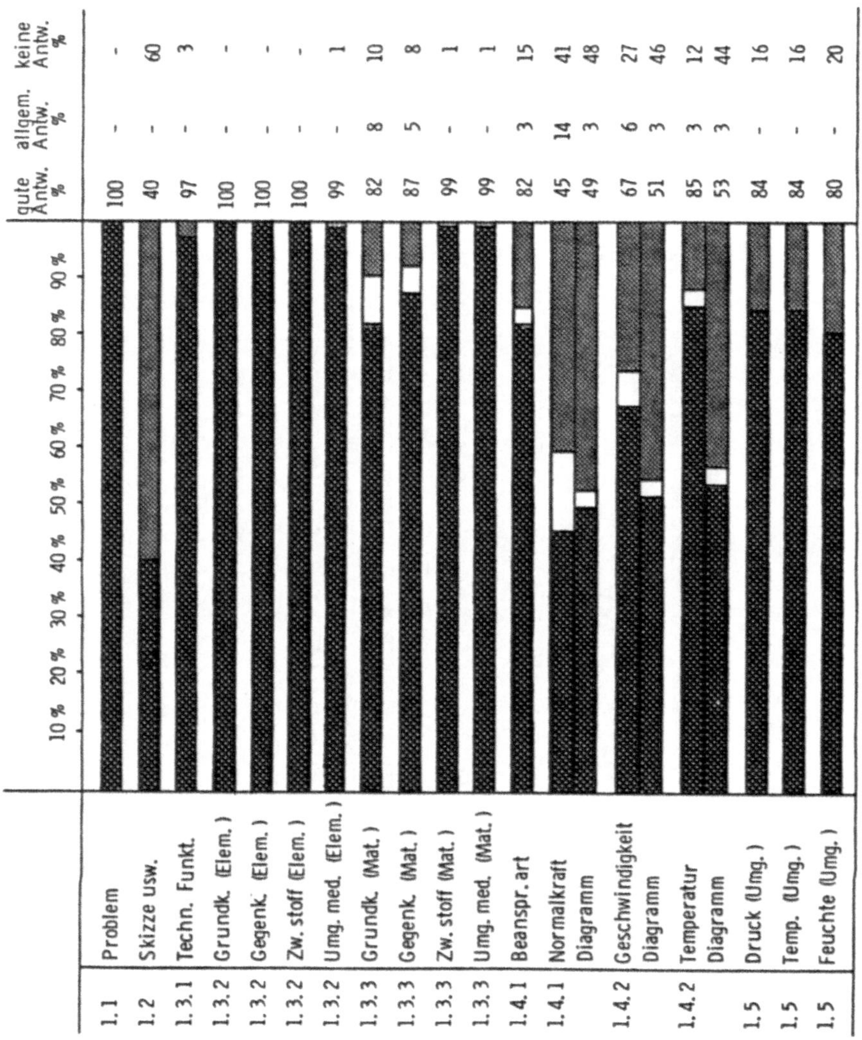

Beantwortungsgrad zum Tribologie-Fragebogen
Fragenkomplex 2 bis 4

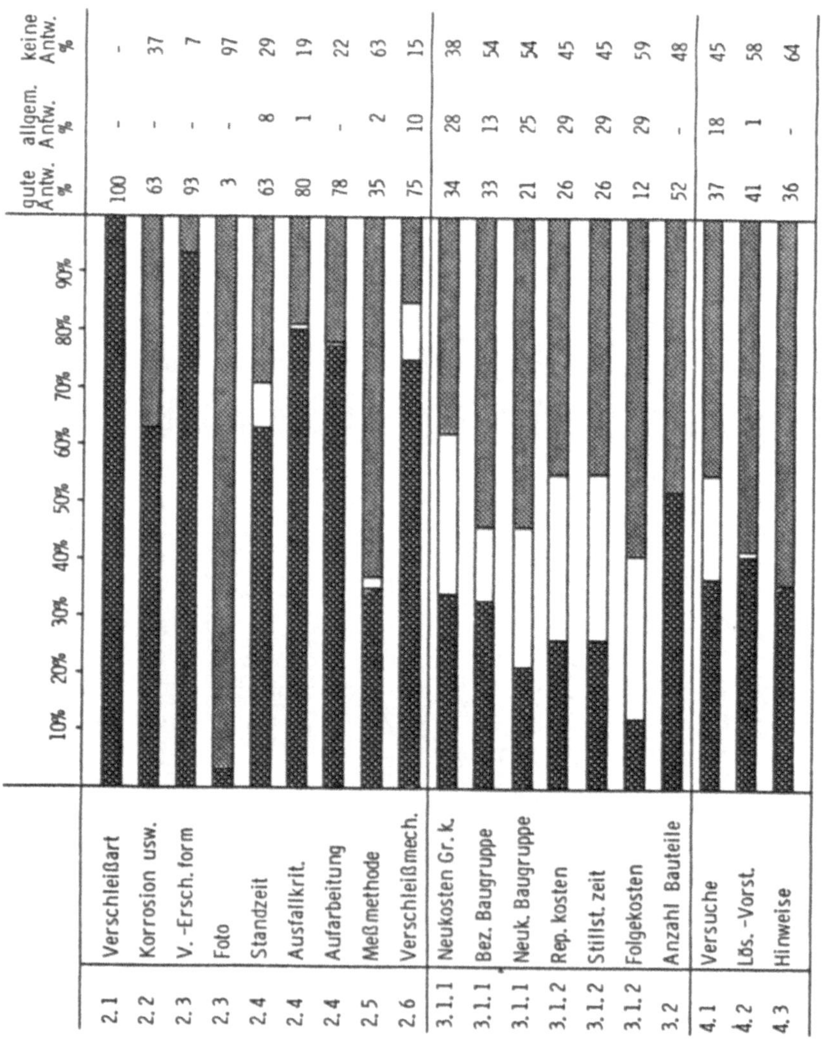

Wenig ergiebig war die Beantwortungsrate beim Fragenkomplex 3, so daß die erhoffte Aussage über die wirtschaftliche Bedeutung eines Verschleißproblems nur in wenigen Fällen möglich war. Auch Fragenkomplex 4 brachte nur wenige wirklich brauchbare Hinweise.

Zusammenfassend zeigt bereits diese Analyse des Beantwortungsgrades deutlich die Schwachstellen des Kenntnisstandes bei Verschleißproblemen auf, die häufig die systematische Bearbeitung verhindern, ohne daß hier Details eines Einzelfalles behandelt wurden.

Um die weitere Auswertung des Datenmaterials zu erleichtern und um eine Übersicht über alle benannten Verschleißfälle zu bekommen, wurden alle Antworten so auf 4 Tabellen übertragen, daß für jeden Einzelfall eine Zeile (insgesamt 110 Zeilen), für jede Fragestellung und deren Antwort eine Spalte (insgesamt 42 Spalten) bereitstanden.

Eine statistische Auswertung der benannten Verschleißfälle nach der Funktion des jeweiligen Verschleißsystems ist in Tabelle 4 dargestellt. Um alle Fälle erfassen zu können, wurden die im DIN-Entwurf vorgeschlagenen Funktionsbereiche entsprechend erweitert. Wie jeder Begriff verstanden wurde, zeigen die Beispiele. In einigen Fällen mußten Mehrfachzuordnungen vorgenommen werden. Zu erkennen ist eine deutliche Bevorzugung der Bereiche Materialtransport, Materialaufbereitung und Bewegungsausübung, die zusammen über 50 % der benannten Fälle ausmachen. In diese Bereiche fallen allerdings auch bevorzugt Verschleißsysteme aus dem Krupp'schen Großanlagenprogramm, bei denen Abrasivverschleiß auftritt.

Tabelle 4

Technische Funktion
der benannten Verschleißfälle (prozentuale Verteilung)

Funktion	%-Satz	Kurz-zeichen	Beispiele
Materialtransp.	23,2	MT	Pumpen, Gebläse, Greifer, Rohrleitungen, Pipelines, Rollgänge usw.
Materialaufber.	16,8	MB	Zerkleinern, Zerquetschen, Mischen, Trennen, Trocknen, Sichten usw.
Bewegungsausübung	14,4	BA	Laufräder, Gelenke, Lager, Gleitschienen, Tragrollen usw.
Führung	8,8	FR	Spurkränze, Gleitführungen, Führungsrohre, Führungsrollen, Führungszähne usw.
Materialumformung	7,2	MU	Strangpressen, Walzen, Schmieden, Ziehen usw.
Materialabtragung	6,4	MA	Baggern, Drehen, Schleifen usw.
Energieübertragung	4,8	EÜ	Kupplungen, Getriebe, (Gebläse) usw.
Abdichtung	4,8	AD	Stopfbuchsen, Ventile usw.
Materialverarb.	4,0	MV	Steinpressen, Ölpressen usw.
Steuerung und Regelung (mech.)	3,2	SR	Drosselklappen, Schieber, Weichen, Ventile usw.
Kühlung	2,4	KL	Kühlrohre
Bewegungshemmung	1,6	BH	Bremsen, Prallbleche
nicht zuzuordnen	2,4		
	100,0		

Die weitere Betrachtung galt schließlich dem Versuch einer Zusammenstellung vergleichbarer oder ähnlich zu behandelnder Verschleißfälle, um bereits an dieser Stelle des Vorhabens auch für die einzelnen Anwender Querverbindungen zum Erfahrungsaustausch zu schaffen. Andererseits sollte eine gewisse Konzentration der Einzelprobleme für die spätere Bearbeitung erreicht werden. Das Ergebnis dieser Analyse ist in Tabelle 5 dargestellt. Insgesamt ergaben sich 3 Gruppen zu 4 Problemen, 10 Gruppen zu 3 Problemen und 11 Gruppen zu 2 Problemen. Die 46 übrigen Probleme konnten u.E. nicht weiter zusammengefaßt werden.

3.1.2. Bewertung und Auswahl der Verschleißprobleme

Es sollten zunächst aus der Gesamtheit der 110 benannten Verschleißprobleme etwa 8 Verschleißkomplexe ausgewählt werden, die stellvertretend für alle anderen Probleme zur weiteren Erarbeitung der systematischen Vorgehensweise und zur Entwicklung eines technischen Lösungskonzeptes geeignet erschienen. Arbeitsschwerpunkt bei der Auswertung der detaillierten Fragebogenaktion war deshalb der Aufbau eines Bewertungsschemas und die Durchführung der Bewertung.

Mit den vorliegenden Erfahrungen wurde in Anlehnung an die Methoden der Nutzwert-Analyse [25] eine Kriterien-Hierarchie mit einer Gewichtung der über- und untergeordneten Kriterien entwickelt, Bild 5. Das Schema ist wie folgt zu lesen:

Oberstes Ziel (Gesamtziel) ist die Auswahl von ca. 8 Verschleißkomplexen, die aufgrund der vorhandenen Daten

Tabelle 5/1

Vergleichbare oder ähnlich zu behandelnde Verschleißfälle

3. Mühlenwalzen-Oberfläche
26. Flockierwalzen für Saatpartikel

---

9. Schake und Bolzen bei Eimerketten
10. Eimerkettengleitführung
33. Eimerketten

---

11. Eimerzähne
34. Eimer und Zähne von Baggern
72. Baggerzähne

---

12. Austragsarme von Bunkerentl.masch.
37. Auskleidungen von Schurren, Bunkern usw.
109. Schlagpanzer-Lamellen im HO-Gichthals

---

16. Seewasser-Pumpenläufer (Kav. u. Kor.)
17. Seewasser-Ejektoren (Kav. u. Kor.)

---

18. Seewasser-Pumpenläufer (Sand u. Kor.)
19. Seewasser-Pumpengehäuse (Sand u. Kor.)
20. Seewasser-Rohrleitungen (Sand u. Kor.)

---

22. Kastenplatten von Kalksandstein-Formwerkz.
23. Kastenplatten von Betonstein-Formwerkz.
75. Hartmetallwerkzeuge beim Steinpressen

---

27. Schneckengänge von Seiherschneckenpressen
28. Seiherstäbe von Seiherschneckenpressen

Tabelle 5/2

32. Kugel- und Rollendrehverb. (Bagger)
106. Laufring und Laufrolle (HO-Drehtrichter)
110. Rollen und Laufringe (Drehkonverter)

---

38. Kolbenstange von Hydraulikzyl.
79. HM-Plunger für Hochdruckverdichter

---

39. Brechwerkzeuge (Kreisel- u. Kegelbrecher)
55. Heißbrecher (Sinteranlage)

---

40. Exzenterbuchse der Exz.lagerung (Brecher)
41. Laufbuchse der Exz.lagerung (Brecher)

---

49. Kreiselpumpen (hydr. Asche-Transport)
50. Krümmer in Rohrleit. (hydr. Transport)
60. Sinterwasserpumpen

---

51. Mischbirne (pneum. Kohlenst. transp.)
52. Verblasungsrohr

---

53. Mahlwerkzeuge der Kohlemahlanlage
103. Mahlwerkzeuge einer Walzenschüsselmühle
104. Innenteile einer Walzenschüsselmühle

---

54. Gebläse für pneumatischen Transport
57. Gebläseschaufeln für Entstaubungsanl.
97. Lüfterrad eines Turbo-Windsichters
98. Sichtergehäuse eines Turbo-Windsichters

---

58. Regel- und Drosselklappen (Entst.anl.)
89. Rohrweichen
90. 2-Wege-Schieber in Förderleitungen

---

64. Rollgangsrollen I
65- Führungsrollen in Rollenführungen
67. Rollgangsrollen II
68. Treibrollen

---

66. Drahtwalzen
77. HM-Warmwalzen (Drahtstraße)

---

84. Triebräder (Panzerkette)
87. Kettenendverbinder (Panzerkette)

---

91. Materialeinlaßventil (Druckgef.förd.)
92. Entspannungsventil (Druckgef.förd.)

---

93. Planetenkühler-Einläufe
94. Planetenkühler-Einbauten (Hubelem.)
95. Stützkonsolen und Schleppplatten

---

99. Hammer am Hammerbrecher (Kalkstein)
100. Roststäbe am Hammerbrecher (Kalkstein)
101. Hammer am Hammerbrecher (Klinker)
102. Prallwände am Hammerbrecher (Klinker)

---

107. Kleine Glocke am HO-Gichtverschluß
108. Glockensitzring am HO-Gichtverschluß

Bild 5

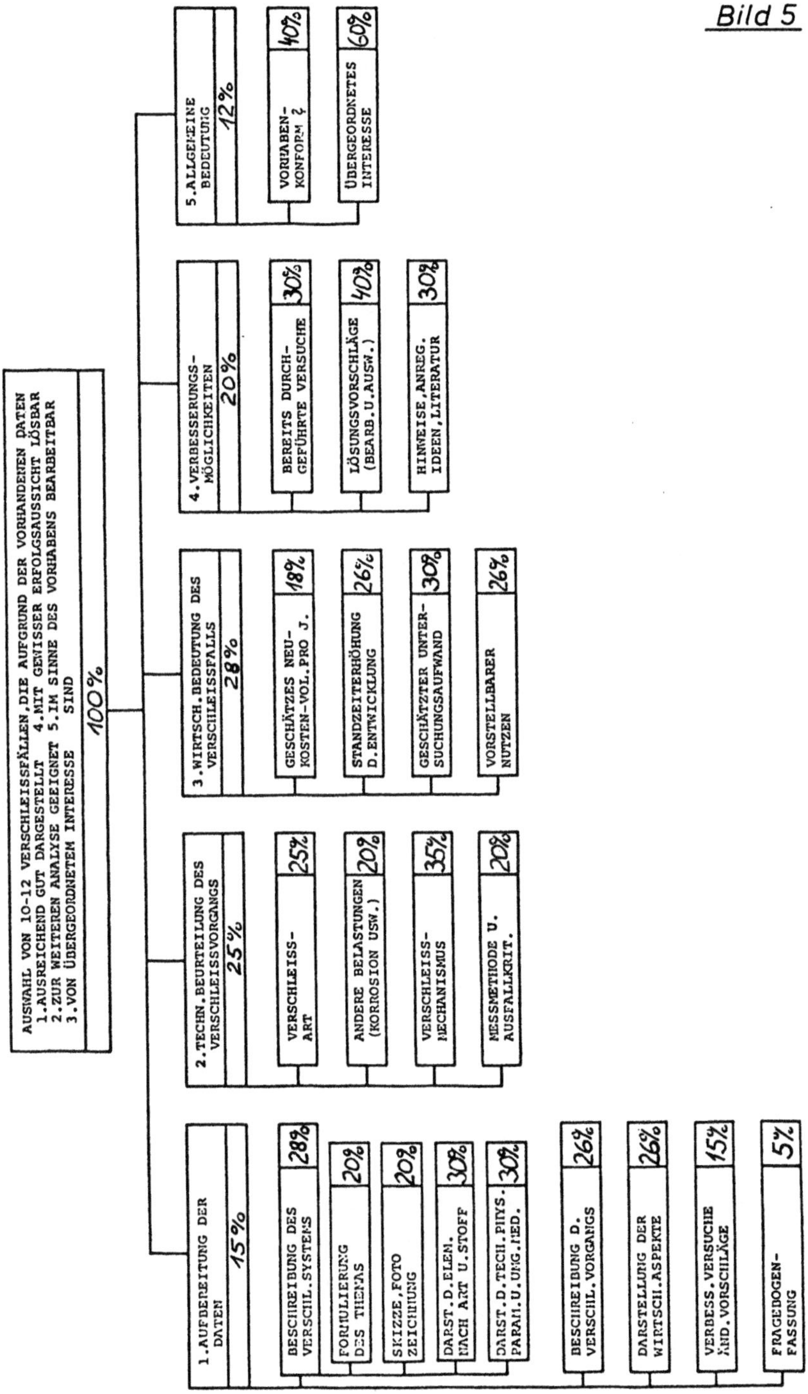

AUSWAHL VON 10-12 VERSCHLEISSFÄLLEN,DIE AUFGRUND DER VORHANDENEN DATEN
1.AUSREICHEND GUT DARGESTELLT 4.MIT GEWISSER ERFOLGSAUSSICHT LÖSBAR
2.ZUR WEITEREN ANALYSE GEEIGNET 5.IM SINNE DES VORHABENS BEARBEITBAR
3.VON ÜBERGEORDNETEM INTERESSE SIND
100%
1.AUFBEREITUNG DER DATEN
15%
BESCHREIBUNG DES VERSCHL.SYSTEMS
28%
FORMULIERUNG DES THEMAS
20%
SKIZZE,FOTO ZEICHNUNG
20%
DARST.D.ELEM. NACH ART U.STOFF
30%
DARST.D.TECH.PHYS. PARAM.U.UMG.MED.
30%
BESCHREIBUNG D. VERSCHL.VORGANGS
26%
DARSTELLUNG DER WIRTSCH.ASPEKTE
26%
VERBESS.VERSUCHE UND.VORSCHLÄGE
15%
FRAGEBOGEN-FASSUNG
5%
2.TECHN.BEURTEILUNG DES VERSCHLEISSVORGANGS
25%
VERSCHLEISS-ART
25%
ANDERE BELASTUNGEN (KORROSION USW.)
20%
VERSCHLEISS-MECHANISMUS
35%
MESSMETHODE U. AUSFALLKRIT.
20%
3.WIRTSCH.BEDEUTUNG DES VERSCHLEISSFALLS
28%
GESCHÄTZES NEU-KOSTEN-VOL.PRO J.
18%
STANDZEITERHÖHUNG D.ENTWICKLUNG
26%
GESCHÄTZTER UNTER-SUCHUNGSAUFWAND
30%
VORSTELLBARER NUTZEN
26%
4.VERBESSERUNGS-MÖGLICHKEITEN
20%
BEREITS DURCH-GEFÜHRTE VERSUCHE
30%
LÖSUNGSVORSCHLÄGE (BEARB.U.AUSW.)
40%
HINWEISE,ANREG. IDEEN,LITERATUR
30%
5.ALLGEMEINE BEDEUTUNG
12%
VORHABEN-KONFORM ?
40%
ÜBERGEORDNETES INTERESSE
60%

1. ausreichend gut dargestellt
2. zur weiteren Analyse geeignet
3. von übergeordnetem Interesse
4. mit gewisser Erfolgsaussicht lösbar
5. im Sinne des Vorhabens bearbeitbar

sind. Dieses oberste Ziel vereinigt in sich alle Unterziele und wird folglich mit 100 % angesetzt. Die 2. Zielebene bringt die Aufteilung in mehrere Hauptkriterien, deren Gewicht im Hinblick auf das Gesamtziel festzulegen ist. Im vorliegenden Fall wurden folgende Hauptkriterien erarbeitet und gewichtet:

1. Aufbereitung der Daten	15 %
2. Technische Beurteilung des Verschleißvorgangs	25 %
3. Wirtschaftliche Bedeutung des Verschleißfalls	28 %
4. Verbesserungsmöglichkeiten	20 %
5. Allgemeine Bedeutung	12 %

Die Gesamtsumme der Gewichte ergibt wieder 100 %. Aus Bild 5 ist weiterhin zu ersehen, wie die Hauptkriterien der 3. und 4. Zielebene unterteilt worden sind und wie hier die Gewichte vergeben wurden. Die Summe der Unterkriteriengewichte, die zu einem übergeordneten Kriterium gehören, ergibt wieder 100 %. Am Ende der Hierarchie stehen die sog. Bewertungskriterien, deren Erfüllungsgrad für jedes Verschleißproblem festzustellen ist. Das endgültige Gewicht eines Bewertungskriteriums ergibt sich aus der Multiplikation seiner eigenen Wichtung als Unterkriterium mit den Wichtungen seiner übergeordneten Kriterien. So beträgt z.B. die endgültige Wichtung des Bewertungskriteriums "Darstellung der Elemente nach Art und Stoff"

$$30\ [\%] \times 28[\%] \times 15[\%] \mathrel{\hat{=}} 0{,}30 \times 0{,}28 \times 0{,}15 = 0{,}0126 \mathrel{\hat{=}} 1{,}26[\%]$$

Da die Wichtung durchaus von subjektiven Betrachtungsweisen beeinflußt sein kann, bietet die Auflistung der Bewertungskriterien in der Reihenfolge der entgültigen Wichtung, Tabelle 6, eine gewisse Kontrolle und Korrekturmöglichkeit. (Um die Vergleichbarkeit zu sichern, wurden nur die Kiterien der 3. Ebene betrachtet.)

Das Bewertungsschema wurde nun so weiterentwickelt, daß jedem Bewertungskriterium 5 Stufen der Erfüllung zuzuordnen waren.

Die Stufe (oder Punktzahl) 0 bedeutet "keine Erfüllung" des Kriteriums, die Stufe (oder Punktzahl) 4 "maximale Erfüllung" des Kriteriums. Der weitere Schritt war, jedem Erfüllungsgrad für jedes Bewertungskriterium eine qualitative oder quantitative Bedeutung zu geben. Es soll hier darauf verzichtet werden, das gesamte Schema der je 5 Erfüllungsgrade für 21 Bewertungskriterien darzustellen, 3 Beispiele mögen es jedoch verdeutlichen.

Bewertungskriterium 1.1.2: Skizze, Foto, Zeichnung

0 = nein
4 = ja

Bewertungskriterium 2.1 : Verschleißart

0 = keine Angaben,Verschleißart nicht erkennbar
1 = komplex und unübersichtlich
2 = -
3 = mit Einschränkung erkennbar
4 = eindeutig erkennbar

Bewertungskriterium 4.2 : Lösungsvorschläge

0 = keine Angaben
1 = allgemeine Angaben
2 = allgemeine Lösungvorschläge
3 = Lösungsvorschläge als Anregung
4 = gute Lösungsvorschläge für gezieltes Arbeitsprogramm

Tabelle 6

Gewichtung der Bewertungskriterien

	Gewichtung %
Verschleißmechanismus	8,75
Geschätzter Untersuchungsaufwand	8,40
Lösungsvorschläge	8,00
Standzeiterhöhung durch Entwicklung	7,28
Vorstellbarer Nutzen	7,28
Übergeordnetes Interesse	7,20
Verschleißart	6,25
Bereits durchgeführte Versuche	6,00
Hinweise, Anregungen, Ideen, Literatur	6,00
Geschätztes Neukosten-Volumen pro Jahr	5,04
Andere Belastungen (Korrosion usw.)	5,00
Meßmethode und Ausfallkriterien	5,00
Vorhabenkonform?	4,80
Beschreibung des Verschleißsystems	4,20
Beschreibung des Verschleißvorgangs	3,90
Darstellung der wirtschaftlichen Aspekte	3,90
Darstellung d.Verbesserungsvers. und Änderungsvorschläge	2,25
Fragebogen-Fassung	0,75
	100,00

Für jedes Bewertungskriterium und für jedes Verschleißproblem war der Erfüllungsgrad zu bestimmen. Die Summe der mit den jeweiligen Gewichtungen (in Prozentzahlen) multiplizierten Erfüllungsgrade der Bewertungskriterien ergab schließlich eine Zahl zwischen 0 und 400 für jeden Verschleißfall. Diese Zahl wurde als Maß für die Einstufung und Auswahl der ersten 10 Verschleißfälle verwendet.

Die erreichten Zahlenwerte lagen für die 110 Verschleißfälle zwischen 114 und 288, so daß der am schlechtesten bewertete Fall 28,5 %, der am besten eingestufte 72,0 % des Maximalwertes von 400 erreichte. Rang 10 wurde bereits mit 259 bewertet, so daß das Verfahren insbesondere in der Spitzengruppe noch ausreichend grob wirkte, um merkliche Unterschiede zwischen den einzelnen Rängen herauszuarbeiten. In Tabelle 7 sind die ersten 10 Verschleißfälle, geordnet nach der erzielten Bewertungssumme, dargestellt.

Um nun die in der Bewertung enthaltenen subjektiven Einflüsse auszugleichen, wurde das Ergebnis mit der in Tabelle 5 aufgeführten Gruppenbildung vergleichbarer oder ähnlich zu behandelnder Verschleißfälle gekoppelt. Damit können im weiteren Verlauf des Vorhabens auch solche Probleme mit berücksichtigt werden, die aufgrund des unzureichenden Kenntnisstandes und des nicht verfügbaren Datenmaterials in der Bewertung schlechter abgeschnitten haben, die aber dennoch als Einzelfall Bedeutung haben. Die so gebildeten 8 Verschleißkomplexe sind in Tabelle 8 zusammengestellt. Sie sollen die Basis der weiteren Analysen bilden.

Mit Hilfe der in Tabelle 4, Tabelle 9 und Tabelle 10 eingeführten Kurzzeichen für

a) Technische Funktion, Tabelle 4,
b) Verschleißart, Tabelle 9,
und

Tabelle 7

Rangfolge der Verschleißfälle
(Ergebnis der Bewertung)

Rang	Bewert. Summe	Thema	Kennziffer
1.	287.87	Auskleidungen von Schurren,Bunkern usw.	(37)
2.	282.47	Führungen von Draht- walzwerken	(45)
3.	280.32	Eimer und Zähne von Baggern	(34)
4.	277.74	Schake und Bolzen bei Raupen- und Eimerketten	(9)
5.	274.82	Eimerketten	(33)
6.	266.92	Kugel- und Rollen- drehverbindungen	(32)
7.	264.87	Tragrollenmäntel (Förderbänder)	(36)
8.	262.90	Eimerketten-Gleitführung	(10)
9.	261.33	Flockierwalzen für Saatpartikel	(26)
10.	259.51	Führungsrollen in Rollenführungen	(65)

Tabelle 8

Auswahl der weiter zu untersuchenden Verschleißprobleme

		Kennziffer ( )
1.	Auskleidungen von Schurren, Bunkern, Gutübergabestellen usw.	(37)
	Austragsarme von Bunkerentleerungsmaschinen	(12)
	Schlagpanzerlamellen im Hochofen-Gichthals	(109)
2.	Führungen von Drahtwalzwerken	(45)
3.	Eimer und Zähne von Baggern	(34)
	Eimerzähne	(11)
	Baggerzähne	(72)
4.	Schake und Bolzen bei Eimerketten	(9)
	Eimerketten	(33)
	Eimerketten-Gleitführungen	(10)
5.	Kugel- und Rollendrehverbindungen von Baggern	(32)
	Laufring und Laufrolle des Drehtrichters am Hochofen-Gichtverschluß	(106)
	Rollen und Laufringe eines Drehkonverters	(110)
6.	Tragrollenmäntel von Förderband-Tragrollen	(36)
7.	Flockierwalzen für Saatpartikel	(26)
	Mühlenwalzen-Oberflächen für Zellstoff	(3)
8.	Führungsrollen in Rollenführungen	(65)
	Rollgangsrollen	(67)
	Treibrollen	(68)

Tabelle 9

Gliederung des Verschleißgebietes nach der Art der tribologischen Beanspruchung

Verschleiß-paarung	Tribologische Beanspruchung (Symbole)	Verschleiß-art	Kurz-zeichen
Festkörper -Festkörper (ohne oder mit Schmierung)	Gleiten	Gleitverschleiß	GL
	Rollen Wälzen	Rollverschleiß Wälzverschleiß	RW
	Bohren	Bohrverschleiß	BV
	Prallen Stoßen	Stoßverschleiß	ST
	Oszillieren	Schwingungs-verschleiß	SW
Festkörper -Festkörper und Partikel	Gleiten	Korngleit-verschleiß	KG
	Wälzen	Kornwälz-verschleiß	KW
Festkörper -Flüssigkeit	Strömen	Kavitations-verschleiß	KA
	Prallen	Tropfenschlag-verschleiß	TS
Festkörper -Flüssigkeit mit Partikeln	Strömen	Spülverschleiß (Erosionsverschleiß)	SP
Festkörper -Gas	Strömen	Ablativ-verschleiß	AV
Festkörper -Gas mit Partikeln	Strömen	Gleitstrahl-verschleiß (Erosionsverschleiß)	GS
	Prallen	Prallstrahl-, Schrägstrahl-verschleiß	PS

<u>*Tabelle 10*</u>

Verschleißerscheinungsformen

Die Veränderungen der Oberflächen von Grundkörper (1) und Gegenkörper (2) durch einen Verschleißvorgang werden durch verschiedene „Verschleißerscheinungsformen" charakterisiert. In Anlehnung an DIN 4761 (Begriffe, Benennungen und Kurzzeichen für den Oberflächencharakter) können unter anderem die nachfolgend aufgeführten makroskopischen Verschleißerscheinungsformen unterschieden werden (siehe Bilder 3 bis 10). (Bei Überlagerung der dargestellten Verschleißerscheinungsformen sind gegebenenfalls die Flächenanteile der einzelnen Verschleißerscheinungsformen innerhalb der gesamten Verschleißflächen anzugeben.)

Makroskopische Verschleißerscheinungsformen:

– Löcher: örtlich begrenzte Vertiefungen der Oberfläche von geringen Maßen (Beispiel: durch Heraustrennen lokaler Adhäsionsbindungen entstandene Oberflächenschäden)

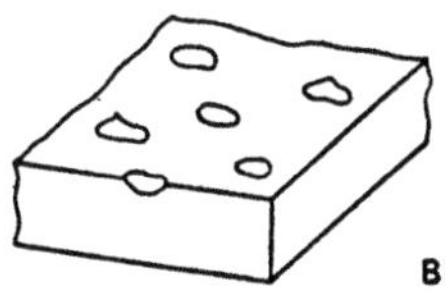

Lö

Bild 3. Löcher

– Mulden: flache Vertiefungen der Oberfläche von rundlicher oder eckiger Form (Beispiel: Oberflächenschäden durch strömende Flüssigkeiten infolge Kavitation)

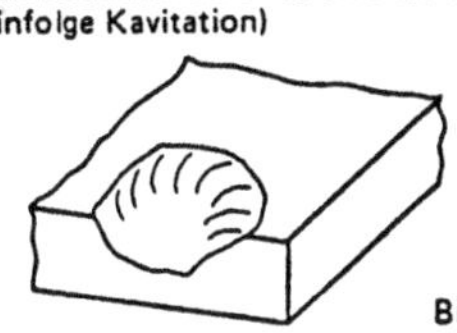

Mu

Bild 4. Mulde

– Wellen: periodische Anordnung von muldenähnlichen Oberflächenveränderungen (Beispiel: Oberflächenschäden durch strömende Flüssigkeiten infolge Erosion)

We

Bild 5. Wellen

– Kuppen: örtlich begrenzte rundliche oder kantige Erhöhungen der Oberfläche in festhaftender oder loser Form (Beispiel: durch Adhäsionsprozesse übertragene Werkstoffpartikel)

Ku

Bild 6. Kuppe

– Schuppen: Überschichtungen der Oberfläche (Beispiel: bei adhäsivem Verschleiß übertragene Werkstoffbereiche)

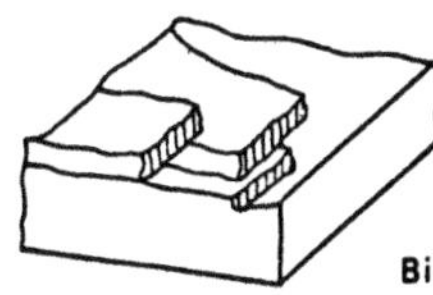

Sp

Bild 7. Schuppen

– Riefen: grabenförmige Verschleißspur großer Länge (Beispiel: Spur eines abrasiv wirkenden harten Partikels)

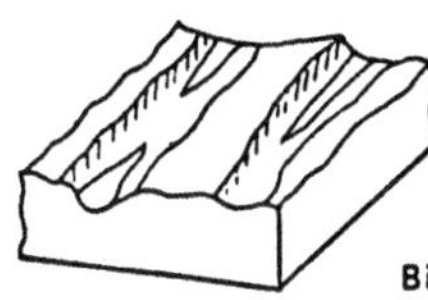

Rf

Bild 8. Riefen

– Kratzer: riefenähnliche Oberflächenbeschädigung geringer Ausdehnung und Länge (Beispiel: lokale Wirkung eines abrasiv wirkenden harten Partikels)

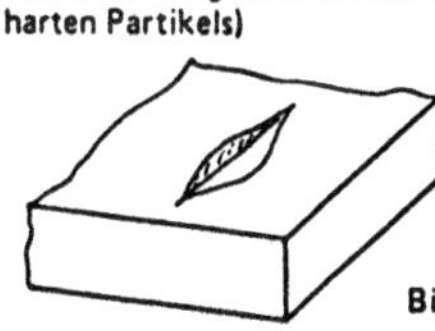

Kr

Bild 9. Kratzer

– Risse: örtlich begrenzte Trennung des Werkstoffgefüges von geringer Breite aber oft beträchtlicher Länge und Tiefe (Beispiel: Erscheinungsform der Oberflächenzerrüttung)

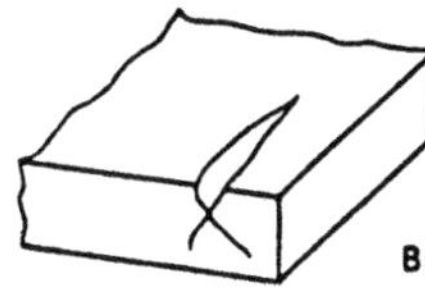

Rs

Bild 10. Riß

allgemein: MA ≙ Materialabtrag

c) Verschleißerscheinungsform, Tabelle 10,

wurden alle Einzelverschleißfälle in ihrer tribologischen Analyse vorgestellt. Eingeschlossen sind die Beurteilung der wirtschaftlichen Aspekte sowie die Abschätzung der realen Lösungsmöglichkeiten der einzelnen Verschleißfälle. Die Mitteilung der tribologischen Darstellung eines Verschleißfalles erfolgt stellvertretend und beispielhaft für alle 110 untersuchten Fälle in Tabelle 11 mit dem Verschleißproblem "Schake und Bolzen bei Eimerketten".
Das hier vorgestellte und durchgeführte Verfahren zur Auswahl geeigneter Verschleißthemen wurde eingehend im Krupp-internen Arbeitskreis diskutiert und kritisiert, letztlich aber als brauchbare Lösung anerkannt, um die Fülle des z.T. sehr komplexen Antwortmaterials auswählend zu beurteilen. Das Verfahren enthält ohne Zweifel sehr viele subjektive Einflüsse der Bearbeiter, was sich in der Aufstellung der Kiterienhierarchie, in der Gewichtung der Kriterien, in der Festlegung und Bedeutung der Erfüllungsgradstufen und letztlich in der Abschätzung des Erfüllungsgrades selbst äußert. Dabei muß anerkannt werden, daß andere Bewerter - die Bewertung wurde von den beiden Verfassern dieses Berichtes durchgeführt - potentiell andere Gesichtspunkte in den Vordergrund stellen können und damit zu anderen Ergebnissen kommen können. Die Diskussion im Arbeitskreis zeigte jedoch, daß die vorliegende Beurteilung sachlich-fachlich zu vertreten ist und anerkannt wurde, so daß eine ausreichende Objektivität als gesichert gelten kann. Insbesondere liegt hier ein Instrumentarium vor,das in jeder Einzelentscheidung nachprüfbar und dessen Ergebnis im Zweifelsfall auch korrigierbar ist. Es ist trotz des erhöhten Aufwandes einer pauschal abschätzenden, in ihren Einzelheiten nicht nachvollziehbaren Betrachtung zur Auswahl geeigneter Verschleißfälle vorzuziehen.

Tabelle 11

FRAGEBOGEN ZUR BESCHREIBUNG VON VERSCHLEISSFÄLLEN.

VERSCHLEISSPROBLEM : SCHAKE UND BOLZEN BEI EIMERKETTEN (9)

TECHN. FUNKTION : MT, MB, BA, FR, MU, MA, EÜ, AD, MV, SR, KL, BH

ELEMENTE/MATERIAL :
GRUNDKÖRPER : BOLZEN/STAHL
GEGENKÖRPER : SCHAKE/STAHL
ZWISCHENSTOFF : FÖRDERGUT (KOHLE, KREIDE)/ = TROCKEN BIS NASS
UMGEBUNGSMEDIUM : LUFT OD. WASSER/=

TECH. PHYSIK. BEANSPRUCHUNGSPARAMETER: BEANSPRUCHUNGSART: DAUERBETR. 5-20 h/d

NORMALKRAFT: 40 - 70 N/mm²

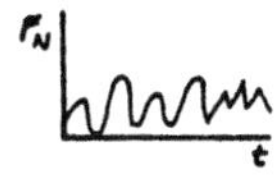

$F_N = f(t)$?

GESCHWINDIGKEIT:

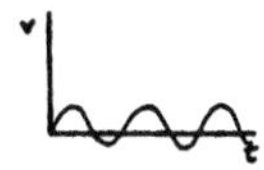

$v = f(t)$?

TEMPERATUR: -10 BIS +30°C

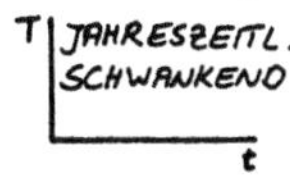

$T = f(t)$?

UMGEBUNGSMEDIUM:
DRUCK : NORMAL
TEMPERATUR : UMGEB.
FEUCHTIGKEIT: TROCKEN, BEI REG. NASS

BESCHREIBUNG DES VERSCHLEISSVORGANGES:
VERSCHLEISSART: GL, RW, BV, ST, SW, KG, KW, KA, TS, SP, AV, GS, PS
KORROSION, ZUNDERUNG, SUBLIMATION : K, Z, S
VERSCHLEISSFORM : Lö, Mu, We, Sp, Rf, Ku, Rs, (MA) Kr
LEISTUNGSGRÖSSE: STANDZEIT : 10% DER GERÄTELEBENSDAUER
STANDWEG :
STANDMENGE :
AUSFALLKRITERIUM: ZU GROSSES SPIEL BEI 15-20% ABNUTZ. AUSW. O. BOLZEN

AUFARBEITUNG : JA/NEIN
MESSMÖGLICHKEIT: 15-20% ABNUTZ. = STANDZ. ENDE
VERSCHLEISSMECHANISMEN : α) ; β) ; γ)

WIRTSCHAFTLICHE ASPEKTE :
NEUKOSTEN GRUNDKÖRPER :
BEZEICHNUNG BAUGRUPPE: } ZU UNTERSCHIEDLICHE WERTE
NEUKOSTEN BAUGRUPPE:
REP. U. EINFAHRKOSTEN:
STILLSTANDZEIT/-KOSTEN: } KOSTEN TRÄGT BETREIBER
FOLGEKOSTEN :
ANZAHL BAUTEILE : $< 10^3$

VERBESSERUNGSVERSUCHE/ÄNDERUNGSVORSCHLÄGE:
VERSUCHE ZUR VERSCHLEISSMINIMIERUNG: JA, 12%-IGE Mn-STÄHLE AM BESTEN BEWÄHRT

VORSTELLUNGEN ZUR LÖSUNG : STANDZEITVERL. DURCH BESSERE WERKSTOFFE

HINWEISE ZUR LÖSUNG:

Teil II des Forschungsvorhabens soll sich speziell mit den folgenden Schwerpunktthemen - einer sinnvollen Auswahl aus den in Tabelle 8 näher erläuterten Verschleißfällen - befassen:

1. Verschleiß an Auskleidungen von Schurren, Bunkern, Gutübergabestellen sowie an Austragsarmen von Bunkerentleerungsmaschinen.

2. Verschleiß an Schlagpanzerlamellen im Hochofen-Gichthals.

3. Verschleiß an Zähnen und Eimern von Großraumbaggern im Braunkohlentagebau.

4. Verschleiß an Schake und Bolzen von Raupen- und Eimerketten sowie an Raupen- und Eimerketten-Gleitführungen.

3.2. Daten- und Informationsbeschaffung, Literaturstudium

Wie bereits erwähnt, kommt den zuvor aufgeführten vier Verschleißfällen bei der weiteren Bearbeitung des F.- + E.-Vorhabens besondere Priorität und Bedeutung zu. Somit galt diesen Themen auch eine besondere Aufmerksamkeit bezüglich Daten- und Informationsbeschaffung in den entsprechenden Betrieben und bezüglich des Literaturstudiums. Die Bemühungen zur Daten- und Informationsbeschaffung gestalteten sich bisweilen unerwartet schwierig, da viele sachliche und gelegentlich auch persönliche Vorurteile abzubauen bzw. zu überwinden waren.

3.2.1. Verschleiß an Auskleidungen von Schurren, Bunkern, Gutübergabestellen sowie an Austragsarmen von Bunkerentleerungsmaschinen

Der Verschleiß an den Auskleidungen der genannten

Elemente der Förder- und Lagertechnik erfolgt im wesentlichen durch Prall- bzw. Stoßverschleiß und nach "Beruhigung" des Fördergutes durch Gleitverschleiß [26, 27], die im Sinne von DIN 50.320 als eine Variante des Abrasivverschleißes zu werten sind. In Bild 6 werden in schematischer Darstellung die genannten Elemente der Förder- und Lagertechnik sowie das Prinzip des Prallverschleißes skizziert. Bild 7 zeigt eine Schurre in Kombination mit einem Schaufelrad. Um den Prallverschleiß generell zu mindern, ist man bemüht, das Fördergut auf glatten und hoch verschleißbeständigen Flächen zu führen, ohne daß eine Gelegenheit zum Anbacken des Fördergutes geboten wird. Dieses ist für die verschiedenen Industriezweige von Wichtigkeit, wie zum Beispiel für die Erzbunker eines Hochofenwerkes [28]. Hier wird nachgewiesen, daß die Auskleidung von Erzbunkern mit Schmelzbasalt eine beträchtliche Standzeit-Erhöhung liefert. Der beschriebene Prallverschleiß (≙ Stoßverschleiß im Sinne von DIN 50.320) und auch der erwähnte Gleitverschleiß sind Furchungsvorgänge, bei denen das meist mineralische Korn des Fördergutes durch ritzende oder schneidende Beanspruchung auf der Lauffläche (≙ Auskleidung) eine Mikrozerspanung verursacht, die zum einen einen tief in den Grundkörper hineinreichenden Vorgang darstellt, Bild 8, und die zum andern nicht nur an der Meßstelle, sondern an beliebig vielen anderen Stellen (bei offenen Verschleißsystemen) erfolgen kann [29, 30]. Ein weiterer wichtiger Parameter beim Prallverschleiß ist darin zu sehen, unter welchem Winkel das mineralische Fördergut auf die Schurre bzw. auf die Gutübergabestelle aufgegeben bzw. aufgebracht wird [22, 31]. Bild 9 zeigt den Strahlverschleiß von zwei Stählen als Grundkörpern in Abhängigkeit vom Anstrahlwinkel des Blasgutes Quarzsand (Gegenkörper): bei einem Anstrahlwinkel von 30° überwiegt der Abrasivverschleiß als dominierender Verschleißmechanismus, während

Bild 6

Schematische Darstellung von Rutschen, Schurren und Rinnen

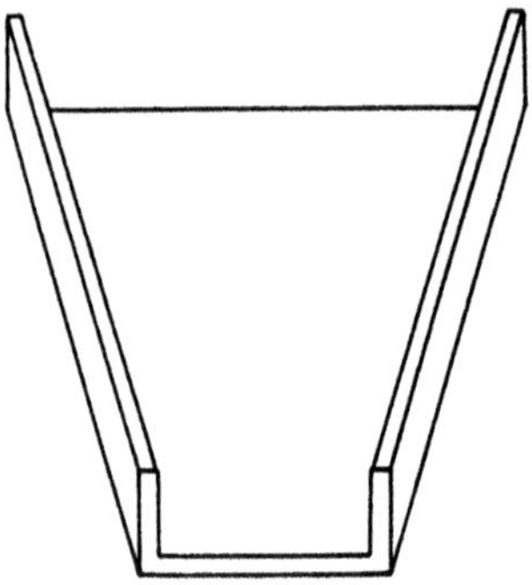

Rutsche

Drehschurre

Art der tribologischen Beanspruchung:

Systemstruktur:

Festkörper
- Festkörper
  (bei Festkörperreibung,
  Grenzreibung,
  Mischreibung)

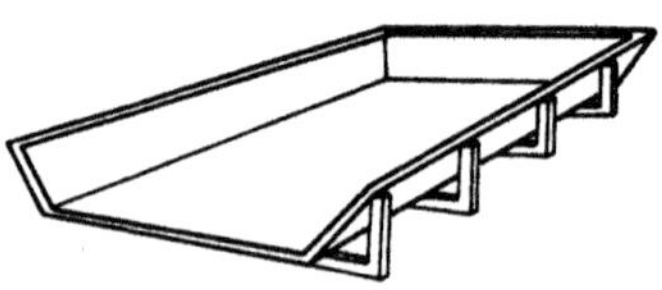

Rinne

Tribologische Beanspruchung (symbolisch) und Verschleißart:

Prallen Stoßen	Prallverschl. Stoßverschleiß
Gleiten	Gleitverschl.

Bild 7

Schurre in Kombination mit einem Schaufelrad

Photo:

KRUPP INDUSTRIE- und STAHLBAU, Rheinhausen

Bild 8

Schematische Darstellung von Furchungsverschleiß (Abrasiv-Verschleiß) [nach H. Uetz]

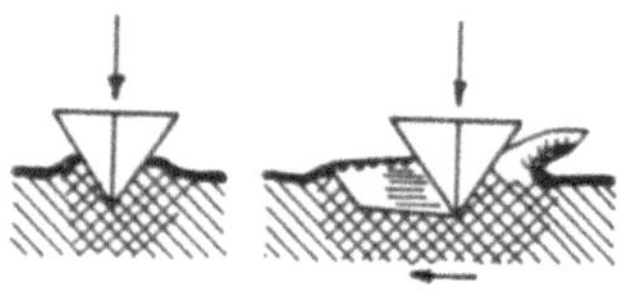

Bild 9

Strahlverschleiß von zwei Stählen in Abhängigkeit vom Anstrahlwinkel [nach K. Wellinger und H. Uetz]

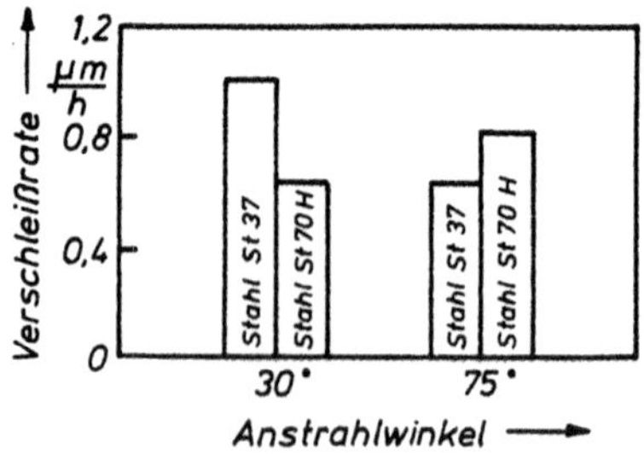

bei einem Anstrahlwinkel von 75° die Oberflächenzerrüttung als Verschleißmechanismus dominant wird und somit der härtere der beiden Stähle schneller verschleißt.

Auch bei der Förderung von Braunkohle und Abraum aus dem Deckgebirge lassen sich durch konstruktive Gestaltung der Geometrie der Übergabe-Schurren deutliche Vorteile im Betriebsverhalten nachweisen [32]. Diese Vorteile konnten durch eigene Beobachtungen bestätigt und weiter gesteigert werden. Auf die bessere, gezieltere Nutzung sog. verschleißkonformer Konstruktionen bei der Auskleidung von Rutschen und Schurren sollte verstärkte Aufmerksamkeit gerichtet werden [18], als es bislang dem "Verschleiß"-Bewußtsein von Werkstoffkundlern, Verfahrenstechnikern und Konstrukteuren entsprochen hat.

Bei der Auskleidung von Schurren, Rinnen, Bunkern und Gutübergabestellen werden in neuerer Zeit [33, 34] zunehmend Kautschuk- bzw. Spezialgummi-Werkstoffe mit großem Erfolg für die Verschleißminderung eingesetzt. In besonderen Fällen werden Gummi-Beläge mit Gewebeverstärkung verwendet. Als Fördergut (Gegenkörper) werden Erze, Hartkeramik, Schotter und Quarz genannt. Neben dem Vorteil der Verschleißminderung ist weiterhin eine Reduzierung des Lärm-Pegels zu nennen. Zudem ist hervorzuheben [28], daß die Aufprall-Energie des Fördergutes bei Einsatz von Spezialgummi-Auskleidungen merklich gemindert werden kann, was wiederum Vorteile für die Gestaltung der Tragkonstruktionen von Schurren, Rinnen, Bunkern und Gutübergabestellen liefert. In diesem Zusammenhang sei auch auf den Einsatz von höchstmolekularem Niederdruck-Polyäthylen [35] verwiesen, welches verschleiß- und lärmmindernd zur Auskleidung von Förder- und Lager-Elementen für die folgenden Güter (Gegenkörper) eingesetzt wird:

Steinkohle;
Braunkohle;
Zementrohstoffe;

Steinsalz;

Schwefelkies;

Formsand;

Kraftfutterrohstoffe, Weißzucker, Phosphat, Soda, Borax;

Gips;

Titanerz;

Rohphosphat;

Rohkali;

Das in Rede stehende höchstmolekulare Niederdruck-Polyäthylen hat ein Molekulargewicht von $\sim 10^6$, verhält sich gegenüber den genannten Fördergütern sehr abrasivverschleißmindernd, weist auch im Tieftemperaturbereich gute Zähigkeitseigenschaften auf und hat sehr günstige Gleiteigenschaften.

3.2.2. Verschleiß an Schlagpanzerlamellen im Hochofengichthals

Gegenüber dem im wesentlichen reinen Abrasiv-Verschleiß, den das Fördergut auf/in Rutschen, Schurren, Bunkern und Gutübergabestellen auf dem jeweiligen Grundkörper verursacht (im wesentlichen mit Luft bei Raumtemperatur als Umgebungsmedium), ist der Verschleiß an den Schlagpanzerlamellen im Hochofengichthals vergleichsweise komplexerer Natur. Grundkörper sind die Schlagpanzerlamellen, sog. innere und äußere Lamellenplatten, die beispielsweise aus GS-17 CrMo 55 bestehen können. Als Gegenkörper ist die Möller-Beschickung zu sehen, die im wesentlichen aus Koks, Erz (Pellets) und diversen Zuschlagstoffen (Kalkstein etc.) besteht. Neben Prall- ($\hat{=}$ Stoß-) und Gleitverschleiß sind die Schlagpanzerlamellen, die bekanntlich das Mauerwerk im Hochofengichthals beim Herabfallen des chargierten Möllers schützen sollen, weiterhin thermischen und atmosphärischen Beanspruchungen ausgesetzt. An der Unterkante der Schlagpanzerlamellen können durchaus Temperaturen bis zu 900°C auftreten, und die Atmosphäre in diesem Bereich des Hochofens kann neben Kohlenmonoxid, Kohlendioxid, Wasserdampf noch Kohlenwasserstoffe und evtl. auch Spaltungskohlenstoff aufweisen. Zudem dürften die thermischen und atmosphärischen Bedingungen während des Hochofenbetriebs laufenden Schwankungen unterworfen sein, so daß eine Definition bzw. Fixierung dieser Parameter nicht möglich sein kann. Die Bilder 10 und 11 zeigen eine innere bzw. äußere Schlagpanzerlamelle, die i.a. als Stahlgußkörper ausgebildet werden und die zudem durch Auftragsschweißen zusätzlich zum Zweck der Verschleißminderung auf den beanspruchten Flächen "gepanzert" werden. Bild 12 dokumentiert anhand einiger Verschleißbilder an Schlagpanzerlamellen, daß diese bereits innerhalb eines Jahres starken Verschleißangriff zeigen.

Neben den genannten thermischen und atmosphärischen Beanspruchungen werden in den geplanten weiterführenden Arbeiten neben den werkstoffkundlichen Fragen [28] vor allem konstruktive Aspekte [18] zu beachten sein, die u.a. auch den Einfluß des Anstrahlungswinkels (d.h.: des Aufprallwinkels) des Möllergutes berücksichtigen. Der Parameter Anstrahlwinkel bei Strahlverschleiß wurde unter Verwendung verschieden harter Strahlmittel, also unterschiedlicher Gegenkörper, wie diese Bestandteile des Möllers sind, systematisch untersucht [31]. Für verschiedene Beanspruchungsformen werden optimierte Werkstoffempfehlungen gegeben. Bei geringen Anstrahlwinkeln sind dieses Hartkeramik (z.B.: Schmelzkorund) und der relativ zähe, hochlegierte Hartguß. Bei Anstrahlwinkeln zwischen 30 und 70° empfiehlt sich der Einsatz von hochfesten Baustählen. Geht jedoch der Anstrahlwinkel gegen 90°, so werden Hartstoff-haltige Verbundwerkstoffe, z.B. Hartmetalle, in Betracht zu ziehen sein. Auch hier zeigt sich recht deutlich, daß es "an sich" gar keine

Bild 10

Skizze einer sog. inneren Schlagpanzerlamelle

Maßstab: ungef. 1:20

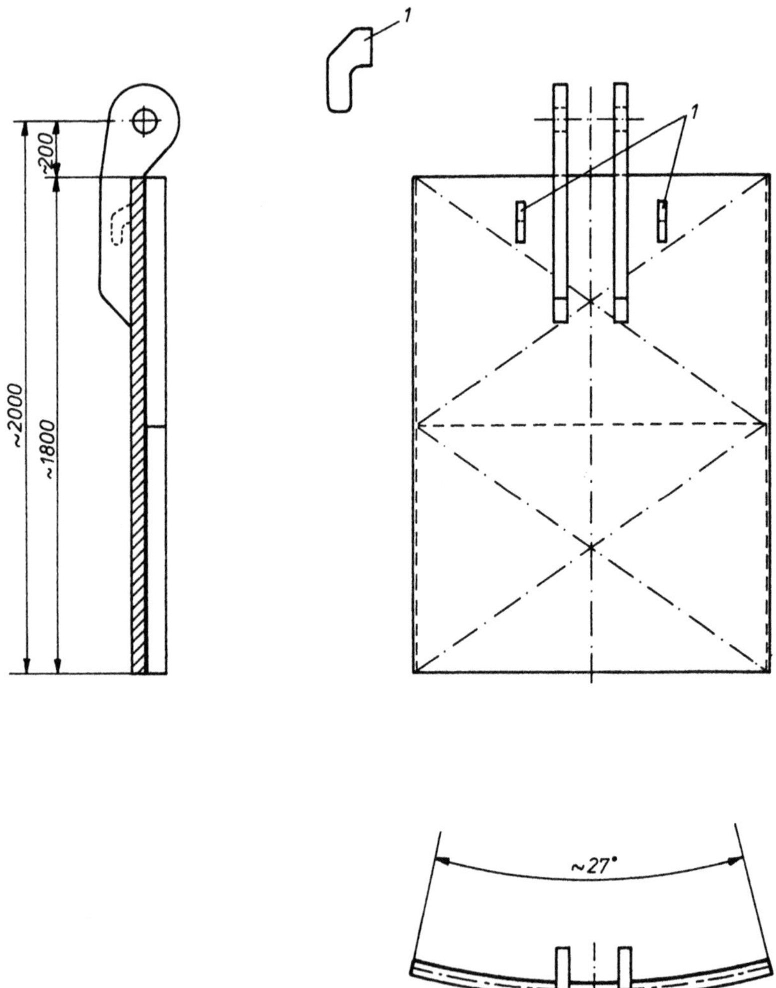

Bild 11

Skizze einer sog. äußeren Schlagpanzerlamelle

Maßstab: ungef.: 1:20

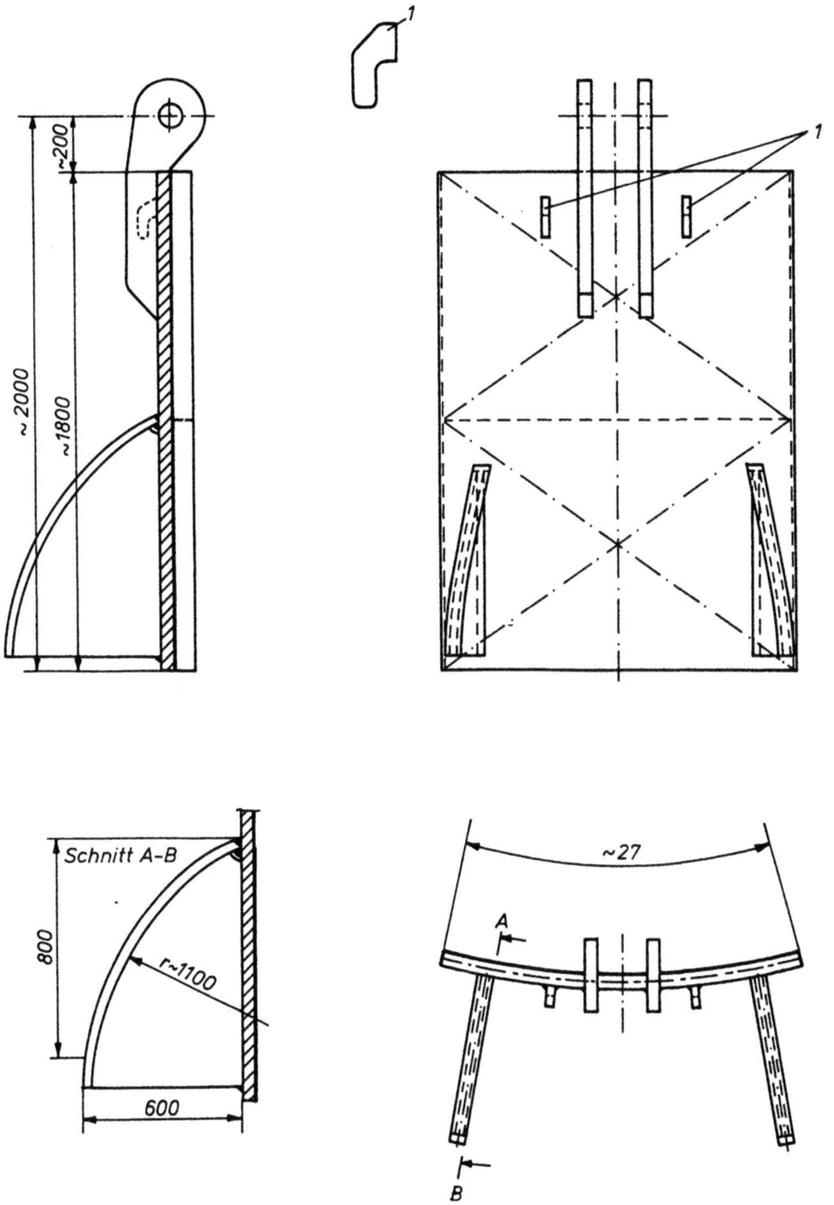

Bild 12

Verschleißbild an Schlagpanzerlamellen im Hochofengichthals

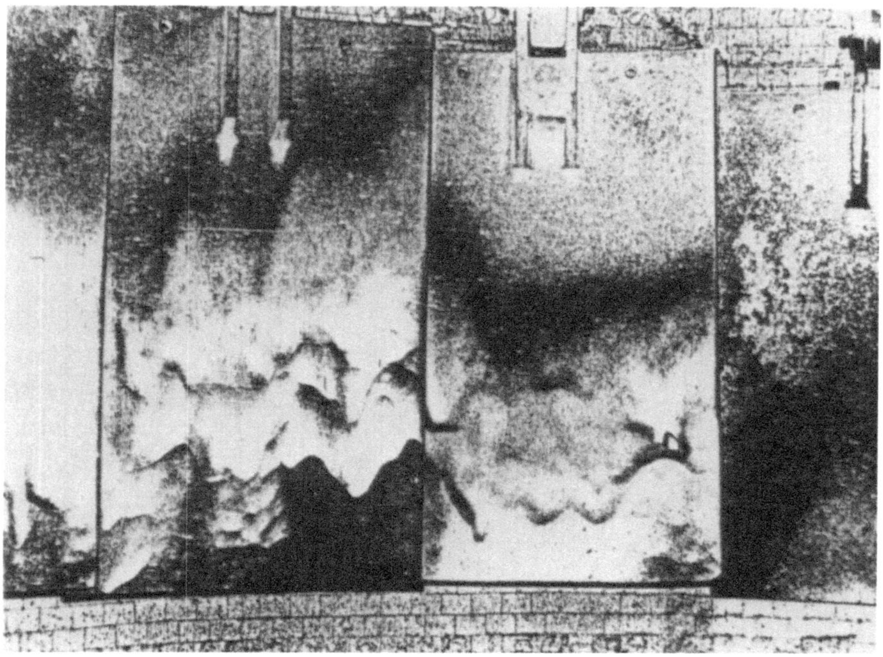

verschleißfesten Werkstoffe gibt, sondern daß die Verschleißbeständigkeit eines Werkstoffes letztlich nur eine Systemgemeinschaft des betreffenden Tribosystems ist.

3.2.3. Verschleiß an Zähnen und Eimern von Großraumbaggern

Der Verschleiß an Zähnen und Eimern von Baggern, insbesondere von Großraumgeräten, im Tagebaubetrieb der Braunkohlenförderung ist ein zentrales und zugleich komplexes Problem, welches sich in erster Näherung als Abrasivverschleiß darstellt und welches zunächst eine werkstoffkundliche Lösung fordert [36-40]. Aber auch die Bedeutung der Geometrie- und Kraftverhältnisse im Tribosystem Baggerzahn/Abbaumaterial während des Betriebes darf nicht unbeachtet bleiben [41].

Bild 13 zeigt die mit Zähnen bestückten Eimer auf dem Rad eines Schaufelrad-Baggers. In Bild 14 wird die Form eines einzelnen Eimers im wesentlichen skizziert. Grundsätzlich sind jedoch auch andere Ausführugnsformen des Eimers möglich [39], die einmal im Zuge der Weiterentwicklung zu sehen sind und die zum anderen auch den örtlichen geologischen bzw. mineralogischen Gegebenheiten angepaßt sein können.

Der Verschleiß an Zähnen und Eimern von Gewinnungsgeräten ist in erster Linie Korn-Gleitverschleiß, im wesentlichen mit dem Gegenstoff Mineral. Bei Betrachtung der Gleitverschleiß-Mechanismen ist zunächst zu beachten, daß Verschleißbeständigkeit und Härte beim Gleitverschleiß eines Grundkörpers nicht gleichartig bzw. nicht analog zu sehen sind. Dieses wird aus Bild 15 ersichtlich, das nach Angaben der BAM [42] zusammengestellt worden ist. Bei logarithmischer Auftragung von Härte und spezifischem Verschleißvolumen verschiedener Werkstoffe über der Werkstoffart zeigt

Bild 13

Zähne und Eimer mit Rad eines
Schaufelrad-Baggers

Photo:

KRUPP INDUSTRIE- und
STAHLBAU, Rheinhausen

Bild 14

Skizze eines Schaufelradeimers

Maßstab: ungef.: 1:15

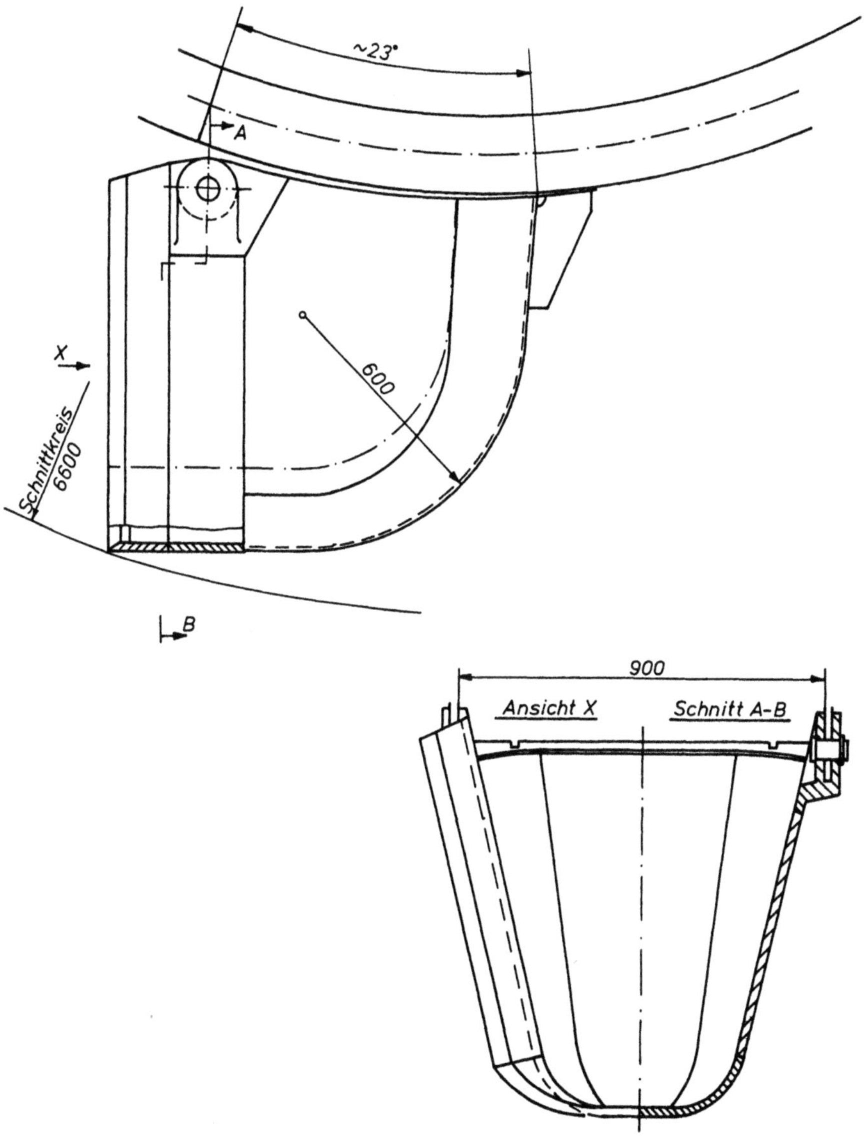

Bild 15

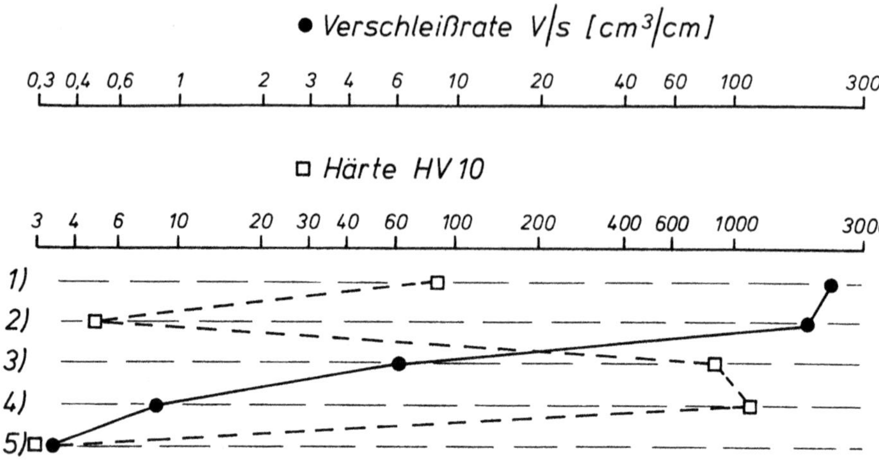

Härte und Verschleißrate verschiedener Werkstoffe beim Gleiten auf Werkzeugstahl (Verbindungslinien nur zur besseren Ablesbarkeit)

Verschleißender Werkstoff:

1) MS 60
2) Teflon
3) Werkzeugstahl
4) WC-Co-HM
5) Polyäthylen

Bild 16

Aufbau einer metallischen Oberfläche (nach H. Uetz)

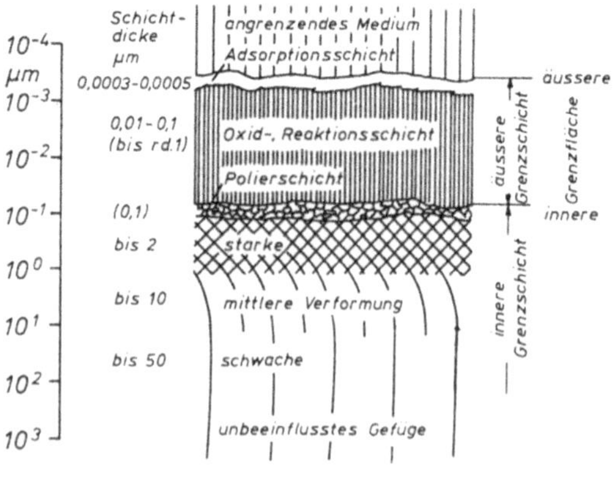

sich für das Gleiten dieser verschiedenen Werkstoffe auf Werkzeugstahl u.a. folgendes: bei einem Härteunterschied, in HV10 gemessen, von fast drei Zehnerpotenzen liegt das spezifische Verschleißvolumen des relativ weichen Polyäthylens nur etwa um die Hälfte höher als das spezifische Verschleißvolumen des härteren Wolframcarbid-Basis-Hartmetalls. Derartige Ergebnis-Relationen sind allerdings nur möglich, wenn unterschiedliche Verschleißmechanismen im Tribosystem "wirksam" sind.Beim Gleitverschleiß, insbesondere beim Korn-Gleitverschleiß mit mineralischem Gegenstoff, steht der Abrasiv-Verschleiß deutlich im Vordergrund. Da die Korn-Gleitverschleiß-Vorgänge im wesentlichen auf der Oberfläche der - hier metallischen - Grundkörper (Zähne, Eimerkanten) ablaufen, sei einiges zum Verständnis des Feinbaus der metallischen Oberfläche gesagt. Bild 16 zeigt, wie in Wirklichkeit eine metallisch blank aussehende Oberfläche bei logarithmisch gespreiztem Maßstab aufgebaut ist [43]. Bei dieser Art der Darstellung ist die äußerste Schicht adsorbierter Gase noch erkennbar. Es schließen sich eine Reaktionsschicht, meist als Oxidfilmschicht ausgebildet, die eigentliche polierte Schicht sowie die mehr oder minder stark kaltverformten Grenzschichten des Grundkörpers an. Wie unter 3.2.1. bereits für den Prallverschleiß ausführlicher beschrieben, kann der hier in Rede stehende Korn-Gleitverschleiß letztlich auch als Furchungsverschleiß angesprochen werden, vgl. Bild 8.

Die bisherige Weiterentwicklung der Zähne und Eimer erfolgte im wesentlichen unter werkstoffkundlichen und konstruktiven Aspekten [37]. Neben dem Verschleiß der einzelnen Zähne verschleißt das Eimermesser vornehmlich in den Rundungen. Die Bagger arbeiten verfahrensbedingt im Blockbetrieb, d.h. der Baggereimer schneidet nur mit einer Seite, nach der der Ausleger schwenkt. Aus diesem Grunde nutzen sich die Eimermesser vornehmlich in der Rundung ab. Aus dieser Erkenntnis wurden die Ecken des

Stahlguß-Eimermessers Verschleiß-konform verstärkt ausgebildet.

Eckmesser aus hochlegiertem Stahlguß hatten aus der Sicht der Verschleißminderung nur geringe Vorteile. Eine Panzerung der Eckmesser mit hochverschleißbeständigen Auftragsschichten auf der Basis von Chrom- und Wolframcarbiden lieferte signifikant bessere Standzeiten der Baggereimer.

Die Grundkörper der Eckmesser bestehen im wesentlichen aus Stahlguß GS-45 und GS-52. Da die Auftragsschweißung i.a. ohne Vorwärmung des Grundmaterials erfolgt, sind feine Risse in der Auftragsschweißung nicht auszuschließen, was sich im betrieblichen Einsatz allerdings nicht als nachteilig erwiesen hat.

Mit dieser Maßnahme konnten die Standzeiten verdoppelt werden, nämlich von $2\text{-}4 \times 10^6$ m^3 auf $6\text{-}9 \times 10^6$ m^3 Abraum/Eimergarnitur. Z.Z. laufen verschiedene Versuche, die Eckmesser mit Hartmetall bzw. mit sog. Verbundhartmetall zu bestücken, wobei gewisse konstruktive Gesichtspunkte [41], insbesondere bezüglich der Ausbildung und Abstützung der Messerspitze, parallel laufen. Mit weiteren deutlichen Verbesserungen wird zu rechnen sein.

3.2.4. Verschleiß an Schake und Bolzen von Raupen- und Eimerketten sowie an Raupen- und Eimerketten-Gleitführungen

Bei dem Verschleiß im Tribosystem: Raupenkettenbolzen gegen Schakenbüchse handelt es sich um Korn-Gleitverschleiß [36], wobei Grund- und Gegenkörper Metalle sind und der Zwischenstoff ggf. mineralischer Natur ist. Dieser Verschleißfall ist Gegenstand vieler spezieller

Untersuchungen [36,37,40,44-49] und hat mit der Einführung der sog. Großgeräte im Braunkohlentagebau eine neue, nicht unbedingt erwünschte Aktualität erhalten.

In Bild 17 wird eine verschlissene, weitgehend plastisch verformte Schake gezeigt, die i.a. gemäß Bild 18 durch Auftragsschweißung gepanzert und dann wieder eingesetzt wird [37]. Dabei ist die Härte der Auftragsschicht der der Laufräder des Fahrwerkes angepaßt.

Die Härte des Schweißgutes liegt zwischen 250 und 500 HV10, während das eigentliche Schakenmaterial ohne Nachbehandlung zum Einsatz gelangt. Allerdings wirkt sich die Grobzeiligkeit des aufgetragenen Schweißgutes nachteilig auf die Mantelfläche der Laufräder, Bild 19, aus. Die entsprechenden Untersuchungen zeigen, daß gefügelokal die zulässige spezifische Flächenpressung weit überschritten wird [50]. Ein egalisierendes Schleifen der Auftragsschweißung in den Laufbahnen der Schaken hat eine gewisse Verschleißminderung auf den Mantelflächen der Laufräder zur Folge, kann aber den Verschleiß nicht in gewünschtem Maße reduzieren.

Inzwischen werden - vor allem bei den Großgeräten - die Raupenkettenglieder aus Manganhartstahlguß (G-X120 Mn12) hergestellt. Dieser bei Beanspruchung durch Martensit-Bildung sich verfestigende Stahl [51] liefert aus tribologischer Sicht eine echte Leistungsverbesserung. Die Raupenkettenglieder sind durch Bolzen miteinander verbunden. Dabei sind in die Kettenglied-Bohrungen Büchsen eingepaßt, wie dieses aus den Bildern 15 und 16 zu entnehmen ist.Das Bild 20 zeigt einen stark verschlissenen Bolzen, dessen Verschleißzustand das Ergebnis von Korn-Gleitverschleiß ist, wie dieser zuvor eingangs dieses Kapitels näher beschrieben worden ist. Im allgemeinen werden die Bolzen aus Manganhartstahl hergestellt, während die Büchse aus Federstahl vom Typ 46 Mn 7 bzw. 50 Mn 7 gefertigt wird [37]. Es finden sich allerdings auch Hinweise dafür, daß die Büchse sinnvoll aus

Bilder 17 und 18

(nach B. Schweins
bzw.
nach R. Oettel)

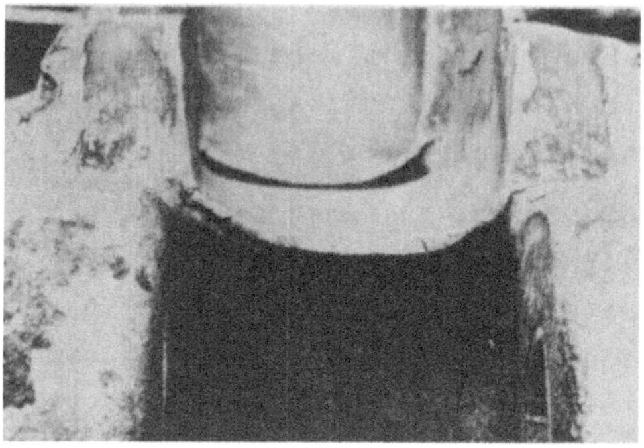

Bild 17

Verschlissenes
Raupen-Kettenglied
(Bereich der Schake)

Bild 18a

Auftragsgeschweißte
Schake

Bild 18b

Auftragsgeschweißte
Schake
(andere Ansicht!)

Bilder 19 und 20

(nach R. Oettel
bzw.
nach B. Schweins)

Bild 19

Schuppen auf der Mantelfläche des Laufrades eines Raupenfahrzeuges

Bild 20

Verschlissener Bolzen

42 CrMo 4 im gehärteten und vergüteten bzw. sogar "nur" im normalgeglühten Zustand eingesetzt werden kann [44].

Bei der Behandlung dieser Thematik dürften künftig werkstoffkundliche Untersuchungen und Arbeiten im Vordergrund stehen.

3.3. Orientierende Vorversuche zum grundsätzlichen Verständnis von Abrasiv-Verschleiß

Bei der Auswertung der Fragebögen, bei der nachfolgenden Daten- und Informationsbeschaffung, insbesondere bei der Literaturauswertung zeigte sich, daß bei der Weiterführung des Vorhabens über die Minderung des Abrasiv-Verschleißes im Großanlagenbau zwei grundsätzliche Fragen neben den genannten vier Verschleißkomplexen vorrangig zu behandeln sein werden:

a) Werkstofffragen vor konstruktiven und verfahrenstechnischen Fragen
sowie

b) Fragen zur Klärung und Beschreibung des Mechanismus von Furchungsverschleiß.

3.3.1. Entwicklung von hochverschleißbeständigem Verbund-Hartmetall

Das Ziel der meisten Werkstoffentwicklungen zur Minderung von Verschleiß - vornehmlich von Abrasiv-Verschleiß - ist es, möglichst hohe Härte <u>und</u> Verschleißbeständigkeit mit möglichst hoher Zähigkeit - bei guter Schweißbarkeit und bei niedrigen Werkstoff-Gestehungskosten - zu kombinieren. Das Ergebnis einer jeden Werkstoffentwicklung mit den genannten Zielsetzungen muß aber notwendigerweise immer ein Kompromiß bleiben.

Im Rahmen dieses Vorhabens wurde eine Entwicklungsarbeit durchgeführt, die bereits zu einem neuen Stahl-Hartmetall-Verbundwerkstoff geführt hat. Dieser Verbundwerkstoff besteht aus einer Metallmatrix, in die Hartstoffe bzw. Hartmetalle in körniger Form eingelagert sind, und besitzt im Gegensatz zu den nach dem Stande der Technik bekannten Verbundwerkstoffen insbesondere eine hohe Härte sowie eine gute Verschleißbeständigkeit und weiterhin günstige Zähigkeitseigenschaften. Für dieses neue Werkstoffkonzept wurden Schutzrechte beantragt [52].

Aus Bild 21 wird das Gefüge des neuen Verbundwerkstoffes ersichtlich, der in der Verschleiß-beanspruchten Oberfläche Hartmetall-ähnliche Härte- und Verschleißbeständigkeits-Eigenschaften erwarten läßt bzw. der aufgrund der Stahlmatrix im Kern oder besser im Grundkörper hohe Zähigkeitseigenschaften und günstige Schweißbedingungen besitzt. Diese Art der "Panzerung" der Verschleiß-beanspruchten Oberflächen zeigt vor allem gegenüber Abrasiv-Verschleiß aufgrund der eingelagerten Hartstoff-Teilchen eine signifikante Beständigkeit.

Aufgrund verschiedener Versuchseinsätze von Verbund-Hartmetall-Körpern im Steinkohlenbergbau und im Braunkohlen-Tagebau wird das Leistungsverhalten von Verbund-Hartmetall in Abbau-Werkzeugen und auch als sog. "wear-resistant-part" positiv bewertet. Gegenüber entsprechenden Hartmetall-Formteilen,
welche im Einsatz durch grobe Ausbrüche bald

Bild 21

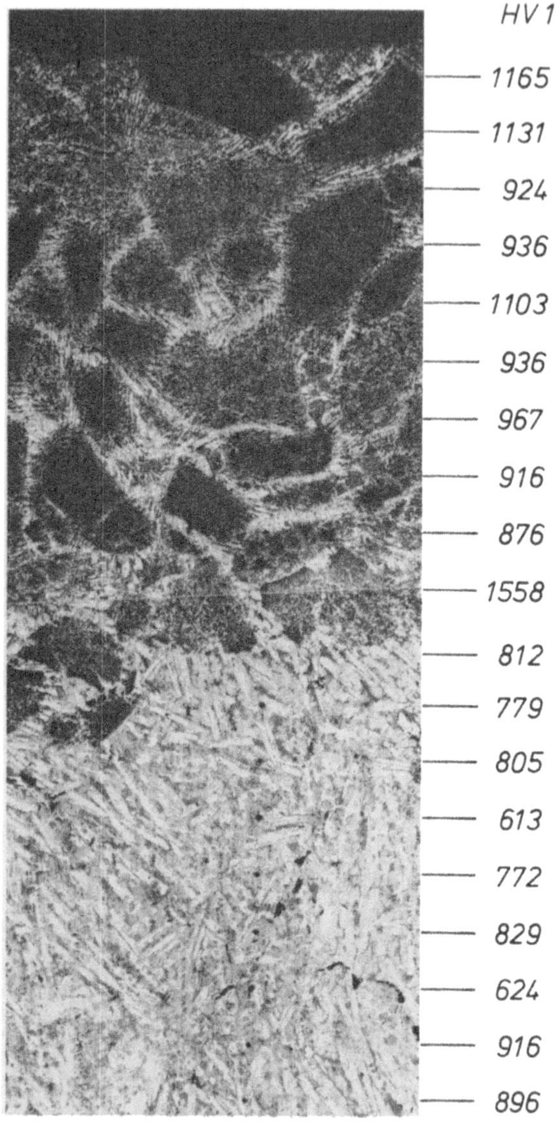

Übersicht

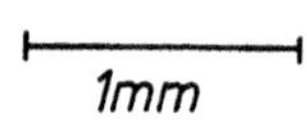

Stahl-Hartmetall-Verbundwerkstoff,
Übergangsgefüge
Ätzung: 4 min Murakami + 20 sec HCl

ausfallen, zeigen Formkörper aus Verbund-Hartmetall nur kleinere Ausbröckelungen, bei denen die Werkzeuge aber voll funktionsfähig bleiben. Dieses günstige Leistungsverhalten von Verbund-Hartmetall-Einsätzen wurde beispielsweise an einem Kohlehobel mit starker Schlagbeanspruchung eingehend studiert. Weiterhin wurden im Wettbewerb mit Manganhartstahl auf den Schlagarmen einer Müllzerkleinerungsanlage deutliche Vorteile für Verbund-Hartmetall festgestellt. Mit Verbund-Hartmetall gepanzerte Schlagarme zeigten eine um ein Mehrfaches gesteigerte Standzeit im Vergleich mit Schlagarmen, die herkömmlich mit Manganhartstahl bestückt waren.

3.3.2. Abriebuntersuchungen beim Mahlen in technischen Rollmühlen

Beim Mahlen von technischen Hartstoffen in Rollmühlen, auch Kugelmühlen genannt, handelt es sich um ein sog. geschlossenes Tribosystem, bei dem in einer Mahltrommel sog. Mahlkugeln Hartstoffteilchen (Zwischenstoff) zerkleinern, wobei das Umgebungsmedium Luft, Schutzatmosphäre oder Flüssigkeit ist. Man könnte die Mahltrommel als Grundkörper und die Mahlkugeln als Gegenkörper bezeichnen, man muß aber auch berücksichtigen, daß der Zerkleinerungsvorgang für den zu zerkleinernden Hartstoff (Zwischenstoff) zwischen zwei Mahlkugeln erfolgen kann.Insofern kann beim Mahlvorgang die Zuordnung der Begriffe Grundkörper und Gegenkörper im Sinne der Norm [4] nicht eindeutig definiert werden.

Der vorherrschende Abrasiv-Verschleiß dürfte mit einiger Sicherheit aus Stoß- (vornehmlich zwischen zwei benachbarten Mahlkugeln) und aus Rollverschleiß (vornehmlich zwischen Mahltrommelinnenwand und Mahlkugel) aufgebaut sein. Der Verschleiß erfolgt sowohl auf der Innenwand der Mahltrommel als auch auf der Oberfläche der Mahlkugeln durch Hartstoffteilchen [53].

Da die Hartstoffzerkleinerung von großem technischen Interesse ist, sind diesem Thema viele Untersuchungen gewidmet. Zur Verschleißminderung an Mahl-Körpern und am Mühlengehäuse sind sowohl werkstoffkundliche als auch verfahrenstechnische und konstruktive Möglichkeiten gegeben [54, 55]. Viele Maßnahmen gelten der Erhöhung von Wirtschaftlichkeit sowie von Betriebssicherheit und damit letztlich der Minderung des Abrasiv-Verschleißes [56-59]. In vielen betriebsnahen und auch in Grundlagen-Untersuchungen werden die Gesetzmäßigkeiten zum Verschleiß beim Mahlen von Hartstoffen diskutiert [29,48,60-67]. Das spezielle Gebiet der Hartstoffzerkleinerung in Prallmühlen und in Fein-Kreiselbrechern wirft sowohl werkstoffkundliche Fragen als auch verfahrenstechnische und konstruktive Problemstellung auf [68,69]. Insbesondere wird in diesen beiden Arbeiten versucht, Ansätze für eine kalkulierbare Gewährleistung von Lebensdauer bzw. Standzeit der auf Abrasiv-Verschleiß beanspruchten Mühlenelemente zu gewinnen.

In den eigenen Abriebuntersuchungen, vgl. Tabelle 12, wurden das Wolfram als Meßgröße für den Abrieb der Mahlkugeln und das Eisen als Meßgröße für den Abrieb an der Innenwand der Mahltrommel verwendet, da man davon ausgehen kann, daß die zerkleinerten Hartstoffe und auch die Mahlflüssigkeit praktisch Wolfram- und Eisen-frei eingesetzt worden sind. Aus der hier gezeigten Prinzip-

Tabelle 12

Abriebuntersuchungen beim Mahlen
von technischen Hartstoffen
in Rollmühlen

Technische Hartstoffe:	Titancarbid; Aluminiumoxid; Quarzsand;
Mahlbedingungen :	Mahltrommel aus rostfreiem Stahl (Werkstoff-Nr.1.4301) Volumen: 8 l Mahlkugeln: 30 kg Hartmetallkugeln WC-Co-Basis, 10 - 15 mm Ø Mahlflüssigkeit: 2,5 l Aceton Mahlgut: Einwaage: a) Titancarbid HCST 1738 3,0 kg b) Aluminiumoxid (Bikorit) 3,64 kg c) Quarzsand F 36 3,0 kg

Prinzip:

Bezeichnung	Symbol
$n=55[min^{-1}]$	
Betriebsversuch zur Untersuchung des Abrasiv-Verschleißes an Mahltrommel u. Mahlkugeln	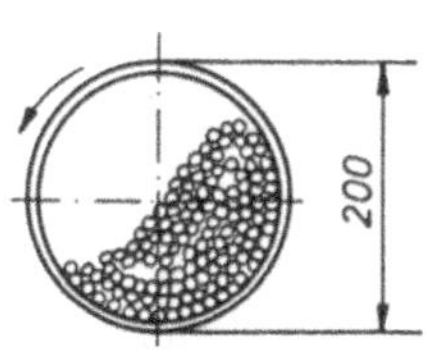

Skizze des Mahlvorganges sind auch der Kugelfall- und Kugelschleppe-Bereich zu erkennen, wobei erwähnt sei, daß im Kugelfall-Bereich der Stoßverschleiß und im Bereich der Kugelschleppe der Rollverschleiß dominieren dürften. Für den Kugelfall-Bereich (vornehmlich Stoßverschleiß) wird bereits in älteren Arbeiten festgestellt, daß der durch Abrasiv-Verschleiß bedingte Gewichtsverlust der Mahlkugeln direkt proportional der Kugelmasse ist [70]. Dagegen ist in der Kugelschleppe der Abrieb-Betrag in gewissen Grenzen als direkt proportional zur spezifischen Oberfläche der einzelnen Kugeln zu sehen [71, 72].

Aufgrund der Abriebuntersuchungen ist zu bemerken, daß die Gewichts-Verhältnisse (W : Co) für die Mahlkugeln und (Fe : C : Ni) für die Mahltrommel in etwa konstant den vom jeweiligen Werkstoff her zu erwartenden Verhältniswerten entsprechen. Dieses bedeutet jedenfalls, daß Mahltrommel und auch die Mahlkugeln gleichmäßig verschleißen, daß also bei den Hartmetallkugeln beispielsweise die Cobalt-reiche Bindephase, die im Vergleich zum eingelagerten Wolframcarbid weich ist, nicht bevorzugt verschleißt.

In den Bildern 22, 23 und 24 werden die Wolfram- (Meßgröße für den Abrieb der Mahlkugeln) und die Eisengehalte (Meßgröße für den Abrieb der Mahltrommel) über der Mahldauer (Zeit-Achse logarithmisch geteilt) aufgetragen. Man erkennt beim Vergleichen der drei letztgenannten Bilder, daß das Titancarbid mit der höchsten Härte (Mikrohärte: HV 0,05 = 3200) sowohl an Mahlkörpern und auch an der Mahltrommel am stärksten abrasiv wirkt. Der Abrasiv-Verschleiß des Aluminiumoxides (Schmelzkorund, HV 0,05 = 2300) und des Quarzsandes (HV 0,05 = 1200) an den Mahlkugeln ist in etwa vom Abriebbetrag her der gleiche, zumindest nach 200 h Mahldauer. Bezüglich des erzielten Abriebs in der Mahltrommel sind die Eisengehalte im Quarzsand vergleichbar hoch wie im relativ wesentlich härteren Titancarbid.

Bild 22

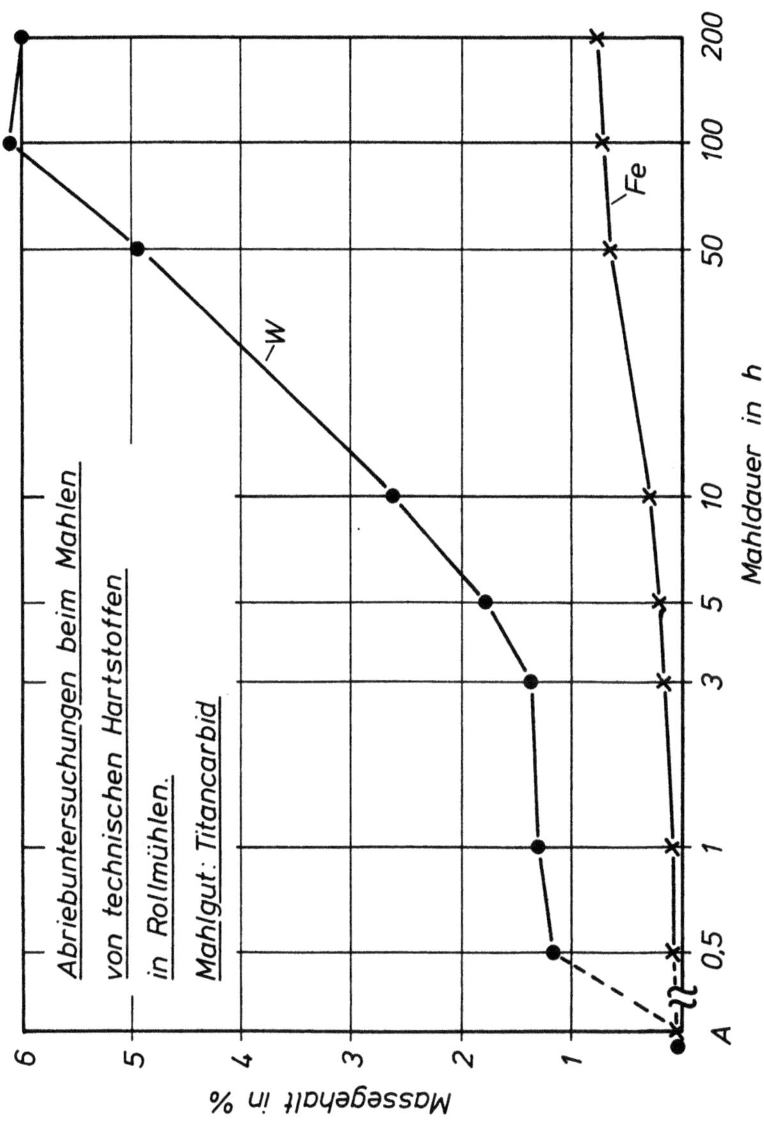

Abriebuntersuchungen beim Mahlen
von technischen Hartstoffen
in Rollmühlen.
Mahlgut: Titancarbid
W
Fe
Massegehalt in %
6
5
4
3
2
1
Mahldauer in h
A
0,5
1
3
5
10
50
100
200

Bild 23

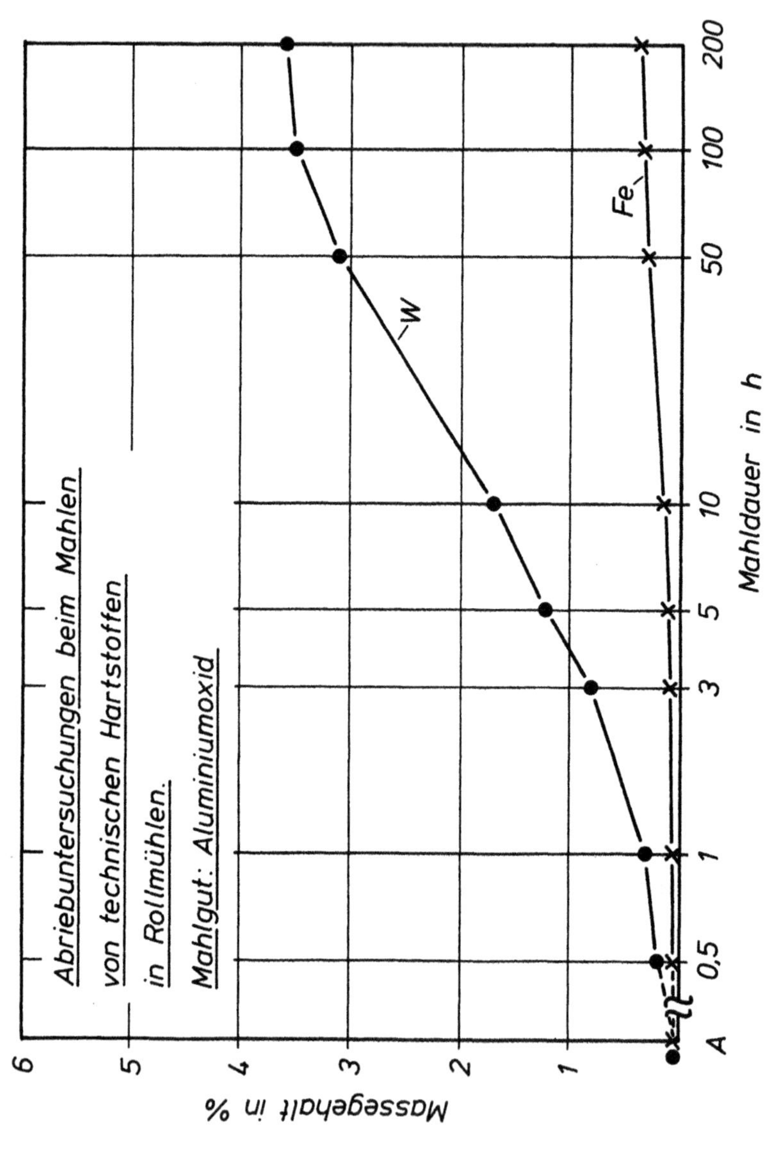

Abriebuntersuchungen beim Mahlen
von technischen Hartstoffen
in Rollmühlen.
Mahlgut: Aluminiumoxid
Massegehalt in %
Mahldauer in h
6
5
4
3
2
1
A
0,5
1
3
5
10
50
100
200
W
Fe

Bild 24

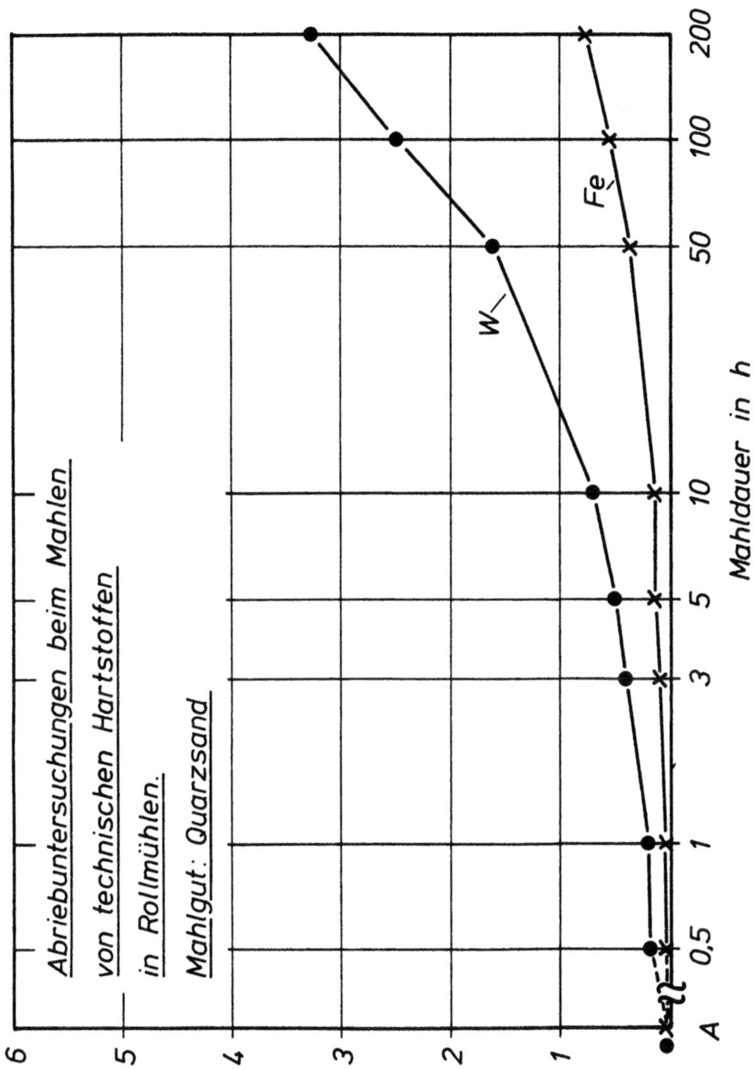
Abriebuntersuchungen beim Mahlen
von technischen Hartstoffen
in Rollmühlen.
Mahlgut: Quarzsand
Massegehalt in %
Mahldauer in h
W
Fe
1
2
3
4
5
6
A
0,5
1
3
5
10
50
100
200

Die Veränderung von Teilchengröße und Temperatur des Mahlgutes in Abhängigkeit von der Mahldauer (Zeit-Achse logarithmisch geteilt) ist den Bildern 25,26 und 27 zu entnehmen. Vor allem für die Hartstoffe Titancarbid und Aluminiumoxid nimmt bis zu 200 h Mahldauer die Teilchengröße stetig ab und nähert sich offensichtlich asymptotisch der Zeit-Achse. In weiterführenden Untersuchungen müßte hier die Veränderung der spezifischen Oberfläche des Mahlgutes miteinbezogen werden, um weitere Folgerungen über den Zerkleinerungsvorgang ziehen zu können. Für alle drei untersuchten Hartstoffe wird nach 5 bis 10 h Mahldauer zwischen 50 und 55^{0}C das "thermische Gleichgewicht" des Tribosystems erreicht.

In Tabelle 13 werden zur weiterführenden Diskussion der Ergebnisse der Abriebuntersuchungen die Daten aus den Bildern 22,23 und 24 zusammengefaßt und nach entsprechender Aufbereitung als Übersicht mitgeteilt. Es zeigt sich beispielsweise, daß die Verhältniszahlen aus den Härtewerten der Hartstoffe sich qualitativ vergleichbar verhalten wie die Verhältniszahlen, die sich aus den Werten des Mahlkugelabriebs ergeben. Daß der Quarz-Sand, wie aus Tabelle 13 ersichtlich ist, besonders auf die Innenwand der Mahltrommel verschleißend wirkt, wurde bereits zuvor erwähnt. Insgesamt liegen die Abriebwerte für den Mahltrommel- und auch für den Mahlkugel-Verschleiß nach 200 h Mahldauer deutlich unter 1 %, bezogen auf die jeweiligen Ausgangswerte. Dieses bedeutet, daß während der Mahldauer von 200 h für alle drei untersuchten Hartstoffe die Mahlkörper-Durchmesser und damit die spezifische Oberfläche der Mahlkörper praktisch unverändert bleiben und daß auch bei der Mahltrommel deren Innendurchmesser und deren Innenoberfläche über die gesamte Versuchsdauer für alle drei Hartstoffe als praktisch konstant betrachtet werden können.

Bild 25

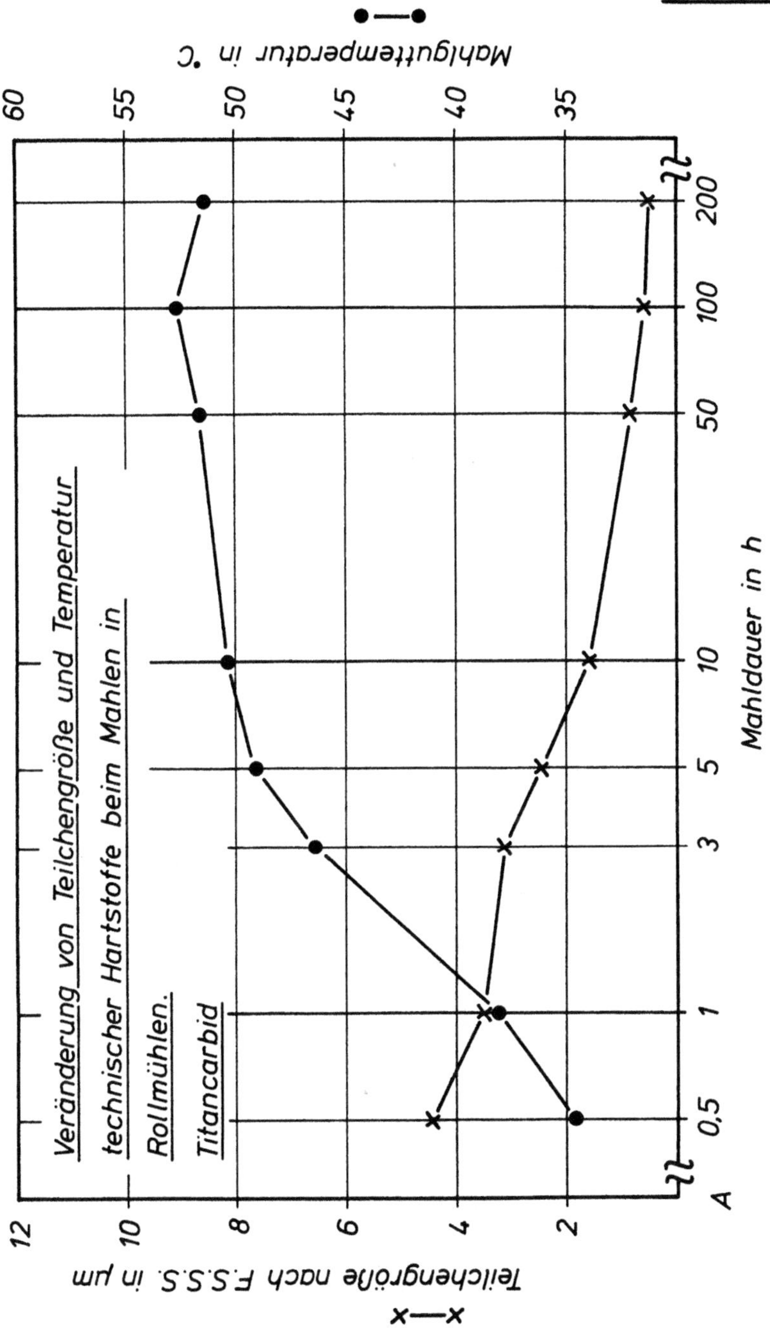
Veränderung von Teilchengröße und Temperatur
technischer Hartstoffe beim Mahlen in
Rollmühlen.
Titancarbid
Mahlguttemperatur in °C
60
55
50
45
40
35
Teilchengröße nach F.S.S.S. in µm
12
10
8
6
4
2
Mahldauer in h
A
0,5
1
3
5
10
50
100
200

Bild 26

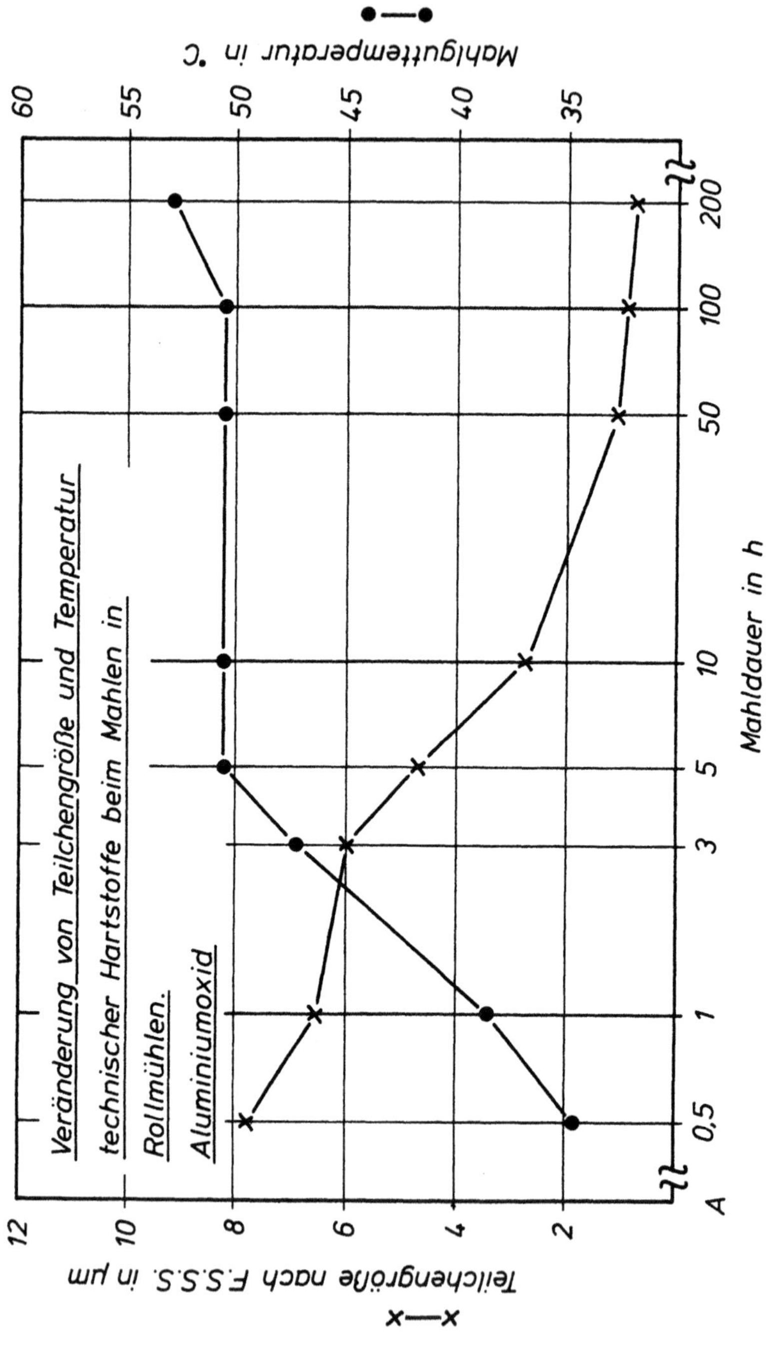

Veränderung von Teilchengröße und Temperatur technischer Hartstoffe beim Mahlen in Rollmühlen.
Aluminiumoxid
Mahlguttemperatur in °C
60
55
50
45
40
35
Teilchengröße nach F.S.S.S. in µm
12
10
8
6
4
2
Mahldauer in h
A
0,5
1
3
5
10
50
100
200

Bild 27

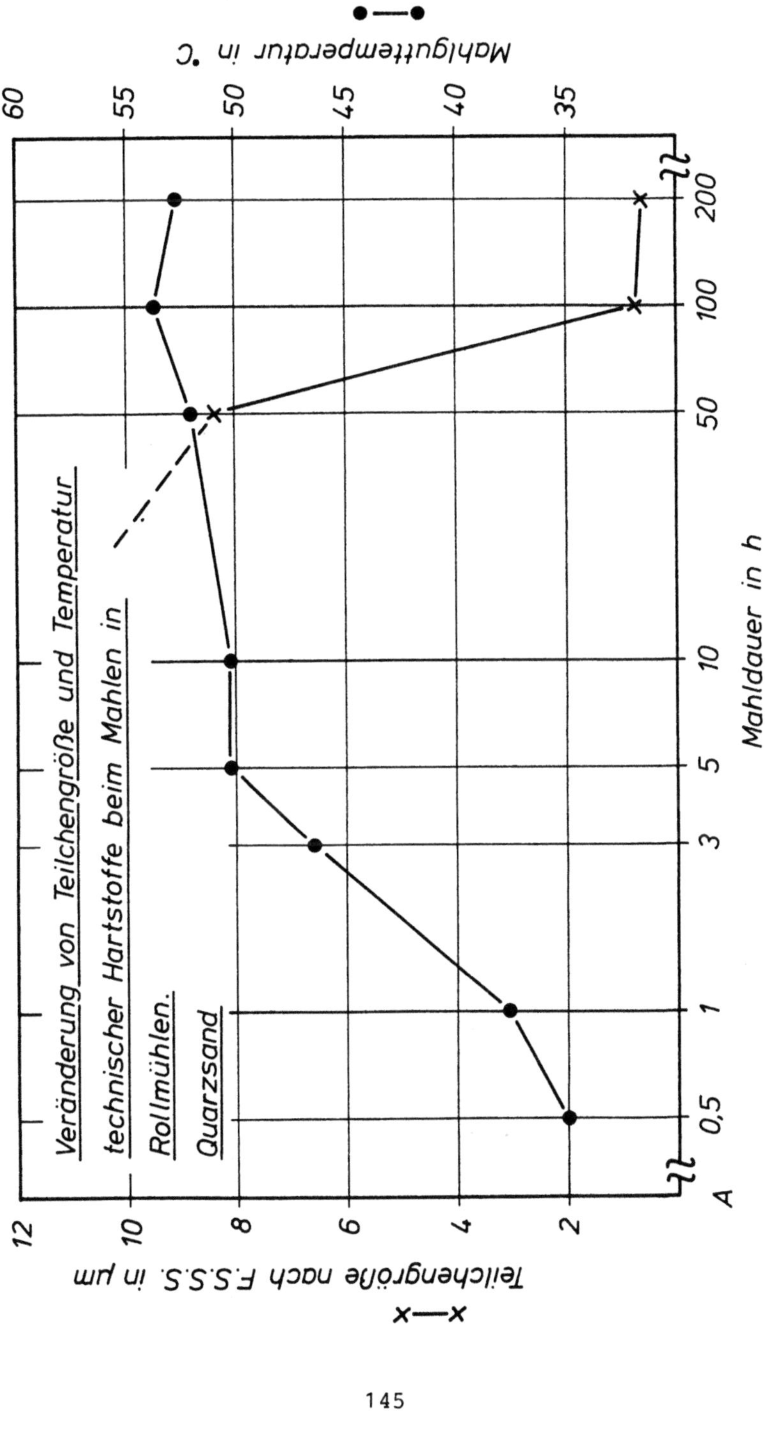

Veränderung von Teilchengröße und Temperatur
technischer Hartstoffe beim Mahlen in
Rollmühlen.
Quarzsand
Mahlguttemperatur in °C
60
55
50
45
40
35
Teilchengröße nach F.S.S.S. in µm
12
10
8
6
4
2
Mahldauer in h
A
0,5
1
3
5
10
50
100
200

Tabelle 13

Übersicht über die Abriebergebnisse bei den Mahlversuchen in technischen Rollmühlen (nach 200 h Mahldauer)

Mahlkugel-Einsatz : 30 kg

Mahltrommelmasse : 7,1 kg (bei Versuchsbeginn!)

Anmerkung: Alle Verhältniszahlen beziehen sich auf Quarz-Sand ≙ 1.

Hartstoff	Härte HV0,05	Verhältnis-zahl	Abriebbeträge Mahltrommel-Abrieb	Verhältnis-zahl	Mahlkugel-Abrieb	Verhältnis-zahl
Titancarbid (TiC)	3.200	2,7	35 g (≙0,49 %)	1,21	253 g (≙ 0,85 %)	2,1
Schmelzkorund (Al_2O_3)	2.300	1,9	18 g (≙0,25 %)	0,62	160 g (≙ 0,53 %)	1,3
Quarz-Sand (SiO_2)	1.200	1	29 g (≙0,41 %)	1	120 g (≙ 0,4 %)	1

Diese Folgerung war nicht unbedingt naheliegend, sie gab allerdings Veranlassung dazu, die Abriebwerte für Eisen (Meßgröße für den Abrieb aus der Innenwand der Mahltrommel) und für Wolfram (Meßgröße für den Abrieb der Mahlkugeln) in ihrer Abhängigkeit von der Mahldauer in einem Koordinatensystem mit logarithmisch geteilten Achsen darzustellen. Die Bilder 28, 29 und 30 vermitteln als Ergebnis, daß sich die Meßdaten in 1. Näherung auf Geraden anordnen. Dieses wiederum bedeutet für die vorliegenden Untersuchungen, daß zwischen Abrieb und Mahldauer folgender grundsätzlicher parabolischer Zusammenhang gegeben ist:

$$A_{Me} = b_{Ha} \cdot t^{n}_{mahl}.$$

Die Bedeutung der Symbole wird in Tabelle 14 mitgeteilt. Weiterhin sind aus Tabelle 14 die aus den Bildern 28, 29 und 30 für die untersuchten Hartstoffe ableitbaren Parabel-Funktionen zu entnehmen.

Während der Exponent n keinerlei Trendrichtung erkennen läßt, zeigt sich für die b_{Ha}-Werte im Vergleich mit den Härtewerten der Hartstoffe, daß sich die jeweiligen Verhältniszahlen für die Härtewerte, b_{Ha}-Werte Eisenabrieb und b_{Ha}-Werte Wolframabrieb, qualitativ vergleichbar verhalten, Tabelle 15. Man kann folgern, daß in dem hier beschriebenen Tribosystem Rollmühle die Härte des zu mahlenden Hartstoffes relativ und qualitativ über die Konstante b_{Ha} in die Abrieb-Mahldauer-Funktion eingeht. Dieses gilt sowohl für den Abrieb aus der Innenwand der Mahltrommel als auch für den Abrieb aus der Oberfläche der Mahlkugeln.

4. Weiteres Vorgehen und Ausblick

Ziel der weiteren Arbeiten wird es sein, die an den genannten ausgewählten Bauteilen des Großanlagenbaus auf-

Bild 28

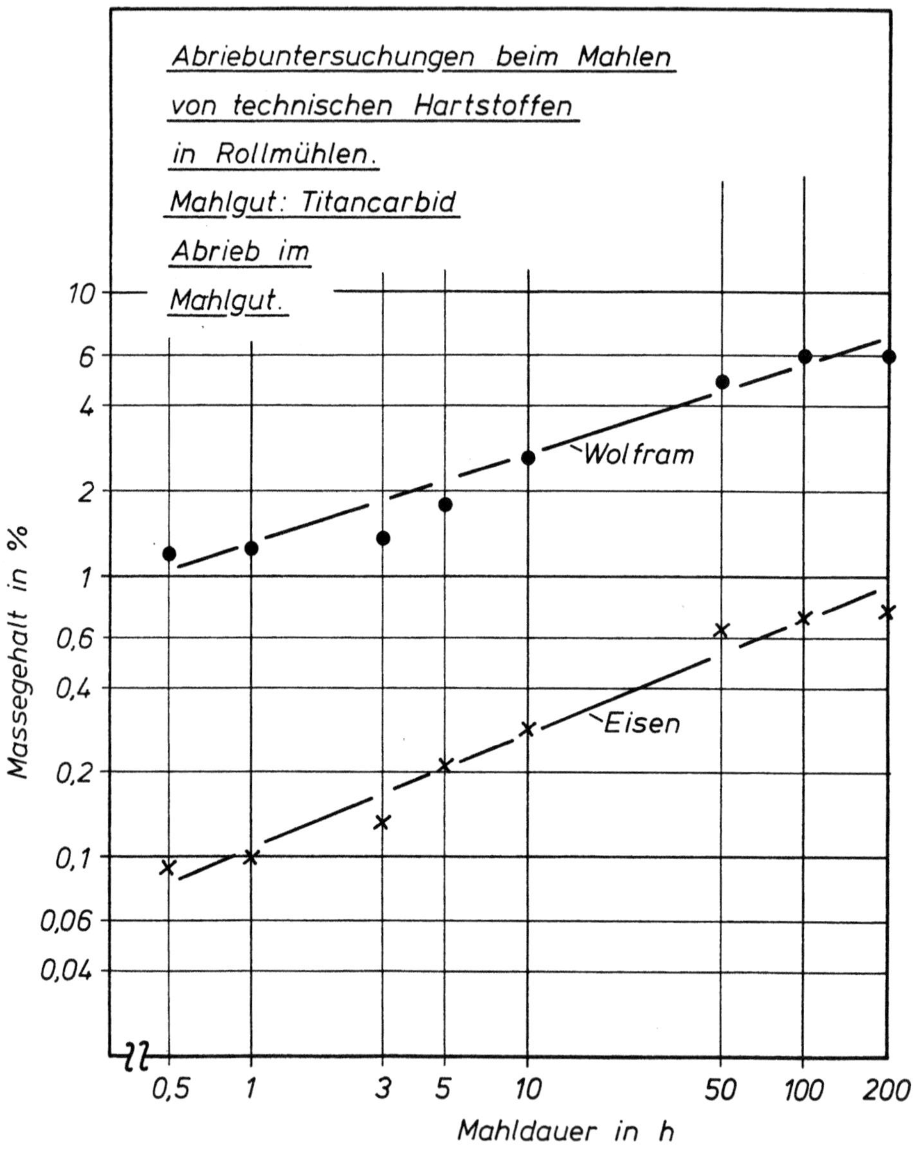

Abriebuntersuchungen beim Mahlen
von technischen Hartstoffen
in Rollmühlen.
Mahlgut: Titancarbid
Abrieb im
Mahlgut.
Wolfram
Eisen
Massegehalt in %
10
6
4
2
1
0,6
0,4
0,2
0,1
0,06
0,04
0,5
1
3
5
10
50
100
200
Mahldauer in h

Bild 29

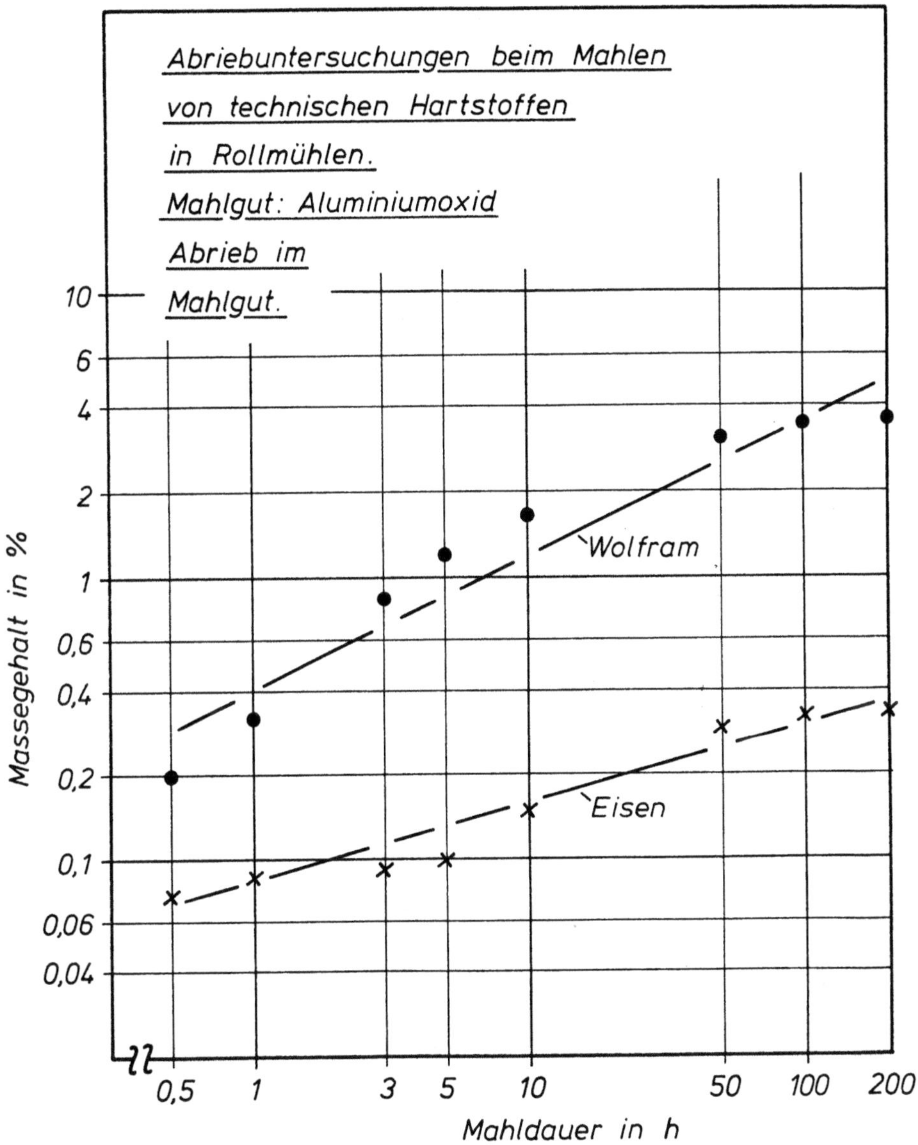

Abriebuntersuchungen beim Mahlen
von technischen Hartstoffen
in Rollmühlen.
Mahlgut: Aluminiumoxid
Abrieb im
Mahlgut.
10
6
4
2
1
0,6
0,4
0,2
0,1
0,06
0,04
Massegehalt in %
Wolfram
Eisen
0,5
1
3
5
10
50
100
200
Mahldauer in h

Bild 30

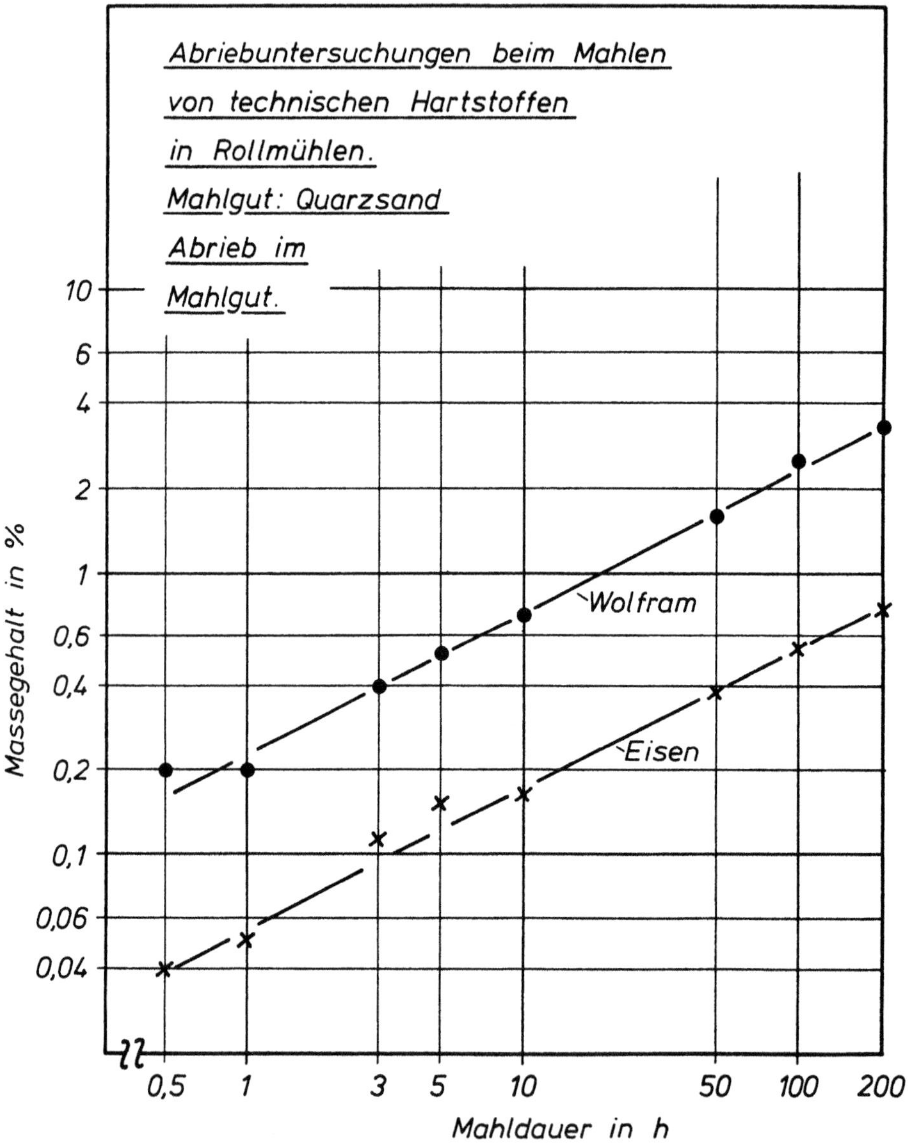

Abriebuntersuchungen beim Mahlen
von technischen Hartstoffen
in Rollmühlen.
Mahlgut: Quarzsand
Abrieb im
Mahlgut.
10
6
4
2
1
0,6
0,4
0,2
0,1
0,06
0,04
Massegehalt in %
Wolfram
Eisen
0,5
1
3
5
10
50
100
200
Mahldauer in h

Tabelle 14

Übersicht über die Abrasiv-Verschleiß-Kinetik aufgrund von Abriebuntersuchungen beim Mahlen von technischen Hartstoffen in Rollmühlen

Vgl.: Bilder 28, 29 und 30

Folgender grundsätzlicher funktionaler Zusammenhang ist gegeben:

$$A_{Me} = b_{Ha} \cdot t_{mahl}^{n}.$$

Hierin bedeuten:

A_{Me} → Abrieb des Metalls im Mahlgut in %;

b_{Ha} → Hartstoff-spezifischer Faktor;

t_{mahl} → Mahldauer in h;

n → Kinetik-spezifischer Exponent;

Hartstoff (Mahlgut)		Meßgröße Eisen	Meßgröße Wolfram
Titancarbid (TiC)	:	$A_{Fe} = 0{,}11 \cdot t^{0,42}$;	$A_W = 1{,}3 \cdot t^{0,33}$;
Schmelzkorund (Al_2O_3)	:	$A_{Fe} = 0{,}08 \cdot t^{0,31}$;	$A_W = 0{,}4 \cdot t^{0,47}$;
Quarz-Sand (SiO_2)	:	$A_{Fe} = 0{,}056 \cdot t^{0,49}$;	$A_W = 0{,}23 \cdot t^{0,52}$;

Tabelle 15

Übersicht über das Abrasiv-Verschleiß-Verhalten der untersuchten Hartstoffe bei den Abriebuntersuchungen beim Mahlen in technischen Rollmühlen

Anmerkung: Alle Verhältniszahlen beziehen sich auf Quarz-Sand ≙ 1.

Hartstoff (Mahlgut)	Härte HV0,05	Verhältnis-zahl	b_{Ha}-Werte Mahltrommel		Mahlkugel	
			Meßgröße Eisen	Verhälnis-zahl	Meßgröße Wolfram	Verhältnis-zahl
Titancarbid (TiC)	3.200	2,7	0,11	1,96	1,3	5,7
Schmelzkorund (Al_2O_3)	2.300	1,9	0,08	1,43	0,4	1,7
Quarz-Sand (SiO_2)	1.200	1	0,056	1	0,23	1

tretenden - vornehmlich mineralisch bedingten - Verschleißvorgänge systematisch zu analysieren, die zugrunde liegenden Verschleißmechanismen zu erkennen und auf dieser Basis Verbesserungsmöglichkeiten für die Einzelprobleme vorzuschlagen. Die notwendigen praktischen Versuche sollen vornehmlich im großtechnischen Rahmen (Betriebsversuche) durchgeführt werden.

Bei sicherlich notwendigen orientierenden Vorversuchen im Labormaßstab oder auch im "pilot-plant-scale" ist darauf zu achten, daß die jeweils simulierten Tribosysteme bei vergleichbarer Energiekonzentration im beanspruchten Bereich arbeiten. Die Übertragbarkeit der Ergebnisse dieser Vorversuche auf die Funktion des tribo-beanspruchten Elements im Großgerät muß in jedem Fall kritisch geprüft werden.

Zu berücksichtigen ist weiterhin, daß das Verschleißgeschehen nicht nur werkstofftechnisch, sondern auch verfahrenstechnisch und konstruktiv zu beeinflussen ist. Neben der Auswahl und Entwicklung geeigneter Werkstoffe sind deshalb auch Änderungen des Prozeßablaufs oder der konstruktiven Voraussetzungen als potentiell verschleißmindernde Parameter in Erwägung zu ziehen.

Neben den Arbeiten zur Deutung und Beherrschung der physikalischen Verschleißerscheinungen muß der Versuch unternommen werden, eine kalkulierbare Verschleißvorhersage zu entwickeln.

Die analytische Beschreibung der technischen Funktion des jeweiligen Tribosystems, die Definition des vorliegenden Beanspruchungskollektivs, die Kennzeichnung der Struktur des betreffenden Tribosystems sowie die Aufnahme der entsprechendn Verschleißkenngrößen erfolgen in Anlehnung an DIN 50.320.

5. Literatur

1. D. Meadows, D. Meadows, E. Zahn und P. Milling: "Die Grenzen des Wachstums", Rowohlt-Taschenbuch, Hamburg, 1973, S. 45 - 57

2. D. Meadows u.a.: "Wachstum bis zur Katastrophe?", Deutscher Taschenbuch-Verlag, München, 1976, S. 13 - 34

3. Bundesministerium für Forschung und Technologie: Forschungsbericht T 76 - 38 ü. Tribologie, Forschungs-Kuratorium Maschinenbau e.V., Frankfrut/M., Juli 1976, S. 13 - 20

4. DIN 50.320: Verschleiß (Begriffe; Systemanalyse von Verschleißvorgängen; Gliederung des Verschleißgebietes), Ausgabe Dezember 1979

5. H. Wahl: Die Technik, Bd. 3, Nr. 5, Mai 1948, S. 193 - 204

6. H. Wahl: "Querschnitt durch das Verschleißgebiet", Metalen, Jaarg. IX, 1954, Nrs. 4, 5 en 6

   Vortrag anl. des 87. Bondsdag des Bond voor Materialenkennis in Utrecht am 9. Okt. 1952

7. K. Wellinger, H. Uetz und G. Gommel: Materialprüfung 9, 1967, Nr. 5, S. 153 - 160

8. P. Eßlinger und H. Uetz: Materialprüfung 9, 1967, Nr. 5, S. 161 - 165

9. W. König: Materialprüfung 9, 1967, Nr. 5, S. 170 - 174

10. G. Gommel:
Materialprüfung 9, 1967, Nr. 5, S. 175 - 178

11. H. Uetz:
Braunkohle, Wärme und Energie, H. 11, November 1968, S. 365 - 376

12. U. Uetz und J. Föhl:
Braunkohle, Wärme und Energie, H. 1, Januar 1970, Bd. 22, S. 1 - 9

13. D. Mouth:
Braunkohle, Wärme und Energie, H. 1, Januar 1970, Bd. 22, S. 21 - 24

14. W. Faust:
Neue Bergbautechnik, 1. Jg., H. 3, 1971, S. 224 - 226

15. H. Motek:
Aufbereitungs-Technik, Nr. 10,1974, S. 547 - 549

16. P. Hainke:
Stahl u. Eisen 94, 1974, Nr, 23, S. 1173 - 1174

17. E. Schaffhirt:
Neue Bergbautechnik, 5. Jg., H. 11, 1975, S. 881-888

18. W. Wahl:
Stahl u. Eisen 96, 1976, Nr. 19, S. 899 - 904

19. P. Hainke:
Stahl u. Eisen 98, 1978, Nr. 24, S. 1301 - 1303

20. H.K. Tönshoff und R. Schulze:
IDR, 13, 1979, H. 1, S. 14 - 20

21. H. Uetz, K. Sommer und M.A. Khosrawi:
VDI-Ber., Nr. 354, 1979, S. 107 - 124

22. H. Czichos:
Z. Metallkde., Bd. 71, 1980, H. 7, S. 421 - 426

23. K.H. Habig:
Verschleiß und Härte von Werkstoffen, Carl Hanser-Verlag, München u. Wien 1980, S. 11 - 15

24. DIN 50.320:
Verschleiß (Begriffe; Systemanalyse von Verschleißvorgängen; Gliederung des Verschleißgebietes), Entwurf November 1976

25. P. Rinza und H. Schmitz:
"Nutzwert-Kosten-Analyse",
Eine Entscheidungshilfe, VDI-Taschenbuch T51

26. Firmenschrift:
Kalenborn Handbuch f. Verschleißschutz, Gleitförderung, Kunststoffanlagenteile, hrsgegeben v. Schmelzbasaltwerk Kalenborn, Dr. Ing. Mauritz K.G., 1978

27. N.N.:
Aufbereitungs-Technik, Nr. 11, 1977, S. 620 - 621

28. R. Stelzer:
Stahl u. Eisen, 99, 1979, Nr. 18, S. 978 - 984

29. H. Uetz:
Aufbereitungs-Technik, Nr. 3, 1969, S. 130 - 141

30. W. Wahl:
VDI-Ber., Nr. 243, 1975, S. 171 - 187

31. K. Wellinger und H. Uetz:
VDI-Forschungsheft 449, Bd. 21, Ausgabe B, 1955, S. 1 - 40

32. A. Schikora:
Braunkohle, H. 1/2, Jan./Febr. 1977, S. 28 - 32

33. N.N.:
Stahl u. Eisen, 97, 1977, Nr. 1, S. 50

34. N.N.:
Aufbereitungs-Technik, Nr. 8, 1970, S. 516 - 518

35. W. Soth:
Aufbereitungs-Technik, Nr. 10, 1969, S. 595 - 598

36. J. Hoffzimmer:
Braunkohle, H. 1/2, Jan./Febr. 1976, S. 11 - 17

37. B. Schweins:
Braunkohle, H. 8, August 1969, S. 257 - 265

38. H. Berns:
Braunkohle, H. 4, April 1979, S. 90 - 94

39. E. Jungbluth:
Braunkohle, H. 10, Oktober 1979, S. 334 - 339

40. D. Kujau:
Neue Bergbautechnik, 2, 1972, H. 1, S. 63 - 67

41. K. Kramer:
Neue Bergbautechnik, 3. Jg., H. 6, Juni 1973, S. 443 - 449

42. Bundesanstalt für Materialprüfung:
Festkörpergleitreibung und Verschleiß, BAM-Ber. Nr. 13, 1972

43. H. Uetz:
Metalloberfläche, 23. Jg., 1969, H. 7, S. 199 - 211

44. H.-F. Hach und R. Mangler:
Neue Bergbautechnik, 6. Jg., H. 7, Juli 1976,
S. 525 - 531

45. E. Jungbluth:
Braunkohle, H. 6, Juni 1973, S. 175 - 182

46. E. Jungbluth:
Braunkohle, H. 3, März 1976, S. 1 - 11

47. H. Wahl:
Aufbereitungs-Technik, Nr. 6, 1969, S. 305 - 322

48. H. Uetz und J. Föhl:
VDI-Berichte Nr. 243, 1975, S. 127 - 142

49. R. Mangler:
Neue Bergbautechnik, 8. Jg., H. 9, Sept. 1978,
S. 495 - 501

50. R. Oettel:
Braunkohle, H. 1, Jan. 1967, S. 14 - 19

51. H. Pacyna:
Manganhartstahl, KEM-Literaturdienst,
Sonderdruck aus den Heften Nr. 8, 9 + 10, 1966

52. DE-OS 29 19 477 vom 15.V.'79,
Offenlegungstag 4. XII.'80

53. D. Ocepek, D. Salatić und M. Grujić:
Erzmetall, Bd. 32, 1979, H. 11, S. 481 - 488

54. W. Wahl:
Aufbereitungs-Technik, Nr. 9, 1971, S. 573 - 578

55. G. Schneider:
Aufbereitungs-Technik, Nr. 9, 1971, S. 537 - 549

56. H.A. Buchmüller:
Aufbereitungs-Technik, Nr. 4, 1971, S. 179 - 190

57. L. Wenz und H.J. Feik:
Aufbereitungs-Technik, Nr. 11, 1970, S. 669 - 678

58. T.E. Norman und K. Röhrig:
Aufbereitungs-Technik, Nr. 6, 1970, S. 356 - 364

59. G. Brüssow:
Aufbereitungs-Technik, Nr. 6, 1970, S. 353 - 355

60. E.W. Hanke:
Zement-Kalk-Gips 11, 1971, S. 520 - 526

61. R. Köhling:
Aufbereitungs-Technik, Nr. 2, 1972, S. 101 - 104

62. H. Motek:
Aufbereitungs-Technik, Nr. 10, 1973, S. 658 - 663

63. M. Clement, J. Voigt und H. Uetz:
Erz-Metall, Bd. 26, 1973, H. 12, S. 583 - 592

64. J.P. Blombled:
Revue des Materiaux de Construction, Nr. 651
S. 387 - 410

65. H.A. Buchmüller:
Aufbereitungs-Technik, H. 5, 1972, S. 305 - 312

66. H. Motek:
Aufbereitungs-Technik, H. 2, 1972, S. 73 - 78

67. H. Motek:
Aufbereitungs-Technik, H. 10, 1974, S. 547 - 549

68. H. Stöckmann:
Aufbereitungs-Technik, H. 10, 1977, S. 511 - 514

69. H. Schmidt:
Aufbereitungs-Technik, H. 10, 1978, S. 484 - 491

70. E.W. Davis:
Trans. AIME 61, 1919, S. 250 - 296

71. H.E.T. Haultain and F.C. Dyer:
Trans. Canad. Inst. of Min. and Metallurgy,
1922, vol. XXV, S. 276 - 292

72. K.A. Rasumov:
"Rationelle Beschickung von Mühlen mit Kugeln"
(russ.); Gorn. Ž. 3 (1947), S. 31 - 36

Prüfung und Optimierung von Hartauftragschweißungen mit Fülldrahtelektroden bei vorwiegend abrasiver Beanspruchung

von

Dipl.-Ing. Ingo Kretschmer

Dr. Ing. Hans Wahl

Dr. Ing. Wolfgang Wahl

Verschleiß-Technik

Dr. Ing. Hans Wahl GmbH & Co

Stuttgart-Ruit

Förderkennzeichen: 01 ZT 097

Laufzeit: 1.1.1978 - 30.6.1980

Zusammenfassung:
Im vorliegenden Untersuchungsvorhaben wurden, unter Verwendung von 8 verschiedenen Fülldrahtelektroden, die Eigenschaftsprofile verschiedenlagiger Panzerungen und reiner Schweißgüter hinsichtlich Abrasionswiderstand, mechanisch-technologischer Eigenschaften, Wärmehärte und Verzunderungsneigung untersucht. Dabei gelang es unter Entwicklung einer Kokillentechnik ausreichend große Biegeproben aus reinem Schweißgut hochharter Legierungen herzustellen. Nach vorangehenden Übertragbarkeitsversuchen konnten sämtliche Proben, mit Ausnahme der Proben für die Gummiradprüfung und der Panzerplatten für den Heißsiebbereich, mittels repräsentativer Kokillenschweißtechnik erstellt werden.

Der Einfluß der Lagenanzahl, der Lagendicke, des Grundwerkstoffes und der Vorwärmung auf das Eigenschaftsprofil der Panzerungen wurden ermittelt. Zusätzlich wurden die Wärmhärte und die Verzunderungsneigung der reinen Schweißgüter ermittelt und ein Schweißzusatz für den Heißbereich im Kohlenstoffgehalt optimiert. Untersuchungen über Gefügeveränderungen und REM-Oberflächenbetrachtungen beim Heißsiebeinsatz runden den Untersuchungsumfang ab.

Modellhafte Verschleißkostenberechnungen wurden auf Grundlage der gemessenen Verschleißgeschwindigkeiten erstellt. Sie ergeben deutliche Vorteile der Hartauftragschweißung im Härtebereich von Glas und Flint als Gegenstoff.

Trotz eines ausgeprägten Optimums der Panzerung mit Schweißzusatzwerkstoff-Nr. 3 im untereutektischen Bereich bei einem Kohlenstoffgehalt von ca. 1,5 % bietet dieser Legierungstyp im übereutektischen Bereich mit Kohlenstoffgehalten von über 4 %, sowohl verschleißmäßige, wie auch verfahrenstechnische Vorteile und ist daher in dieser Version wirtschaftlich sinnvoller.

Inhaltsverzeichnis:

1. Einleitung:

Die generelle Feststellung "der Verschleiß ist das Sorgenkind im Maschinenbau" /1/ gilt in besonders krasser Weise für Betreiber von Anlagen, die abrasivem Verschleiß ausgesetzt sind. Abrasive Verschleißschäden treten vor allem dort auf, wo mineralische Rohstoffe und ähnlich harte, körnige Stoffe in großem Umfang gewonnen, transportiert, zerkleinert und verarbeitet werden /2, 3/. Besonders betroffene Industrien sind demnach

- der Bergbau
- die eisenschaffende Industrie
- Steine und Erden incl. Bauwirtschaft
- kunststoff- und gummiverarbeitende Industrie

Eine kostenmäßige Abschätzung der volkswirtschaftlichen Gesamtschäden durch Abrasivverschleiß ist naturgemäß schwierig und mit großen Unsicherheiten verbunden. Nach /3/ werden sie mit ca. 10 % des geschätzten Instandhaltungsaufwandes von 60 - 80 Md. DM für 1974 in der Bundesrepublik angesetzt. Eine neuere Schätzung /4/ setzt den jährlichen Gesamtverlust aufgrund abrasiver Verschleißschäden für die Bundesrepublik (Materialverlust incl. Kosten für Ausfallzeiten und Instandhaltung) auf ca. 5 Md. DM/Jahr an. Für die Zukunft ist zu erwarten, daß die durch Abrasivverschleiß verursachten volkswirtschaftlichen Schäden noch deutlich steigen /5, 6/, da: -
mit der weiterhin zu erwartenden Steigerung der absoluten und spezifischen Durchsatzleistung die Verschleißschäden überproportional steigen.
- Das zunehmende Arbeiten im Verbundsystem verschleißbedingte Ausfallkosten ebenfalls überproportional steigen läßt und
- bei der Rohstoffgewinnung mit steigendem Anteil von Ballaststoffen gerechnet werden muß.

Durch die kombinierte Ausnutzung von konstruktiven, betrieblichen, werkstofflichen und sonstigen Möglichkeiten /7/ zur Minderung der Verschleißschäden ist die Steigerung der Anlagengröße und deren spezifischen Durchsatzleistung vielfach erst technisch und wirtschaftlich sinnvoll geworden. Verschleißschutzmaßnahmen durch Entwicklung und Einsatz optimaler Werkstoffe an den verschleißgefärdeten Stellen nehmen im Rahmen dieses Maßnahmebündels einen besonders hohen Stellenwert ein.

Die werkstofflichen Möglichkeiten der Verschleißminderung sind extrem vielfaltig und erstrecken sich bei der Bekämpfung abrasiven Verschleißes über eine weite Werkstoffpalette /8, 9/. Ein Kennzeichen der modernen Technik, auch bei der Verschleißbekämpfung, ist ein deutlicher Trend zum Einsatz von Verbundwerkstofflösungen /10, 11/.

Unter den vielen Beschichtungsmöglichkeiten /12, 13/ steht bei abrasiver Bauteilbeanspruchung der Verschleißschutz durch das Hartauftragschweißen an vorderster Stelle und hat durch die Entwicklung von Fülldrahtelektroden, die das Auftragschweißen beliebiger Legierungen kostengünstiger automatisierter Form ermöglichen, an Bedeutung noch entscheidend hinzugewonnen. /14, 15, 16/

Fülldrahtelektroden sind Drahtelektroden, die aus einem metallischen Mantel und einem Kern aus nichtmetallischen und / oder metallischen Pulvern bestehen. Wie Bild 1 zeigt /17/ kann ihre Querschnittsform sehr unterschiedlich sein, jedoch hat sich aus Wirtschaftlichkeitsgründen und aus Gründen des hohen Füllvolumens bei der Hartauftragschweißung die Form b als Standard durchgesetzt.

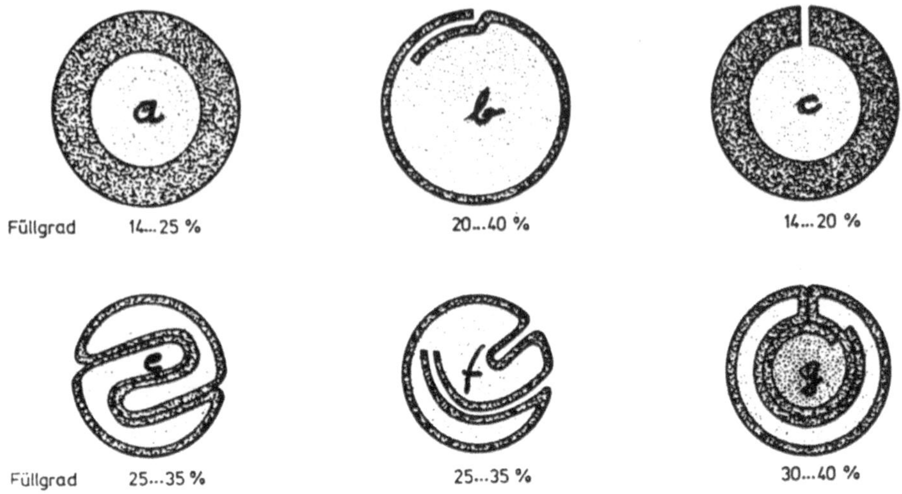

Bild 1. Querschnittsformen von Fülldrahtelektroden.

Ihre Entwicklung in "selbstschützender" Form für das Metall-Lichtbogenschweißen hat zum steigenden Einsatz der Hartauftragschweißung auch in Form großflächiger und großvolumiger Panzerungen entscheidend beigetragen da sie

a) leicht teil- und vollmechanisiert verarbeitbar sind und dadurch die Wirtschaftlichkeit und Gleichmäßigkeit der Auftragschweißung erhöhen, sowie

b) durch Hartstoffzugabe in der Drahtfüllung höchst abrasionsbeständige Legierungen in Drahtform erbringen die aus Massivdrähten nicht herstellbar sind.

2. Aufgabenstellung:

Mit der aussagefähigen und reproduzierbaren Prüfung von Hartauftragschweißungen hat man sich seit jeher schwer getan und die Richtlinien dazu, soweit überhaupt vorhanden, sind noch wenig ausreichend /18/ oder umgehen die vorhandenen Schwierigkeiten unter

Inkaufnahme des weitgehenden Verlustes der Aussagekraft. So schreibt z.B. /19/ vor, Hartauftragschweißungen für den Faltversuch bis auf einen Restbestand von 0,5 mm abzuarbeiten und in vorsichtiger Weise vorab in die Panzerung Risse mit einem Abstand von 8 - 10 mm einzuleiten, um auf diese Weise die vom TÜV Regelwerk für Verbindungsschweißungen ausgelegten Biegewinkel zu erreichen.

Der Schwierigkeit der Herstellung reiner Hartstoff-Schweißgutproben in Abmessungen wie sie für mechanisch-technologische Untersuchungen nötig sind, ist man bisher dadurch ausgewichen, daß man Gußproben aus gleichartigen Legierungen hergestellt und deren Ergebnisse alsauch für Schweißgutproben gültig betrachtet hat. Auswirkungen der Überhitzung im Lichtbogen und der Tropfenreaktion mit der Umgebung sind dabei naturgemäß nicht erfasst und vernachlässigt worden.

Demgemäß sollte bei diesem Vorhaben ein Beitrag zur Prüftechnik von Fülldraht-Hartauftragschweißungen geleistet werden, unter Berücksichtigung der Eigenschaften des reinen Schweißgutes und der Kombinationseigenschaften verschiedenlagiger Schweißgüter mit unterschiedlichen Grundwerkstoffen. Ein 2. Ziel dieses Vorhaben war es, eine im Heißsinterbereich bereits erfolgreich eingesetzte übereutektische Chrom-Kohlenstoff-Legierung durch Reduzierung des Kohlenstoffgehaltes im untereutektischen Bereich für den Heißsinterbereich zu optimieren. Dahinter stand die Überlegung, daß

1. die Verzunderungsneigung von Chrom-Kohlenstoff-Legierungen mit fallendem Kohlenstoffgehalt sinkt und

2. die Auftragschweißung bei niedrigerem Kohlenstoffgehalt in erwünschter rissfreier Ausführung möglich ist.

Zusätzlich zur meßtechnischen Erfassung der Eigenschaftsprofile waren kostenmäßige Überlegungen zum vorteilhaften Einsatz der untersuchten Fülldrahtelektroden anzustellen.

3. Herstellung der Schweißproben:

Im Rahmen dieses Vorhabens wurden die in Tabelle 1 aufgelisteten 8 ausgewählten Fülldrahtelektroden verarbeitet.

Schweißzusatz Nr.	Durchmesser (mm)	Schweißgut - Richtanalyse (reines Schweißgut) C %	Si %	Mn %	Cr %	Mo %	Nb %	Sonstige	Rest	Bezug zu DIN 8555
1	2.8	5.0	0.5	0.7	22	-	7	--	Fe	~ Legierung Gr.10 weiterentwickelt
2	2.4	5.8	0.7	1.0	22	7	6	2 % W 1 % V	Fe	"
3	2.4	1.9	1.5	1.7	28.5	1	-	3 % Ni	Fe	~ Leg. Gr. 10
4	2.8	1.9	0.3	1.5	7.0	1.4	-	5 % Ti	Fe	Leg.Gr.4 weiterentwickelt
5	2.4	1.0	0.4	13.7	4.7	-	-	-	Fe	Leg.Gr. 7 weiterentwickelt
6	2.4	0.3	0.6	16.5	16.5	-	-	-	Fe	
7	2.8	1.0	--	--	28.0	-	-	4	Co	Leg. Gr. 20
8	2.4	60 % W_2C							40 % Fe	Leg. Gr. 22

Tabelle 1

Der große Probenanfall, die schwierige Ausarbeitung höchstverschleißbeständiger Auftragschweißgüter und deren Rissanfälligkeit machten die Entwicklung einer repräsentativen Methode der schweißtechnischen Herstellung kleiner Proben nötig, die auch unter Verwendung formstützender Kupferkokillen gelang. Diese in Bild 2 und 3 abgebildeten Kupferkokillen wurden eingesetzt bei den

a) 1, 2- und dreilagigen Schleiftellerproben
b) 1, 2- und dreilagigen Schleiftopfproben
c) 1, 2- und dreilagigen Fallhammerproben
d) reinen Schweißgutproben zur Biegeprüfung

Bild 2

Bild 3

Bild 4 skizziert die Vorgangsweise bei dieser formgebenden Auftragschweißtechnik und für die Verschleißprüfung nach dem Schleifteller- und Schleiftopfverfahren eingesetzten Probeformen.

Bei sämtlichen Schweißungen mittels Kokillentechnik wurden folgende Parometer eingehalten

a) Kokillentemperatur 130 - 200° C
b) Schweißstrom 260 - 290 A für 2,4 mm Elektroden Ø
300 - 330 A für 2,8 mm Elektroden Ø
c) Schweißspannung 26 - 28 V für 2,4 mm Elektroden Ø
27 - 29 V für 2,8 mm Elektroden Ø

Die Einhaltung dieser relativ niedrigen Fülldrahtelektrodenbelastung in Kombination mit der Kokillenvorwärmung und deren formmäßigen Abstimmung auf die gewünschte Probeform entsprechend Bild 4, führte zu gleichen Härte- und Verschleißwerten, wie bei nach DIN 8555 - Empfehlung geschweißten Panzerungen auf 20 mm Grundwerkstoff.

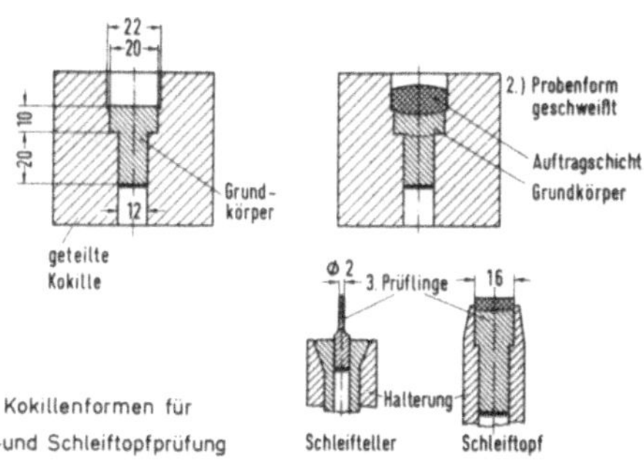

Proben- und Kokillenformen für Schleifteller-und Schleiftopfprüfung

Bild 4

Mit im Durchschnitt 30 - 35 % Grundwerkstoffaufmischung bei 4,5 mm dicken Panzerungen auf St 37 ergeben sich, bei Anwendung dieser Kokillenschweißmethode, auch von der Analysenverdünnung der Panzerung Werte, wie sie beim Metall-Lichtbogenschweißen mit Fülldrahtelektroden an großen Teilen bekannt und üblich sind.

Die Panzerungen für die Prüfung nach dem Schleifteller-verfahren, mit einer zu prüfenden Panzerungsoberfläche von 70x25 mm, wurden unter Einhaltung der DIN 8555 Empfehlung auf St 37 als Grundwerkstoff aufgebracht und daraus herausgearbeitet. Desgleichen wurden sämtliche Platten für den Betriebseinsatz im Heißsinterbereich vollmechanisiert geschweißt unter Anwendung der unter 6.3. näher beschriebenen Legierungs-Kompensationstechnik durch zusätzliche Pulverzugabe.

4. Verschleißprüfung der hartauftraggeschweißten Proben

Bei sämtlichen im folgenden beschriebenen Verschleißprüfungen handelt es sich um Modellversuche, mit einfachen Probekörpern die nach neuerem Vorschlag /20/ der Kategorie VI zu zuordnen sind. Man muß sich dementsprechend über die Einschränkungen bei der Ergebnisübertragung dieser Werte auf komplexere Verschleißvorgänge bewußt sein /21/22/23/ und die Beanspruchung wie auch die Verschleißerscheinungsform /24/25/ bei der Abschätzung der Übertragbarkeit im Einzelfall mitberücksichtigen. Trotz dieser Einschränkung ist jedoch der Wert solcher einfachen Versuche insbesondere bei der Werkstoffoptimierung unbestritten /26/ und somit auch für die Prüfung von Hartauftragschweißungen.

4.1. Beschreibung der Prüfverfahren

Die 1, 2- und 3-lagigen Auftragschweißungen wurden mit den in Bild 5 skizzierten Schleifteller- Schleiftopf- und Gummiradverfahren unter Verwendung der im folgenden beschriebenen Parameter und verschiedener Gegenkörper verschleißmäßig überprüft.

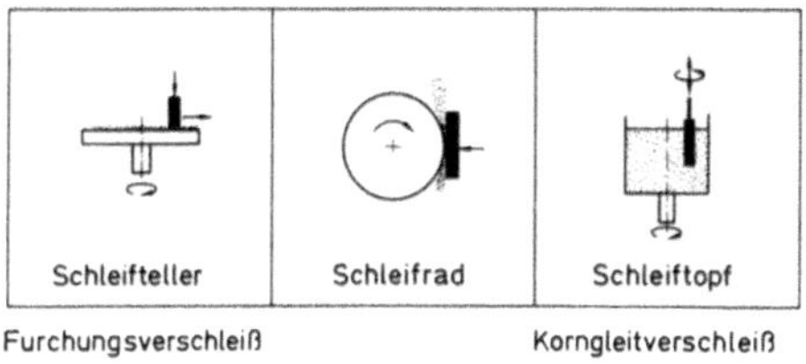

Benutzte Prüfmethoden zur Beurteilung des Abrasionswiderstandes
Bild 5

Bei dem eingesetzten Schleiftellerverfahren wird eine mit 5 N belastete zylindrische Probe des Durchmessers 2 mm derart über ein Schleifteller bestimmter Körnung und Härte geführt, daß sie bei jeder Schleifpapierumdrehung jeweils um die Probebreite in Form einer archimedischen Spirale nach außen gedrückt wird. Sie wird somit ständig von neuen mineralischen Gegenstoffen verschlissen.

Als Gegenstoff wurde dabei eingesetzt:

Gegenstoff	korrigierte Härte (HV_{10})
Glas	500
Flint	930
Granat	1420
Korund	1780
Siliziumkarbid	2700

Bei den verschiedenen Schleifpapieren handelt es sich durchwegs um Schleifpapier der Körnung 60. Der von der Probe pro Schleifpapierblatt abgefahrene Schleifweg beträgt 14 m und wird in 90 sec. zurückgelegt.

Beim Prüfen des Verschleißwiderstandes von Werkstoffen mit der Schleiftopfmaschine werden zylindrische Proben des Durchmessers 15 mm entsprechend Bild 5 außermittig in mit dem Gegenstoff gefüllten Werkstoffbehälter auf und ab geführt. Gleichzeitig wird der Schleiftopfbehälter mit einer Drehzahl von 3,5/min. und die Probe mit einer Drehzahl von 1000/min. gegensinnig zueinander gedreht. Die Auf- und Abbewegung der Probe erfolgt 42 Mal/min. über eine Strecke von jeweils 60 mm, wobei der untere Druckpunkt der Probe 20 mm vom Boden des Schleiftopfes entfernt ist. **Die Versuchsdauer pro** ermittelten Einzelverschleißwert beträgt 4 min. Der Prüfling wird bei dieser Art der Prüfung sowohl auf Ritzung wie auch insbesondere in der Nähe des unteren Todpunktes auf Stoß beansprucht. Als Gegenstoff dienten

<u>Mischung 1:</u> 5 kg feiner Quarzsand H34 mit folgender Körnungsanalyse:

3 %	< als 0.063 mm
60 %	0.063 mm - 0.1 mm
32 %	0.1 - 0.2 mm
5 %	> 0.3 mm

<u>Mischung 2:</u> grobkörniger, scharfkantiger Quarzsand (0.9 - 1.2 mm)

<u>Mischung 3:</u> rundlicher Quarzsand 0.7 - 1.2 mm (Mit dieser, weniger differenzierfähigen Mischung als Mischung 2 wurden nur vereinzelte Werkstoffe geprüft).

<u>Mischung 4:</u> 2,5 kg Neckarkies (3 - 7 mm)
2,5 kg Neckarsand (0 - 7 mm)

<u>Mischung 5:</u> = Mischung 4 + 1 Liter Wasser

Vor der endgültigen Festlegung der Probeformen des Prüflings beim Schleiftopfverfahren war im Rahmen eines Grundsatzversuches versucht worden, den Einfluß einer Probenanphasung auf das Verschleißgeschehen zu erfassen. Dabei zeigte sich, daß die Verschleißbeträge unabhängig von vorhandener oder nicht vorhandener Anphasung proportional zu der bei der Abwärtsbewegung stark druckbelasteten Stirnfläche sind und die auf diese Stirnfläche bezogene Verschleißbeträge somit konstant ausfielen. Bei den Folgeversuchen wurde daher auf die Probenanphasung verzichtet.

Beim Schleifradverfahren nach Haworth wird nach Bild 5 zwischen eine sich mit 250 U/min. drehende Gummiradscheibe (Ø 233 mm, Dicke 15 mm) und dem Prüfling (70x25 x10 mm) ein spezifizierter mineralischer Gegenstoff zugegeben. Der Prüfling wird über einen Hebelarm mit 20 N an die Gummiradscheibe gepresst und vom sich in die Gummischeibe einpressenden Gegenstoff vorwiegend ritzend beansprucht.

Die Prüfzeit beträgt jeweils 4 min. Ausgehend von der glattgeschliffenen Probe wurden jeweils 3 hintereinander folgende 4 min. Prüfperioden in Schweißrichtung und Gegenschweißrichtung erfaßt.

4.2. Ergebnisse der Modellverschleißprüfungen

4.2.1. Ergebnisse der Schleiftellerprüfung:

Die bei der Schleiftellerprüfung ermittelten Werte 1-, 2- und 3-lagiger Panzerungen sind in Tabelle 2 zusammen mit Werten von Referenzwerkstoffen wieder gegeben.
Generell kann festgelegt werden, daß die Verschleißbeträge sämtlicher Legierungen, mit Ausnahme der Legierungen entsprechend Legierungsgruppe 10 tendenzmäßig mit steigender Lagenanzahl nur leicht fallen.

Das heißt, daß im Rahmen der auftretenden Schweißzusatzwerkstoffverdünnungen durch die Aufmischung von Grundwerkstoff zwar eine Tendenz zu höherem Ritzwiderstand bei geringerer Schweißzusatzwerkstoffverdünnung zu erkennen ist, die Abhängigkeit jedoch nur mäßig ist. Eine Ausnahme von dieser Regel bilden die im reinen Schweißzustand stark übereutektischen Legierungen Nr. 1 und 2 bei der einlagigen Auftragschweißung. Insbesondere im Bereich von Flint und Granat als Gegenstoff schwanken hier die Verschleißbeträge um bis zu 300 %, abhängig davon, ob die Kokillenschweißung 4 mm oder 5 mm Auftragshöhe erbracht hat.

Prüfling				Verschleißbetrag (mg/Blatt bei Gegenstoff				
Schweißzusatz Nr.	Lagenanzahl	Lagendicke	Härte HV	Glas	Flint	Granat	Korund	Silizium Karbid
1	1	4	707	0.1	4.75	5.8	6.4	7.9
		5	733	0.1	2.45	4.6	7.7	8.7
	2	4.5	824	0.2	1.5	3.6	7.0	8.0
	3	(4-5)	847	0.1	1.05	2.4	5.8	8.9
2	1	4	762	0.3	4.1	7.5	8.1	8.6
		5	762	0.2	1.5	2.8	6.5	6.7
	2	4.5	894	0.2	0.4	2.0	5.5	8.4
	3	(4-5)	946	0.2	0.2	1.1	5.1	8.7
3	1	4.5	342	4.7	8.8	11.5	9.3	9.8
	2	(4-5)	370	4.7	9.6	10.7	10.3	8.8
	3		390	3.4	10.4	12.5	12.5	10.5
4	1	4.5	733	0.1	7.5	8.8	10.4	9.4
	2	(4-5)	762	0.3	5.5	7.1	7.3	7.4
	3		762	0.2	5.1	6.2	6.4	7.7
5	1	4.5	215	6.0	9.6	10.1	10.4	9.9
	2	(4-5)	224	5.6	7.3	10.2	9.0	10.1
	3		227	4.9	8.7	10.3	9.5	9.6
6	1	4.5	217	9.9	10.2	13.5	11.5	10.9
	2	(4-5)	227	8.5	9.6	12.2	11.1	10.2
	3		239	8.5	10.3	11.9	10.7	9.9
7	1	4.5	363	4.6	11.2	13.2	11.4	12.0
	2	(4-5)	420	0.7	10.6	14.1	12.9	13.3
	3		446	0.4	11.5	13.8	11.4	13.2
8	1	4.5	974	0.5	1.8	1.9	6.4	14.1
	2		-	-.--	-.-	-.-	-.-	-.-
	3	(4-5)	-	-.--	-.-	-.-	-.-	-.-
St37			120	17	19	19	20.4	17
Hartmanganstahl				3.8	8.3	10.0	9.4	8.8
GX210Cr12			780	0.3	3.5	5.5	5.8	7.5

Tabelle 2
Schleiftellerwerte

Prüfling				Verschleißbetrag (mg/Versuch) bei Gegenstoff				
Schweißzusatz Nr.	Lagenanzahl	Lagendicke	Härte (HV)	feinkörnig Quarzsand H34	scharfk. Quarzsand (0.9-1.2mm	rundlicher Quarzsand 0.7-1.2mm	Sand-Kies	Sand Kies Wasser
1	1	4		(1.8) 0.8	(28.8) 22.6		(8.6)8.3	(13.8)11.
		5		(1.5) 0.7	(17.4) 15.2		(4.4)3.4	(13.0) 9.
	2	4-5		(1.3) 0.7	(8.3) 7.5		(3.9)4.1	(9.8) 8.
	3	4-5		(1.6) 1.2	(5.4) 5.7		(2.8)2.3	(6.8) 7.
2	1	4		(0.9) 0.9	(23) 19.2	(11.7)10.8	(5.7)5.2	(9.3) 6.8
		5		(1.8) 0.7	(18.9)17.9	(10.7) 9.0	(3.4) 2.6	(12.4)7.5
	2	4-5		(1.0) 1.0	(5.1) 4.3	(7.9) 3.6	(2.8) 2.3	(6.4) 5.0
	3	4-5		(1.8) 0.9	(5.4) 3.4	(6.8) 3.6	(1.6) 1.4	(8.5) 5.4
3	1	4.5		(3.6) 3.1	(32.5)32.7		(31.6)29.8	48.7
	2	(4-5)		(4.7) 3.9	(47.3)43.4		(36.7)27.8	52.0
	3	"		(4.0) 3.0	(43) 41.5		(30.9)27.7	43.8
4	1	"		(2.8) 2.6	(34.6)27.3	(10.5) 9.6	(7.6) 5.2	(14.4)12.
	2	"		(2.9) 2.5	(31.5)26.9	(8.8) 8.0	(5.3) 4.2	(14.8)12.
	3	"		(2.5) 2.5	(24.4)21.8	(8.2) 7.0	(3.8) 3.6	(11.7)10.
5	1	"		(6.3) 5.1	(96.1)88.5	(52.0)49.8	(22.7)20.4	(46.6)42.
	2	"		(5.5) 5.5	(83.9)79.0	(50.6)50.1	(19.1)18.5	(40.4)33.
	3	"		(6.4) 5.9	(91.5)81.2	(45.0)43.5	(20.0)19.5	(40.4)36.
6	1	"		(12.8)11.7	(94.3)88.5		(27.5)25.3	(47.1)39.
	2	"		(12.0)10.8	(85.7)76.9		(27.3)25.3	(37.1)33.
	3	"		(10.0) 8.4	(62.0)60.7		(25.9)24.5	(42.6)36.
7	1	"		(7.3) 5.8	(191.2)130.6	(57.8)51.	(29.6)27.2	(70.5)61.
	2	"		(6.3) 4.7	(159.2)138	(53.5)52.	(33.7)28.9	(60.2)49
	3	"		(5.7) 3.8	(134)120.0	(57.5)52.5	(27.3)27.4	(61.6)50
8	1			(2.5) 2.1	(10.5) 7.9	(6.8) 2.7	(4.7) 2.2	(9.1)6.0
	2	-		-	-	-	-	-
	3	-		-	-	-	-	-

Tabelle 3
Schleiftopfwerte

Bild 6

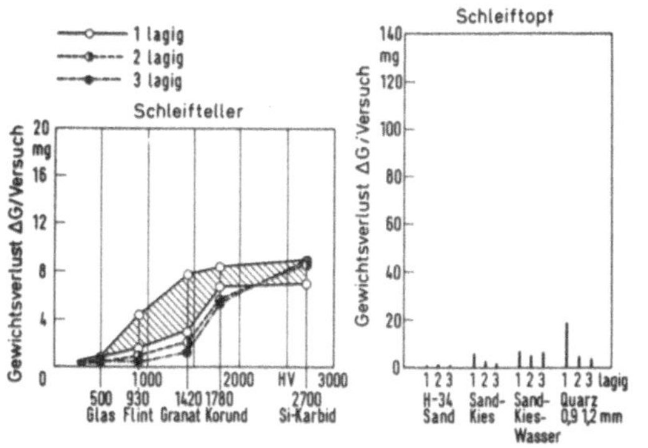

Verschleißbeträge an Auftragschweißungen mit Schweißzusatz Nr. 2

Dementsprechend zeigt die auszugsweise grafische Darstellung der Ergebnisse von Schweißzusatz-Nr. 1 bei der Einlagenschweißung, entsprechend Bild 6, ein breites Streufeld deren hohe Verschleißwerte den angefallenen 4 mm dicken Panzerungen und deren niedrige Verschleißwerte den 5 mm dicken Panzerungen zuzuordnen sind. Mit anderen Worten erreicht man bei gleicher Einbrandtiefe und einer von 4 auf 5 mm erhöhten Auftragdicke nicht nur eine Erhöhung der Verschleißdicke um die 20 Relativ-Prozent, die mehr aufgetragen worden sind, sondern eine zusätzliche Erhöhung des Verschleißwiderstandes bei ritzender Beanspruchung durch Flint und Granat um das 2- bis 3-fache. Bei einlagiger Auftragschweißung dieser Legierungen ist demnach auf das Erreichen eines geringen Aufmischungsverhältnisses bzw. hoher Auftragdicken besonders zu achten. Die Abweichung der Verschleißbeträge von S-Linienverlauf /27/ bei hohen C-Gehalten entspricht dem an Grußproben bereits von /26/ ermittelten Werkstoffverhalten.

Als einzige sonstige Fülldrahtelektrode zeigte die Fülldrahtelektrode Nr. 7 (Stellit-6) eine starke Abhängigkeit der Verschleißbeträge von der Lagenzahl und zwar bei Glas als Gegenstoff. Während bei der Einlagenschweißung der Verschleißbetrag mit nur 4,6 mg/Blatt bereits in der Hochlage liegt, ist bei der Zwei- und Dreilagenschweißung ein ritzarmes Verschleißen in der Tieflage zu beobachten.

Hier handelt es sich um ein Problem nicht ausreichender Matrixhärte in der ersten Lage. Die einlagige Stellitschweißung liegt mit HV 363 unter der beim Gegenstoff Glas für den beginnenden Steilabfall benötigte Härte von 400 HV.

Die 60 % Wolfram-Karbid enthaltende Fülldrahtelektrode Nr. 8 zeigt als einzige Fülldrahtelektrode ganz offensichtlich deutliche Unterschiede im Verhalten des Schweißgutes gegenüber dem Schweißgutverhalten entsprechender umhüllter Stabelektroden. Beim Verarbeiten dieser Fülldrahtelektroden ist ein so großer Teil des Wolfram-Karbides unerwünschtermaßen aufgeschmolzen, daß das Schweißgut extrem hart und spröde wird, mit der Folge, daß es beim Auftragen einer zweiten Lage zu Abplatzungen kommt. Dies ist vor allem die Folge davon, daß bei der Fülldrahtelektrode aus Platzgründen die mittlere Korngröße der Wolframkarbidfüllung viel feiner ist als bei entsprechend dickeren Stabelektroden. Eine summarische Darstellung der Schleifteller-Verschleißwerte der einlagigen Auftragschweißungen zeigt Bild 7, zusammen mit den Werten der Referenzwerkstoffe. Diese Werte sind auch Grundlage für die unter 7.1. aufgeführten wirtschaftlichen Folgerungen.

Bild 7

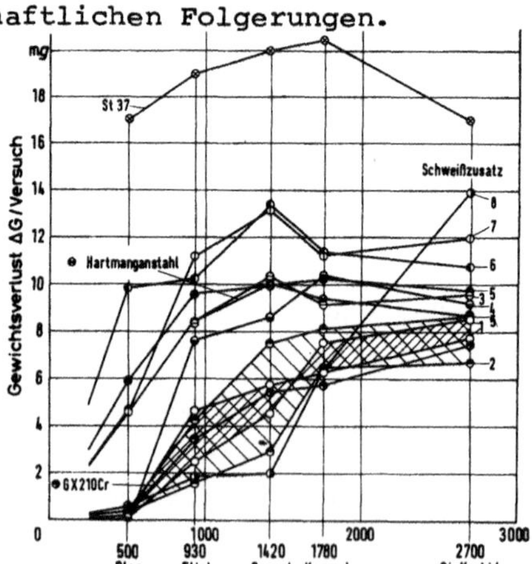

Verschleißbeträge von einlagigen Auftragschweißungen und Vergleichswerkstoffen beider Schleiftellerprüfung

4.2.2. Ergebnisse der Schleiftopfprüfung:

Die in Tabelle 3 aufgeführten Verschleißbeträge stellen Mittelwerte von mindestens zwei Versuchsreihen dar, wobei eine Versuchsreihe sich über vier Einzelversuchsmessungen von 4 min. erstreckt. Da bei jeder Versuchsreihe in der Regel der erste Wert deutlich über den folgenden lag, wurden die ersten Werte jeweils als Einlaufwerte getrennt erfaßt und erschienen in der Aufstellung zusätzlich als Klammerwerte.

Ähnlich wie beim Schleifteller ist die Differenzierung zwischen Einlagen-, Zweilagen- und Dreilagenschweißung in der Gruppe hochkohle-, hochchromhaltige Zusatzwerkstoffe (Legierungsgruppe 10) insgesamt am deutlichsten. Am deutlichsten ist die Differenzierung ausgeprägt bei Verwendung des scharfkantigen Quarzsandes der Körnung 0.9 - 1.2 mm, am geringsten beim feinkörnigen Quarzsand " H 34 ".
Wie beim Schleiftellerverfahren zeigt sich der Einfluß geringer werdender Schweißzusatzwerkstoffverdünnung mit steigender Lagenanzahl bei den übrigen Schweißzusatzwerkstoffen tendenzmäßig in der Form, daß die Verschleißbeiträge mit steigender Lagenanzahl im allgemeinen geringfügig zurückgehen.

Die Gütefolge unter Verwendung verschiedener Gegenstoffe ist, wie auch beim Schleiftellerverfahren, nicht generell einheitlich, besonders jedoch sticht das extrem schlechte Abschneiden von Stellit-nr. 6 bei sämtlichen grobkörnigen Gegenstoffen hervor der oberflächlich weniger verfestigt als die weicheren

Manganaustenite.

4.2.3. Ergebnisse der Gummiradprüfung:

Die bei der Gummiradprüfung ermittelten Werte zeigen Bild 8 für den feinkörnigen Quarzsand und Bild 9 für groben, scharfkantigen Quarzsand als Gegenstoff.

Bild 8

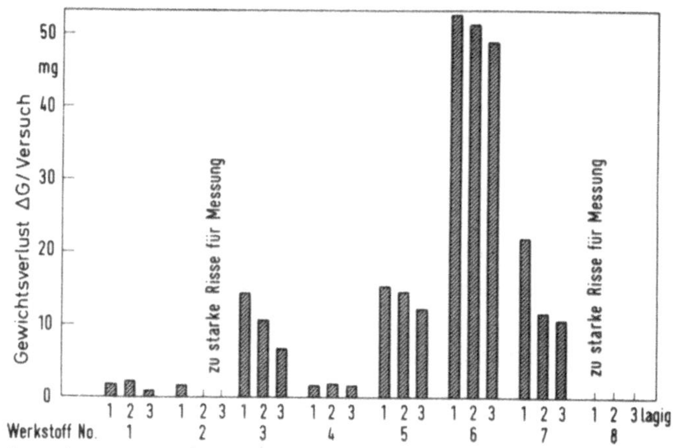

Versuchsergebnisse Gummiradversuche feinkörnigem Quarzsand (H-34)

Bild 9

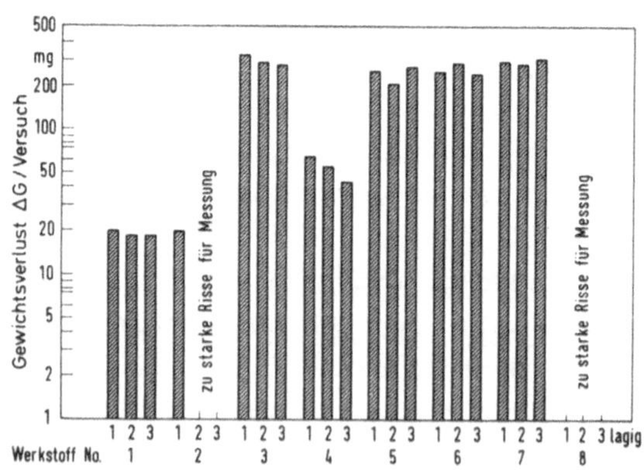

Versuchsergebnisse Gummiradversuch mit scharfkörnigem Quarz (0,9-1,2 mm)

Wie auch schon beim Schleiftopfversuch festgestellt, ist bei Verwendung des groben, scharfkantigen Quarzsandes in der Regel die Differenzierung eindeutig und die relative Streubreite geringer als bei Verwendung des feinkörnigen H 34-Quarzsandes. Insbesondere bei den hochkohlenstoffhaltigen Schweißzusatzwerkstofftypen führt die Rißbehaftung und teilweise auch die Ausbruchneigung der Karbide bei den an sich niedrigen Verschleißbeträgen unter Verwendung von H 34-Quarzsand als Zwischenstoff zu einer erheblichen Streuung der Werte.

In der Regel sind die mit H 34-Quarzsand als Zwischenstoff gemessenen Verschleißwerte zumindest eine Zehnerpotenz geringer als die Werte mit Quarzsand 0.9 - 1.2 mm.

Einen deutlichen Unterschied beim Vergleich der Verschleißgeschwindigkeitsrelationen der Auftragschweissungen unter Verwendung der zwei verschiedenen Quarzsande als Gegenstoff zeigt der Auftragschweißwerkstoff Nr. 6, der bei Verwendung von H 34-Quarzsand besonders ungünstig abschneidet. Dieser Chrom-Mangan-Austenit-Werkstoff mit nur 0.4 % Kohlenstoff verliert unter den gegebenen Bedingungen offensichtlich jede Aufhärtungsmöglichkeit und schneidet hier sogar schlechter ab als St 37. Bei dem ähnlich aufgebaute Schweißzusatzwerkstoff Nr. 5 mit 1 % Kohlenstoff ist dieses extrem schlechte Verhalten gegen feinkörnigen Quarzsandes nicht mehr zu beobachten.

5. Prüfung mechanisch-technologischer Eigenschaften

Werden auftraggeschweißte Bauteile zusätzlich zur abrasiven Beanspruchung auf Stoß beansprucht, wie z.B. Prallbarren bei Prallmühlen, so kann eine ungenügende Zähigkeit der Auftragschweißung insbesondere

an Kanten leicht zu Ausbrecherscheinungen führen und somit zu einem Verlust des gewünschten Schutzes gegen abrasiven Verschleiß. Eine mit steigender Stoßbeanspruchung steigende Zähigkeit der Verbundlösung Auftragschweißung ist somit in vielen Fällen Nebenbedingung für den erfolgreichen Einsatz von Hartauftragschweißungen. Ein möglicher zweiter Schadensmechanismus bei Stoß- und Druckbeanspruchung ist das Einbrechen der Auftragschweißung, wobei dieser Schadensmechanismus in der Regel ein Problem des Verbundwerkstoffcharakters der Auftragschweißung darstellt oder - anders ausgedrückt - bei mangelnder Druckfestigkeit des Grundwerkstoffes bevorzugt auftritt.
Ziel der folgenden Untersuchungen an reinen Schweißgutproben und an Auftragschweißungen auf verschiedene Grundwerkstoffe ist es somit, solchen Schadensmechanismen durch gezielte Auswahl des Auftragwerkstoffes und der Auftragwerkstoff-Verbundlösung vorzubeugen bzw. das Energieaufnahmevermögen von Auftragschweißproben meßtechnisch zu erfassen und so zu straffen, daß die nächst günstigere Lösung gezielt ausgewählt werden kann.
Bild 10 zeigt die im Rahmen dieser Untersuchungen hergestellten Proben und die Prüfanordnung bei den Biegebruch- und Fallhammerversuchen. Nach vorhandener Literaturkenntnis ist es dabei durch den Einsatz der in Absatz 3 gezeigten Doppelkokillentechnik erstmals gelungen ausreichend groß reine Schweißgutproben aus hochverschleißfestem Schweißgut mittels Metall-Lichtbogenschweißens schweißtechnisch herzustellen. Eine Probenherstellung in Größe der verwendeten Biegeproben führt bei konventioneller Viellagenschweißung infolge der Schweißeigenspannungen bei solchen Schweißgütern unweigerlich zur Rißbildung. Die Proben müssen daher in einer Hitze und einer Lage hochgezogen werden.

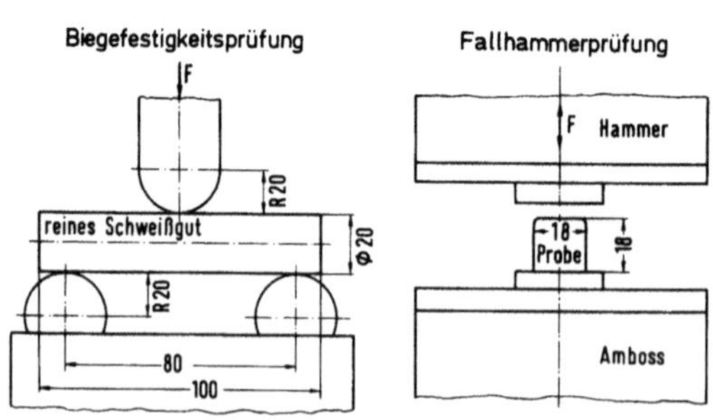

Bild 10

Ermittlung reiner Schweißgüter mechanisch technologischer Eigenschaften

Die in Bild 3 gezeigte auf- und zuklappbare Doppelkokille ermöglicht, in der Kombination der Fähigkeit von Fülldrahtelektroden sich auch mit einem freien Drahtende von 5 - 6 cm gut verarbeiten zu lassen, die Herstellung solcher bis zu 10 cm langen Schweißgutproben ohne Unterbrechung.

5.1. <u>Biegefestigkeitsprüfung:</u>

Bild 11 zeigt sowohl die jeweils 3 ermittelten Einzelwerte der Biegefestigkeitsprüfung der reinen Schweißgüter, sowie deren arithmetischen Mittelwert.

Bild 11

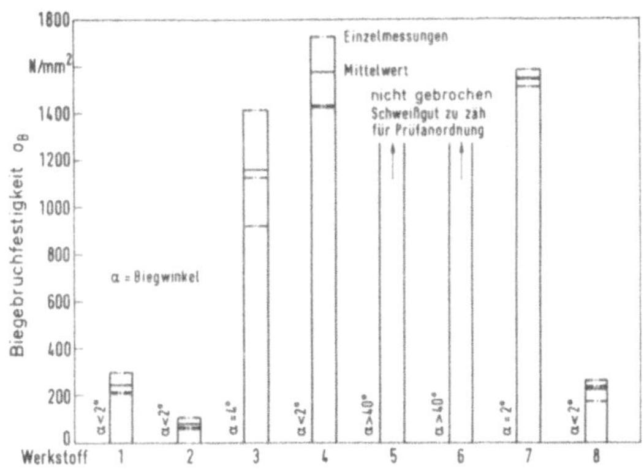

Erwartungsgemäß wird die Prüfanordnung der hohen Zähigkeit der hartmanganstahlähnlichen Schweißgüter Nr. 5 und Nr. 6 nicht gerecht, zeigt jedoch bei den harten Hartauftragschweißgütern eine gute Reproduzier- und Differenzierbarkeit. Im Rahmen eines Ausweitungsversuches der Biegebruchprüfung auf den Standardwerkstoff der Legierungsgruppe 10 mit 5 % C und 30 % Cr konnte in der Literatur erstmals in /28/ aufgestellte These, daß der teilweise Ersatz von Chrom durch Niob (entsprechend Schweißzusatz Nr. 1) zähigkeitsfördernd sei nicht bestätigt werden, da sowohl der Durchschnittswert der Bruchfestigkeit mit 358 N/mm² wie auch sämtliche Einzelwerte bei der 5 % C - 30 % Cr-Version höher lagen als bei Schweißzusatz Nr. 1. Verschleißmäßig positiv kann sich demnach allenfalls der Härtesteigerungseffekt nicht aber eine angenommene Steigerung der Zähigkeit auswirken.

5.2. Fallhammerprüfung:

Aus der Reihe der Fallhammerprüfung zeigt das folgende Bild 12 die an reinen Schweißgutproben ermittelten Verformungswerte bis zum Bruch.

Bild 12

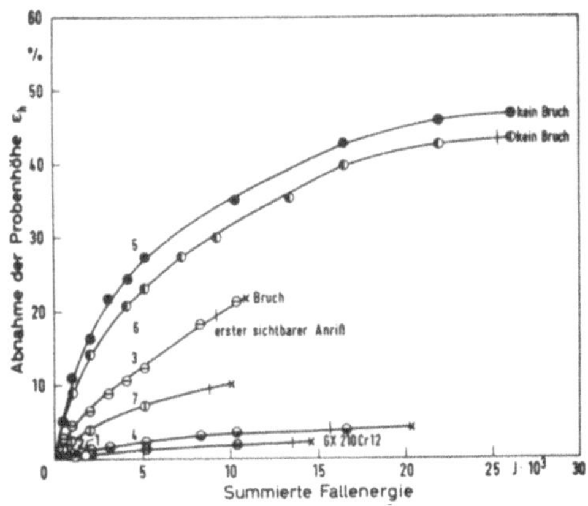

Verhalten reiner Schweissgutproben bei der Fallhammerprüfung

Dabei wurde der 23,5 kg schwere Fallhammer, entsprechend Bild 10, beginnend mit einer Fallhöhe von 10 cm, von Fall zu Fall um 5 cm höher, bis zur gerätemäßig (275 cm) maximal vorhandenen Fallhöhe angehoben und die Fallenergie summiert aufgetragen. Zusätzlich wurden die verschiedenen Schweißzusätze auf St 37, hochfestem Stahl und hartmanganähnlichem Grundwerkstoff 1, 2- und 3-lagig aufgeschweißt, der Fallhammerprüfung unterzogen und ausgewertet. Dabei zeigten sich tendenzmäßig folgende Einflüße:

a) die Grundwerkstoffe erhöhen den Bruchwiderstand einlagiger, hochharter Panzerungen generell

b) das Energieaufnahmevermögen der auftraggeschweißten Kombination ist mit St 37 als Grundwerkstoff relativ am niedrigsten und mit hartmanganstahlähnlichem Grundwerkstoff am höchsten.

c) Bei mehrlagigen Auftragschweißungen wirkt sich eine Erhöhung der Vorwärmtermeratur spannungsabbauend und daher günstiger aus.

6. Verhalten von Auftragschweißungen bei hohen Temperaturen:

Auftragsgemäß ist, in Zusammenarbeit mit dem BFI-Vorhaben "Verschleißminderung an Heißsintersieben" das Verhalten von Auftragschweißungen bei hohen Temperaturen als zusätzlicher Schwerpunkt mit in das Programm genommen worden.

Dabei war, aufbauend auf die Ergebnisse von /29/, die dort bewährte übereutektische Panzerungslegierung auf Chrom-Kohlenstoffbasis mit austenitstabilisierenden Molybdän- und Nickelzugaben versuchsweise im untereutektischen Kohlenstoffbereich zu optimieren. Durch geringeren Kohlenstoffgehalt sollte

a) die Zunderbeständigkeit erhöht, und
b) die Rißanfälligkeit der Panzerung verringert werden.

Dies ist in Form von Kohlenstoffvariationen der Panzerung mit der Fülldrahtelektrode Nr. 3 in der in 6.3. beschriebenen Weise erfolgt.

Sämtliche untersuchten Schweißzusätze wurden zusätzlich auf ihren Warmhärteverlauf und ihr Verzunderungsverhalten bei hohen Temperaturen untersucht.

6.1. Verzunderungsverhalten:

Das Verzunderungsverhalten der 8 untersuchten Schweißzusatzwerkstoffe ist durch Glühung im elektrisch beheizten Ofen unter Luftatmosphäre an reinen Schweißgutproben der Abmessung Ø = 20 mm, h = 8 mm ermittelt worden und die Bilder 13 - 16 zeigen die bei 2- und 20-stündiger Glühung infolge Verzunderung erfolgten absoluten und oberflächenbezogenen Gewichtszunahmen im Bereich von 600-900° C.

Bild 13

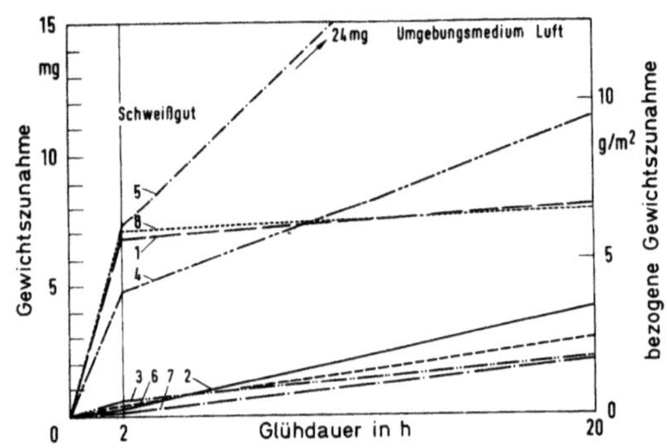

Gewichtzunahme von Schweißgutproben durch Zunderbildung bei 600 °C

Bild 14

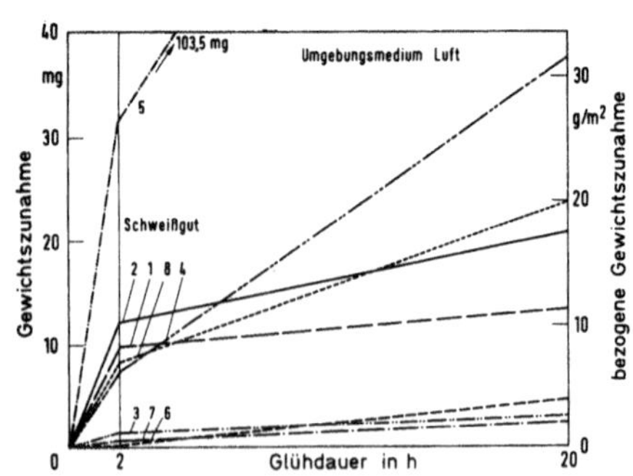

Gewichtszunahme von Schweißgutproben durch Zunderbildundung bei 700°C

Bild 15

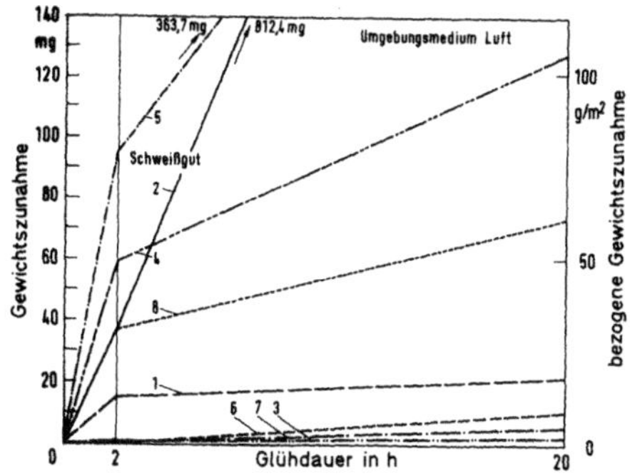

Gewichtszunahme von Schweißgutproben durch Zunderbildung bei 800 °C

Bild 16

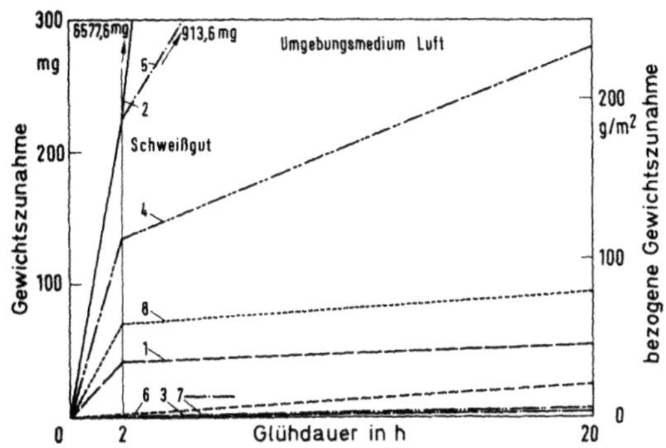

Gewichtszunahme von Schweißgutproben durch Zunderbildung bei 900 °C

Dabei fällt insbesondere das sehr gute Verhalten von Schweißzusatz Nr. 3 (ähnlich wie der Co-Basis-Schweißzusatz Nr. 7) im gesamten Temperaturbereich auf, sowie das infolge zu geringer Chromgehalte schlechte Abschneiden von Schweißzusatz Nr. 5. Von besonderer Bedeutung ist das extrem schlechte Abschneiden des im Kaltbereich optimal verschleißbeständigen Schweißzusatz Nr. 2, beginnend im Temperaturbereich über 700° C.

Die Bilder 17 und 18 zeigen das Verzunderungsverhalten bei unterschiedlich langer Glühung nochmals übersichtlich in logarithmischer Form.

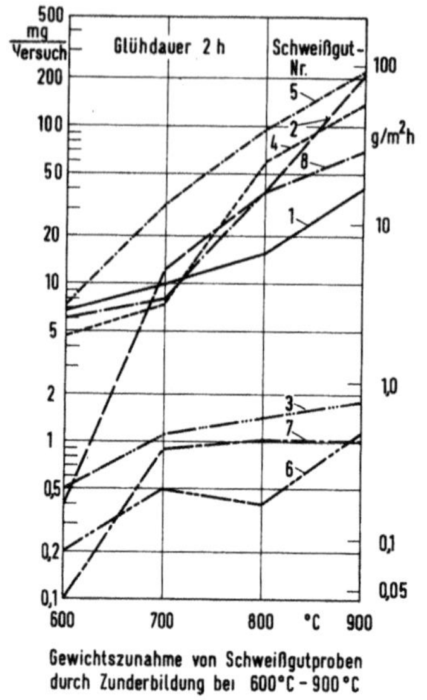

Bild **17**

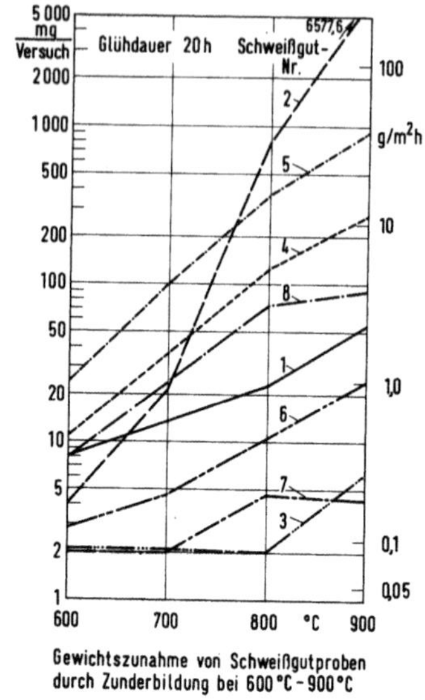

Bild **18**

Setzt man als oberen erlaubten Verzunderungswert für 5 mm Panzerungen 5 g/m^2h, entsprechend etwa 1 mm Legierungsverzunderung/Monat, an so kann man aus diesen Bildern den von der Verzunderungsseite her noch akzeptablen Temperatur-Werkstoffbereich für den Einsatz im Hochtemperaturbereich ablesen.

Dreidimensionale Darstellungen in den Bildern **19** und **20** verdeutlichen, das günstige Verzunderungsverhalten von Schweißzusatz Nr. 3 im gesamten Temperaturbereich und das Einsetzen sogenannter katastrophaler Oxidation bei Schweißzusatz Nr. 2 mit Wegfall des Deckschichtbildungseffektes über 700° C.

Bild **19**

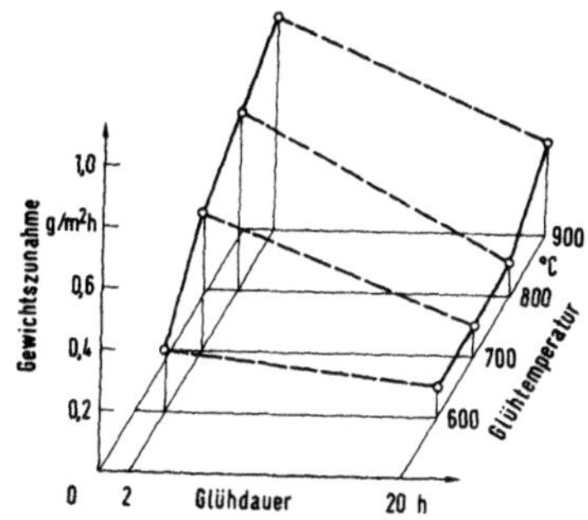

Verzunderungsgeschwindigkeit im Zeit/Temperaturfeld (Schweißzusatz Nr. 3)

Bild **20**

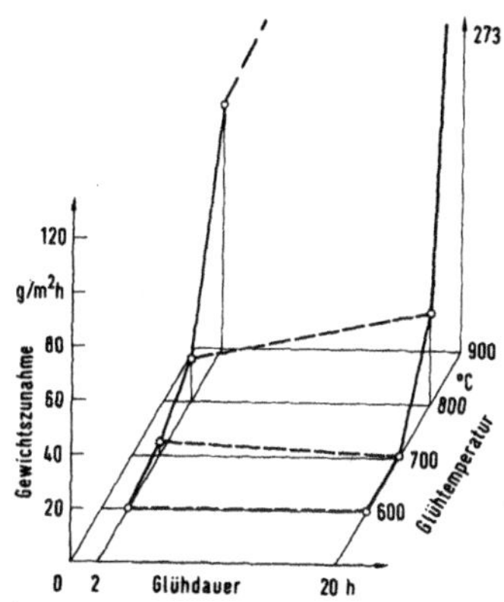

Verzunderungsgeschwindigkeit im Zeit / Temperaturfeld (Schweißzusatz Nr. 2)

6.2. Warmhärten:

Die Übersicht der an reinen Schweißgutproben ermittelten Warmhärten in Bild **21** zeigt als Besonderheit bei Probe-Nr. 2 im Bereich 700° C einen ausgeprägten Anstieg, verursacht durch Sekundärkarbidausscheidungen. Relativierend dazu muß erwähnt werden, daß die dynamische Warmhärte bei hohen Werten, auf Vickershärten umgerechnet, vergleichsweise zu hohe Werte bringt und zusätzlich ein Absinken dieses Härteanstieges bei längerer Temperaturbelastung zu erwarten ist. Die Warmhärteverläufe wurden an der RWTH Aachen unter Schutzgasatmosphäre ermittelt, d.h. Einsatzbegrenzungen die durch zu schnelle Verzunderung hervorgerufen werden, sind daraus nicht ersichtlich.

Bild **21**

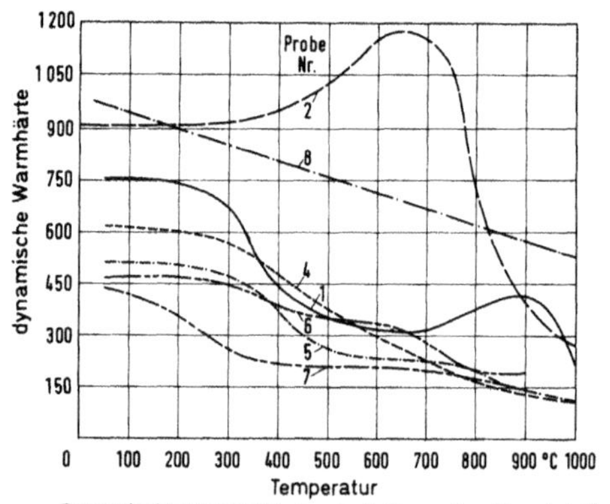

Dynamische Warmhärten von Auftragschweißwerkstoffen

6.3. Optimierungsversuche für den Heißsinterbereich:

Für die Optimierungsversuche von Schweißzusatz Nr. 3, hinsichtlich eines optimalen Kohlenstoffgehaltes mußte eine Auftragmethode gefunden werden, bei der bewußt nur der Kohlenstoffgehalt der Panzerung variiert wird, ohne die sonstigen Legierungszusätze zu beeinflußen. Dies wurde im Bereich von 1 %-**2,5%** Kohlenstoff dadurch erreicht, daß stets der gleiche Röhrchendraht (Schweißzusatz Nr. 3) in Kombination mit Ferro-Chrom-Zusatzpulver gleicher Menge aber unterschiedlichen Kohlenstoffgehaltes verschweißt wurde. Die Ferro-Chrom-Pulverzugabe erfolgte stets konstant über eine elektronische Dosiereinrichtung mit der Draht-Pulvereinstellung 1 : 0.4, d.h. pro Kilogramm abgeschmolzenen Röhrchendraht wurde insgesamt 0.4 kg Ferro-Chrom unterschiedlichen Kohlenstoffgehaltes zugesetzt. Bei der gemessenen mittleren Einbrandtiefe von 1.2 mm und der Auftraghöhe von 5 mm erhielt man die effektive Zusammensetzung der jeweiligen Auftragschweißung in der Form der Mischrechnung:

20 % Grundwerkstoff + 57 % reines Schweißgut des Röhrchendrahtes + 23 % Ferro-Chrom-Pulver

In anderen Worten wird bei dieser Verfahrensweise die Legierungsverdünnung des reinen Röhrchendrahtschweißgutes durch die aufgeschmolzenen 20 % Grundwerkstoff hinsichtlich Chrom durch die zusätzlcihe Ferro-Chrom-Pulverzugabe (Chromgehalt jeweils 67 %) stets in gleicher Weise kompensiert.

Als Ferro-Chrom-Pulverzusatz standen zwei Sorten mit extrem verschiedenen Kohlenstoffgehalten zu Verfügung und zwar:

Sorte A:	Kohlenstoffgehalt	0.1 %
Sorte B:	Kohlenstoffgehalt	9.0 %

die in unterschiedlichem Mischungsverhältnis zueinander zugegeben wurden.

Im folgenden werden die aufgrund der Vorversuche eingestellten Schweißdaten sowie die bei unterschiedlichen Kohlenstoffwerten erhaltenen Härten der einlagigen Schweißpanzerung:

1. Schweißzusätze

1.1. Fülldrahtelektrode Nr. 3, Ø 2,4 mm

1.2. Ferro-Chrom-Pulver (Körnung 0 - 0.6 mm, Analyse: 67 % Cr, Rest = Fe + C)

Kohlenstoffversionen des Ferro-Chrom-Pulvers:

Version A:	C = 0 %	Version B:	C = 1.5 %
Version C:	C = 3.0 %	Version D:	C = 5.0 %
Version E:	C = 9.0 %		

2. Schweißdaten:

Schweißstrom	=	420 A
Schweißspannung	=	33 V
freies Drahtende	=	35 mm
Draht : Pulver	=	1 : 0.4
Penzelfrequenz	=	60 min.
Pendelbreite	=	30 mm
Raupenbreite überlappt	=	30 mm
Brennvorschub	=	200 mm/min.
Brenneranstellung	=	85° (schleppend)

3. Erzielte Werte:

Zusatzversion des Chrompulvers	% C in Panzerung	Panzerungshärte HV
A	0.98	360
B	1.31	446
C	1.50	464
D	1.86	503
E	2.40	540

Mit dieser Schweißmethode wurden daraufhin Platten aus 13 CrMo 44 der Abmessung 600x450 mm mit unterschiedlichen Panzerungs-Kohlenstoffgehalten geschweißt und unter Überwachung des BFI im Projekt Verschleißminderung an Heißsinter-Siebmaschinen, sowohl als Blindplatten wie auch als Siebplatten eingesetzt. Als Kohlenstoffzielwerte wurden alternativ 1 %, 1,5 %, 2,0 % und 2,5 % angestrebt.

6.4. Ergebnisse im Heißsinterbereich:

Mit der vorab beschriebenen Methode wurden insgesamt drei Platten wie folgt gefertigt:

Kohlenstoffzielgehalt %

	Teilbereich 1,	Teilbereich 2,	Teilbereich 3
Platte 1	1	1,5	2,0
Platte 2	1,5	2,0	2,5
Platte 3	1,5	2,0	2,5

Platte 1 wurde in gelochter Form eingesetzt (Siebplatte), die Platten 2 und 3 als Blindplatten.
Als Maximaltemperatur wurde durch in Bohrungen eingebrachte Schmelzpulver etwa 750°C ermittelt. Anfritten des Pulvers mit Schmelzpunkt 774°C, Schmelzen des 704°-Pulvers.
Die Abhängigkeit des Verschleißes vom Kohlenstoffgehalt war bei der Siebplatte besonders deutlich, wie Bild 22 eindringlich zeigt.
Eine Nachanalyse aus den verbliebenen Panzerungen der 3 Teilbereiche ergab folgende Werte:

C = 2,4 % ... Teilbereich der praktisch fast zur Gänze verschlissen ist.
C = 1,92 % ... mittlerer Teilbereich
C = 1,45 % ... Teilbereich mit geringstem Verschleiß.

Bild 23 zeigt den mittleren Härteverlauf der Panzerungen im Schweißzustand und im Zustand nach dem Betriebseinsatz. Demnach wird die Härte der Panzerungen im Heißeinsatz und in den Grenzen C = 1,5 % bis C = 2,5 % auf einheitlich 440-450 HV abgebaut mit der Konsequenz, daß der höhere Kohlenstoffgehalt verschleißmäßig an Einfluß verliert.

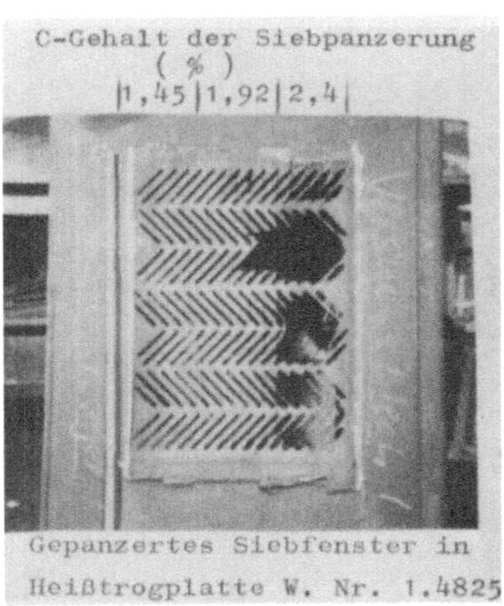

Bild 22

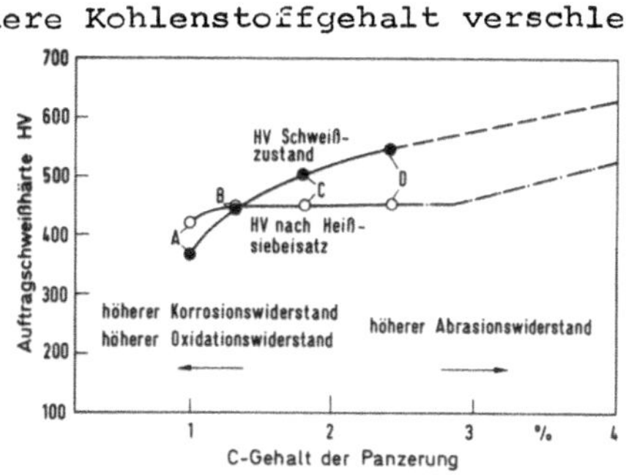

Einfluß des Kohlenstoffgehaltes auf die Schweißguthärte (Schweißzusatz-Nr. 3)

Bild 23

Daraus und aus dem günstigeren Zunderverhalten der Panzerungen mit niedrigem Kohlenstoffgehalt ist das Optimum bei C = 1,5 % entsprechend Bild 24 erklärbar:

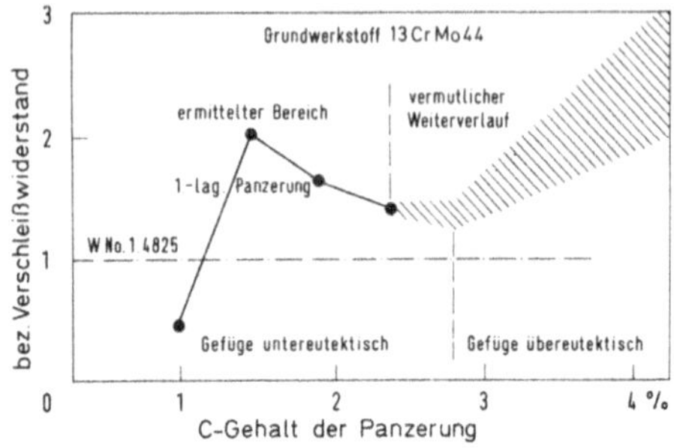

Einfluß des C-Gehaltes auf den Heißabrieb (Schweißzusatz-Nr. 3)

Bild 24

Im Bereich unter 1,5 % Kohlenstoff verliert das verschleißmäßig günstige Austenitgefüge im Hochtemperatureinsatz an Stabilität und wandelt sich zu schnell um, wodurch das rapide Abfallen des Verschleißwiderstandes erklärt werden kann. Panzerungsvariationen mit unter 1 % C bilden bereits im Schweißzustand teilweise ferritisches Gefüge.
Bild 25 zeigt vergleichsweise REM-Oberflächenaufnahmen der verschlissenen Siebplatte im Teilbereich C = 1,45 % und C = 2,4 %. Beide weisen keine, beim Abrasiv-Verschleiß im Kaltbereich üblichen, Ritzspuren auf. Die Deckschichtbildung der Probe mit C = 1,45 % ist einheitlicher und zeigt bereits bei der Probenvorbereitung eine extrem gute Haftung.

C = 1,45 %

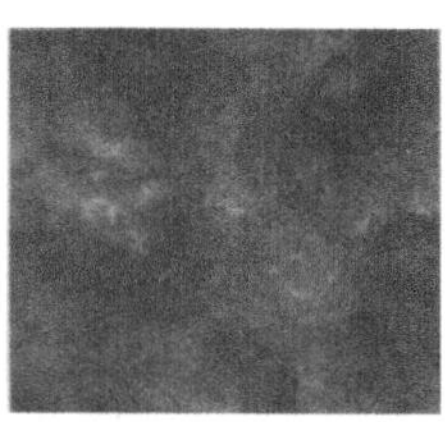

C = 2,4 %

Heißsieb-Panzeroberfläche
REM-Aufnahme, Vergrößerung 4000-fach

Bild 25

7. <u>Schlußfolgerungen wirtschaftlicher Art:</u>

7.1. Einsatz Kaltbereich

Bild 26 zeigt die grafische Auswertung einer vergleichsweisen Berechnung direkter Verschleißkosten. Verglichen werden einfache Formteile einlagig gepanzert und ungepanzert aus St 37. Die Auswertung gilt genau nur auf der Basis der Verschleißgeschwindigkeiten, wie sie bei der Schleiftellerprüfung gemessen worden sind und unter der Annahme der Übertragbarkeit dieser Verschleißgeschwindigkeiten auf große Flächen. Da bei der Verschleißprüfung sehr scharf (hohe Flächenbelastung, stets frischer Gegenstoff) geprüft wird, ergeben sich für die Praxis bei großflächigem Verschleißangriff zu hohe Werte, jedoch sind die Kostenrelationen für übertragbare Abrasivbeanspruchungsfälle repräsentativ.

Bei St 37 als Vergleichswerkstoff ergibt sich demnach der größte Kostenvorteil bei den Panzerungen mit Schweißzusatz Nr. 1 und Nr. 2 im Bereich von Glas bis Granat als Gegenstoff und bei Schweißzusatz Nr. 4 im Bereich Glas/ Flint. Der Kostenvorteil gegenüber St 37 ist naturgemäß abhängig vom Ausnutzungsvolumen des Bauteiles das bei St 37 modellhaft mit alternativ 10 und 30 % angenommen

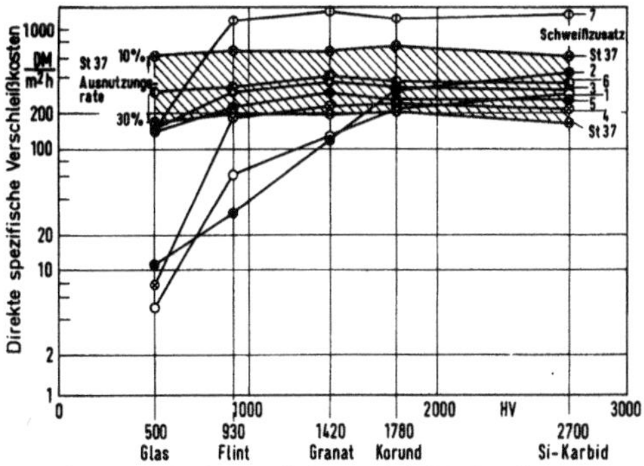

Direkte spezifische Verschleißkosten bei einlagigen Panzerungen und bei St 37 (Verschleißgeschwindigkeit entsprechend Schleiftellerprüfung)

Bild 26

wurde und bei Panzerungen als regenerierbar.

7.2 Einsatz Heißbereich:

Untereutektische Legierungsvariationen von Schweißzusatz Nr.3 haben ein Verhaltensoptimum im Sinter-Heißbereich bei Kohlenstoffgehalten um 1,5 % mit gegenüber dem Standardwerkstoff 1.4825 deutlich niedrigeren Verschleißwerten. Diese optimierte Form erreicht jedoch bei in etwa gleichen Panzerungskosten nicht den Verschleißwiderstand der gleichartigen übereutektischen Legierung mit C-Werten über 4 %. Eine weitere Legierungsoptimierung sollte sich daher bei Panzerungen auf hochkohlenstoffhaltige Typen konzentrieren.

8. Literaturübersicht:

(1) GENATH, B.: Der Verschleiß ist das Sorgenkind im Maschinenbau - VDI-Nachrichten 52 (1971) Nr.47, Seite 1-7

(2) WAHL,H.: Allgemeine Verschleißprobleme bei der Zerkleinerung - Techn.Mitteilungen 59 (1966) 5, Seite 228-232

(3) KOTTSIEPER,H.: Instandhaltung dynamisch beanspruchter Bauteile/Fachtagung Instandhaltung 75 des Deutschen Komitee Instandhaltung

(4) LANG,O.et al: Tribologie-Reibung, Verschleiß, Schmierung Forschungsbericht 76-38, erstellt im Auftrag des BMFT, Forschungsk. Maschinenbau e.V. (1976)

(5) WAHL,W.: Materialeinsparung durch aktiven Verschleißschutz, VDI-Bericht Nr. 277 (1977) Seite 219-233

(6) WASSERMANN, R.: Instandhaltung aus Unternehmersicht, Vortrag DKIN/Okt. 1980 Wiesbaden

(7) WAHL,W.: Abrasive Verschleißschäden und ihre Minderung, VDI-Bericht Nr. 243, 1975, Seite 171-187

(8) WAHL,W.: Werkstoffauswahl für abrasiv beanspruchte Bauteile, VDI-Bericht Nr. 410, 1980, Seite 161-177

(9) HABIG, K.H.: Verschleiß und Härte von Werkstoffen, Carl-Hauser-Verlag, München/Wien 1980

(10) WAHL, W.: Verschleißkostensenkung durch Einsatz von Verbundlösungen. Aufbereitungstechnich 20 (1979) Heft 10, Seite 559-566

(11) KEMPE,P.C.: Laminated Martensitic White Irons ...Properties

and Application Transaction, Volume LXX, 1967,pp.225-230

(12) RAZIM,C., DÜLL,CHR., RÄUCHLE,W.: Über die Beeinflussung von Bauteilen-Grundeigenschaften durch Verschleißschutzschichten, VDI-Berichte Nr. 333 (1979), Seite 11-22

(13) DANNÖHL,H.D.: Verschleißschutzschichten - Auswahlkriterien für den speziellen Verschleißfall. VDI-Berichte Nr.333 (1979), Seite 159-165

(14) WAHL,W., KRETSCHMER,I.: Verschleißkostensenkung in der Aufbereitungstechnik durch mechanisiertes Hartauftragschweißen. Aufbereitungstechnik, Jahrgang 18(1977), Heft 10, Seite 623-642

(15) HENNEMANN,K., MONDRE,J., SOMMER,P.: Verbesserte Lichtbogenschweißverfahren für Werkstatt und Baustelle. DVS-Bericht, Band 46 (1977), Seite 1-7

(16) GURDET,F., PERTENEDER,E., WABNEGGER,F.: Neuentwickelte Fülldrahtelektroden zum Schweißpanzern verschleißbeanspruchter Oberflächen. DVS-Bericht Nr. 65, Seite 197-201 Deutscher Verlag für Schweißtechnik GmbH, Düsseldorf

(17) N.N. Fachbuchreihe "Schweißtechnik" Band 55, Teilbericht 2.17 - Metall-Lichtbogenschweißen mit Fülldrahtelektroden Deutscher Verlag für Schweißtechnik

(18) DIN 8555: Teil 2 : Schweißzusätze zum Auftragschweißen Beuth Verlag GmbH, Berlin

(19) VdTÜV-Merkblatt "Verfahrensprüfung für die Hartauftragschweißung" - Schweißtechnik 1156-11.78,Maximilian-Verlag, 4900 Herford

(20) UETZ,H., SOMMER,K., KOSHRAWI: Übertragbarkeit von Versuchs- und Prüfungsergebnissen bei abrasiver Verschleißbeanspruchung auf Bauteile, VDI-Bericht Nr.354(1979),S.107-124

(21) HEINKE,G.: Verschleiß - eine Systemeigenschaft, Auswirkungen auf die Verschleißprüfung. Werkstofftechnik 6 (1975), Seite 164/169

(22) WAHL, H.: Allgemeine Ordnungsprobleme bei der Bearbeitung von Verschleißfragen, insbesondere durch Verschleißprüfung. Giesserei 53 (1966), Heft 13, Seite 421-ff.

(23) DIN 50 320: Verschleiß-Begriffe Systemanalyse und Verschleißvorgänge - Ausgabe Dez. 1979, Beuth-Verlag GmbH, Berlin

(24) UETZ,H., FÖHL,J.: Erscheinungsformen von Verschleiß-schäden - VDI-Bericht Nr. 243 (1975) "Methoden der Schadensuntersuchung", Seite 127/142

(25) UETZ, H., FÖHL,J.: Wear as Energy Transformation process Wear 49 (1978), Seite 253-264

(26) WAHL,W.: Untersuchungen über die Bestimmung der Verschleißbeständigkeit und Bruchneigung von harten Werkstoffen, insbesondere Hartguß - Diss. Uni.Stgt. 1970

(27) WAHL,H.: Praktische Verschleißabwehr - Archiv für Metallkunde (1949) Heft 4, Seite 121-128

(28) SCHRAMM, H.: Chrom-Niob legierte Hartlegierungen für das Auftragschweißen - Industrieanzeiger Essen, Jahrgang 1975, Seite 1567-1569

(29) CORDES, u.a.: Verschleißminderung durch Anwendung der Schweißtechnik in Sinteranlagen - VdEH, Fachausschußbericht Nr. 5.017 - Düsseldorf (1977)

Untersuchungen
über
Herstellung und Eigenschaften von Hartstoffschichten
die durch reaktives Aufdampfen abgeschieden werden

von
Dr.rer.nat. U. König
Dr. Ing. H. Grewe

Fried.Krupp GmbH
Krupp Forschungsinstitut
Essen

Förderkennzeichen: 01 ZT 277
Laufzeit: 1.1.78 bis 31.12.80

Kurzfassung

Die Standzeit von Hartmetallschneidplatten zur Zerspanung von Eisenwerkstoffen kann durch das Aufbrigen einer etwa 10 µm dicken Harststoffschicht wesentlich erhöht werden. Solche Hartstoffschichten aus TiC, TiN, Al_2O_3 u.a. werden industriell durch chemische Gasphasenabscheidung (Chemical Vapour Deposition, CVD) bei relativ hohen Temperaturen (1000 bis 1100°C) hergestellt. Die vorliegende Untersuchung befaßt sich mit der Herstellung und den Eigenschaften von Hartstoffschichten, die durch reaktive Aufdampfverfahren abgeschieden werden. Diese sog. PVD-Verfahren (Physical Vapour Deposition) bieten gegenüber CVD-Verfahren den Vorteil der geringeren Substrattemperatur und der höheren Beschichtungsraten. Am Beispiel der Abscheidung von Titannitrid wurden verschiedene Herstellmethoden zur Erzeugung von haftfesten und verschleißbeständigen Schichten auf Hartmetallsubstraten nach PVD-Verfahren erprobt. Die Hartstoffschichten wurden hinsichtlich ihrer Zusammensetzung, Härte, Phasenreinheit, Gefüge und Verschleißverhalten mit verschiedenen Methoden untersucht. Es gelang, bei Substrattemperaturen von 600°C Schichten zu erzeugen, die sich in ihrem Verschleißverhalten als ebenbürtig zu CVD-beschichteten Proben erwiesen. Ähnlich günstige Ergebnisse wurden auch mit TiC, HfN- und HfC-Schichten sowie mit Schichten aus den Doppelnitriden Ti-Cr-N und V-Ti-N erzielt.

Gliederung

1. Einleitung

Die Beschichtung von Verschleißteilen mit Hartstoffen gewinnt immer größere Bedeutung in allen Bereichen der Technik. Die technische Bedeutung liegt darin, daß man durch Beschichtung eines relativ zähen Grundmaterials mit einem Hartstoff die an sich gegenläufigen Eigenschaften Härte (Verschleißbeständigkeit) und Zähigkeit miteinander paart.

Im vorliegenden Vorhaben wurden neue Verfahren der Beschichtung von Hartmetallen mit Hartstoffen untersucht und erprobt. Beschichtete Hartmetalle finden heute z.B. in Form von sog. Wendeschneidplatten für die spandende Bearbeitung von metallischen Werkstoffen ausgedehnte Verwendung. Durch eine Hartstoffbeschichtung erhöht sich die Schneidleistung der Hartmetallwerkzeuge bei der Bearbeitung von Eisenbasiswerkstoffen um ein Mehrfaches. Man kann die Werkzeugkosten durch Verlängerung der Schnittzeiten senken und/oder die Produktivität durch Erhöhung der Schnittgeschwindigkeit erhöhen.

Eingang in die industrielle Praxis haben bis heute mit verschiedenen Hartstoffen beschichtete Hartmetall-Schneideinsätze gefunden, die nach dem Verfahren der chemischen Abscheidung aus der Gasphase (Chemical Vapour Deposition, CVD) beschichtet werden. Ein Beispiel dafür ist die Beschichtung von Stählen und Hartmetallen mit Titancarbid, die aus einem Gasgemisch von Titantetrachlorid, Methan und Wasserstoff bei Temperaturen von 1000 bis 1100°C durchgeführt werden kann /1/. Eine Übersicht über den Stand der Technik der CVD-Hartstoffbeschichtungen wird in Abschnitt 2 gegeben. Obwohl sich diese Produkte z.T. hervorraggend bewährt haben, haben die CVD-Verfahren einige, den Anwendungsbereich einschränkende Nachteile. Die benötigten hohen Substrattemperaturen engen den Kreis der geeigneten Substratwerkstoffe ein. Die erreichbaren Beschichtungsraten von ca.1 bis 2 µm/h sind im Hinblick auf die benötigten Schichtdicken von ca. 10 µm gering. Es erschien daher aussichtsreich, die sog. physikalischen Aufdampfverfahren (Physical Vapour Deposition, PVD) auf ihre Eignung zur Hartstoffbeschichtung von Hartmetallen zu untersuchen.

Prinzipiell lassen sich mit PVD-Verfahren die Substrate bei niedrigeren Temperaturen als beim CVD-Verfahren beschichten. Man kann somit auch Substrate mit Hartstoffschichten überziehen, deren Eigenschaften sich bei höheren Temperaturen bereits wesentlich ändern, wie z.B. Stähle. Es sollten sich auch gelötete oder eingeschrumpfte Werkzeuge ohne Schädigung der Verbindung zwischen Hartmetall und Stahl beschichten lassen. Ein weiterer potentieller Vorteil ist es, daß aufgrund der Leistungsfähigkeit von modernen Elektronenstrahlverdampfern wesentlich höhere Abscheidungsraten von mehr als 60 µm/h möglich sind. Grundsätzlich sollte mit Hilfe der Elektronenstrahltechnologie als Fernziel auch die Entwicklung einer kontinuierlich arbeitenden Beschichtungsanlage möglich sein, wie sie z.B. für die Aluminiumbeschichtung von Stahlblechen und für die Chrombeschichtung von Turbinenschaufeln großtechnisch eingeführt sind. Es wurde bewußte von "potentiellen Vorteilen" gesprochen. Es ist nämlich von vornherein nicht ohne weiteres geklärt, ob die wesentlichen Vorteile des PVD-Verfahrens, nämlich niedrige Substrattemperaturen und hohe Beschichtungsraten, wie sie von der Herstellung von einfachen Metallüberzügen bekannt sind, auch beim Erzeugen von Hartstoffschichten durch reaktives Verdampfen erhalten bleiben können. Hartstoffe können nämlich nicht als Phase verdampft werden, sondern sie bilden sich letztlich auch durch eine chemische Reaktion auf der Substratoberfläche. Die Substrattemperatur hat somit auch beim PVD-Verfahren einen großen Einfluß. Das Metalloid des Hartstoffes muß aus der Gasphase im Verdampfungsraum bereitgestellt werden. Zur Bildung eines stöchiometrisch zusammengesetzten Hartstoffes muß der Druck des reaktiven Gases genügend hoch sein. Auf der anderen Seite darf der Dampfstrom der Metallatome durch Streuung an den Gasatomen bzw. Molekülen nicht zu sehr geschwächt werden. Die erzielbaren Beschichtungsraten werden daher auch begrenzt sein.

Aus der Literatur /2/ ist bekannt, daß mit Hilfe eines PVD-Verfahrens harte TiC-Schichten hergestellt werden können. Es war jedoch bei Beginn der vorliegenden Untersuchung nicht sicher, ob die so auf Hartmetallsubstraten erzeugten Schichten den extremen mechanischen, thermischen und chemischen

Beanspruchungen, wie sie beim Einsatz als Schneidwerkzeuge auftreten, gewachsen sein würden. Hier erschienen Bedenken am Platze, da bekannt war, daß PVD-Schichten andersgeartete Gefüge haben /3/ und daß wegen verlangsamter Diffusionsvorgänge die Haftfähigkeit der Schichten möglicherweise nicht ausreichend sein würde. Wegen der großen technischen Bedeutung solcher Hartstoffbeschichtungen und auch wegen neuartiger, durch CVD-Beschichtungen nicht zu erfassender Anwendungen wurde es notwendig, die Eignung von PVD-Hartstoffschichten für die genannten Einsatzfälle unter praxisnahen Bedingungen in direktem Vergleich zu durch CVD hergestellten Hartstoffen zu testen. Über Herstellung und Eigenschaften der letztgenannten Verschleißschutzschichten lagen im Hause der Verfasser bereits umfangreiche Erfahrungen vor.

2. Zur Bedeutung der Beschichtung von Hartmetallen mit Hartstoffen

Durch die breite industrielle Einführung der Hartmetalle auf WC-Co- und WC-TiC-TaC-Co-Basis wurden Anfang der 30iger Jahre revolutionierende Fortschritte in der Zerspanungstechnik erzielt. Es stand nun ein Werkstoff zur Verfügung, dessen Härte wesentlich höher als die der zu zerspandenden Metalle - insbesondere der zahlreichen Eisenwerkstoffe - war und der gleichzeitig eine ausreichende Festigkeit und Zähigkeit hatte, um den hohen Zerspanungskräften zu widerstehen. Durch die Einführung der Hartmetalle konnten z.B. die bis dahin beim Zerspanen mit Schnellstählen zulässigen Schnittgeschwindigkeiten nahezu verzehnfacht werden. Doch auch die Hartmetallwerkzeuge zeigten Verschleißerscheinungen, so insbesondere die durch den ablaufenden Span entstehende Kolkmulde und den sog. Freiflächenverschleiß an der dem Werkstückzugekehrten Seite (s. Bild 9). Man versuchte daher, den Verschleiß der Hartmetallwerkzeuge zu vermindern, indem man einen zähen Grundkörper mit einer harten, verschleißfesten Oberflächenschicht kombinierte. Hierzu gehören Versuche mit sog. Sandwich-Hartmetallen, bei denen auf sintertechnischem Wege ein feinkörniges, bindemetallarmes Hartmetall auf einen zähen Grundkörper aufgebracht wird /4/. Wegen der komplizierten Herstellung kam ihr Einsatz jedoch nicht über das Versuchsstadium hinaus. Andere Verfahren zur Erzeugung dünner, verschleißfester Oberflächen sind das Nitridieren /5/ und das Borieren /6/. Auch sie haben bis heute keine größere Bedeutung erlangt. Ein wesentlicher Fortschritt wurde erst durch die Einführung von mit dünnen Schichten aus Titancarbid überzogenen Hartmetallen Mitte der 60er Jahre erzielt. Die dafür benutzte Methode der chemischen Abscheidung aus der Gasphase (CVD-Verfahren) war ursprünglich zur Beschichtung von Eisenwerkstoffen entwickelt worden /7/. Die weitere Entwicklung führte zu Schichten aus TiC/TiN, TiC/Al_2O_3 und schließlich zu kompliziert aufgebauten Mehrstoffschichten aus den Elementen Ti, Al, C, N und O /8/.

Bild 1 zeigt in einem Schneidhaltigkeitsvergleich (Kolktiefe KT und Verschleißmarkenbreite VB nach einer Schnittzeit von 5 min) beim Drehen von Stahl C60, welche Leistungen heute mit beschichteten Hartmetallschneidplatten erzielt werden. Eine unbeschichtete Schneidplatte wäre bei den gewählten Testbedingungen schon nach etwa 1 min zu Bruch gegangen.

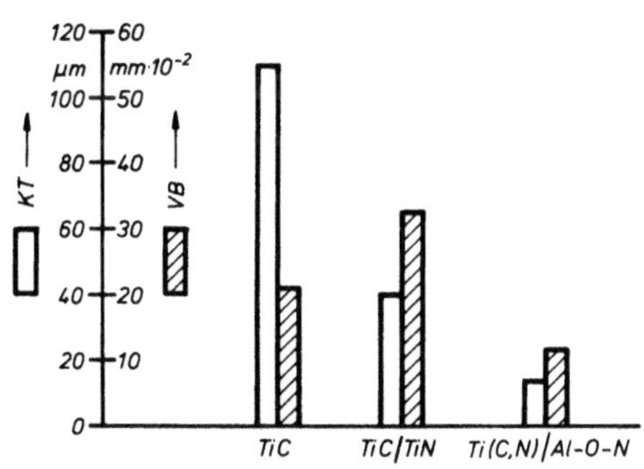

Bild 1: Verbesserung der Schneidhaltigkeit von Hartmetall-Wendeschneidplatten SNUN 120408, die mit Hartstoffschichten aus TiC, TiC/TiN und Ti(C,N)/Al-O-N beschichtet wurden. KT Kolktiefe und VB Verschleißmarkenbreite beim Drehen im glatten Schnitt an Stahl C60 mit v = 240 m/min, s = 0,28 mm/U, a = 1,5 mm, (vgl. Bild 9). Drehzeit 5 min.

In manchen Anwendungsfällen haben sich die Schnittzeiten durch die Einführung der Hartstoffbeschichtungen verzehnfacht, oder die Produktivität konnte durch höhere Schnittgeschwindigkeiten bedeutend erhöht werden. Der Anteil der beschichteten Hartmetalle bei der zerspanenden Drehbearbeitung von Eisenbasiswerkstoffen dürfte heute die 50%-Grenze überschritten haben.

3. Übersicht über PVD-Verfahren

Bild 2 gibt eine Übersicht über die sog. physikalischen Beschichtungsverfahren (PVD-Verfahren). Nach der Art der Atomisierung der Feststoffkomponente kann man die Verfahren einteilen in solche, die auf Verdampfung beruhen (Aufdampfen) und in Verfahren der Kathodenzerstäubung (Sputtern).

Mit den Aufdampfverfahren lassen sich die größten Beschichtungsraten erzielen. Zur Diskussion der verschiedenen Verfahren sei auf die prinzipielle Anordnung von Verdampfer, Substrat und ggf. Hilfselektroden in Bild 3 hingewiesen.

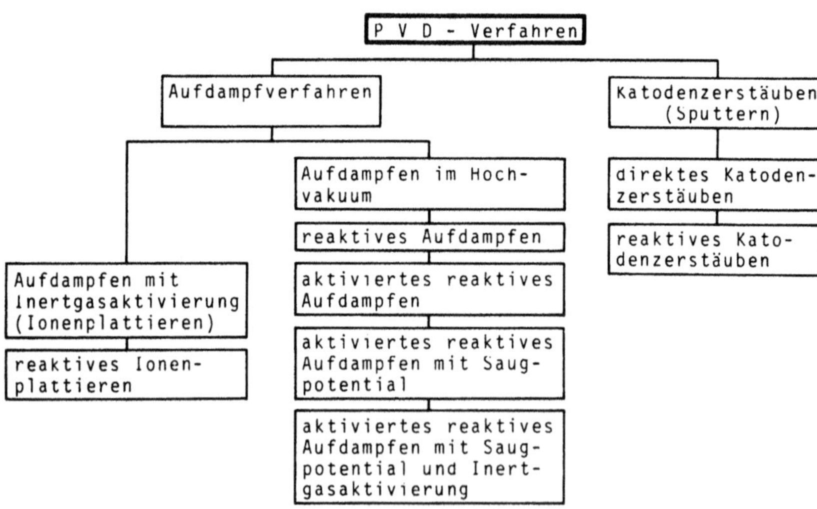

Bild 2: Übersicht der PVD-Verfahren.

Das einfache <u>Aufdampfen im Hochvakuum</u> (Schalter S_1 und S_2 in Bild 3 geöffnet, Druck < 10^{-5} mbar) kann jedoch im wesentlichen nur auf reine Metalle und auf Verbindungen angewendet werden, die beim Verdampfen nicht dissoziieren. Bei Legierungen haben infolge der unterschiedlichen Dampfdrücke der Komponenten die aufgedampften Schichten andere Zusammensetzungen als das eingesetzte Verdampfungsgut (von wenigen Sonderfällen abgesehen). Die hier interessierenden Hartstoffe können - abgesehen von der zu erwartenden thermischen Dissoziation - wegen ihrer hohen Schmelzpunkte nicht direkt verdampft werden.

Mit dem Verfahren des reaktiven Aufdampfens /9/ (Schalter S_1 und S_2 in Bild 3 geöffnet, Druck des reaktiven Gases $\sim 10^{-3}$ mbar) kann man dicke Schichten aus Metalloxiden herstellen. Man verdampft das Metall (z.B. Silizium) in Gegenwart eines Teildrukkes von Sauerstoff (z.B. 10^{-4} mbar). Das auf dem Substrat abgeschiedene Silizium oxidiert dann zu SiO_2. Bei Versuchen, nach der Methode des reaktiven Aufdampfens TiC-Schichten durch Verdampfen von Titan in Gegenwart von C_2H_2 herzustellen, erhielten Bunshah und Raghuram /2/ nur metallische Titanschichten.

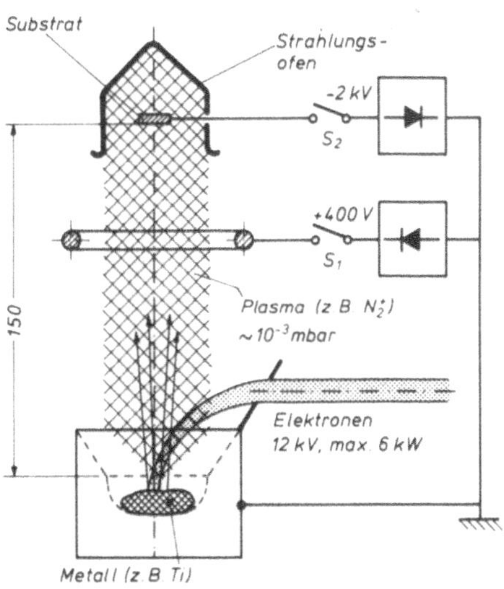

Bild 3: Prinzip der Verfahren des reaktiven Aufdampfens.

Titancarbidschichten entstanden, wenn das reaktive Gas im Rezipienten durch eine positive Hilfselektrode ionisiert wurde (Schalter S_1 in Bild 3 geschlossen). Die Ionisierung des Gases entsteht durch die Beschleunigung der durch Glühemission aus der Schmelze austretenden Elektronen. Bunshah prägte dafür den Begriff aktiviertes reaktives Aufdampfen. Nach dieser Methode konnten u.a. Schichten aus TiC, TiN, ZrC, VC, NbC, HfC, HfN und Al_2O_3 /10-13/ auf metallischen Substraten hergestellt werden. Die Schichten hatten im wesentlichen die für den jeweiligen Hartstoff typischen Eigenschaften (Härte, Stöchiometrie, Farbe u.a.).

Ein weiteres auf Verdampfung beruhendes Verfahren ist das Ionenplattieren /14/. Hierbei wird ein negatives Potential (ca. -2 bis -5 kV) an das Substrat gelegt (Schalter S_1 offen, S_2 geschlossen, s. Bild 3). Im Unterschied zum aktivierten reaktiven Aufdampfen werden wesentlich höhere Gasdrücke (10^{-2} bis 10^{-1} mbar) eingestellt. Es kann daher in einer Argonatmosphäre durch Stoßionisation eine stabile Glimmentladung entstehen. Die ionisierten Argonatome werden gegen die aufwachsende Schicht beschleunigt und bewirken dort Diffusionsvorgänge, die das Schichtgefüge stark beeinflussen. Da die freie Weglänge klein ($\sim$ 1 mm) ist, können die Dampfteilchen nicht mehr direkt auf das Substrat gelangen. Es bildet sich eine Art Dampfströmung zwischen Schmelzbad und Substrat aus. Die Gasstreuung bewirkt, daß auch Seitenflächen und mit verminderter Rate die Rückseite des Substrates beschichtet werden. Fügt man der Argonatmosphäre einen Anteil eines reaktiv wirkenden Gases hinzu, so gelangt man zum Verfahren des reaktiven Ionenplattierens, mit dem sich auch Hartstoffschichten herstellen lassen /15/.

Im Verlaufe der vorliegenden Untersuchungen wurde vorwiegend nach den Verfahren des aktivierten reaktiven Aufdampfens gearbeitet, wobei noch, wie weiter unten beschrieben wird, Modifizierungen vorgenommen wurden. Diese Verfahren können dann durch aktiviertes Aufdampfen mit Saugspannung (Schalter S_1 und S_2 in Bild 3 geschlossen, Druck $\sim$ 10^{-3} mbar) und durch aktiviertes reaktives Aufdampfen mit Saugspannung und Inertgasaktivierung

(Schalter S_1 und S_2 geschlossen, Druck des Reaktivgases $\sim 10^{-3}$ mbar, Inertgasbeimengung) bezeichnet werden. Eine andere Linie von PVD-Verfahren (s. Bild 2) beruht auf dem Vorgang des Kathodenzerstäubens (Sputtern). Bild 4 zeigt das Prinzip der sog. Magnetronkathode. Als Kathodenmaterial, das entmischungsfrei auf das Substrat gestäubt wird, kann man praktisch alle Materialien einsetzen, z.B. Legierungen aus Metallen mit sehr verschiedenen Schmelzpunkten (z.B. TiAl) oder auch - bei Anwendung der HF-Zerstäubung - keramische Materialien (z.B. Al_2O_3). Es können Schichtwachstumsraten von 3 µm/h und mehr erzielt werden. Man kann die Hartstoffe direkt (d.h. nur unter Zuhilfenahme von Argon mit einem Druck von etwa 10^{-2} mbar) und durch Einsatz eines entsprechenden Targets aufstäuben (direktes Sputtern). Man kann aber auch das Metall der Hartstoffverbindung (z.B. Ti) als Target einsetzen und durch Zusatz eines reaktiven Gases zum Sputtergas (z.B. Ar + N_2) das reaktive Sputtern anwenden. Da die abgestäubten Atome energiereicher als beim Aufdampfen sind, ergibt sich schon bei relativ niedrigen Temperaturen ein gutes Haftvermögen. Das Verfahren der Kathodenzerstäubung ist bereits seit langem bekannt, jedoch konnten nur geringe Beschichtungsraten ($\sim 0,1$ µm/h) erzielt werden. Erst durch die Entwicklung der Magnetronkathode konnten die Raten bedeutend erhöht werden (3 bis 10 µm/h), so daß auch diese Verfahren zur Erzeugung von verschleißmindernden Hartstoffschichten interessant werden. Die Sputtertechnik wurde jedoch in dem vorliegenden Vorhaben nicht untersucht.

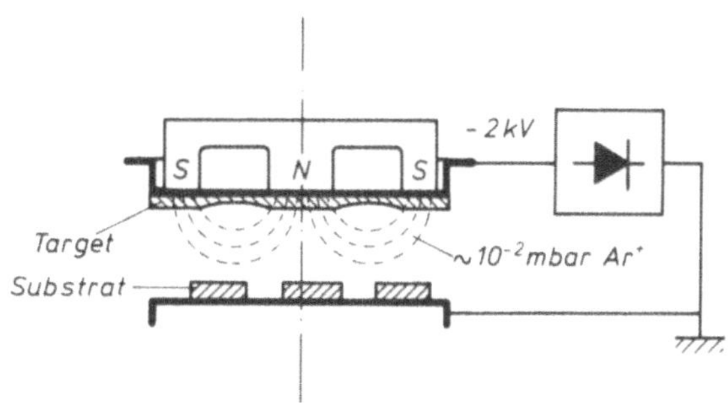

Bild 4: Prinzip des Magnetron-Kathodenzerstäubens.

4. Vorgänge beim reaktiven Aufdampfen

Es sei im folgenden die in Bild 3 dargestellte Anordnung diskutiert. Die Metallkomponente der auf dem Substrat aufzubringenden Hartstoffschicht wird durch eine Elektronenstrahlkanone aus einem Tiegel verdampft. Der Druck des im Verdampfungsraum befindlichen Reaktivgases kann genau geregelt werden. Es ist eine Ringelektrode mit positiver Vorspannung und eine negative Vorspannung des geheizten Substrates vorgesehen.

4.1 Reaktives Aufdampfen

Der Prozeß der Bildung einer Hartstoffschicht sei am Beispiel von Titannitrid TiN betrachtet. Das verdampfte Titan scheidet sich auf der Substratoberfläche ab, deren Temperatur weit unterhalb des Schmelzpunktes des Titans liegt. Es wird daher durch Kondensation nahezu jedes auftreffende Atom haften bleiben. Die Anzahl ν_{Ti} der Stöße der Titanatome pro Flächen- und Zeiteinheit berechnet sich aus Kondensationsrate K und der Anzahl N_{Ti} der Titanatome pro Volumeneinheit der Schicht zu

$$\nu_{Ti} = K \cdot N_{Ti}$$

Für TiN ist $N_{Ti} = 5{,}3 \cdot 10^{22} cm^{-3}$. Die Wandstoßzahlen des Reaktionspartners Stickstoff N_2 sind nach der kinetischen Gastheorie (M Molekulargewicht)

$$\nu_{N_2} = \frac{P_{N_2} \cdot L}{4RT} \sqrt{\frac{8RT}{\pi M}}$$

Wenn sich stöchiometrisches TiN an der Substratfläche bilden soll, dann muß das Verhältnis N : Ti mindestens gleich 1 sein. Schon eine einfache Betrachtung zeigt jedoch, daß das Metalloidatom-Angebot aus dem Gasraum sehr viel größer als das Metallatom-Angebot sein muß (s. Bild 5). Nimmt man an, daß nur die Stickstoffmoleküle, die auf nicht mit Stickstoff abgesättigte Bereiche der aufwachsenden Schicht treffen, chemisorbiert werden, so wird die Wahrscheinlichkeit für ein Stickstoffmolekül aus dem Gasraum mit zunehmender Absättigung der Oberfläche einen unge-

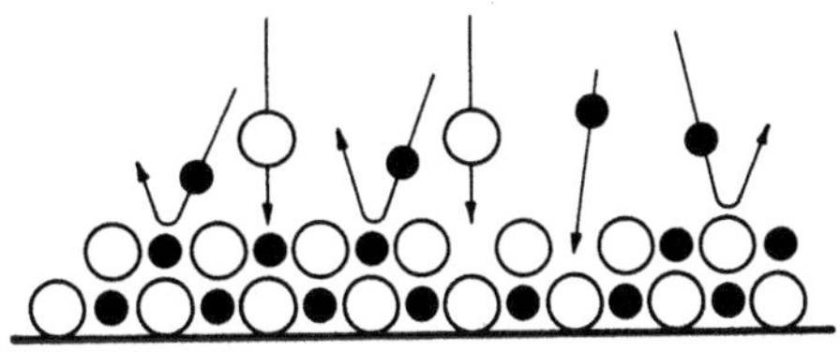

Bild 5: Einfaches Modell des Schichtwachstums. Das Metalloid (schwarz) haftet beim Auftreffen auf ungesättigte Gitterplätze und wird von abgesättigten Schichtbereichen reflektiert. Die Metallatome (weiß) haften an gesättigten und ungesättigten Plätzen.

sättigten Platz zu finden, immer ungünstiger. Zur gleichen Zeit treffen aber auch Titanatome auf die Schicht, deren Wahrscheinlichkeit für das Haftenbleiben für gesättigte und nicht gesättigte Bereiche ungefähr gleich groß ist. Daher entstehen im Schichtgitter zwangsläufig auch Stickstoff-Leerstellen, die durch Diffusion von der Oberfläche her aufgefüllt werden müssen. Aus dieser Überlegung folgt, daß das reaktive Aufdampfen eines Metalls mit einer Komponente aus der Gasphase (z.B. N_2, O_2, C_2H_2) nur dann zu stöchiometrischen Verbindungen führt, wenn das Verhältnis der Wandstoßzahlen ν_{Gas}/ν_{Metall} hinreichend groß ist. So ergibt z.B. beim reaktiven Aufdampfen von SiO_2 erst ein Verhältnis von $\nu_{O_2}/\nu_{Si} \sim 100$ eine stöchiometrische Schicht /16/.

Bild 6 gibt eine Übersicht über Stoßzahlenverhältnisse ν_{N_2}/ν_{Ti}, die sich in Abhängigkeit von der Aufdampfrate und dem Gasdruck ergeben. Danach kann bei hohen Beschichtungsraten ein großes Stoßzahlenverhältnis nur bei relativ hohen Gasdrücken erreicht werden. Z.B. muß, um $\nu_{N_2}/\nu_{Ti} = 50$ bei $K = 10$ µm/h zu erreichen, der Druck $p = 3{,}3 \cdot 10^{-3}$ mbar sein. Dann kommt man jedoch schon in einen Druckbereich, in dem die mittlere freie Weglänge nicht mehr groß im Vergleich zum Abstand zwischen Verdampfer und Substrat ist. Die mittlere freie Weglänge für Titanatome in nichtionisiertem N_2 ist in Bild 6 ebenfalls mit eingezeichnet.

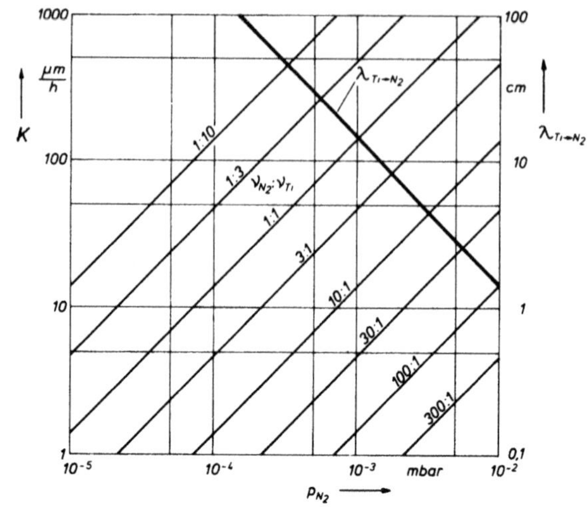

Bild 6: Stoßzahlenverhältnis ν_{N_2} : ν_{Ti} in Abhängigkeit von der Aufdampfrate K und dem Druck des Reaktivgases p_{N_2} bei 600 °C λ_{Ti-N_2} freie Weglänge von kollimierten Titanatomen in Stickstoff in Abhängigkeit vom Druck.

Ähnliche Größenverhältnisse, wie sie hier am Beispiel der Abscheidung von TiN diskutiert wurden, gelten auch für andere Hartstoffe. In Wirklichkeit sind die physikalischen und chemischen Vorgänge beim Entstehen der Hartsoffschicht viel komplizierter, insbesondere bei Berücksichtigung von Aktivierungsvorgängen. Dennoch zeigt diese einfache Überlegung bereits, bei welchen Drücken (des Reaktivgases) und Schichtwachstumsraten man etwa stöchiometrisch zusammengesetzte Hartstoffe erwarten kann.

4.2 Verdampfungs- und Abscheideraten

Nach der kinetischen Gastheorie beträgt die Masse G, die aus der Oberfläche einer Schmelze pro Zeiteinheit in einen Vakuum-Halbraum verdampft /17/

$$G = \alpha \, p_s \sqrt{\frac{M}{2\pi RT}}$$

In dieser Gleichung bedeuten: T absolute Temperatur, R Gaskonstante, M Molekular- bzw. Atomgewicht, p_s Sättigungsdampfdruck.

Der Koeffizient $\alpha \leq 1$ berücksichtigt Abweichungen vom idealen Verhalten. Der Sättigungsdampfdruck der Metalle kann Tabellenwerken mit thermodynamischen Daten /18/ entnommen werden. Über der Schmelzfläche des Verdampfers bildet sich beim Verdampfen eine keulenförmige Dampfstromdichteverteilung aus, die ungefähr durch die Funktion $\phi(r,\Theta) = \phi_o(r)\cos^n\Theta$ mit $n = 3$ bis 4 beschrieben werden kann, wobei Θ der Winkel zur Normalen ist /19/. Im Zentrum der Dampfkeule erhält man etwa die vier bis fünffache Stromdichte, bezogen auf den isotropen Fall mit $\phi(r) = G/(2\pi r^2)$. Bei idealer Verdampfung und vollständiger Kondensation ($\alpha = 1$) kann man die Abscheideraten in Abhängigkeit von der Temperatur des Schmelzbades berechnen. Bild 7 zeigt das Ergebnis für einen Abstand von 15 cm zwischen Schmelzfläche und Substratfläche für die Metalle Al, Co, Ti, V und Hf. Im Falle des Titans sind Oberflächentemperaturen von etwa 2500 K erforderlich, um Beschichtungsraten von 100 µm/h zu erreichen.

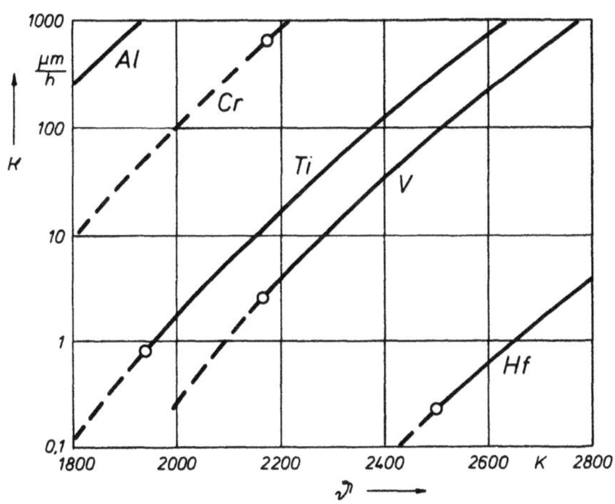

Bild 7: Aufdampfraten verschiedener Metalle in Abhängigkeit von der Schmelzbadtemperatur. Fläche der Quelle 1 cm^2, Kondensationskoeffizient = 1.

4.3 Gleichförmigkeit der Verdampfung

Bild 7 zeigt, daß jeweils 100 K Temperaturerhöhung etwa eine Verdopplung der Verdampfungsrate zur Folge haben. Zur Erzielung einer gleichförmigen Verdampfungsrate sollte also die Oberflächentemperatur so konstant wie möglich gehalten werden. Bei der Verdampfung aus dem wassergekühlten Kupfertiegel sind folgende Größen von Einfluß auf die Oberflächentemperatur der Schmelze:

1. Masse des Schmelzkörpers
2. Wärmeübergang: Schmelzkörper → Tiegel (Formanpassung, Schlackenreste)
3. Bildung von TiN auf den kälteren Teilen des Schmelzkörpers
4. Einstelldaten der Elektronenkanone (Leistung, Fokussierung, Landungspunkt des Strahles, Wobbelamplitude).

Während die letztgenannten Daten genau einstellbar sind, entziehen sich die Einflüsse nach 2. und 3. der unmittelbaren Kontrolle. Letztere verändern sich auch während des Verdampfens. Während es also einerseits nach dem Gesagten wichtig ist, die Oberflächentemperatur des Schmelzbades konstant zu halten, ist es andererseits schwierig, letztere reproduzierbar einzustellen. Eine pyrometrische Temperaturmessung ist wegen des mit der hohen Verdampfungsrate verbundenen Beschlagens der Fenster und wegen verfälschender Einflüsse durch die Strahlung des ionisierten Gases (N_2, C_2H_2) nur mit sehr großem Aufwand möglich. Als wirksamste Methode der Kontrolle der Verdampfungsrate bietet sich ein Schichtdickenmeßgerät mit einem Schwingquarz im Dampfstrom an, obwohl dessen Einsatz wegen der hohen Beschichtungsraten, der hohen Temperatur am Ort des Substrates und der Anisotropie des Dampfstromes ebenfalls problematisch ist. Wegen der großen Anisotropie der Dampfstromdichteverteilung erhält man eine von der Ablage der Position des Substrates von der Symmetrielinie stark ahängiges Schichtdickenwachstum /19/. Bei einem Abstand von 150 mm zwischen Verdampfer und Substrat ergeben sich bei

Ablagen von 30 mm etwa 87% und bei 60 mm etwa 60% der Rate im Zentrum. Den gleichen Effekt bewirkt eine Neigung der Schmelzfläche, die man bei Elektronenstrahlverdampfern mit schräg auftreffenden Elektronen beobachten kann. Man erhält dann ein von der mehr oder weniger zufälligen Neigung der Schmelzfläche abhängiges Schichtdickenwachstum, wodurch systematische Versuchsreihen, bei denen das Verhältnis der Metallatome zu den Metalloidatomen vorgegeben werden soll, erschwert werden.

5. Herstellung von Hartstoffschichten

Für die Beschichtungsexperimente wurde eine PVD-Anlage entworfen und gebaut, deren prinzipieller Aufbau bereits in Bild 3 gezeigt wurde. Mit dieser Anlage (Bild 8) konnten alle in Abschn. 3 genannten Aufdampfverfahren durchgeführt werden. Der Vakuumkessel besitzt einen Innendurchmesser von 495 mm und wird durch eine Diffusionspumpe (Saugleistung 1000 l/s) evakuiert. Der Kessel hat mehrere Beobachtungsfenster. Zum Verdampfen des Schmelzgutes wird eine Elektronenkanone mit 90°-Umlenkung und

Bild 8: Ansicht der Apparatur.

einer max. Strahlleistung von 6 kW verwendet. Um auch bei höheren Gasdrücken im Rezipienten verdampfen zu können, hat die Elektronenkanone eine eigene Pumpe. Es wird aus einem wassergekühlten Kupfertiegel mit einem Fassungsvermögen von etwa 11 ml verdampft. Der Tiegel liegt auf Erdpotential. Über dem Tiegel befindet sich eine ringförmige Elektrode ("Sonde") für das positive Aktivierungspotential. Das isoliert aufgehängte Substrat kann an ein negatives Potential gelegt werden. Zur Versorgung dient ein 5 kV / 1 kW - Gleichspannungsgerät. Ein weiteres Gleichspannungsgerät 600 V / 0,5 A versorgt die als Ring ausgebildete Hilfselektrode. Die Heizung des Substrates erfolgt indirekt durch die Strahlung eines Ofens aus Molybdänblech. Die Temperatur des Substrates wird an einem in äquivalenter Position angeordneten Substrat gemessen, mit dem ein Pt-PtRh-Thermoelement kontaktiert ist. Die Substratheizung ist handgeregelt und ermöglicht eine Temperaturkonstanz von etwa 10°C. Der Druck des Reaktivgases wird über ein Ventil mit automatischer Druckregelung eingestellt. Es wird hochreines Stickstoffgas verwendet, bei dem sich eine anfangs eingesetzte Gasreinigungsanlage als entbehrlich erwies. Zur weiteren Ausrüstung gehören ein Vakuummeßgerät und ein Massenspektrometer zur Gasanalyse. Die Hartmetallsubstrate wurden vor dem Beschichten in einem Ultraschallbad mit Trichloräthylen und mit Alkohol gereinigt. Nach dem Einbau in die Apparatur wurde der Rezipient auf etwa 10^{-6} mbar abgepumpt und die Probe auf die Beschichtungstemperatur gebracht. Danach wurde der Rezipient bis auf 0,5 mbar mit Argon gefüllt und das Substrat durch Zünden einer Glimmentladung etwa 10 min gesäubert. Nach erneutem Abpumpen wurde der Rezipient mit dem Reaktivgas gefüllt, das zur Verdampfung gelangende Metall (Titan) mit der Elektronenstrahlkanone aufgeschmolzen und die Leistung eingestellt. Die Abhängigkeit der Abscheiderate von der Leistung wurde durch Vorversuche bestimmt. Danach wurde das Reaktivgas durch Anlegen einer positiven Spannungvon etwa 400 V an die Hilfselektrode ionisiert, ggf. die Substratspannung eingeschaltet und eine Blende geöffnet. Nach Beendigung der Beschichtung (typische Dauer: 20 min) und dem

Abschalten von Elektronenkanone und Hilfselektrodenspannung wurde der Rezipient bis zur Beendigung der Abkühlung mit 5 bis 10 mbar Stickstoff gefüllt. Für Versuche mit Gasgemischen (z.B. N_2 und Ar) als Reaktivgas wurde das benötigte Gemisch in einer besonderen Gasflasche hergestellt und über ein Nadelventil eingespeist.

6. Untersuchung der Hartstoffschichten

Die abgeschiedenen Hartstoffschichten wurden in der Regel röntgenographisch, metallographisch und mit Hilfe eines Rasterelektronenmikroskopes untersucht. Die Verschleißbeständigkeit wurde durch Schneidhaltigkeitsuntersuchungen geprüft. An ausgewählten Proben wurden auch elektronenmikroskopische Untersuchungen in Durchstrahlung vorgenommen.

6.1 Röntgenographische Untersuchungen

Die Röntgenbeugungsdiagramme wurden mit Cu-Kα -Strahlung auf einem Zählrohrgoniometer aufgenommen. Hieraus wurden folgende Informationen erhalten:

(1) Die in der Schicht vorhandenen Phasen sowie eine ungefähre Abschätzung des Mengenanteiles dieser Phasen.

(2) Gitterkonstante der kfz-TiN_x -bzw. TiC_x-Phase, die bei kompakten Proben vom stöchiometrischen Verhältnis x abhängt /20/. Wegen der Eigenspannung der Schichten konnte jedoch der Parameter x nicht aus der gemessenen Gitterkonstante bestimmt werden.

(3) Aus den relativen Intensitätsverhältnissen der Linien einer Phase Aussagen darüber, ob eine Vorzugsorientierung (Textur) oder eine isotrope Kristallitausbildung vorliegt. Da die relativen Linienintensitäten nicht nur von einer evtl. vorhandenen Textur, sondern auch vom nicht genau bekannten Stöchiometrieparameter x abhängen, ist nur eine nicht sehr genaue Abschätzung der Vorzugsorientierung möglich.

(4) U.U. Aussagen über die Korngrößen anhand der Linienbreiten.

6.2. Metallographische Untersuchungen

Nach der Röntgenbeugungsanalyse wurden die Proben mit einem 5.7°-Schrägschliff versehen. Unter dem Mikroskop wurden dann die Schichtdicken bestimmt, Veränderungen in der Randzone (η-Phasenbildung, Poren) beobachtet, Gleichmäßigkeit und Porosität der Schicht beurteilt und schließlich noch die zuvor eingedrückten Vickers-Mikrohärteeindrücke (durchweg 0,5 N Last) ausgemessen. An Hand des Schliffbildes konnte auch gut die Haftung der Schicht am Substrat beurteilt werden.

6.3 Rasterelektronnenmikroskopische Untersuchungen

Die Zusammensetzung der Schicht wurde in einem Rasterelekronenmikroskop mit einem energiedispersiven Röntgenstrahl-Analysensystem (EDAX) und einem Kristallspektrometer bestimmt. Die Genauigkeit bei den leichten Elementen C,N und O ist nicht sehr hoch. Als Standard bei der TiN-Beschichtung wurde eine durch CVD hergestellte Probe gewählt. Einige Proben wurden nach dem Schneidhaltigkeitsversuch zerbrochen, um das Bruchgefüge mit dem Rasterelektronenmikroskop aufnehmen zu können.

6.4 Schneidhaltigkeitsuntersuchungen

Eine Anzahl der beschichteten Hartmetall-Wendeschneidplatten wurde auf einer Prüfdrehbank durch Drehen von Stahl C60 im glatten Schnitt geprüft. Bild 9 zeigt die Anordnung der Prüf-

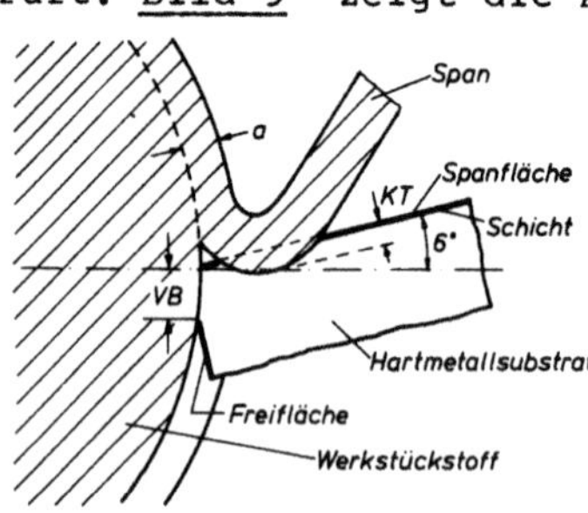

Bild 9: Schneidhaltigkeitsprüfung. KT Kolktiefe, VB Verschleißmarkenbreite, a Spantiefe.

platte beim Eingriff in das Werkstück. Die verschiedenen geometrischen Parameter der Schneidplatten (s. Bild 9) wurden für alle Versuche beibehalten. Da die Zerspanbarkeit auch derselben Stahlsorte von Charge zu Charge verschieden sein kann und sich die Tests über mehrere Jahre erstreckten, wurden zum Vergleich stets durch CVD mit TiC oder TiC/TiN beschichtete Wendeschneidplatten gleicher Qualität mitgeprüft. Als Kriterium für die Verschleißfestigkeit wurde der Kolkverschleiß, gekennzeichnet durch die Kolktiefe KT (s. Bild 9) gewählt, da bei den PVD-Beschichtungen nur die obere Spanfläche beschichtet wurde. Durch Versuche mit allseitig durch CVD beschichteten Wendeschneidplatten wurde nachgewiesen, daß durch Abschleifen der Schicht auf der Freifläche sich der Kolkverschleiß nicht erhöht (etwa durch Erhöhung der Schneidentemperatur durch vermehrte Reibung der unbeschichteten Freifläche). Üblicherweise wurde 5 Minuten lang mit einer Schnittgeschwindigkeit von 220 m/min, einer Schnittiefe von 1,5 mm und einem Vorschub von 0,28 mm/U gedreht. Der Kolkverschleiß hängt von der Schichtdicke d ab. Durch empirische Untersuchungen von CVD-beschichteten Proben mit verschiedenen Schichtdicken wurde die Abhängigkeit des Kolkverschleißes von der Schichtdicke bestimmt und bei der Auswertung der Schneidhaltigkeitsversuche entsprechend berücksichtigt.

Am Zustand der Probe nach dem Schneidhaltigkeitsversuch konnte auch die Haftfestigkeit der Schicht beurteilt werden (Abplatzungen).

7. Aktiviertes reaktives Aufdampfen

Wie bereits in der Übersicht erwähnt, gelang es Anfang der 70iger Jahre Bunshah und Mitarbeitern /2/, nach dem Verfahren des aktivierten reaktiven Aufdampfens "dicke" Hartstoffschichten (d.h. im Bereich von 10 µm und mehr) abzuscheiden. Die Erzeugung von TiC-Schichten durch Verdampfung von Ti in Gegenwart von C_2H_2 ist nur durch "Aktivierung" möglich, d.h. beim Anliegen eines Potentials an der Hilfselektrode. Bei Drücken von

$p_{C_2H_2} = 5 \cdot 10^{-4}$ Torr und bei Substrattemperaturen von 450°C wurden TiC -Schichten auf Tantal- und V2A-Substraten mit x = 0,8 bis 0,95 und Mikrohärten von HV 0,05 = 2775 erhalten. Beachtlich sind die angegebenen Verdampfungs- und Beschichtungsraten. Mit einer 10 kV/3 kW-Elektronenkanone verdampft man 0,67 g/min und erreicht ein Schichtwachstum von 4 µm/min. In der Arbeit von Raghuram und Bunshah /21/ wurde ein lineare Abnahme der Beschichtungsrate mit zunehmender Beschichtungstemperatur beobachtet. Dieses wird dadurch erklärt, daß bei höherer Temperatur die Wiederverdampfung zunimmt, weil die Lebenszeit eines auf der Substratoberfläche vagabundierenden Atomes, bevor es in einer Potentialmulde haften bleibt, ebenfalls zunimmt. Bis zu Substrattemperaturen von 830°C zeigen die Schichten eine zunehmende (220)-Textur, bei 1120°C eine Textur senkrecht zu den (200)-Flächen und bei 1450°C eine kristallographische Gleichverteilung. Die Mikrohärte (Vickershärte HV 0,05) nimmt mit der Substrattemperatur zu. Bei 1080°C und 1120°C wurden HV 0,05-Werte von 4110 bzw. 4160 gefunden, also erheblich höhere als aus der Literatur für TiC bekannte HV-Werte (∿ 3200). Bei Substrattemperaturen von 520°C bis 830°C liegen die HV 0,05-Mikrohärtewerte relativ hoch zwischen 2330 und 2955.

Über mit dem Verfahren des aktivierten reaktiven Verdampfens hergestellte Hartstoffschichten wurde in der Folgezeit noch verschiedentlich berichtet. Von Raghuram u.a. /10/ wurden nach dem ARE-Verfahren hergestellte TiC-, ZrC- und HfC-Schichten durch Linienprofilanalysen der Röntgenbeugungslinien auf Korngröße und Eigenspannungen hin untersucht. Es wurde keine Korrelation zwischen der Mikrohärte und der Korngröße und auch den Eigenspannungen festgestellt. Die bei 500°C abgeschiedenen TiC-Schichten hatten Korngrößen von nur etwa 20 nm. Über Schneidhaltigkeitsversuche an mit ca. 5 bis 8 µm dicken, durch den ARE-Prozeß hergestellten TiC-Schichten auf Schnellarbeitsstahl (M-43 nach AISI-Standard) berichten Bunshah u.a. /22/ . Bei der Bearbeitung von Baustahl (4340 nach AISI-Standard, etwa 40 NiCoMo6) wurde eine Standzeitverlängerung von 300-800% und um 50% niedrigere Schnittkräfte gegenüber unbeschichteten Werkzeugen festgestellt. Dabei dominierte der Freiflächenverschleiß. Eine Variante des ARE-Verfahrens wurde von K. Nakamura u.a. /23/ vorgeschlagen, die als "Low pressure plasma depostition"-Verfahren bezeichnet wurde. Bei diesem Verfahren wird eine positive Gleichspannung oder eine HF-Spannung an das Substrat angelegt. Die Höhe der Spannung wurde nicht angegeben. Beim TiC-Beschichten von Hartmetallproben erwies sich ein C_2H_2-Druck von $4 \cdot 10^{-4}$ Torr als günstig.

Die Herstellung von Schichten aus Titannitrid durch aktiviertes reaktives Aufdampfen wird erstmalig in der Arbeit von Bunshah und Raghuram /2/ erwähnt. Danach sollen bereits bei Raumtemperatur und Drücken von $5 \cdot 10^{-4}$mbar N_2 bzw. NH_3 TiN-Schichten entstehen, ohne daß hierzu nähere Angaben gemachten wurden. Hill u.a. /24/ und Zega u.a. /25/ berichteten 1977 über die Abscheidung von TiN-Schichten nach dem Verfahren des Ionenplattierens. Schulze /26/ stellte 1978 TiN-Schichten nach den Verfahren des aktivierten reaktiven Aufdampfens und erstmalig auch bei gleichzeitiger Inertgasaktivierung her. In der Folgezeit wurde noch verschiedentlich über die Herstellung von TiN-Schichten auf metallischen Substraten durch Aufdampfverfahren berichtet /26,27,28,29/.

7.1 Abscheidung von TiN-Schichten

Die Herstellung von Titannitridschichten wird durch die Existenz zweier Phasen im Bereich zwischen 25 und 38 Atom-% Stickstoff erschwert. Für das System Ti-N wurde von Toth /30/ das in Bild 10 dargestellte Zustandsdiagramm vorgeschlagen. Neben der TiN-Phase gibt es noch die ε-Ti_2N-Phase. Daneben besitzt der Stickstoff noch eine größere Löslichkeit in α- und β-Titan. Die Phase TiN_x mit B1-Struktur existiert in einem weiten Bereich von $x \sim 0{,}25$ bis $x \sim 1{,}16$. Die Gitterkonstante dieser Phase nimmt von a = 0,4220 nm bei x = 0,45 auf a = 0,4238 bei x = 1 mit anwachsenden Stickstoffgehalt zu /20/. Die Messung der Gitterkonstanten eignet sich aber nicht zur Bestimmung der stöchiometrischen Verhältnisse x, da in dünnen Schichten die Gitterkonstante durch vom Substrat verursachte Spannungen beeinflußt ist. Die ε-Ti_2N-Phase ist nach Einkristallmessungen von Holmberg /31/ tetragonal mit der Raumgruppe $P4_2/mm$ und den Gitterkonstanten a = 0,4945 nm und c = 0,3034 nm. Sie ist röntgenographisch gut von der Phase mit B1-Struktur zu unterscheiden.

Erste eigene Versuche, mit aus der Literatur bekannten experimentellen Parametern harte TiN-Schichten auf Hartmetallsubstraten abzuscheiden, führten nicht zum Erfolg. In einer größeren Versuchsreihe wurden daher die Verfahrensparameter Druck des reaktiven Stickstoff-Gases p_{N_2}, Substrattemperatur θ_s und Verdampfungsrate K variiert. Als Substrate wurden durchweg Wendeschneidplatten der Größe SNUN 120408 (Abmessung 12,7 x 12,7 x 4,8 mm) aus einem Hartmetall der ISO-Sorte M15 (Zusammensetzung 80,5% WC, 5% TiC, 8% TaC, 6,5% Co) eingesetzt. Diese Sorte hat sich als Grundwerkstoff für durch CVD beschichtete Schneidplatten besonders bewährt. In einigen Versuchen wurde festgestellt, daß mit Beschichtungsraten, die über 60 µm/h hinausgingen, keine harten und verschleißfesten Schichten erhalten werden konnten, offenbar, weil das Stoßzahlverhältnis zu ungünstig wurde. Daher wurden die Beschichtungen vorwiegend

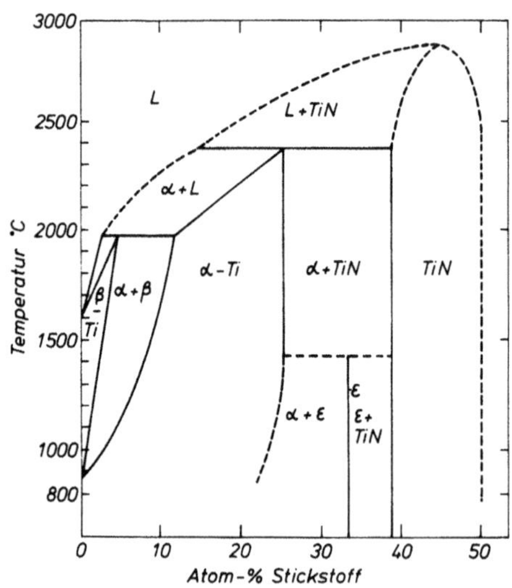

Bild 10: Zustandsdiagramm des Systems Ti-N nach Toth /30/.

mit Raten von etwa 20 µm/h durchgeführt. Nach dem Beschichten (Versuchsablauf s. Abschnitt 5) wurden die Platten in der in Abschnitt 6 beschriebenen Art und Weise untersucht. Aus zahlreichen Einzelergebnissen wurde folgendes Bild über die Randbedingungen der Entstehung von Titannitridschichten gewonnen. Bei Substrattemperaturen von θ_s = 600°C und Reaktivgasdrucken unterhalb von etwa $6 \cdot 10^{-4}$ mbar wurden Schichten mit Härten von 600 bis 1000 HV 0,05 erhalten. Röntgenographisch bestanden die Schichten aus TiN und etwas α-Ti. Bei höheren Temperaturen (ab etwa 650°C) zeigten sich in den Beugungsbildern neben den Linien der TiN_x -Phase die der ε-Ti_2N-Phase. Die Härte nahm im allgemeinen zu, obwohl der analysierte Stickstoffgehalt zu höheren Temperaturen hin geringer wurde. Bei Drücken der Reaktivgase von 10^{-3} mbar bestanden die Schichten etwa bis zu Substrattemperaturen von 700°C aus TiN , darüber teilweise oder zum überwiegenden Teil aus ε-Ti_2N.

<u>Bild 11</u> zeigt als Beispiel die Beugungsdiagramme von Schichten, die bei Drücken p = $2 \cdot 10^{-3}$ mbar und Substrattemperaturen von θ_s = 600°C bzw. 870°C hergestellt wurden. Bei θ_s = 870°C überwiegt die ε-Ti_2N-Phase. Die Härten der bei θ_s = 600°C erzeugten Schichten erreichten jedoch, wie die Schliffbilder in <u>Bild 12</u> zeigen,

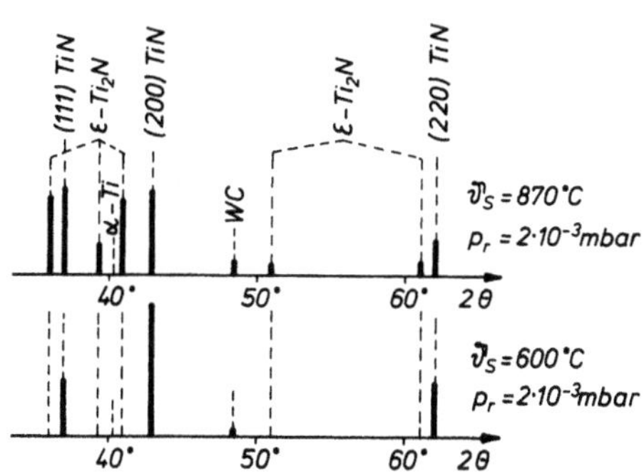

Bild 11: Röntgendiagramme von Titannitrid-Schichten bei verschiedenen Substrattemperaturen θ_S.

nicht das von kompaktem Titannitrid her bekannte Niveau von ca. 1900 HV 0,05. Die ε-Ti_2N-haltige Schicht ist etwas härter als die phasenreine TiN_x-Schicht. Ingesamt ergab eine Sichtung von mehr als 50 Einzelversuchen, die mit Beschichtungsraten von 10 bis 20 µm/h durchgeführt wurden, die in Bild 13 dargestellten Grenzen in der θ_s-p_r-Ebene für das Auftreten von α-Ti, TiN_x und ε-Ti_2N. Bei N_2-Drücken unterhalb von $4 \cdot 10^{-4}$ mbar wurden immer größere Anteile von α-Ti in der Schicht gefunden. Mit anwachsendem N_2-Druck findet man, abhängig von der Substrattemperatur, entweder TiN_x oder TiN_x und kleinere Anteile von ε-Ti_2N oder fast ausschließlich ε-Ti_2N. Nach diesen Beobachtungen begünstigen hohe Substrattemperaturen und niedrige Stickstoffdrücke die Entstehung von ε-Ti_2N, während umgekehrt bei hohen Stickstoffdrücken ($>10^{-3}$ mbar) und Substrattemperaturen von etwa 600°C das Auftreten der ε-Phase nicht festgestellt wurde. Ebenfalls wurde ein Anwachsen des Stöchiometrieparameters x in der angezeigten Richtung festgestellt. Bei höheren Beschichtungsraten (>50 µm/h) verschieben sich die Grenzen für das Auftreten der Phasen in Richtung höherer Drücke und kleinerer Substrattemperaturen.

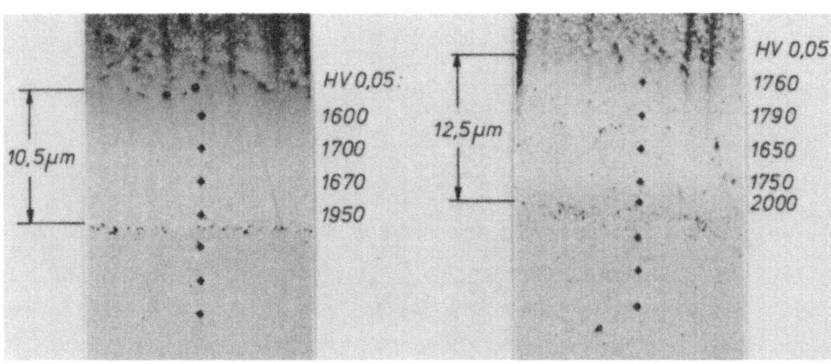

Bild 12: Schliffbilder von Titannitridschichten. $p_r = 2 \cdot 10^{-2}$mbar, links θ_s = 600 °C, rechts θ_s = 870 °C.

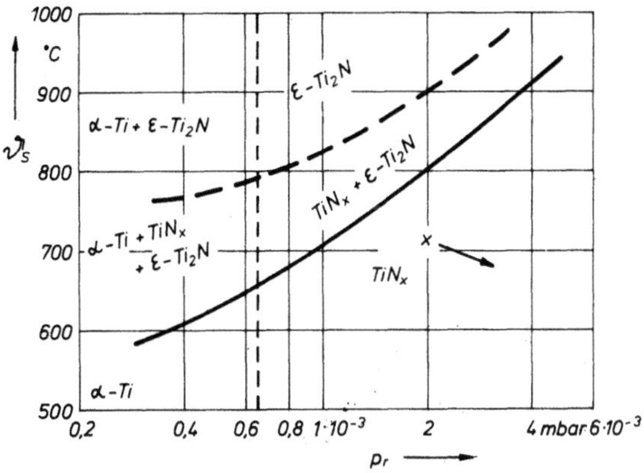

Bild 13: Phasen Ti-N der Schichten in Abhängigkeit von Stickstoffdruck p_r und Substrattemperatur θ_s bei Aufdampfraten von etwa 20 µm/h.

Diese Beobachtungen sind in Übereinstimmung mit Ergebnissen, die von Suri u.a. /28/ sowie von Jacobson u.a. /29/ mitgeteilt wurden. Diese Autoren finden bei der Abscheidung von Titannitrid auf rostfreiem Stahl oder Tantal bei θ_s = 550°C TiN, bei 735°C und 1000°C jedoch TiN_x und ε-Ti_2N. Bei θ_s=550°C und gleichbleibendem Druck von $2 \cdot 10^{-4}$ mbar, jedoch zunehmender Beschichtungsrate, erscheinen nacheinander die Phasen TiN_x, TiN_x + ε-Ti_2N und ε-Ti_2N+α-Ti. Ähnliche Ergebnisse erhielten - bei Berücksichtigung der unterschiedlichen Beschichtungsraten - auch Nakamura u.a. /23/, die bei θ_s=250°C, p_{N_2} = $2 \cdot 10^{-3}$ mbar TiN_x, bei θ_s= 500°C jedoch TiN_x+ε-Ti_2N erhielten. Bei sehr hohen Beschichtungsraten wird ε-Ti_2N und ggf. α-Ti, jedoch kein TiN_x erhalten /15/.

Die gefundenen Zusammenhänge schränken die Wahl der Verfahrensparameter ein, wenn man möglichst stöchiometrisches TiN_x erhalten will. Der Druck des reaktiven Gases sollte möglichst hoch gewählt werden. Weiterhin gilt: Je höher die Beschichtungsrate, um so niedriger sollte die Substrattemperatur sein. Auf der anderen Seite wirkt sich eine niedrigere Substrattemperatur nachteilig auf eine gute, durch Diffusion zwischen Substrat und Schicht vermittelte Haftung aus. Die Gefüge aller durch aktiviertes reaktives Aufdampfen erhaltenen Schichten zeigten ausgeprägte Stengelstrukturen (vgl. Abschn. 7.3).

7.2 Abscheidung von TiN-Schichten mit NH_3 als Stickstoffträger

Bereits Bunshah /2/ erwähnt, daß es möglich ist, TiN-Schichten nach dem Verfahren des aktivierten reaktiven Aufdampfens mit NH_3 als Stickstoffträgergas herzustellen. Beim CVD-Verfahren werden bei Verwendung von NH_3 sehr feinkörnige Schichten erhalten /32/.

Beim Auftreffen eines NH_3-Moleküls auf die erwärmte Substratfläche werden, sofern es nicht bereits in der Gasphase dissoziiert, ein freies Stickstoffatom und Wasserstoff entstehen. Man sollte annehmen, daß der Stickstoff in statu nascendi die Bildung von TiN erleichtert. Der Wasserstoff könnte diffusionserleichternd wirken. Nachteilig ist jedoch, daß pro auftreffendes Gasmolekül nur ein Stickstoffatom angeboten wird. Es wurden einige Versuche mit NH_3 als Stickstoffträger durchgeführt. Die Ergebnisse können wie folgt zusammengefaßt werden:
Bei $p_r = 1 \cdot 10^{-3}$ mbar und $\theta_s = 800°C$ wurde nur wenig Stickstoff (5,5 Gew.-%) in die Schicht eingebaut. Im Röntgendiagramm sind schwache TiN- und ε-Ti_2N-Linien vorhanden. Die Vickers-Mikrohärten betragen etwa 1000. Es entstand eine breite Diffusionszone an der Grenzfläche zwischen Schicht und Hartmetallsubstrat. Im Röntgenbeugungsdiagramm ist die intermetallische Phase Co_7W_6 besonders intensiv. Bei höherem NH_3-Druck ($p_r = 3 \cdot 10^{-3}$ mbar) erhöhte sich der Stickstoffgehalt der Schicht auf 8,1%. Es bildete sich bevorzugt ε-Ti_2N mit Härtezahlen von 1500 bis 1700 HV 0,05. Insgesamt konnte somit kein Vorteil durch die Verwendung von NH_3 als Stickstoffträger festgestellt werden.

7.3 Abscheidung von TiN-Schichten mit Saugpotential

Um das Angebot der reaktiven Gasmoleküle auf der zu bedampfenden Substratoberfläche zu erhöhen, wurde in Abänderung des aktivierten reaktiven Aufdampfverfahrens an das Substrat ein negatives Potential von -1 bis -2 kV gelegt (s. Bild 3, Schalter S_1 und S_2 geschlossen). In dem entstehenden, negativen Potentialfeld werden die positiven Stickstoffionen gegen das Substrat beschleunigt. Dabei ionisieren sie noch zusätzlich neutrale Moleküle durch Stoßionisation. Für die Entstehung einer selbständigen Gasentladung ist der gewählte Gasdruck von $2 \cdot 10^{-3}$ mbar noch zu klein. Die Ionisierung durch die Beschleunigung der Glühemissionselektronen zur positiven Hilfselektrode hin wird also nach wie vor benötigt.

Durch Anwendung dieses Verfahrens konnte der Stickstoffgehalt der Schichten deutlich erhöht werden. Bild 14 zeigt als Beispiel eine bei 600°C aufgedampfte Schicht. Die Schicht besteht nur aus der TiN_x-Phase. Auch bei der höheren Temperatur von 850°C entstand, durch das erhöhte Stickstoffangebot kein ε-Ti_2N. Die Härtewerte der Schichten lagen deutlich höher als beim einfachen reaktiven Aufdampfen und erreichten z.T. das Niveau von kompaktem TiN. Die in Bild 13 angegebenen Grenzen für das Auf-

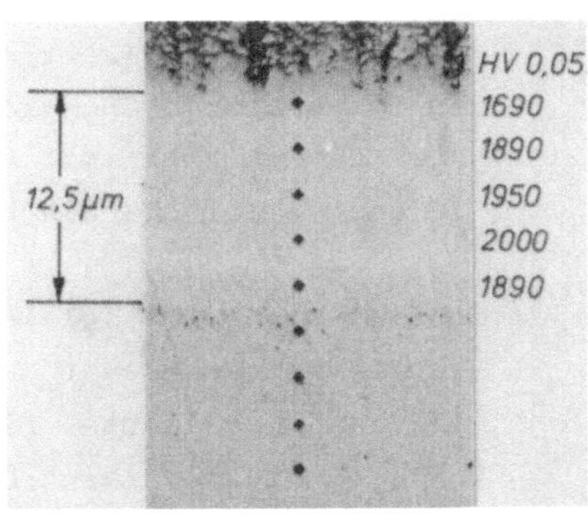

Bild 14: Schliffbild einer Titannitridschicht, die bei 600 °C mit angelegtem Saugpotential erhalten wurden.

treten von ε-Ti_2N dürften sich unter diesen experimentellen Bedingungen zu höheren Temperaturen verschieben. Das Schichtgefüge ist auch bei diesem Verfahren durch stengelförmiges Wachstum geprägt. Bei höheren Substrattemperaturen neigten die Schichten zur Rißbildung.

7.4 Gefüge der TiN-Schichten

Nach dem Zonenmodell von Movchan und Demchishin /33/ kann die Kornstruktur von mit hoher Rate aufgedampften Schichten der unterschiedlichsten Materialien in Abhängigkeit von T/T_m, wobei T die Kelvin-Temperatur des Substrates und T_m die des Schmelzpunktes des aufgedampften Materials ist, wie folgt klassifiziert werden:

1. Zone $0<T/T_m<0{,}3$: Zapfenförmige Kristallite und kuppelförmige Spitzen, getrennt, durch z.T. leere Grenzflächen. Hohe Fehlstellendichte. Geringe laterale Festigkeit.

2. Zone $0{,}3<T/T_m<0{,}5$: Stengelförmige Körner mit aneinandergefügten interkristallinen Grenzflächen. Mattes Aussehen oder Oberfläche. Korngröße nimmt mit T/T_m zu. Bei metallischen Schichten werden die Eigenschaften des kompakten Materials erreicht. Die aus der Temperaturabhängigkeit des Korngrössenwachstums bestimmte Aktivierungsenergie deutet auf Oberflächendiffusion hin.

3. Zone $0{,}5<T/T_m<1{,}0$: Gröbere Körner mit gleichgerichteten Hauptachsen senkrecht zur Oberfläche, glänzende Oberfläche. Die Temperaturabhängigkeit des Korngrößenwachstums deutet auf eine Aktivierungsenergie, die konsistent mit einer Volumen-Selbstdiffusion ist, hin. Die Eigenschaften der Schicht entsprechen voll dem Grundmaterial.

Will man die von uns hergestellten Titan-Hartstoffschichten nach diesem Modell klassifizieren, hat man zunächst die Größe T/T_m festzulegen. Dafür sind zwei Annahmen möglich. Zum ersten kann man annehmen, daß sich bei der Kondensation des Titans an der Substratoberfläche in Mitwirkung der Gasphase direkt die Phase TiN bildet. Dann ist mit T_m = 3223 K bei den von uns gewählten Substrattemperaturen zwischen 600°C und 1000°C

$$0,27<T/T_m<0,39$$

Es ist aber auch vorstellbar, daß zunächst das verdampfte Titan als α-Ti kondensiert und erst durch einen zeitlich nachfolgenden Diffusionsprozeß die Phase TiN gebildet wird. In diesem Falle würde die Kornstruktur im wesentlichen die Morphologie von aufgedampftem Titan mit T_m = 1948 K haben. Dann ist für Substrattemperaturen von 600°C bis 1000°C

$$0,45<T/T_m<0,65$$

Während man im ersten Fall einen Schichtaufbau mit den Eigenschaften der 2. Zone bekommen sollte, hat man im zweiten Fall mehr die Charakteristika der 3. Zone zu erwarten.

Die Härte nimmt bei keramischen Stoffen mit T/T_m zu, bei Metallen dagegen ab /3/. In der Zone 2 erfolgt das Wachstum noch vorwiegend lokal-epitaktisch an den am weitesten dem Dampfstrom entgegenstehenden Flächen. Als Folge davon ist die Bindung zwischen den Körnern noch nicht besonders fest. Die 3.Zone mit T/T_m >0,5 ist gekennzeichnet durch Volumen-Diffusion und beginnende Rekristallisation, durch die die Schicht auch in lateraler Richtung hin eine bessere Festigkeit bekommt.

Zur Charakterisierung der hergestellten TiN-Schichten wurden REM-Aufnahmen von Bruchgefügen und von Draufsichten auf die Schichten aufgenommen (Bild 15 und 16). Die Schichttextur wurde an Hand des Röntgenbeugungsbildes abgeschätzt.

Im folgenden werden Bruchgefüge besprochen, die nach den Verfahren des aktivierten Aufdampfens ohne und mit Saugspannung

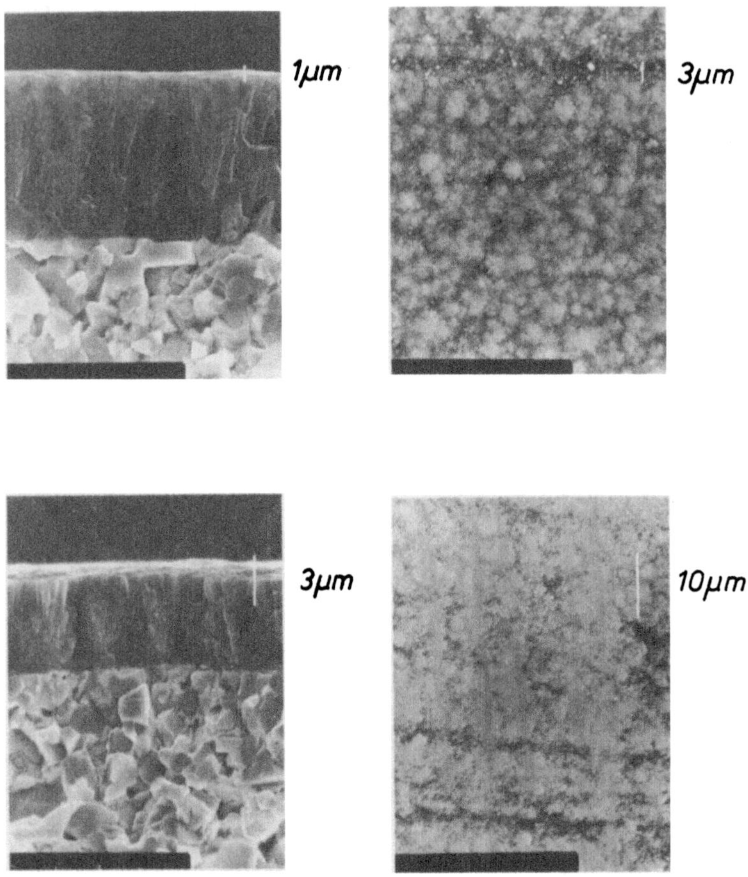

Bild 15: Bruchgefüge und Schicht-Draufsichten von Titannitridschichten. Substrattemperatur 600 °C (oben) und 800 °C (unten).

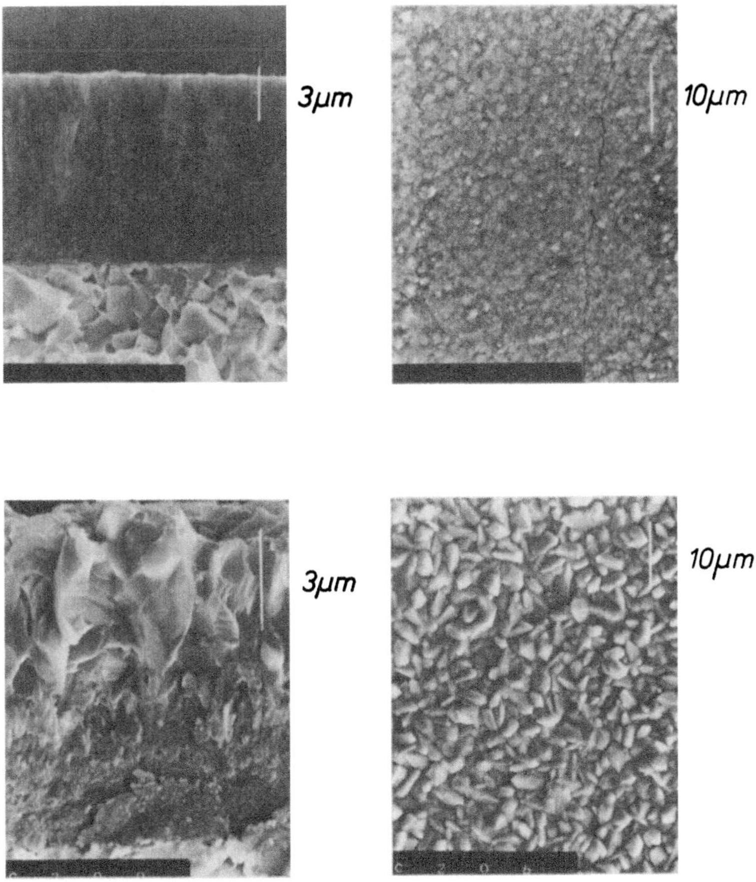

Bild 16: Bruchgefüge und Schicht-Draufsichten von Titannitrid-schichten. Oben: PVD-Schicht bei θ_S = 900 ^{o}C. Unten: CVD-Schicht (θ_S ~ 1000 ^{o}C).

hergestellt wurden. Zwischen diesen beiden Methoden konnten im Bruchgefüge keine Unterschiede erkannt werden.

In den Bildern 15 und 16 sind die Bruchgefüge von drei TiN-Schichten, die bei den Temperaturen 600°C, 800°C und 870°C (T/T_m = 0,27; 0,33 und 0,39) unter sonst annähernd gleichen Bedingungen beschichtet wurden, gezeigt. Alle drei Schichten haben Stengelstruktur. Die Probe bei θ_s = 600°C hat besonders scharfkantig ausgebildete Stengel. Das ist ein Hinweis auf die noch schlechte Verbindung der Stengel untereinander. Das Gefüge ist auf jeden Fall viel feiner als bei einer CVD-Schicht, die in Bild 16 gezeigt ist. Die Draufsichten aller drei Proben zeigen noch keine einzelnen Körner.

Die Stengelstruktur der Bruchgefüge läßt erkennen, daß es sich bei den TiN-Schichten wahrscheinlich um Gefüge der 2. Zone handelt, deren Wachstum noch durch Oberflächendiffusion gekennzeichnet ist. Im Unterschied zu CVD-Schichten haben die PVD-Schichten eine kristallographische Vorzugsorientierung längs (100). Lediglich bei der Substrattemperatur von 870°C war auch der (220)-Reflex stärker ausgeprägt.

Bei der abgebildeten CVD-Schicht (Bild 16) wurde zwischen dem Hartmetall und der TiN-Schicht noch eine Zwischenschicht aus TiC aufgebracht. Beim CVD-Verfahren entsteht die TiN-Schicht nach der Reaktion

$$2TiCl_4+N_2+4H_2 \rightarrow 2TiN+8HCl$$

bei Gasdrücken um 700 mbar und Temperaturen von 1000 bis 1100°C. Das Bruchgefüge zeigt in der TiN-Schicht ein grobkörniges Gefüge mit Korngrößen bis zu 3 µm. Die Draufsicht zeigt meist längliche Körner. Kristallographisch haben die Körner eine sehr ausgeprägte Vorzugsorientierung mit der Richtung (110) parallel zur Normalen, wie das Röntgendiagramm mit seinem starken (220)-Reflex zeigt.

Zur weiteren Aufklärung der Gefügestrukturen wurden noch Durchstrahlungsaufnahmen mit einem Elektronenmikroskop gemacht. Im Falle der TiN-Schicht (θ_s = 650°C, p_r = $6 \cdot 10^{-4}$mbar) werden Korngrößen von etwa 50 bis 100 nm gefunden. Diese Schicht ist als Folge des niedrigen N_2 -Druckes bei der Herstellung nicht einphasig. Neben den TiN-Körnern liegen sehr feine, helle Teilchen, bei denen es sich auf Grund der Elektronen-Feinbereichsbeugung um ε-Ti_2N oder um Ti_2O_3 handeln könnte. Vergleicht man die in den TEM-Aufnahmen gefundenen Korngrößen mit dem aus den Bruchgefügen abschätzbaren Durchmesser der gewachsenen Stengel (ca. 400 bis 600 nm), so folgt, daß letztere noch kein einkristallines Gefüge haben

7.5 Abscheidung von TiN-Schichten mit Saugpotential und Inertgasaktivierung

Die in Abschn. 7.4 besprochene Stengelstruktur der Hartstoffschichten ist typisch für die bisher besprochenen Aufdampfverfahren. Beim Zerspanen eines Werkstückes mit einer Wendeschneidplatte wird die Spanfläche durch den ablaufenden Span in erheblichem Maße tangential beansprucht. Man kann vermuten, daß die normal zur Oberfläche ausgebildete Stengelstruktur der Schicht den tangential angreifenden Reibungskräften weniger Widerstand als ein isotropes Schichtgefüge entgegenzusetzen vermag. Daher tragen sicherlich Maßnahmen, die die Ausbildung der unidirektionalen Stengelstruktur unterdrücken und die laterale Bindung verbessern, zur weiteren Verbesserung des Schneidverhaltens bei.

Aus der Technik des Sputterns ist bekannt, daß man die bei niedrigen Substrattemperaturen (niedrig bezüglich des Schmelzpunktes des Schichtwerkstoffes) entstehende Stengelstruktur vermeiden kann, wenn während des Beschichtens ein Ionenbombardement der sich bildenden Schicht durch Inertgasatome erfolgt (sog. Bias-Sputtern) /34/. Die gegen das negativ vorgespannte Substrat beschleunigten Edelgasatome bewirken durch Abgabe ihrer kinetischen Energie eine zusätzliche Aktivierung der aufwachsenden Schichtatome (Bildung von Nukleationszentren) und behindern das für die Entstehung der Stengelstruktur erforderliche epitaktische Wachstum. Proben, die nach diesem Verfahren bei θ_s = 800°C mit Gasgemischen von $2 \cdot 10^{-3}$mbar N_2 und $0,6 \cdot 10^{-2}$mbar Ar mit einem Potential von -1,5 kV am Substrat hergestellt wurden, wiesen in der Schicht wiederum hohe Anteile von ε-Ti_2N auf. Die Aktivierung durch das Argonbombardement wirkt sich offenbar in Richtung einer Erhöhung der Substrattemperatur (Begünstigung der Entstehung von ε-Ti_2N) aus, wie ein Vergleich mit dem bis auf den Argonzusatz unter gleichen Bedingungen durchgeführten Versuchen nach Abschn. 7.3 zeigt. Bei einer Substrattemperatur von θ_s = 600°C bestanden die Schichten aus der TiN_x-Phase und hatten hohe Härtewerte bis 2200 HV 0,05 (Bild 17). Die analysierten Stickstoffgehalte lagen bei x = 0,80. Eine wesentliche Veränderung ist zu erkennen, wenn man die Bruchgefügebilder der Proben vergleicht, die ohne und mit Inertgasaktivierung hergestellt wurden (Bild 18). In dem letzteren Falle wurde die Ausbildung der Stengelstruktur weitgehend unterdrückt.

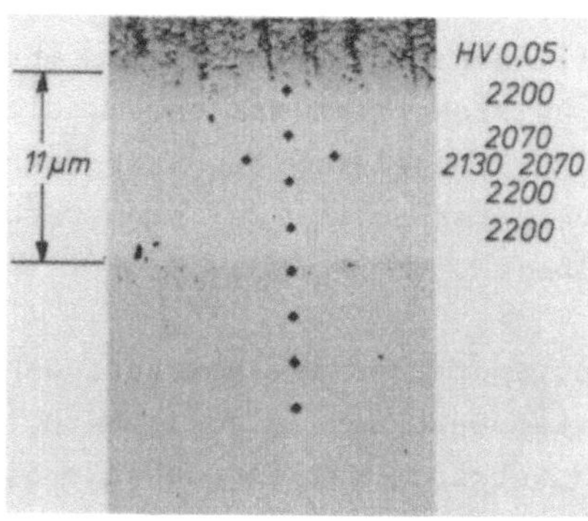

Bild 17: Schliffbild einer bei θ_s = 600 °C mit $p_{N_2} = 2 \cdot 10^{-3}$mbar und $p_{Ar} = 0,6 \cdot 10^{-3}$mbar hergestellten TiN_x-Schicht.

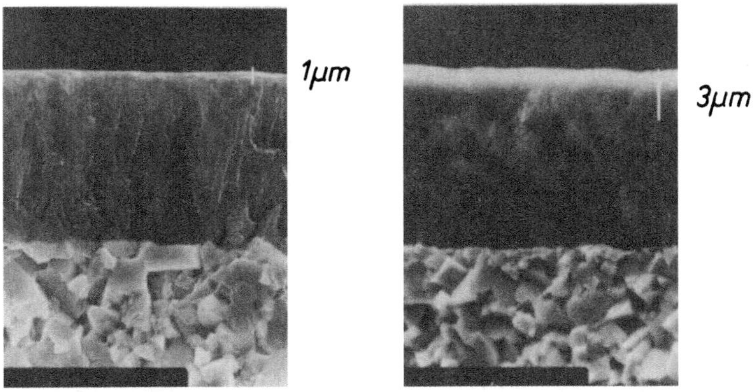

Bild 18: Bruchgefüge von bei θ_S = 600 ^{o}C hergestellten TiN_x-Schichten. Links: p_{N_2} = $2\cdot 10^{-3}$mbar. Rechts p_{N_2} = $2 \cdot 10^{-3}$mbar und p_{Ar} = $0,6\cdot 10^{-3}$mbar.

7.6 Schneidhaltigkeitsuntersuchungen

Als Kriterium für die Verschleißbeständigkeit wurde durchweg der durch Drehen im glatten Schnitt entstandene Kolkverschleiß gewählt. Die Versuchsdurchführung wurde bereits in Abschn. 6.4 angegeben. In Bild 19 ist der Kolkverschleiß von einigen Proben dargesteltt, die durch aktiviertes reaktives Aufdampfen (Abschn. 7.1) bei Temperaturen um 800°C abgeschieden wurden. Es ist klar zu erkennen, daß sich die Anwesenheit der ε-Ti_2N-Phase nachteilig auf den Kolkverschleiß auswirkt. Bei manchen Proben waren auch Abplatzungen und Bröckelungen an der Schneidkante zu beobachten. Die harte ε-Ti_2N-Phase wirkt sich im Sinne einer Versprödung der Schicht aus.

Ergebnisse von Schneidhaltigkeitsuntersuchungen an Proben, die durch aktiviertes reaktives Aufdampfen mit Saugspannung hergestellt wurden, zeigten bessere Ergebnisse, insbesondere bei niedrigeren Substrattmperaturen um 600°C (Bild 20). Während der Kolkverschleiß der durch PVD beschichteten Proben unter optimalen Bedingungen das Niveau der durch CVD beschichteten Proben erreichte, war der Freiflächenverschleiß der PVD Proben stets

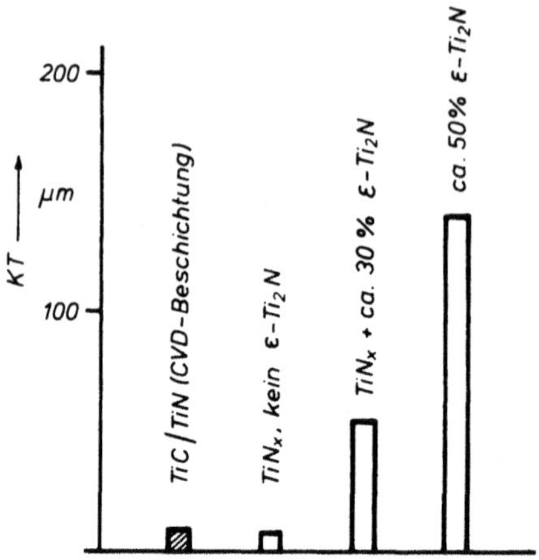

Bild 19: Einfluß von ε-Ti_2N auf den Kolkverschleiß von mit Titannitrid beschichteten Wendeschneidplatten SNUN 120408. Schnittgeschwindigkeit 220 m/min, Vorschub 0,28 mm/U, Spantiefe 1,5 mm, Drehzeit 5 min, Werkstückstoff Stahl C60.

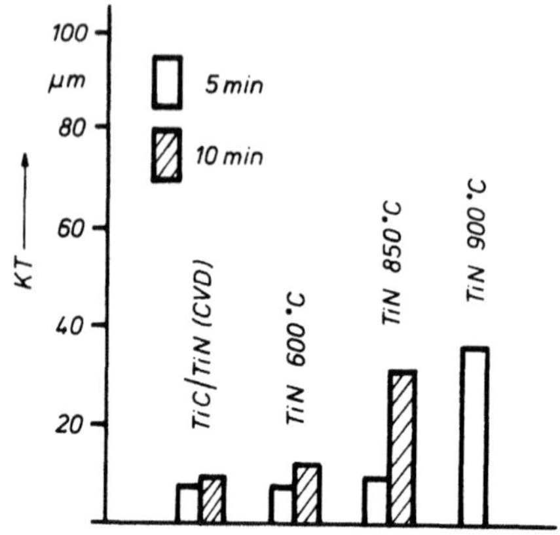

Bild 20: Kolkverschleiß von durch aktiviertes reaktives Aufdampfen mit Saugspannung hergestellte TiN_x-Schichten. Schnittdaten s. Bild 19.

deutlich größer. Bei den CVD beschichteten Vergleichsproben handelte es sich um Doppelschichten von einer ca. 2-4 µm dicken TiC-Schicht und einer ca. 8 µm dicken Deckschicht aus TiN. Um einen besseren Vergleich der nach den beiden Methoden hergestellten TiN-Schichten zu erhalten, wurden bei einigen TiN-Beschichtungen nach dem Verfahren des aktivierten reaktiven Aufdampfens Substrate eingesetzt, die durch CVD mit einer 2,5 µm dicken TiC-Schicht vorbeschichtet waren. Bild 21 zeigt das Ergebnis einer Schneidhaltigkeitsuntersuchung, die bis auf 25 Minuten Drehzeit ausgedehnt wurde. Sowohl die direkt auf das Hartmetall als auch die auf das mit TiC vorbeschichtete Substrat abgeschiedene TiN-Schicht haben bessere Kolkverschleißwerte als die CVD-Vergleichsprobe. Letztere zeigt sich jedoch beim Freiflächenverschleiß überlegen, dicht gefolgt von der vorbeschichteten Probe. Die nur auf der Spanfläche beschichtete Probe zeigt den größten Freiflächenverschleiß. Die bessere Kolkverschleißfestigkeit der PVD-Proben dürfte auf das wesentlich feinkörnigere Gefüge zurückzuführen sein. Der Stickstoffgehalt dieser TiN_x-Schichten lag bei nur

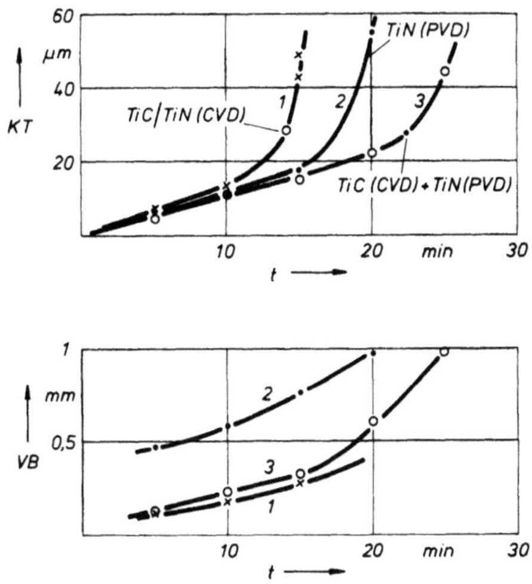

Bild 21: Kolkverschleiß KT und Verschleißmarkenbreite von verschiedenen TiN_x- bzw. TiC/TiN_x-Schichten (s. Text). Schnittdaten s. Bild 19.

x = 0,67. Dieses zeigt, daß für die Verschleißbeständigkeit der Stöchiometrie vielleicht nicht ganz die Bedeutung zukommt, die ihr oft beigemessen wird.

Bild 22 zeigt Ergebnisse, die an durch aktiviertes reaktives Aufdampfen mit Inertgasaktivierung hergestellten Proben erhalten wurden (Abschn. 7.5). Den geringsten Kolkverschleiß hat die bei 600°C hergstellte Probe, auch im Vergleich mit der CVD-Probe. Bei θ_s = 500°C ist der Verschleiß wegen geringerer Härte der Schicht (HV 0,05 ∿ 1500) und bei θ_s = 800°C wegen ε-Ti_2N-Anteilen in der Schicht größer.

Bild 22: Kolktiefe KT von durch aktiviertes reaktives Aufdampfen mit Inertgasaktivierung hergestellten TiN_x-Schichten. Schnittdaten s. Bild 19.

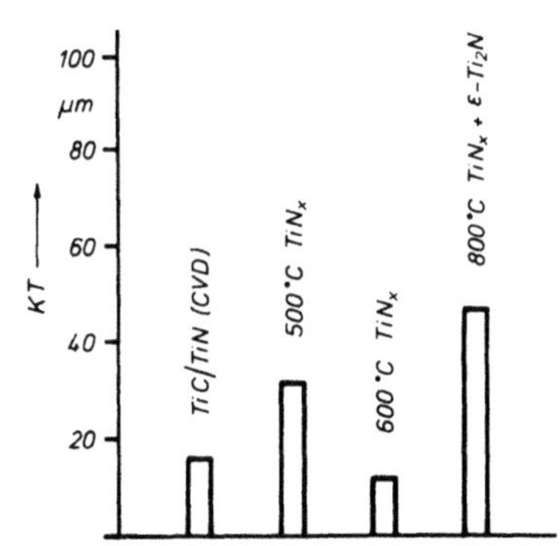

8. Abscheidung von TiC-Schichten

Das Phasendiagramm des Systems Ti-C /30/ zeigt zwischen 0 und 35 Atom-% C α-Ti (bzw. β-Ti bei T >920°C) und TiC. Von 35 bis 49 Atom-% C existiert ausschließlich TiC. Schwierigkeiten, wie sie beim Ti-N durch eine zweite Hartstoffphase (ε-Ti_2N) entstanden, können somit nicht auftreten. Bei mehr als 49 Atom-% C besteht jedoch die Gefahr der Ausscheidung von festem Kohlenstoff. Für die Abscheidung von TiC-Schichten wurde C_2H_2 als Reaktivgas, in einigen Versuchen auch mit Ar-Beimengung, benutzt. Es gelang, harte TiC-Schichten bei θ_s = 900°C und C_2H_2-Drucken von $2 \cdot 10^{-3}$ mbar herzustellen. Die Röntgendiagramme zeigten nur die TiC-Phase mit B1-Struktur. In Bild 23 werden das Bruchgefüge und die Draufsicht einer bei 900°C hergestellten TiC-Schicht gezeigt. Auch hier liegt eine Stengelstruktur mit vermutlich geringer Festigkeit quer zur Stengelrichtung. Die TiC-Schicht hat wie die TiN-Schicht die (100)-Richtung parallel zur Flächennormale als Vorzugsorientierung. Von einer TiC-Schicht wurden auch Durchstrahlungsaufnahmen mit dem Elektronenmikroskop hergestellt. Das Gefüge war mit Korngrößen von 2 bis 3 µm noch feinkörniger als bei

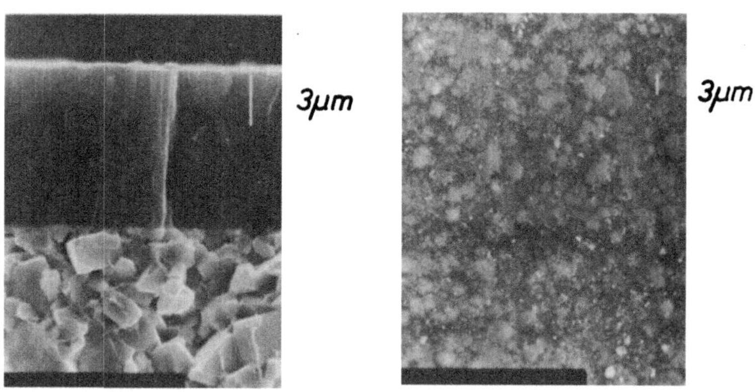

Bild 23: Bruchgefüge und Schicht-Draufsicht einer bei θ_s = 900 °C hergestellten TiC-Schicht.

den TiN-Schichten. Die Feinbereichsbeugung ergab nur die Reflexe des TiC. Schneidhaltigkeitsuntersuchungen ergaben, daß der Kolkverschleiß das Niveau der durch CVD mit TiC beschichteten Vergleichsproben erreichte. Der Freiflächenverschleiß war deutlich höher. Bei Doppelbeschichtungen mit TiC und TiN wurde ein ähnlich geringer Kolk- und Freiflächenverschleiß wie bei entsprechenden CVD-Vergleichsproben festgestellt.

9. Abscheidung von HfN- und HfC-Schichten

HfN-Schichten haben gegenüber TiN-Schichten einen dem Grundmetall ähnlicheren Ausdehnungskoeffizienten und eine höhere Warmhärte /35/. Sie haben von allen 4a-, 5a- und 6a-Metallnitriden die größte thermodynamische Stabilität, ausgedrückt durch die Bildungsenthalpie (ΔH = - 88,2 kcal/mol für HfN, ΔH = - 80,4 kcal/mol für TiN). HfN-Schichten können ebenfalls durch CVD-Verfahren hergestellt werden, jedoch bereitet das Verfahren infolge der höheren Verdampfungstemperatur des Hafniumchlorides größere Schwierigkeiten als bei der Abscheidung von TiN. In der Literatur wurde über die Abscheidung von HfC /10/ und von HfN /13/ auf Tantalsubstraten sowie auf HSS-Substraten durch das Verfahren des aktivierten reaktiven Aufdampfens berichtet. Infolge des höheren Schmelzpunktes von Hafnium (2222°C) ist mit geringeren Raten als bei der Titanverdampfung zu rechnen. Im System Hf-N existiert nur HfN

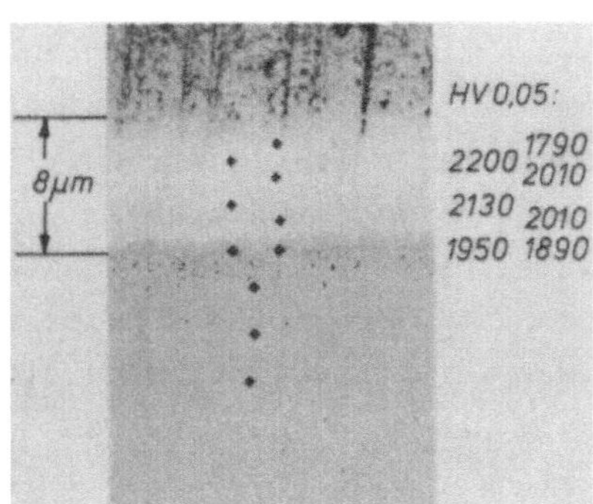

Bild 24: Schliffbild einer HfN-Schicht.

mit B1-Struktur als Hartstoffphase /30/. Ein Beispiel der von uns durch das Verfahren nach Abschn. 7.5 abgeschiedenen HfN-Schichten ist in Bild 24 dargestellt. Die Beschichtungsrate betrug etwa 5 µm/h. Es entstand eine gleichmäßig harte Schicht mit (100)-Textur. Im Schneidhaltigkeitsversuch wurde an den mit HfN beschichteten Proben nach 5 Minuten Drehzeit eine Kolktiefe von 9 µm und an der CVD-Vergleichsprobe eine Kolktiefe von 13 µm gemessen.

Eine Abscheidung von HfC mit B1-Struktur gelang ebenfalls, jedoch mit geringerer Härte als beim HfN. Diese Schichten zeigten keine besonderen verschleißmindernden Wirkungen.

10. Durch aktiviertes reaktives Aufdampfen von Titanlegierungen hergestellte Hartstoffschichten

Von der Hartstofflegierung VC - 25% TiC ist bekannt, daß sie bei hohen Temperaturen (1800°C) eine etwa 10-fach höhere Druckstreckgrenze als die "Eltern" TiC bzw. VC haben /36/. Es ist daher nicht ausgeschlossen, daß Hartstoffschichten solcher Legierungen möglicherweise bessere Verschleißeigenschaften haben. Bisher wurde ein Versuch in dieser Richtung bekannt. R. Nimmagadda und R.F. Bunshah /13/ dampften eine Legierung VTi 22 reaktiv in Gegenwart von C_2H_2 auf Tantalsubstrate auf. Sie erhielten kubisches (Ti,V) C mit Mikrohärten von 2400 KH 0,05.

Im Prinzip können Schichten aus Legierungen von beliebiger Zusammensetzung aufgedampft werden, wenn man die einzelnen Komponenten - die in der Regel sehr unterschiedliche Aktivitäten und Dampfdrücke haben - aus separaten Tiegeln mit einzeln verstellbarer Verdampfer-Leistung zur Verdampfung bringt. Die in Abschn. 4.3 erwähnten Schwierigkeiten bei der Konstanthaltung der Verdampfer-Leistung erhöhen sich hier noch beträchtlich. Um mit der im Rahmen dieser Untersuchung allein möglichen Eintiegelverdampfung einige Ergebnisse zu erhalten, wurden zwei Wege eingeschlagen. Zum einen wurden Titanlegierungen verdampft, bei denen aufgrund der Dampfdruckverhältnisse eine partielle Entmischung zu erwarten war. Die gewählten Legierungen TiMe10 (Anteil des Legierungselementes Me : 10 Gew.-%) sind in Tabelle 1 aufgeführt. Sie wurden in einem Lichtbogenofen erschmolzen. Mit diesen Einsatzlegierungen wurden nach den Verfahren nach Abschnitt 7.3 Hartstoffschichten auf Hartmetall abgeschieden. Tabelle 1 zeigt die Zusammensetzungen der erhaltenen Schichten. Es gelang, Al, Hf, V und Cr in die Schichten einzubauen. Zr und Nb konnten nicht analysiert werden, während Ta nur in geringen Mengen nachgewiesen wurde. Die Schichten mit Al und Cr zeigten von innen nach außen abnehmende Konzentrationen von Al bzw. Cr. Mit Ausnahme der Al-haltigen Schicht hatten alle Schichten zumindest die von den TiN-Schichten her gewohnten Härtewerte (Tabelle 1). Deutlich höhere Härtewerte wurden bei den V- und den Cr-haltigen Schichten festgestellt. Diese Schichten zeigten auch den niedrigsten Kolkverschleiß.

In einer zweiten Versuchsreihe wurden Titanlegierungen verwendet, bei denen eine entmischungsfreie Verdampfung möglich ist. Die Partial-Dampfdrücke der Komponenten einer binären Legierung A_xB_{1-x} sind in erster Näherung proportional zur molaren Konzentration (Molenbruch) der Komponenten (Raoultsches Gesetz):

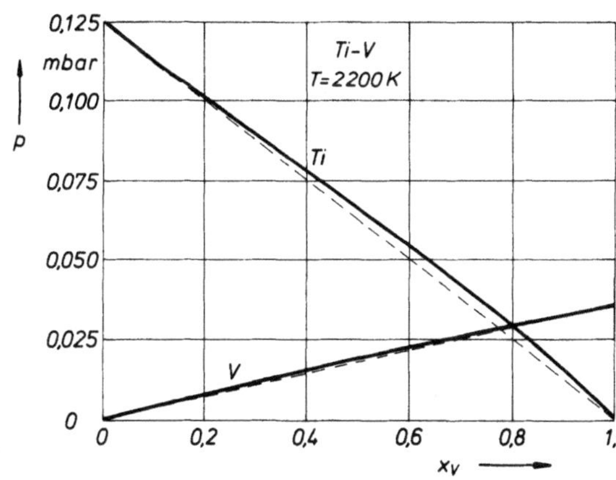

Bild 25: Partialdrücke von Ti und V über der Legierung V_xTi_{1-x} bei 2200 K.

Tabelle 1

Durch Verdampfen von Titanlegierungen hergestellte Hartstoffschichten

Legierung Ti Me	Schichtanalyse (%) Ti	O	N	Me	HV0,05	KT[2] µm
Ti Al 20	81,3	1,2	1,6	15,9	800	n.b.
Ti Zr 10	84,1	0,2	17,7	0	1900	14
Ti Hf 10	82,7	0,6	16,2	0,5	1800	20
Ti V 10	83,4	0,5	13,8	2,3	2300	8
Ti Nb 10	83,7	0,7	15,6	0	2000	11
Ti Ta 10	88,1	0,2	11,7	0,61	1900	n.b.
Ti Cr 10	75,0	0,3	16,8	7,9	2200	9
Ti Cr 2	79,8	0,4	17,1	2,7	1800	29
V Ti 25	30,3	0,9	20,0	48,8	2500	13
V Ti 50	64,5	0,5	19,9	15,0	2000	n.b.
TiC/TiN[1]					2000	16

1) Durch CVD beschichtete Vergleichsprobe.

2) KT Kolktiefe nach 5 min. Schnittdaten wie in Bild 19

$$p_A = x \; p_A^o$$

$$p_B = (1-x) \; p_B^o$$

p_A^o bzw. p_B^o sind die Dampfdrücke der reinen Komponenten. Daraus folgt, daß bei der Konzentration x_1

$$x_1 = \frac{p_B^o}{p_A^o + p_B^o}$$

die Partialdrücke p_A und p_B über der Legierung gleich sind. Für den Materialtransport aus der Schmelze ist außer den Partialdrücken noch die Geschwindigkeit der Dampfatome maßgebend, die bei gleicher thermischer Energie von der Masse M_A und M_B der Atome abhängt. Eine entmischungsfreie Verdampfung ist bei dem Molenbruch

$$x_2 = \frac{p_B^o}{\sqrt{M_B/M_A} \; p_A^o + p_B^o}$$

gegeben, wenn das Raoultsche Gesetz gilt. Bei Abweichungen davon kann die Konzentration x_2 bestimmt werden, wenn der Verlauf der Aktivitäten und somit der Dampfdrücke in Abhängigkeit des Molenbruchs der Legierungskomponenten bekannt ist. Da beträchtliche Abweichungen vom Raoultschen Gesetz vorkommen können /37/, wurden mit dem thermodynamischen Datenbanksystem THERDAS die Aktivitäten und die Dampfdrücke für eine Reihe von für Hartstoffsysteme interessanten Legierungen berechnet und daraus in der geschilderten Weise die Zusammensetzungen x_2 bestimmt, die sich entmischungsfrei verdampfen lassen.

Als Beispiel sind in Bild 25 die Partialdrücke über der Legierung V_xTi_{1-x} bei T = 2200 K dargestellt. Bei der Temperatur 2200 K hat die Legierung mit x_1 = 0,80 gleiche Dampfdrücke. Eine kongruent verdampfende Legierung liegt vor bei x_2 = 0,81, entsprechend einem Titangehalt von 18 Gew.-%. Die Berechnung der Aktivitäten bei T = 2300 K führt auf gleiche Drücke der Komponenten bei x_1 = o,76, also auf die Legierung VTi21. In gleicher Weise wurden die Systeme Ti-Cr, Al-Ti, Zr-Ti, Ti-Hf und Hf-Zr

Tabelle 2

Legierungen A_xB_{1-x} mit Dampfdruckgleichheit ($x = x_1$) der Komponenten und kongruenter Verdampfung ($x = x_2$) sowie deren Zusammensetzung

A - B	Temperatur K	x_1	x_2	Zusammensetzung
V - Ti	2200	0,80	0,81	VTi18
	2300	0,78	0,79	VTi20
Ti - Cr	2000	0,010	0,010	TiCr1,1
	2100	0,012	0,012	TiCr1,3
	2200	0,013	0,014	TiCr1,5
Al - Ti	2000	0,0017	0,0012	TiAl0,07
	2100	0,0024	0,018	TiAl0,10
Zr - Ti	2200	$1{,}8\ 10^{-4}$	$1{,}3\ 10^{-4}$	ZrTi0,006
Ti - Hf	2200	$6{,}5\ 10^{-4}$	$3{,}6\ 10^{-4}$	HfTi0,01
Hf - Zr	3000	0,51	0,60	HfZr25

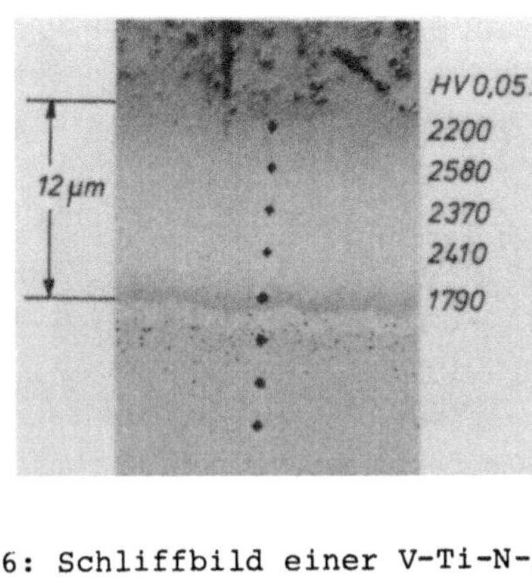

Bild 26: Schliffbild einer V-Ti-N-Schicht.

berechnet. Eine Zusammenstellung der Ergebnisse gibt Tabelle 2. Von den untersuchten Systemen haben alle außer dem Ti-V und dem Zr-Hf eine Dampfdruckgleichheit nur bei sehr kleinen Gehalten des zweiten Legierungspartners.

Für Versuche standen die Legierungen TiCr2, TiV25 und TiV50 zur Verfügung. Hiermit erzeugte Hartstoffschichten sind in Tabelle 1 aufgeführt.

Ein Beispiel einer Schicht aus einem Titan-Vanadium-Nitrid ist in Bild 26 gegeben. Die V-haltigen Schichten hatten Härten HV 0,05 bis 2580, also deutlich höhere Werte als beim TiN. An den Härteeindrücken entstanden z.T. Risse, ein Hinweis auf größere Sprödigkeit dieser Schichten. Die Doppelnitridschichten besaßen eine gute Verschleißfestigkeit, neigten jedoch bei größeren Schichtdicken zu Abplatzungen.

Eine Anwendung könnten diese harten Doppelnitride als dünne Einzelschichten in Verbundschichten (ähnlich dem TiC in der Doppelschicht TiC/TiN) finden.

11. Ionenplattieren

Beim Ionenplattieren /14/ wird nur ein negatives Potential (ca. -2 bis -5 kV) an das Substrat gelegt (Schalter S_1 offen, S_2 geschlossen, s. Bild 3). Im Unterschied zu den verschiedenen im Abschn. 7 beschriebenen Verfahren werden wesentlich höhere Gasdrücke eingestellt. Der Rezipient ist mit einem zu ca. 75 Vol.-% Argon und 25% Vol.-% N_2 (Reaktivgas) bestehenden Gasgemisch mit einem Gesamtdruck von etwa 10^{-2} mbar gefüllt. Da die mittlere freie Weglänge klein ($\sim$ 1 mm) ist, können die Dampfteilchen nicht mehr direkt auf das Substrat gelangen. Es bildet sich eine Art Dampfströmung zwischen Schmelzbad und Substrat aus. Die Gasstreuung bewirkt, daß auch Seitenflächen und mit verminderter Wachstumsrate die Rückseite des Substrates beschichtet werden. Es gelang zwar, nach dieser Methode harte Schichten mit kubischem TiN abzuscheiden, jedoch konnten keine Vorteile gegenüber den nach dem in Abschn. 7.5 beschriebenen Verfahren abgeschiedenen Schichten festgestellt werden.

12. Zusammenfassung

Die vorliegende Untersuchung befaßt sich mit der Herstellung und den Eigenschaften von Hartstoffschichten zur Verschleißminderung nach Verfahren, die auf der Aufdampftechnik beruhen (PVD-Verfahren). Im Prinzip handelt es sich hierbei auch um eine Abscheidung aus der Gasphase. Die Metallatome werden jedoch nicht, wie bei der chemischen Abscheidung (CVD), durch chemische Umwandlung einer thermodynamischen stabilen, gasförmigen Metallverbindung bereitgestellt, sondern durch direkte Verdampfung. Die Atome der Gasphase des Metalles beim PVD-Verfahren sind - bezogen auf die Substrattemperatur - weit vom thermodynamischen Gleichgewicht entfernt und werden beim Auftreffen auf die Substratfläche vollständig kondensieren.Das trifft jedoch nicht für die Metalloidatome zu, die aus der Gasphase angeboten werden. Bei den PVD-Verfahren hat man es daher bei der Ausbildung der Hartstoffschicht immer mit zwei konkurrierenden Prozessen zu tun, nämlich der Kondensation des Metalldampfes und mit der nachfolgenden Reaktion der kondensierten Metallatome mit den Metalloiden aus der Gasphase. Bei den CVD-Verfahren besteht die Gefahr der Kondensation eines Reaktionspartners nicht, da nur das Reaktionsprodukt bei der Temperatur der Beschichtung eine feste Phase bildet. Daher hat man auch nicht das Problem, nichtstöchiometrische Schichten zu erhalten. Bei allen reaktiven Aufdampfverfahren ist es dagegen wegen der Kondensation schwierig, stöchiometrische Hartstoffschichten zu erhalten. Die Bedingungen, mit denen stöchiometrische oder nahezu stöchiometrische Hartstoffschichten erhalten werden können, wurden am Beispiel der Abscheidung von Titannitrid genau untersucht. Ohne zusätzliche Aktivierung, d.h. nur in Anwesenheit des Reaktivgases, werden nur geringe Mengen an Stickstoff in die aufgedampfte Schicht eingebaut. Der von Bunshah /2/ vorgeschlagene und erprobte Weg zur Aktivierung des Abscheidevorganges durch die Ionisierung des reaktiven Gases wurde in Einzelheiten untersucht. Bei den Versuchen zur Abscheidung von Titannitrid wurden je nach den Abscheideparametern die Phasen α-Ti, ε-Ti_2N und/oder TiN_x in den Schichten gefunden. Die tribologisch besonders günstige TiN_x-Phase bildet sich vorzugsweise bei Substrattemperaturen um 600°C und bei N_2-Drücken von mehr als 10^{-2} mbar. Die dabei erhaltenen TiN_x-Schichten mit $x \sim 0,6$ bis 0,7 waren jedoch weit von der stöchiometrischen Zusammensetzung entfernt und hatten noch verhältnismäßig geringe Härten. Eine Verbesserung wurde erzielt durch das Anlegen eines negativen Potentials an das Substrat, welches die ionisierten Moleküle des Gases gegen das Substrat beschleunigt. Der Stickstoffgehalt der Schichten konnte dadurch merklich erhöht werden. Die in dieser Weise beschichteten Hartmetall-Wendeschneidplatten zeigten bei Drehversuchen Schneidhaltigkeitseigenschaften, die an die von CVD-beschichteten Wendeschneidplatten heranreichten. Das Gefüge der Schichten ist charakterisiert durch eine ausgeprägte Stengelstruktur. Zur Verbesserung des lateralen Schichtverbundes wurde die Schicht während des Wachstums durch ein Bombardement von Argonionen zusätzlich aktiviert. Durch die Inertgasaktivierung konnte die Ausbildung der uniaxialen Stengelstruktur unterdrückt werden. Die Schichten erreichten hohe Härtegrade und gute Verschleißeigenschaften. Das Entwicklungsziel des Vorhabens, nach dem PVD-Verfahren Hartstoffschichten von ähnlicher Qualität wie nach dem CVD-Verfahren abzuscheiden, wurde damit erreicht. Gegenüber dem CVD-Verfahren konnten bei gleichzeitiger Verfeinerung des Schichtgefüges die Beschichtungstemperaturen von ca. 1000°C auf 600°C gesenkt und die Beschichtungsraten von ca. 2 - 3 µm/h auf ca. 20 bis 60 µm/h gesteigert werden. Es darf jedoch nicht übersehen werden, daß die reproduzierbare Durchführung des aktivierten reaktiven Aufdampfverfahrens erhebliche Schwierigkeiten infolge der nur schwer kontrollierbaren Verdampfungsgeschwindigkeit bereitet und beträchtliche technische Probleme mit sich bringt. Die genaue Einhaltung der z.T. kritischen und nur schwer kontrollierbaren Verfahrensparameter bereitet nicht unerhebliche Schwierigkeiten. Für eine allseitige Beschichtung ist weiterhin ein komplizierter Bewegungsmechanismus der Substrate erforderlich. Die Entwicklung einer kontinuierlich arbeitenden PVD-Beschichtungsapparatur mit Ein- und Ausschleusung der Proben, Substratheizung, auf Hochspannungspotential liegendem Bewegungsmechanismus der Substrathalterung, für einen Dauerbetrieb geeigneten Beschichtungsratenmesser und anderes mit, bleibt daher eine Herausforderung an die Vakuumtechnik. Dabei ist abzusehen, daß sich durch das leichter beherrschbare Verfahren des reaktiven Hochleistungskathodenzerstäubens Hartstoffschichten ebenfalls mit dem Vorteil der geringen Substrattemperatur erzeugen lassen. Die entwickelten PVD-Verfahren werden aber in jedem Fall ihren Platz bei der Entwicklung neuer, durch CVD-Methoden nicht oder nur mit sehr großem Aufwand herstellbarer Schichten haben. Beispiele dafür sind die hier ebenfalls hergestellten und untersuchten Schichten aus Hafniumnitrid, Hafniumcarbid und den Doppelnitriden Ti-Cr-N und Ti-V-N. Besonders bei den letztgenannten Hartstoffschichten konnten Härten gemessen werden, die höher als die von einfachen Titannitriden waren. Eine Fortsetzung der Arbeiten zur Herstellung neuer Hartstoffschichten wird daher bei der weiteren Entwicklung der Hartstoff-Schichtsysteme von großem Nutzen sein.

13. Schrifttum

1 G. Schumacher, Techn. Zeitschr. prakt. Metallbearbeitung, 63 (1969), S. 275-278

2 R.F. Bunshah, A.C. Raghuram, J.Vac.Sci, Techn. 9 (1972), S. 1385-1388

3 J.A. Thornton, Ann. Rev. Mater. Sci. 7 (1977), S.239-260

4 Deutsche Patentschrift 615 262 (1934), US Patentschrift 2 053 977 (1935), Deutsche Auslegeschrift 1 053 447 (1959)

5 O. Rüdiger, H. Grewe und J. Kolaska, Wear 48 (1948), S. 267-282

6 K.R. Leuschen: Ein Beitrag zur Feststoffborierung von Cobalt, Gußlegierungen auf Cobaltbasis und Sinterhartmetallen, Dissertation TH Aachen (1977)

7 A. Münster und W. Ruppert, Z. Elektrochem. 57 (1953), S. 564-571

8 N. Reiter: Fortschritte bei der spanenden Bearbeitung mit Hartmetall und Schneidkeramik; Vortrag: Symposium Zerspanung der Metalle, Bad Nauheim, 18.-19.9.1980

9 US Patentschrift 2 784 115 (1957)

10 A.C. Raghuram, R. Nimmagadda, R.F. Bunshah und C.N.J. Wagner, Thin Solid Films 20 (1947), S. 187-199

11 R.F. Bunshah und R.J. Schramm, Thin Solid Films 40 (1977), S. 211-216

12 R.J. Hill, G. Scheuermann und R. Lucariello, Thin Solid Films 40 (1977), S. 217-222

13 R. Nimmagadda und R.F. Bunshah, Thin Solid Films 63 (1979), S. 327-331

14 D.M. Mattox, J.Vac.Sci.Technol. 10 (1973) S. 47-52

15 I. Barany, G. Kienel, Vakuum-Technik 28 (1979), S. 168-172

16 E. Ritter, J.Vac.Sci.Techn. 3 (1966), S. 225-226

17 E.H. Kennard: Kinetic Theory of Gases, New York and London 1938

18 I. Barin, O. Knacke, O. Kubaschewski: Thermochemical properties of inorganic substances, Berlin, Heidelberg, New York 1977

19 S. Schiller, U. Heisig, S. Panzer: Elektronenstrahltechnologie, Stuttgart 1977

20 P. Ehrlich, Z. anorg. Chem. 259 (1949), S. 1-41

21 A.C. Raghuram, R.F. Bunshah, J.Vac.Sci.Techn. 9 (1972) S.1389-1394

22 R.F. Bunshah, A.H. Shabiak, R. Nimmagadda, J. Cony, Thin Solid Films 45 (1977), S. 453-457

23 K. Nakamura, K. Inagawa, K. Tsuruoka u. S. Komiya, Thin Solid Films 40 (1977), S. 155-167

24 R.J. Hill, G. Scheuermann, R. Lucariello, Thin Solid Films 40 (1977), S. 217-222

25 B. Zega, M. Kornmann, J. Amiguet, Thin Solid Films 45 (1977), S. 577-582

26 D. Schulze, Die Technik 33 (1978), S. 399-402

27 W. Fleischer, D. Schulze, R. Wilberg, A. Lunk, F. Schrade, Thin Solid Films 63 (1979), S. 347-356

28 A.K. Suri, R. Nimmagadda, R.F. Bunshah, Thin Solid Films 72 (1980), S. 529-533

29 B.E. Jacobson, R.R. Bunshah, R. Nimmagadda, Thin Solid Films 63 (1979), S. 357-362

30 L.E. Toth: Transition Metal Carbides and Nitrides, New York, London 1971

31 B. Holmberg, Acta Chem.Scand. 16 (1962), S. 1255-1258

32 M.E. Sjöstrand, Proceed. 7th Int.Conf.CVD (1979), S. 452-462

33 B.A. Movchan und A.V. Demchishin, Fiz.metal.metalloved. 28 (1969), S. 653-660

34 R.D. Bland, G.J. Kominiak, D.M. Mattox, J.Vac.Sci.Technol. 11 (1974), S. 671-674

35 Manufacturing Engineering, Jan. 1977, S. 40-42

36 G.E. Hollox, Mat.Sci.Eng. 3 (1968/69), S. 121-137

37 G.Zinsmeister, Vakuumtechnik 13 (1964), S. 233-240

Hartstoffbeschichtung durch thermochemische Abscheidung aus der Gasphase

von

Dipl. Ing. Lienhard Paterok

Fa. Bernex GmbH
Oberflächenmetallurgie
Langenfeld
(Geschäftsführer: B. Filthaut)

Förderkennzeichen: 01 ZT 287
Laufzeit: 1. 1. 1978 - 31. 3. 1981

Vorwort

Die vorliegende Arbeit entstand Dank der Subventionierung durch das Bundesministerium für Forschung und Technologie in Bonn.

Unser besonderer Dank gilt Herrn Prof. Dr.-Ing. Hans Krause und seinen Mitarbeitern Herrn Dipl.-Ing. Ch. Schroelkamp und Herrn Dipl.-Ing. A. Klinkenberg aus dem Lehrstuhl für "Die Abnutzung der Werkstoffe" in der RWTH Aachen für durchgeführte röntgenographische Eigenspannungsmessungen, verschleiß- und rasterelektronenmikroskopische Untersuchungen und eine effektive Zusammenarbeit.

Den Herren Prof. Dr.-tech. O. Knotek (Inst. für Werkstoffkunde der RWTH Aachen), Dr.-Ing. M. Pohl (G.f.E. der RWTH Aachen), Dr.-Ing. W.-G. Burchard (G.f.E. der RWTH Aachen), Dr.-Ing. H. Reimann (Inst. für Werkstoffkunde der RWTH Aachen), Dr.-Ing. K. Keller, Dr.-Ing. U. Putzier, Dr.-Ing. H. Mathesius und Herrn B. Filthaut möchte ich für Ihre wertvollen Hinweise und Diskussionen danken.

Mein Dank gilt ferner Frau Gerber, den Herren J. Krejsa, F. Koch, A. Werner, A. Dimmers und H. Zickler für die sehr gute Zusammenarbeit und Unterstützung bei der Durchführung der Untersuchungen.

Inhaltsübersicht

Es wurde eine CVD-Anlage, in der man Metalle der III. - VI. Gruppe des Periodensystems abscheiden konnte, konstruiert und gebaut. Das Vorhaben wurde in Phasen unterteilt, in der ersten Phase wurde ein eingehendes Literaturstudium durchgeführt und der Stand der Technik überprüft. Aufgrund der Ergebnisse wurde in der zweiten Phase eine praxisnahe Normal-Unterdruck-CVD-Anlage mit einem Versuchsreaktor von 300 mm Durchmesser gebaut, in der Abscheideversuche von Titan-, Vanadin-, Zirkonium-, Niob-, Wolfram-, Chrom- und Tantalcarbiden durchgeführt wurden. Ein Teil der Versuche ist mißlungen. Die Ursache dafür waren irrationale Halogenid-Verflüchtigungsmethoden und zu geringe Härte des Schichtträgers. In dieser Phase wurde beschlossen, das Hauptgewicht der Untersuchungen auf die Schnellstahlbeschichtung und Härtung zu verlagern.
Das Probenmaterial (Wendeschneidplatten, Bohrer, Schübeplatten) ist so gewählt worden, daß man es direkt in der Praxis überprüfen konnte. Die Beschichtungsversuche von Bohrern sind mit positivem Resultat beendet worden, es mißlang aber der Standzeitversuch, weil der Reibungskoeffizient zwischen dem Schaft und dem Bohrkopf zu gering war. Eine Lösung des Problems konnte im Rahmen dieses Vorhabens nicht gefunden werden. Auf der Basis der neuen Erkenntnisse wurde die 3. Phase eröffnet, in der eine neue Konzeption einer Zweikammer-Vakuum-CVD-Anlage entstand. In dieser Anlage sind außer CVD-Prozessen Aufkohlungs-, Nitrier-, Carbonitrier-, Inchromierprozesse unter Vakuum-, Normal- und Unterdruckbedingungen durchführbar. Parallel wurden spezielle Halogenid-Verflüchtigungseinheiten gebaut und optimiert. In der 4. Phase wurde eine halbtechnische Zweikammer-Vakuum-CVD-Anlage gebaut und optimiert. Die ersten Beschichtungsversuche wurden auf Titan-, Vánadin-, Zirkonium- und Niob-Hartstoffe beschränkt. Es mißlang die Abscheidung von Zirkonium-Hartstoffen, die Ursache lag in der Verstopfung von Zuführungsleitungen durch Zirkonium-Halogenide. Die Lösung des Problems der Zirkonium-Abscheidung konnte im Rahmen dieses Vorhabens nicht gelöst werden. Andere Abscheidungen wurden positiv beendet. In der 4. Phase gelang es, die Vorversuche, bezogen auf Zerspanung und Trockenreibung zu beenden.
Strukturuntersuchungen mittels REM, Mikrosonde, Ψ-Goniometer, Makro- und Mikrogefüge ergaben Erkenntnisse über den Schichtaufbau und die Haftfestigkeit zu Schichtträgern. Die Ergebnisse werden als Grundlage für weitere Optimierung und Rationalisierung der:
- Beschichtungsprozesse

- Geometrie der Wendeschneidplatten
- Gasdosiersysteme
- Abschrecksysteme der Härtekammer
- Schiebersysteme

in der zukünftigen 5. Industrialisierungsphase dienen.

II. Genaues Thema des Forschungsvorhabens und die Aufgabenstellung

A. Thema

Entwicklung, Untersuchung und Erprobung hochverschleißfester Hartstoffoberflächen, beschichtet durch thermochemische Gasphasen-Abscheidung von Carbiden, Nitriden und Carbonitriden der Metalle der III. - VI.-Gruppe des period. Systems auf Hochleistungsschnell- und Vergütungsstählen.

B. Aufgabenstellung

a) Herstellung bzw. Anpassung einer thermochemischen Behandlungsanlage zur thermochemischen Gasphasenabscheidung der Hartstoffverbindungen der Metalle der III. - VI.-Gruppe des period. Systems auf Hochleistungsschnell- und Vergütungsstählen mit integrierter Einrichtung zur kontrollierten thermische Nachvergütung der Beschichtungsteile direkt aus der Beschichtungshitze.

b) Entwicklung bestgeeigneter Hartstoffbeschichtungen unter Berücksichtigung der Substratwerkstoffeigenschaften von Hochleistungsschnellstählen (HSS).

c) Untersuchung der Schichteigenschaften, insbesondere in ihrer Wechselbeziehung mit den Substratwerkstoffen und deren thermischer Umwandlung bzgl. residueller Spannungen, Hafteigenschaften, Maßänderungen.

d) Erprobung und Optimierung der Schichten und Erprobung der beschichteten Teile in der Praxis.

e) Erarbeitung der Grundlagen für die Beschichtungsindustrialisierung.

III Mechanismus und Chemie der Abscheidung von Metallverbindungen aus der Gasphase (AMG)

Gegenwärtig unter dem Begriff der Abscheidung von Metallverbindungen aus der Gasphase (Chemical Vapour Deposition) versteht man physikalisch-chemische Reaktionen, die in pyrolysierten Gasen und Dämpfen bei einem exakt definierten Druck durchgeführt werden. Die Pyrolyse der Reagenzmedien läßt die gasförmigen Stoffe katalysiert an der Substratoberfläche in feste Substanzen umsetzen. Parameter wie Druck, Temperatur, Konzentration, Strömungsart, Reaktionszeit determinieren die Abscheidungsgeschwindigkeit, Quantität (Schichtdicke), Qualität und Haftfestigkeit der abgeschiedenen Schicht.

Druck	- hat qualitativen und quantitativen Einfluß auf die Schicht
Temperatur	- mit steigender Abscheidetemperatur nehmen Abscheideraten und Homogenität der Schicht zu
Konzentration	- mit zunehmender, z.B. Titantetrachlorid, Konzentration in der Gasphase, bis zu einem Grenzwert steigt die Abscheiderate, oberhalb der Grenze fällt sie stark ab
Strömungsart	- hat einen Einfluß auf die Feinkörnigkeit der Schicht.

Die Anzahl der Grundparameter, die einen qualitativen und quantitativen Einfluß auf den Beschichtungsprozeß haben, bereiten im Optimierungsprozeß gewisse Schwierigkeiten. Es ist festgestellt worden, daß Beschichtungsparameter aus einem CVD-Beschichtungsprozeß (z.B. Chromcarbidabscheidung) sich in einen anderen Beschichtungsprozeß (z.B. Titancarbidabscheidung) nicht übertragen lassen. Jeder Abscheideprozeß muß individuell optimiert werden. In der Literatur sind CVD-Verfahren beschrieben worden, die sich grundsätzlich in zwei Richtungen entwickelten. Das erste ging in die Richtung der Temperatur (Tieftemperatur-, Mitteltemperatur- und Hochtemperaturverfahren) und das zweite in die Richtung des Drucks (Hochdruck-, Normaldruck und in den letzten Jahren durch LSRH-Bernex industrialisiertes Vakuumverfahren).
Der Beschichtungsprozeß beginnt durch das Einführen von einer definierten Metall-Halogen-Dampfkonzentration mittels Wasserstoff und Methan auf die zu beschichtende Oberfläche des Substrats. Die Beschichtungstemperatur liegt im Bereich 1000$\pm$ 50°C. Das Metallhalogenid zersetzt sich in Gegenwart des Substrats ab ca. 200°C in Chlorionen und ableitende Titanchloride. Die Abscheidetemperatur wird von den Verdampfungstemperaturen der Eisenhalogenide

(Eisentrichlorid und Eisendichlorid) bestimmt. Eisentrichlorid bildet sich als feste Verbindung schon bei 200°C an der Substratoberfläche aus. Bei 307°C geht es in die flüssige Phase über und verdampft bei 319°C. Das Eisendichlorid bleibt im festen Zustand an der Substratoberfläche bis zu ca. 677°C, bei dieser Temperatur geht es in die flüssige Phase über, die an der Oberfläche des Werkstückes einen Eisendichlorid-Film bildet, der die Abscheidung der metallischen Komponente der Hartstoffschicht am Substrat hindert. Erst bei 1020°C verdampft das Eisendichlorid und die Oberfläche des zu beschichtenden Substrats ist von der flüssigen Phase frei. Jetzt setzen sich an der Oberfläche reine Atome des hochschmelzenden Partners am Substrat fest, und ein Teil der Atome geht mit dem Substrat eine Reaktion ein und bildet Verbindungen, wie z.B. Ferrotitaniden bei einer Titancarbidabscheidung oder Ferrochromiden bei einer Chromcarbidabscheidung. Der größte Teil der Atome entzieht dem Substrat ein bestimmtes Gehalt an Kohlenstoff und bildet eine reine Carbidphase der Modifikationen $Me_{23}\,C_6$, $Me_7\,C_3$, $Me_3\,C_2$ aus. Die Phase $Me_{23}\,C_6$ ist ungesättigt, nimmt Kohlenstoff aus der Gasphase auf und geht in die Phase $Me_7\,C_3$ über. Man kann folgendes Reaktionsschema im Abscheideprozeß beobachten

$$Me_{23}\,C_6 \xrightarrow[T,\ p]{+\ C} Me_7\,C_3 \xrightarrow[T,\ p]{+\ C} Me_3\,C_2 \qquad (1)$$

Diese Umwandlungen sind in der Schicht im Bereich bis 4 µm von der Substratoberfläche aus zu beobachten. Sie sind durch eine absichtliche unterstöchiometrische Prozeßführung in weiteren Bereichen der Schicht auch noch feststellbar. In der Zone bis 4 µm bleiben auch vorher erwähnte Ferro-Verbindungen bestehen. Der größte Teil der Ferro-Verbindungen verbindet sich mit Kohlenstoff und bildet Carbide der Modifikationen $(Me_x\,Me_y)_{23}\,C_6$ und $(Me_x\,Me_y)_7\,C_3$. Die Mischung von reinen Karbiden und Mischcarbiden in der Zone bis 4 µm wurde fast bei 90% der abgeschiedenen Schichten festgestellt. Der Verdampfungspunkt der Eisendichloride läßt sich mittels Druck und Temperatur so verschieben, daß eine Hartstoffschichtabscheidung unter 1000°C möglich ist. Die Art der Prozeßführung ist immer von den Beschichtungsausgangsmaterialien determiniert (siehe Tabelle 1).

Tabelle 1

	Titan-Carbid	Titan-Carbonitrid	Titan-Nitrid	Zirkonium-Carbid	Vanadin-Carbid	Niob-Carbid
Metall-spender	Titan-Metall Ferro-Titan (Fe Ti 70) Ti Cl_4	Titan-Metall Ferro-Titan (Fe Ti 70) Ti Cl_4	Titan-Metall Ferro-Titan Fe Ti 70) Ti Cl_4	Zirkon-Metall Ferro-Zirkon (Fe Zr 85) Zr Cl_4; Zr Br_4	V-Metall-Späne (99,4% V) Ferro-Vanadium (Fe V 90) V Cl_4	- Ferro-Niob (Fe Nb R) Nb Cl_5
Kohlen-stoff-spender	Methan 99,95 Äthan Propan Dekalin	Methan Äthan Propan Dekalin	Methan Äthan Propan Dekalin	Methan Äthan Propan Dekalin	Methan Äthan Propan Dekalin	Methan Äthan Propan Dekalin
Stick-stoff-spender	—	Stickstoff (N_2) Ammoniak (NH_3)	Stickstoff (N_2) Ammoniak (NH_3)	—	—	—
Träger-gas	Wasserstoff 99.93% Argon 99,998%	Wasserstoff 99.93% Argon 99,998%	Wasserstoff 99.93% Argon 99,998%	Wasserstoff 99,93% Argon 99,998%	Wasserstoff 99.93% Argon 99.998%	Wasserstoff 99,93% Argon 99,998%
Chlor-ionen-spender	—	—	—	Chlorwasser-stoff 99,8%	Chlorwasser-stoff 99,8%	Chlorwasser-stoff 99,8%

IV. Allgemeines über Beschichtungstechnologie

Es sind gegenwärtig drei Beschichtunsprozesse in bezug auf das Kriterium Druck bekannt, wovon nur zwei an großtechnischer Bedeutung gewonnen haben, und zwar Normaldruck- und Vakuumprozeß. Überdruckprozesse haben die Entwicklungsphase nicht verlassen. Jeder Prozeß determiniert eindeutig das technologische und konstruktive Anlagengebilde. Anlagen für Hartstoffabscheidung unter Normaldruck sind einfach und bestehen aus einem Ofen, einem Reaktor und einer Reagenzdosiereinheit (siehe Bild 1). Weitere Beschreibungen erfolgen im Abschnitt VII.: Herstellung einer thermochemischen Behandlungsanlage. Normaldruckanlagen sind für Stahlbeschichtungen vorgesehen, während Vakuumanlagen grundsätzlich für Hartmetallbeschichtungen gebaut wurden. Eine CVD-Vakuumanlage (Bild 2) besteht aus: Ofen, Reaktor, Reagenzdosiereinheit, Stromversorgungseinheit und zusätzlich aus einer Vakuumpumpeneinheit, die zur kontinuierlichen Vakuumüberwachung und -haltung während des Beschichtungsprozesses bestimmt ist.

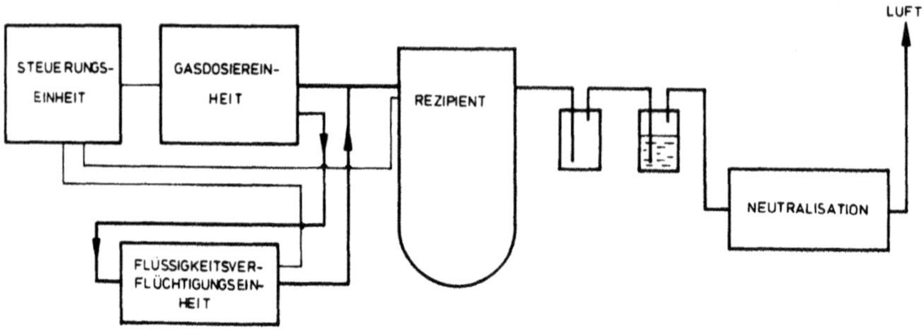

Bild 1 Normaldruck CVD-Anlage (Schema)

Für Schnellstahlbeschichtung wurde eine spezielle CVD-Vakuumanlage gebaut, in der nach dem Beschichten das Härten und Anlassen durchführbar ist. Diese CVD-Anlage (siehe Bild 3) ist so konzipiert, daß man Prozesse unter Normaldruck, Vakuum- oder kombiniert Vakuum- und Normaldruck durchführen kann.

Gegenwärtig werden in der Industrie Stähle nach folgendem Schema beschichtet und thermisch nachbehandlet:

a) Reinigung der zu beschichtenden Teile (Perbad)
b) Chargen-Zusammenbau (definierte Zuordnung der zu beschichtenden Teile zu einer Chargiervorrichtung)
c) Einsatz der Charge in den Rezipienten
d) Evakuierung des Rezipienten
e) Aufheizen des Ofens und Rezipienten lt. Stahlbehandlungsvorschriften
f) Reduzieren der an der Substratoberfläche übriggebliebenen Oxyde mittels Wasserstoff im Temperaturbereich von 700 bis 1000°C
g) Aufkohlen - nur im Falle, daß der Kohlenstoffpegel im zu beschichtenden Substrat unter 0,9% liegt
h) Hartstoffschichtabscheidung
i) Abschrecken der Charge
j) Anlassen
k) Auseinanderbau der Charge

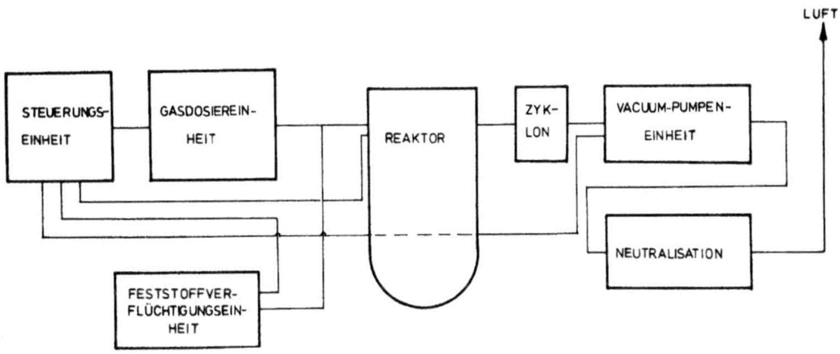

Bild 2 Vakuum CVD-Anlage (Schema)

Heute werden Hochleistungsschnellstähle und einige Werkzeugstähle nach dem CVD-Beschichten aufgrund unzureichender Härte einem thermischen Nachbehandlungsprozeß mit Anwärmen, Vorwärmen, Härten und Anlaßstufen in einem externen Vakuumofen unterzogen.
In von uns hergestellten halbtechnischen Vakuum CVD-Anlagen lassen sich Hartstoffschichten wie: Titancarbid, Titannitrid, Titancarbonitrid, Vanadincarbid, Niobcarbid und Titansilizid nach folgenden Reaktionen abscheiden:

Titancarbid

$$TiCl_4 + CH_4 \xrightarrow[Fe]{H_2;\ T;\ p} \downarrow TiC + \uparrow 4\ HCl \qquad (2)$$

Titannitrid

$$2\ TiCl_4 + N_2 + 2H_2 \xrightarrow[Fe]{H_2;\ T;\ p} \downarrow 2\ TiN + \uparrow 4\ HCl \qquad (3)$$

Titancarbonitrid

$$TiCl_4 + CH_4 + {}^{y}/_{2}N_2 \xrightarrow[Fe]{H_2;\ Ar;\ T;\ p} \downarrow Ti\ (C_x\ N_y) + \uparrow 4\ HCl \qquad (4)$$

Vanadiumcarbid

$$VCl_4 + CH_4 \xrightarrow[Fe]{H_2;\ T;\ p} \downarrow VC + \uparrow 4\ HCl \qquad (5)$$

Niobcarbid

$$2NbCl_5 + 2CH_4 + H_2 \xrightarrow[Fe]{H_2;\ T;\ p} \downarrow 2NbC + \uparrow 10HCl \qquad (6)$$

Titansizid

$$TiCl_4 + SiCl_4 + 4H_2 \xrightarrow[Fe]{H_2;\ T;\ p} \downarrow TiSi + \uparrow 8HCl \qquad (7)$$

V. Werkstoffwahl

1. Allgemeines über SS, HS und HSS Stähle

In der Maschinenbauindustrie werden die Werkstoffe nach DIN-Normzeichen und Werkstoffnummern eingekauft. Schnellstähle für Schneidwerkzeuge sind mit einem Kurzzeichen wie SS, HS, HSS versehen, deren Bedeutung nicht klar umrissen ist. Die physikalischen Eigenschaften und chemischen Zusammensetzungen können aus den allgemeinen Bezeichnungen SS, HS oder HSS nicht erkannt werden. Die chemische Zusammensetzung ist jedoch die Grundlage zur Beurteilung der gewünschten Eigenschaften, wobei die Qualität, Wärmebehandlung und konstruktive Form des Werkzeuges vorausgesetzt werden muß. In der deutschen Literatur ist die Auslegung der Kurzzeichen SS, HS und HSS anders üblich als in der internationalen Literatur. Vor 1939 wurde die Bezeichnung SS für Schnellstahlwerkzeuge neben dem Fabrikzeichen geführt. Ende 1946 sind sogenannte Friedensqualitäten, B 18 (1.3355), DMo 5 (1.3343), BMo 9 (1.3346) und D (1.3318) eingeführt worden, deren Leistung nicht zufriedenstellend war. Deutsche Hersteller haben das Kurzzeichen HSS (Hochleistungsschnellstahl) eingeführt, um die Qualität der Stähle und die Leistungsdifferenzen hervorzuheben. Im Ausland gilt die Bezeichnung nicht, da dort der Begriff Sparstähle unbekannt ist. Für Verbraucher von Schnellstahlprodukten wurden die üblichen Kurzzeichen SS, HS und HSS beibehalten und im Werkstoffblatt 320/53 des VDEh festgehalten (siehe Tabelle 2).

Tabelle 2

Gruppe	Bez. I	Bez. II	Stoff-Nr.	Kurz-zeichen	Leistung
A	ABC II	S 9-1-2	1.3316	SS	mittlere
	ABC III	S 3-3-2	1.3333	SS	mittlere
B	D	S 12-1-2	1.3318	HS	hoch (high speed)
	DMo 5	S 6-5-2	1.3343	HS	
	BMo 9	S 2-9-1	1.3346	HS	
	BMo 9V	S 18-0-1	1.3355	HS	
	B 18	S 2-9-2	1.3348	HS	
C	EV4Co	S 12-1-4-5	1.3202	HSS	höchste (high speed special)
	EW9Co10	S 10-4-3-10	1.3207	HSS	
	ECo3	S 12-1-2-3	1.3211	HSS	
	E18Co5	S 18-1-2-5	1.3255	HSS	
	EV 4	S 12-1-4	1.3302	HSS	
	EMo5V3	S 6-5-3	1.3344	HSS	
	EMo5Co5	S 6-5-2-5	1.3243	HSS	
	E18Co 10	S 18-1-2-10	1.3265	HSS	

Gruppen:

A - Sparlegierungen

B - Standardschnellstahl

C - Schnellstahl für hohe Zerspanungsanforderungen

Gegenwärtig werden dem deutschen Verbraucher 30 Schnellstähle angeboten (siehe Tabelle 3).

Tabelle 3

Pos.	Werkstoff-Nr.	Kurzname	C-Pegel
1	1.3202	S 12-1-4-5	1,30 - 1,45
2	1.3202 S	S 12-1-4-5 (S=0,08)	1,30 - 1,45
3	1.3207	S 10-4-3-10	1,20 - 1,35
4	1.3211	S 12-1-2-3	
5	1.3222	S 6-5-2-8	1,40 - 1,50
6	1.3243	S 6-5-2-5	0,88 - 0,96
7	1.3245	S 6-5-2-5 (S=0,10)	0,88 - 0,96
8	1.3246	S 7-4-2-5	1,05 - 1,15
9	1.3247	S 2-10-1-8	1,05 - 1,12
10	1.3248	S 2-9-2-5	
11	1.3249	S 2-9-2-8	0,85 - 0,92
12	1.3255	S 12-1-2-5	0,75 - 0,83
13	1.3257	S 18-1-2-15	0,60 - 0,70
14	1.3265	S 18-1-2-10	0,72 - 0,80
15	1.3302	S 12-1-4	1,20 - 1,35
16	1.3316	S 9-1-2	0,78 - 0,86
17	1.3318	S 12-1-2	0,90 - 1,00
18	1.3333	S 3-3-2	0,95 - 1,03
19	1.3335	S 3-3-2 Si (Si=0,7-1,0)	0,95 - 1,05
20	1.3340	SC 6-5-2 S (S=0,12)	0,95 - 1,05
21	1.3341	S 6-5-2 S (S=0,12)	0,84 - 0,92
22	1.3342	SC 6-5-2	0,95 - 1,05
23	1.3343	S 6-5-2	0,86 - 0,94
24	1.3344	S 6-5-3	1,17 - 1,27
25	1.3345	S 6-5-2 Si (Si=0,7-1)	0,84 - 0,92
26	1.3346	S 2-9-1	0,78 - 0,86
27	1.3347		
28	1.3348	S 2-9-2	0,97 - 1,07
29	1.3355	S 18-0-1	0,70 - 0,78
30	1.3357	S 18-0-2	

2. Eliminationskriterien

Nicht alle Stähle lassen sich in einer wirtschaftlichen Zeit optimal beschichten und aus diesem Grund sind folgende Eliminationskriterien für einen optimalen Einsatz eingeführt worden:

2.1 Schwefelgehalt kleiner als 0,03%
2.2 Siliciumgahalt kleiner als 0,45%
2.3 Kohlenstoffgehalt größer als 0,9%
2.4 Anlaßhärte im Bereich 64-67 HRc

2.1 Schwefelgehalt kleiner als 0,03%

Aus der Tabelle 3 sind 4 Stähle gestrichen worden:

1. S 12-1-4-5 S (1.3202) S = 0,08%
2. S 6-5-2-5 S (1.3245) S = 0,10%
3. SC 6-5-2 S (1.3340) S = 0,12%
4. S 6-5-2 S (1.3341) S = 0,12%

2.2 Siliciumgehalt kleiner als 0,45%

Aus den 26 übriggebliebenen Stählen sind weitere 2 eliminiert worden:

1. S 3-3-2 Si (1.3335) Si = 0,7 - 1,0%
2. S 6-5-2 Si (1.3345) Si = 0,7 - 1,1%

2.3 Kohlenstoffgehalt größer als 0,9%

Die übriggebliebenen Stähle sind in der Tabelle 4 zusammengestellt.

2.4 Anlaßhärte 64-67 HRc

Die aufgelisteten Stähle in Tabelle 4 entsprechen diesen Anforderungen.

Tabelle 4

Pos.	Werkstoff-Nr.	Kurzname	A-Härte (HRc)	C-Pegel %
1	1.3202 HSS	S 12-1-4-5	64 - 66	1,30 - 1,45
2	1.3207 HSS	S 10-4-3-10	65 - 68	1,20 - 1,35
3	1.3222 (HSS)	S 6-5-2-8	64 - 69	1,40 - 1,50
4	1.3246 (HSS)	S 7-4-2-5	64 - 66	1,05 - 1,15
5	1.3247 (HSS)	S 2-10-1-8	65 - 68	1,20 - 1,35
6	1.3302 HSS	S 12-1-2	64 - 66	1,20 - 1,35
7	1.3318 HS	S 12-1-2	64 - 66	0,90 - 1,00
8	1.3333 SS	S 3-3-2	64 - 66	0,95 - 1,03
9	1.3342 (HSS)	S 6-5-2	65 - 67	0,97 - 1,07
10	1.3344 HSS	S 6-5-3	65 - 67	1,17 - 1,27
11	1.3348 HS	S 2-9-2	64 - 66	0,97 - 1,07

VI. Festlegung der Beschichtungsvarianten

Nach einer eingehenden Analyse der während des Zerspanungsprozesses an Hochleistungs-Schnellstählen entstehender Verschleißarten und ihrem qualitativen und quantitativen Charakter, wurde bei der Festlegung der Hartstoffe das Kriterium Härte der Schicht als Grundlage genommen. Die härtesten Schichten sind:

Titancarbid (3200 HV_{50-10}),
Vanadincarbid (2944 HV_{50-10}),
Titancarbonitrid (2550 HV_{50-10}),
Titannitrid (2450 HV_{50-10}),
Niobcarbid (2400 HV_{50-10}),

Diese Hartstoffschichten dienten als Basis zur Festlegung folgender Beschichtungsvarianten:

HSS + TiC
HSS + TiC + TiCN
HSS + TiC + TiCN + TiN
HSS + TiN
HSS + VC
HSS + NbC

VII. Herstellung einer Anlage zur thermochemischen Gasphasenabscheidung mit integrierter Einrichtung zur kontrollierten Nachvergütung direkt aus der Beschichtungshitze

In dieser Projektphase sind alle konventionellen Anlagen für Normaldruck- und Vakuumbeschichtungsprozesse analysiert worden unter Berücksichtigung von

1. Konstruktionskriterien
1.1 Beschichtung unter Vakuumbedingungen von 100 bis 1 Torr absolut
1.2 Direktes Abhärten aus der Beschichtungshitze
1.3 Härten in Vakuum und in der Schutzgasatmosphäre
1.4 Abkühlungsgeschwindigkeit kontrollierbar

2. Konstruktionsannahmen
2.1 Dichtheit des Rezipienten und der Abkühlhaube 2×10^{-5} Torr l/s
2.2 Dichtheit der dynamischen Ventile 2×10^{-5} Torr l/s
2.3 Standzeit des Rezipienten 2500 bis 3000 Stunden
2.4 Säurebeständigkeit der Anlage

Es stellte sich heruas, daß es keine CVD-Anlage gab, in der man Hochleistungs-Schnellstähle härten und anlassen kann. Es wurde beschlossen, die Herstellung einer Anlage auf 2 Stufen zu unterteilen. In der ersten wurde ein Umbau einer Normaldruckanlage für Vakuumbeschichtungen und Wasserhärtung vorgenommen (siehe Bild 2 und 3). In der zweiten Stufe wurde aufgrund der Ergebnisse der ersten Phase und aufgrund der Beschichtungsversuche in einer Vakuum CVD-Glockenanlage, Typ Berna-Bernex (siehe Bilder 6 und 7) eine halbtechnische 2 Kammer Vakuum CVD-Anlage, Typ KKZ6P-8 mit einer Turbokühlungseinheit entwickelt.

VIII. Normal- Unterdruck CVD-Anlage, Typ KRR-80

Diese Anlage besteht aus:

a) Schachtofen

Für diese CVD-Anlage wurde ein Schachtofen für eine Dauerwärmeerzeugung bis zu 1200°C gekauft. Dieser Ofen besaß in der ersten Version einen quadratischen Innenraum von 380 x 380 mm und eine Höhe von 1600 mm. In den Ofenraum wurde von oben ein Rezipient von einem Durchmesser 350 mm eingeführt, der sich nach ca. 15 bis 25 Chargen so verzogen hatte, daß man die Chargiervorrichtung und den Rezipienten nicht mehr aus dem Ofen herausnehmen konnte. Die Ursache für diesen Verzug war die Seitentür. In der zweiten Phase wurde dieser Ofen durch einen speziellen runden Sechs-Zonen-Ofen mit einer Bohrung von 380 mm und einer Tiefe von 1500 mm ersetzt (siehe Bild 3, Pos. 1).

b) Rezipienten

Die Wanddicke des Rezipienten (siehe Bild 3, Pos. 2) für den CVD-Normaldruckprozeß ist von der Abkühlungsart abhängig. Es werden drei Abkühlungsprozesse angewandt:
Abkühlung im Ofen (Abkühlungszeit bis zu 36 Stunden)
Abkühlung an der Luft (Abkühlungszeit zwischen 3 und 8 Stunden)
Abkühlung im Wasser (Abkühlungszeit unter einer Stunde).
Der mit der Abkühlung verbundene Thermoschock führt zu:

1. axialem und radialem Verzug des Außenmantels
2. Versprödung des Werkstoffes
3. Rißbildung im Bereich der Schweißnaht
4. Bodenverzug.

c) Chargiervorrichtungen

Im ersten Stadium wurde eine Chargiervorrichtung gebaut, die zur Überprüfung der Strömungsverhältnisse von Reagenzmedien bestimmt war. Aufgrund der Ergebnisse ist im zweiten Stadium eine Chargiervorrichtung für einen zweiseitigen Beschichtungsvorgang konstruiert worden (Beschichtungsrichtung von unten nach oben und von oben nach unten). Diese Chargiervorrichtung wurde mehrmals optimiert. Im dritten Konzeptionsstadium entstand eine universelle, zerlegbare Chargiervorrichtung, siehe Bild 4.

Bild 3 Angepaßte Normal- Unterdruck- CVD-Anlage, Typ KRR-80

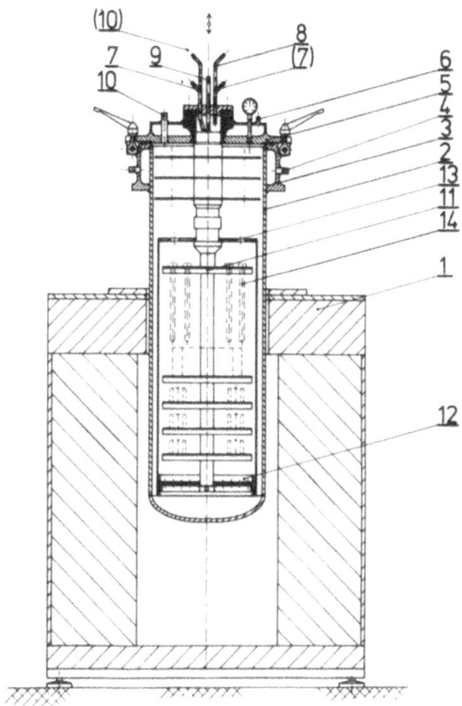

1. Ofen
2. Rezipient
3. Kühlflansch
4. Kühlwassereingang
5. Deckel
6. Kühlwassereingang
7. Reagenzien-Eingangsleitungen
8. Abgasleitungen
9. Thermoelement
10. Warmwasserleitungen
11. Chargiervorrichtung
12. Gasdusche
13. Strömungsglocke
14. Probeteile

d) Deckel

Der Deckel (Bild 3, Pos. 5) ist das zweitwichtigste Element der Normaldruckanlage. Er ist ein Teil der Chargiervorrichtung und durch ihn werden Reagenzien in den Rezipienten ein- und abgeführt. Der Deckel besitzt einen Kühlraum, der die Dichtung im Rezipientenflansch vor einer Überhitzung schützt und die Temperatur (ca. 70°C) konstant hält. Die Kühlung des Deckels im Bereich des Abführrohres unter 70°C führt zur Abscheidung von festen Chloriden am Ausgang (was zu einer Verstopfung und Abbrechung des CVD-Prozesses führt) und zur Verringerung des effektiven Reaktionsraumes.

Der Deckel ist mit konstanter Temperatur für Normal-, Überdruck- und Vakuumprozesse konstruiert worden. Die besten Werkstoffe für Deckel sind säure- und hitzebeständige Stähle.

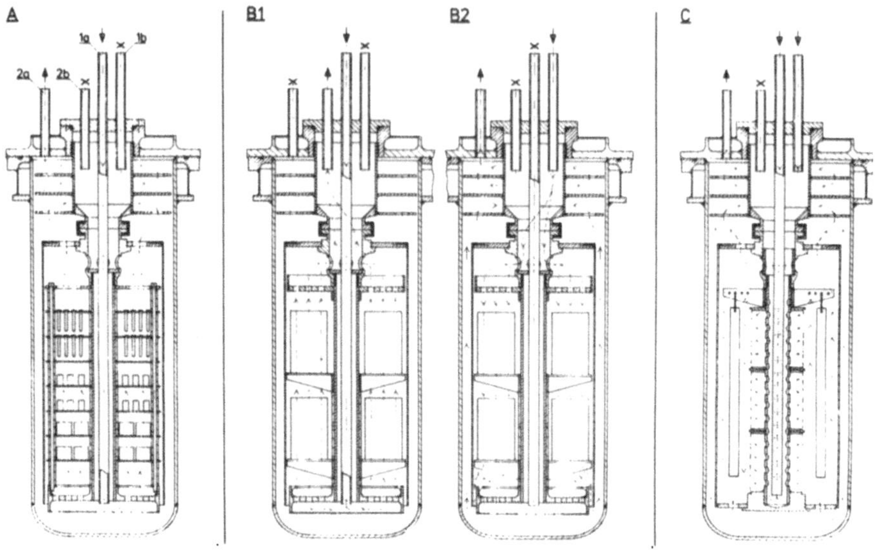

Bild 4 Universelle Chargiervorrichtung, Typ IKR 80

e) Halogenid-Verdampfungseinheit

Es wurden drei Verdampfungsarten von flüssigen Halogeniden überprüft:

a) mittels Gasverdampfer, extern der Anlage
b) mittels Verdampfer, unmittelbar an der Anlage
c) direkte Verdampfung im Rezipienten.

Eine Halogenid-Verdampfungseinheit besteht aus hermetisch dichtem Verflüchtigungsgefäß, Verdichter, Thermostat mit Heizpilz und Temperaturanzeigegerät.

f) Steuerungs- und Überwachungseinheit

Mittels dieser Einheit wurde die Temperatur:

a) des Ofens in Abhängigkeit der Größe der Charge und Beschichtungsart
b) des Kühlsystems im Rezipienten
c) des Abschreckbehälters
d) der Halogenid-Verflüchtigungseinheit

gesteuert und überwacht.

g) Gasdosiereinheit

Diese Einheit besteht aus 5 Rotametern, Feinventilen zur Fein- und Grobregulierung, 5 Gas-Reinigungsanlagen und 5 Druckminderventilen. Die Rotameter sind mit Skalen in Litern/Minute ausgestattet. Ventile, Rotameter und Leitungen wurden aufgrund in der Anlage existierender leicht saurer Atmosphäre aus Edelstahl eingesetzt. Die direkte Ablesegenauigkeit der Instrumente lag bei 0,05 l/min.

h) Wasser-Kühlsystem (WKS)

Das Wasser-Kühlsystem ist auf drei Untersysteme unterteilt worden:

a) im Rezipientenflansch
b) im Deckel
c) an Leitungen.

i) Abschrecksystem

Das System besteht aus Abschreckbehälter, Heizvorrichtung, Intensivmischer und Dampfabführungsvorrichtung.

k) Abscheidekontrolle

Während der Abscheidereaktion im Rezipienten entstehen Nebenprodukte wie Eisendichlorid, Titandichlorid, Titantrichlorid, die mit den Abgasen abgeführt werden. In die Abgasstrecke wurden zwei Glasgefäße eingebaut, im ersten scheiden sich an den Wänden feste Chloride ab und im zweiten befindet sich alkalisiertes Wasser mit einem Indikator für die Beobachtung der Startreaktion.

IX. Einkammer-Vakuum-CVD-Glockenanlage, Typ KZ-77

Die CVD-Glockenanlage (Bild 5) besteht aus:

a) Grundgestell

Der Sockel dient als Basis für den Manipulator, die Rezipientengrundplatte und die Steuerkonsole.

b) Rezipient

Der Rezipient wurde für Vakuumbetrieb ausgelegt und besteht aus der in den

Sockel eingelassenen Grundplatte und der beweglichen Haube. In die Grundplatte sind wassergekühlter Dichtflansch, Durchführungen für Vakuum-, Gas- und Meßleitungen, sowie Sicherheitsklemmvorrichtungen eingebaut. Letztere sichert die Haube gegen Abheben durch Überdruck. Die Haube hat zwecks optimaler Wärmeleitung eine nur wenige Millimeter dicke Wand. Durch Verwendung hoch hitzebeständiger Werkstoffe wurde eine hohe Sicherheit gegen Verformung gewährleistet. Für Hochtemperaturbeschichtung oberhalb 1100°C werden Kaltwandrezipienten verwendet.

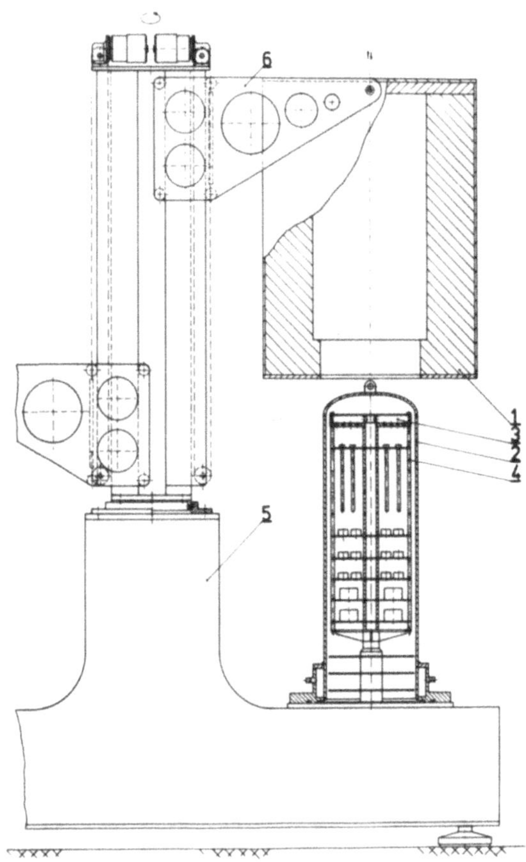

Bild 5
Einkammer-Vakuum-CVD-Glockenanlage, Typ KZ-77

1. Haubenstrahlungsofen
2. Rezipient
3. Gasdusche
4. Chargiervorrichtung
5. Grundgestell
6. Manipulator

Gegenwärtig werden Vakuum-CVD-Glockenanlagen mit folgenden Durchmessern hergestellt: 100, 180, 230, 300, 380 und 430 mm. Die Länge der Rezipienten beträgt: 300, 600, 800, 1200 und 1500 mm. Die Rezipienteneinbauten bestehen aus Meßrohr für die Aufnahme bis zu 10 Thermoelementen, Gasverteilung,

Strömungsleitblechen, Abgasstutzen und Chargiervorrichtungen. Letztere stützen sich entweder auf der Reaktionsgrundplatte oder auf dem Saugrohr ab.

Bild 6 Einkammer-Vakuum-CVD-Glockenanlage, Typ KZ-77 (Hersteller: Berna Olten)

Bild 7 Tandem-Vakuum-CVD-Glockenanlage (Hersteller: Berna Olten)

c) Heizsystem

Die Rezipienten werden bis zu 1200°C mittels Haubenstrahlungsöfen beheizt, die zur Einstellung eines optimalen Temperaturprofils über mehrere getrennte Heizzonen verfügen.

e) Rezipientenkühlung

Zur Abkürzung der Beschichtungszyklen und zur rascheren Abkühlung des Beschichtungsgutes wurden folgende Einrichtungen verwendet:

1. Kühlgebläsehaube zur äußeren Rezipientenkühlung, gleichzeitig als Abhebevorrichtung für die Rezipientenhaube
2. Durchflutungskühlung zur direkten Gaskühlung im Rezipienten mittels Heliumkreislauf und zwischengeschaltetem Wärmetauscher in Ergänzung zu 1.

f) Gasdosiereinheit

Je nach Reinheitserfordernis wurden die Gase einer chemischen Reinigung unterzogen. Die Trocknung z.B. von Wasserstoff erfolgte in Tandem-Molekular-Siebkolonnen. Sauerstoff und Kohlenmonoxyde, Dioxyde und übrige Verunreinigungen wurden mittels geeigneten Absorbern entfernt. Die Gasflußregelung und -messung erfolgte mittels Präzisionsventile und Schwebekörper-Durchflußmesser. Zur Überdruckregelung und -messung dient ein Kontakt-Maximum-Minimum-Manometer. Ein eigens entwickeltes Mikro-Reguliersystem ermöglichte eine exakte Dosierung der Gase in den Reaktor. Sämtliche Gasleitungen, sowie Regel- und Meßgeräte sind je nach Bedarf säurebeständig ausgeführt. Eine vorzeitige Reaktion der Behandlungsgase wurde durch streckenweise Beheizung der Leitungen verhindert.

g) Verdampfungseinheiten

Es wurden sowohl Flüssigkeitsverdampfer als auch Feststoffverdampfer verwendet. Die Verdampfereinheiten sind in einer normalen CVD-Anlage direkt in die Gasversorgungszentrale eingebaut. Das Vakuumpumpsystem wird im nächsten Kapitel beschrieben.

X. Zweikammer-Vakuum-CVD-Anlage, Typ PKZ-80

Aufgrund einer genauen Analyse und Konstruktionsänderungen der Anlagen bei der Berücksichtigung von Probebeschichtungen mit anschließender Substrathärtung entstand eine neue Konzeption einer Zweikammer-Vakuum-CVD-Anlage,

Typ PKZ-80. Diese Konzeption entspricht den festgelegten Konstruktionsannahmen und Kriterien (siehe VII. 1 u. 2), sie ist in Bild 8 schematisch dargestellt.

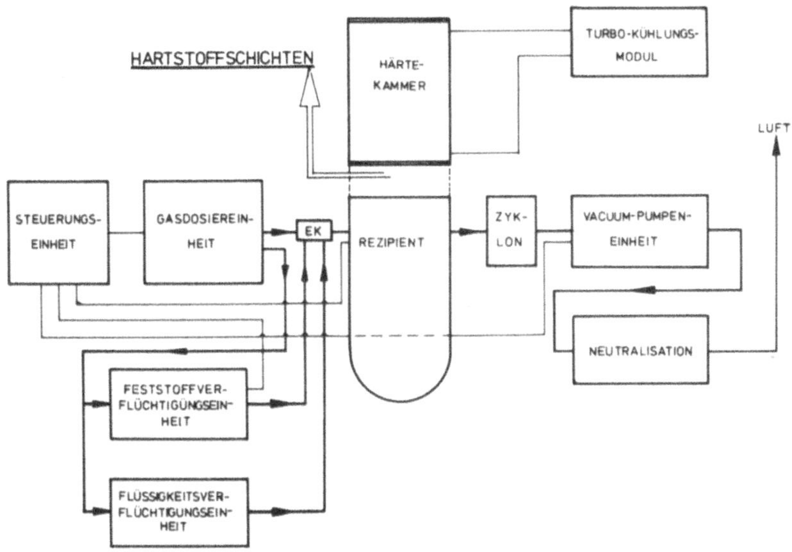

Bild 8 Schema einer Zweikammer-Vakuum-CVD-Anlage, Typ PKZ-80

Eine Zweikammer-Vakuum-CVD-Anlage (siehe Bild 8) besteht aus:

a) Schachtofen (siehe Bild 9, Pos. 1)

Spezieller runder Sechszonen-Ofen mit einer Öffnung für einen Rezipienten Ø 380 mm, Tiefe 1600 mm und einer Wärmedauererzeugung von 1200°C.

b) Rezipient (siehe Bild 9, Pos. 2)

Außer den vorher im Text erwähnten Konstruktionsannahmen und Kriterien mußten zusätzlich folgende physikalische Effekte für den Temperaturbereich von 20°C bis 1200°C berücksichtigt werden.

a) Desorption absorbierter Gase an der Innenfläche
b) Diffusion absorbierter Gase aus der Rezipientenwand in das Innere
c) Permeation von Gasen in den Ofenraum nach der Schrittfolge:
   Absorption an der Rezipientenoberfläche
   Diffusion durch die Wand (bei 1000°C bis 1200°C)
   Desorption im Vakuumraum.

Die drei Effekte zwangen uns zur Anwendung von säure- und hitzebeständigen Werkstoffen mit einer Gasabgaberate von $I \leq 5 \times 10^{-6}$ Torr l/s cm². Die äußere Druckbelastung des Rezipienten während des Beschichtungsprozesses bewegt sich bis zu 1 bar/cm². Diese Belastung führte zur gleichmäßigen radialen Schrumpfung. Der Schrumpfungswert ist von der Belastungszeit und den -zyklen abhängig. Er determiniert im Vakuumprozeß die Lebensdauer des Rezipienten und nicht wie im Normaldruckprozeß der unkontrollierte dreiaxiale, durch Thermoschock verursachte Verzug.

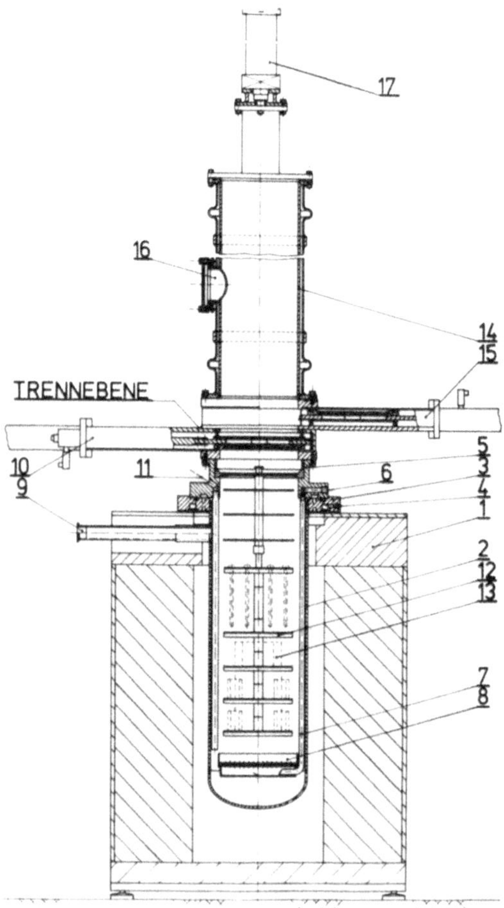

Bild 9
Halbtechnische Zweikammer-Vakuum-CVD-Anlage, Typ PKZ-80

1. Ofen
2. Rezipient
3. Rezipientenflansch
4. Kühlwassereingang
5. Zwischenflansch
6. Reagenzieneingang
7. Gasrohr
8. Gasdusche
9. Absaugstutzen
10. Schieber I
11. Thermoelementenrohr
12. Chargiervorrichtung
13. Proben
14. Härtekammer KKZ-78
15. Schieber II
16. Fenster
17. Fördervorrichtung

c) Chargiervorrichtungen (Bild 9, Pos. 12)

Es wurde eine Chargiervorrichtung entwickelt, mit der man die Strömungsverhältnisse, die Reaktionseffektivität und den nutzbaren Beschichtungsraum untersuchen wollte. Danach sind 6 Chargiervorrichtungen aus säure- und hitzebeständigem Stahl gebaut worden, um ihre Funktionsfähigkeit im direkten Einsatz auszuprobieren.

d) Gasdosiereinheit

Eine Gasdosiereinheit für eine Zweikammer-Vakuum-CVD-Anlage, Typ PKZ-80 besteht aus: Durchflußmengenmessern Typ Rota, (für Wasserstoff, Chlorwasserstoff, Argon, Methan, Stickstoff), Druckminderventilen, Druckhalteventilen, Feindosierventilen, Molekulartrocknungsanlagen, Manometern und Vakuummetern. Diese Einheit ist so konstruiert worden, daß man Beschichtungsprozesse unter Zwischenvakuum-, Normal- und Überdruckbedingungen führen kann, und das Härten unter Schutzgas im Druckbereich von 1 mbar bis 10 bar möglich ist.

e) Schiebersystem (Bild 9, Pos. 10 und 15)

Für diese Anlage wurde ein spezielles Schiebersystem entwickelt. Das System besteht aus 2 Schieberventilen, das erste wurde dem Rezipienten und das zweite der Abkühlkammer zugeordnet. Das Kriterium für den Bau der Ventile ist die Dichtheit im Gehäuse bei 2×10^{-7} Torr l/s und am Ventilsitz bei 2×10^{-5} Torr l/s bei 20°C gewesen. Es sind 2 Konzeptionen entstanden:

a) Schieberventil mit zerlegbarem Gehäuse und Wasserkühlsystem im Flansch
b) Schieberventil mit nicht zerlegbarem Gehäuse, Wasserkühlung in der Schieberplatte und im Flansch, Typ PRZ-79.

Die erste Konstruktion erwies sich aufgrund ihrer 15 abzudichtenden Stellen als nicht optimal und die angesetzte Dichtheit war praktisch nicht zu erreichen. Die Undichtheiten führten zu direkten Sauerstoffeinbrüchen in den Rezipienten und in den Schieberkasten. Jeder Sauerstoffeinbruch führt zu Oxydation und Abplatzungen der abgeschiedenen Hartstoffschichten und damit zum Ausfall der Charge. Daher wurde die Konzeption 2 ausgearbeitet, in der nur 6 Stellen abzudichten waren, davon 3 statisch und 3 dynamisch. Diese Schieberventile sind mit einem doppelt wirkenden Pneumatikzylinder ausgerüstet, der den Ventilteller ohne schädliches Reiben der Dichtung beim Schließen und Öffnen im Schieberkasten bewegt und ihn gegen den Sitz anpreßt und gleichzeitig verriegelt. Diese Schieber sind für Schließ- und

Öffnungszeiten von 2 Sekunden ausgelgt, die Betriebstemperatur liegt unter 100°C, der Differenzdruck am Teller ist 0,35 bar, die Dichtheit ist gewährleistet. Dem Gehäuseflansch auf der Ventilseite ist ein Kühlring zugeordnet, der mit Kühlwasser bis 8 bar geflutet werden kann.

Bild 10 Schiebersystem PRZ-79

f) Härtekammer (Bild 9, Pos. 14 und Bild 11)

Laut erster Konzeption (siehe Bild 11 A, Typ KZ-77) sollte das Chargiergut möglichst schnell nach dem Beschichten im Zwischenvakuum oder unter definierter Schutzgasatmosphäre abgekühlt werden. Als Lösung bot sich eine auf den Rezipienten aufsetzbare Abkühlkammer an, die das Chargiergut direkt unter Grob- oder Zwischenvakuumbedingungen aufnehmen sollte. Es wurde ein Hohlzylinder mit einem definierten Durchmesser d, von unten mit einem Schieberventil und von oben mit einer Fördervorrichtung versehen, der gewährleistete, das Chargiergut in den Rezipienten unter vorher genannten Bedingungen einzuführen und herauszuziehen. Auf die Abkühlhaube sind zusätzlich Ein- und Ausgangsstutzen für entsprechende Kühlmedien angeschweißt worden. Die ersten Versuche bestätigten die richtige Wahl der Konzipierungsrichtung: die beschichteten Teile sind blank geblieben, die Abkühlzeiten lagen je nach Größe der Charge zwischen 10 Minuten und 3,5 Stunden, die Härte war nicht zufriedenstellend (z.B. Stahl 1.2601 26 HRC bis 56 HRC).

Aufgrund dieser Erkenntnisse wurde im Stadium 2 eine Abkühlkammer mit einem Wassermantel gebaut (siehe Bild 11 B, Typ KZ-78).Das innere Rohr wurde aus 3,5 mm Blech gewalzt und konzentrisch zum Außenmantel zusammengeschweißt. Anschließend wurde die Kühlkammer mittels akustischer Prüfmethode (Prüfdruck 0,1 bis 2 bar, Prüfmedium: Luft, kleinste meßbare Undichtheit 1 Torr l/s) und der Blasenbildungsmethode (Prüfdruck 1,5 bis 2 bar, Prüfmedium:

Schaummittel, kleinste meßbare Undichtheit 10^{-3} Torr l/s) auf Dichtheit geprüft. Die Ergebnisse waren positiv, Teile aus den 10 ersten Chargen besaßen die optimale Härte von 60 bis 67 HRC und die Oberfläche war silberglänzend. Beschichtete Teile aus weiteren Chargen wiesen sehr intensive Oxydationsschatten auf den Oberflächen auf, die nur Folge eines konzentrierten Luft- oder Wasserstrahleinbruches in dem Abkühlraum sein konnten. Nach langwierigen Untersuchungen gelang es, mittels Helium-Massenspektrometer (kleinste meßbare Undichtheit 10^{-7} Torr l/s) und einer auf 200°C erwärmten Kühlkammer, Undichtheiten im Bereich der Schweißnaht zu finden. Man stellte fest, daß durch hohe thermische Belastungen beim Herausfahren der Charge im direkten Bereich der Schweißnaht Mikrorisse entstanden sind, durch die Wasserdampf eindrang und zur Oxydation der Hartstoffschicht führte. In der weiteren Phase wurde das geschweißte Innenrohr durch ein nahtloses ersetzt, geschweißt und einer Dichtheitsprüfung mittels Helium-Massenspektrometer unterzogen. Der erforderliche Dichtheitspegel ist erreicht worden; über die Standzeit der Kammer läßt sich zur Zeit noch nichts sagen, gegenwärtig ist sie ca. 2500 Std. im Einsatz. Die Abkühlkammer wurde aus säurebeständigem Stahl gefertigt. An die Abkühlkammer wurde ein pneumatischer Zylinder mit einer Greifvorrichtung zur Einführung und Entnahme des Chargiergutes aus dem Reaktionsraum angebaut. Diese Vorrichtung erlaubt es, eine 450 kg schwere Charge in den Rezipienten einzusetzen. Gegenwärtig wird an einer 3. Konzeption gearbeitet (Bild 11 C, Typ PK-80), in der man Abkühlgeschwindigkeiten $\lambda = 0{,}1$ bis 3 zu erreichen versucht. Es wurde ein Turbo-Kühlmodul entwickelt, welches mit der Härtekammer integriert wird. Als Kühlmedium sind Gase wie Wasserstoff ($\lambda = 0{,}160$ kcal/mh grd), Helium ($\lambda = 0{,}130$ kcal/mh grd), Argon ($\lambda = 0{,}015$ kcal/m h grd) und Stickstoff ($\lambda = 0{,}022$ kcal/m h grd) vorgesehen.
Das Abkühlen soll im Druckbereich 50 mbar bis 10 bar möglich sein.

g) Vakuum-Pumpensystem

Das Pumpensystem besteht aus:

1. Flüssigkeitsringpumpe
2. zwei Drehschiebepumpen
3. Durchgangsventile
4. Sicherheitsventile
5. Belüftungsventile
6. Bourdon-Feder-Vakuummeter 0 bis 1000 mbar
7. Feinmeßmanometer.

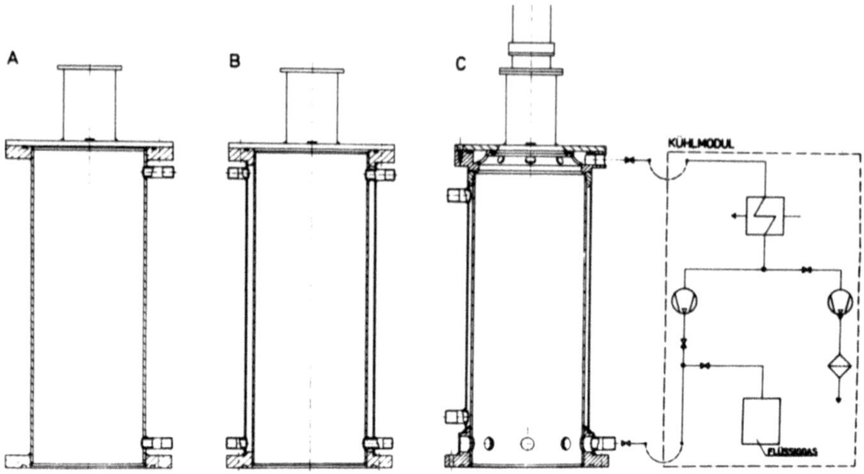

Bild 11 Härtekammer A-Typ KZ-77 (ohne Wassermantel)
B-Typ KZ-78 (mit Wassermantel)
C-Typ PK-80 (mit Turbo-Kühlmodul)

Flüssigkeitsringpumpen werden meist mit Wasser betrieben. In unserem Fall bot sich die direkte Neutralisierung der sich in den Abgasen befindenden Salzsäure durch Austausch des Wassers gegen Natronlauge an. Die Maßnahme führte zu einer Senkung des HCl-Pegels in den Abgasen von 50 ppm auf 0,1 ppm. Die zulässige maximale Arbeitskonzentration beträgt 5 ppm.

h) Verflüchtigungseinheiten

Für die Herstellung von Hartstoffschichten müssen entsprechende Metalle in die Gasphase überführt werden. Die Ausgangssubstanzen können reine Metalle oder ihre Halogenverbindungen in flüssiger oder fester Form sein. Der physikalische Zustand der Ausgangssubstanz determiniert die Art der Verflüchtigungseinheit, und zwar werden bei flüssigen Halogeniden eingesetzt:

a) Glasverdampfer
b) indirekte Einspritzung
c) direkte Einspritzung

bei festen Halogeniden und Metallen:

d) direkte Verflüchtigung im Rezipienten

e) außerhalb des Rezipienten

h.a.) Glasverdampfer (Bild 12)

Die älteste bekannte Art der Verdampfung einer Flüssigkeit ist das direkte Aufkochen in einem geschlossenen Behälter. Sie kann durch definierte Wärmezuführung sehr exakt gesteuert werden. Der Dampf wird mittels Transportgase verdünnt und in den Reaktionsraum durch eine Gasdusche eingeführt (siehe Bild 9, Pos. 8). Der Druck und die Menge werden auf der Strecke zwischen dem Verdampfer und dem Rezipienten kontinuierlich überwacht. Für CVD-Beschichtungesprozesse werden wegen der hohen Toxizität von Halogenverbindungen hermetisch geschlossene Gefäße benutzt.

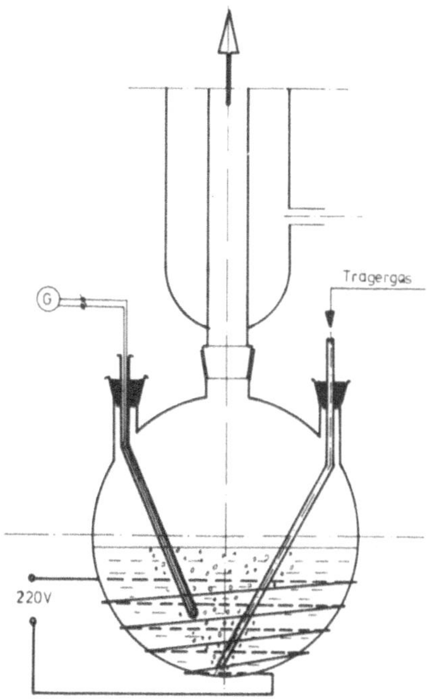

Bild 12 Glasverdampfer

h.b.) Indirekte Einspritzung (Bild 13)

Unter indirekter Einspritzung ist das kontinuierliche Einspritzen und Verdampfen von Flüssigkeiten in einem Glaskolben außerhalb des Rezipienten zu verstehen. Die Verdampfungseinheit wurde mit der Gasdosiereinheit integriert und in einen Steuerschrank eingebaut. Sie besteht aus: Dosierpumpe, Leitungen, Dosierzwischenventilen, Thermostaten, Glaskolben, Temperaturmeßgeräten, Heizpilz und einem Manometer. Die Nachteile dieser Methode sind:

- sehr lange Zuführungsleitungen
- Kondensierung der Dämpfe in den Leitungen
- Beheizung der Leitungen
- Dichtheit der Leitungen.

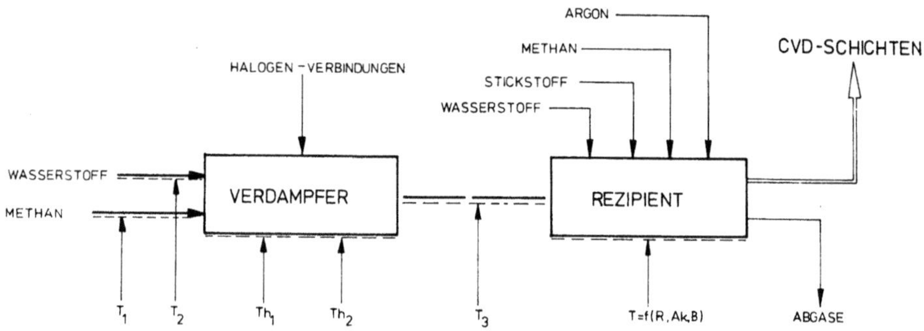

T = Heizband$_{1,2...}$ Ak = Aufkohlen

Th... = Thermostat$_{1,2...}$ B = Beschichten

R = Reduzieren

Bild 13 Schema der indirekten Einspritzung und Verdampfung von Halogenid-Verbindungen in einer CVD-Anlage

h.b. Direkte Einspritzung (Bild 14)

Unter direkter Einspritzung von Flüssigkeiten in einen CVD-Rezipienten ist eine Einheit zu verstehen, die folgenden Voraussetzungen entspricht:

- maximale Entfernung der Einspritzkammer vom Reaktor 100 cm
- direkte Verbindung der Einspritzkammer mit der Gasdusche
- maximale Temperaturdifferenz zwischen Einspritzkammer und Gasdusche 500°C
- direktes Gasmischsystem
- steuerbare Einspritzmengen mit einer Genauigkeit von $\pm$ 0,05%
- Gewährleistung der Vakuumbedingungen.

Die zur Verdampfung bestimmte Flüssigkeit wurde mittels einer sehr genauen Pumpe und sehr präzisen Dosierzwischeneinheiten in konstanten Zeitabständen in eine am Rezipienten befindliche Turbulenzkammer eingeführt. Die minimale Dosiermenge lag bei 0,05 ml, der Zeitabstand bei 0,5 s.

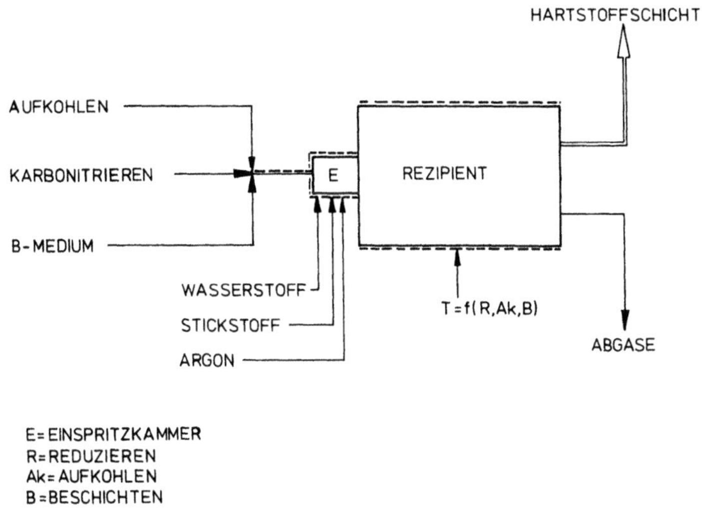

Bild 14 Schema einer direkten Einspritzung und Verdampfung von Halogenverbindungen in einer CVD-Anlage

h.d.) Direkte Verflüchtigung im Rezipienten

Bei dieser Methode werden reine Metalle oder ihre Ferro-Legierungen in den Rezipienten eingesetzt und mittels Chlorwasserstoff bei entsprechender Temperatur in Halogenverbindungen übergeführt. So entstehen feste und flüssige Halogenverbindungen und durch weitere Temperaturerhöhung verdampfen sie mit dem zu beschichtenden Substrat. Parameter wie Temperatur, Druck und Strömungsgeschwindigkeit bestimmen die Verflüchtigungsgeschwindigkeit und die Reaktion mit dem zu beschichtenden Substrat.

h.e.) Direkte Verflüchtigung außerhalb des Rezipienten

Es wurde eine zweite Methode entwickelt, mittels deren sich reine Metalle und bis zu 850°C verdampfende Halogenide in die Gasphase überführen lassen. Es wurde ein spezieller Verdampfer konstruiert, gefertigt und erprobt. Die Führung des Verdampfungsprozesses ist unter Über-, Normal-, Unterdruckbedingungen und Temperatur von 20 bis 850°C möglich. Die Konstruktion des

Verdampfers ermöglicht eine direkte Temperaturmessung des Verdampfungsprozesses und seine Führung. Transportgase werden tangential zur Innenwand des Verdampfers eingeführt, um eine intensive Durchmischung mit entstehenden Dämpfen zu gewährleisten, und durch ein konzentrisches Rohr in die Turbulenzkammer der CVD-Anlage abgeführt.

XI. Beschichtungsversuche

a) Titancarbid

a.a. Eigenschaften

Titancarbid besitzt ein kubisch-flächenzentriertes Gitter, Typ B I mit Gitterkonstanten von 4,280 Å bis 4,316 Å, die charakteristisch für die Phasen $TiC_{1,5}$ und $TiC_{0,22}$ sind. Die maximale Härte beträgt 3200 HV, der Wärmeausdehnungskoeffizient nimmt Werte von 6,7 bis 7,42 x 10^{-6} an, die Wärmeleitfähigkeit beträgt 0,05 cal/cm sC°, Kennzahl für den Adhäsionswiderstand bei TiC gegen GGL 15 (5,1 - 9,9) x 10 N/cm² mg und TiC gegen Cf 53 7,8 x 10 N/cm² mg.

Titancarbid fängt schon bei 450°C an zu oxydieren, die obere Oxydationsgrenze liegt bei 1100°C. TiC ist in Salpetersäure - Flußsäure leicht löslich, wird von Salzsäure und Schwefelsäure kaum angegriffen, zersetzt sich im Vakuum oberhalb 1000°C.

Der Reibungskoeffizient nimmt für folgende Paare die Werte an:

Reibpaarungen	Reibkoeffizient µ geschmiert	trocken
Stahl/Stahl	0,09	0,70
TiC/Al_2O_3	0,10	0,14
$Stahl/Al_2\ O_3$	0,09	0,60
TiC/Stahl	0,10	0,15

a.b. Ausgangsmaterialien

Für die Abscheidung von Titancarbid auf Schnellstählen in der Zweikammer-Vakuum-CVD-Anlage sind folgende Materialien gewählt worden:

1. Titantetrachlorid (flüssig) 99,953%
2. Methan (CH_4) 99,950%
3. Wasserstoff (H_2) 99,993%

4. Argon (Ar) 99,998%

a.c. Abscheidebedingungen

Abscheidedruck: Zwischenvakuum

Abscheidetemperatur: von 950°C bis 1700°C

Abscheidezeit: 2 bis 3,5 Stunden

Gaszusammensetzung: von $TiCl_4 : CH_4 : H_2 = A : 1 : 200)$
bis $(TiCl_4 : CH_4 : H_2 = A : 1 : 500)$

a.d. Titantetrachloridverflüchtigung

In die Zweikammer-Vakuum-CVD-Anlage wurde das Titantetrachlorid mittels Glasverdampfer und direkter Einspritzung eingeführt. Bessere Beschichtungsergebnisse brachte die Anwendung der direkten Einspritzung.

a.e. Versuchsmaterial

Folgende Werkzeuge wurden als Versuchsmaterial zugrundegelegt:

1. Wendeschneidplatten (siehe Bild 15) - 12,7 x 12,7 x 3,2 in einer Anzahl von 500 Stck., wurden aus einem Stab 15 x 15 x 5000 geschnitten und auf eine definierte geometrische Form geschliffen und poliert.

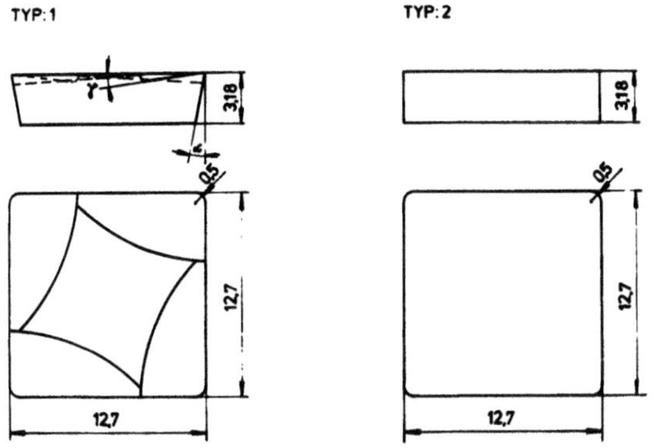

Bild 15 Versuchswendeschneidplatten Typ 1 und Typ 2

2. Schübeplatten - 44 x 29 x 5 in einer Anzahl von 130 Stck., wurden aus einem Flachstahl 40 x 6 x 6000 gesägt und auf vorgeschriebene Form geschliffen und poliert.

Wendeschneidplatten und Schübeplatten wurden hergestellt aus den Werkstoffen:

1.3207	S 10-4-3-10	
1.3222	S 6-5-2-8	
1.3247	S 2-10-1-8	
1.3265	S 18-1-2-10	
1.3343	S 6-5-2	
1.3344	ASP 30	pulvermetallurgisch hergestellt

Wie schon vorher zu ersehen war, entspricht der Hochleistungsschnellstahl 1.3265 dem Eleminationskriterium 3 nicht, ist aber aus Versuchsgründen in den Beschichtungsplan einbezogen worden.

3. Bohrer (Typ N nach DIN 338) 5 h 8 - in einer Anzahl von 200 Stck., davon wurden 100 Bohrer aus dem Stahl S 6-5-2 (1.3343) und 100 Stck. Bohrer aus dem Stahl S 6-5-2-8 (1.3222) gekauft und Probebeschichtungen unterzogen.

a.f. Versuchsdurchführung

Alle zum Beschichten vorgesehenen Proben wurden im Perbad entfettet. Nachher wurden sie einer entsprechenden Chargiervorrichtung zugeordnet, in die Härtekammer eingesetzt, die Härtekammer mit dem Chargiergut auf den Rezipienten aufgesetzt und evakuiert. In der gleichen Zeit wurde der Rezipient durch mehrmaliges Spülen und Evakuieren auf die Aufnahme der Charge vorbereitet. Nach dem Abschluß der Vorbereitungsoperationen wurden der Härtekammer- und Rezipientenschieber geöffnet, das Chargiergut im kalten Rezipienten abgesetzt, der Rezipient geschlossen und die Proben nach dem im Bild 16 dargestellten Temperatur-Zeitdiagramm beschichtet. Das Versuchsmaterial wurde im Vakuum bei 450°C angewärmt, bei 750°C bis 850°C vorgewärmt und in der 2. Vorwärmestufe beschichtet. Im Bereich von 850°C bis 1000°C wurden an den Probenoberflächen entstehende Oxyde mittels Wasserstoff reduziert und im Fall eines zu niedrigen Kohlenstoffpegels anschließend aufgekohlt. Nach dem Beschichten wurde das Probenmaterial auf die Härtetemperatur gebracht, gehärtet und anschließend aus der Härtehitze in die Härtekammer eingeführt und abgeschreckt. Die Proben konnten direkt angelassen und in die CVD-Anlage für den nächsten Beschichtungsprozeß vorbereitet werden.

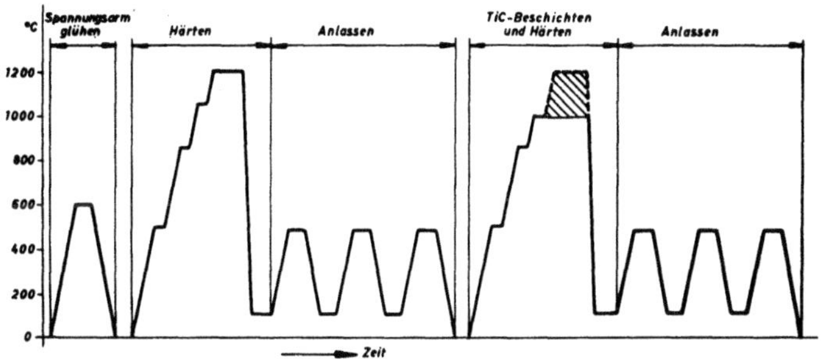

Bild 16 Temperatur-Zeit Diagramm

a.g. Beschichtungsergebnisse

Es wurden insgesamt 30 TiC-Beschichtungsversuche durchgeführt. Die Schichtdicken variierten bei:

1.3207	S 10-4-3-10	zwischen 4 und 6 µm
1.3222	S 6-5-2-8	zwischen 8 und 10 µm
1.3247	S 2-10-1-8	zwischen 8 und 12 µm
1.3265	S 18-1-2-10	zwischen 8 und 11 µm
1.3343	S 6-5-2	zwischen 8 und 11,5 µm
1.3344	ASP 30	zwischen 7 und 10 µm

Die Schichten sind homogen (siehe Bild 17) mit sehr guter Haftfestigkeit (siehe Bild 21). In Bild 21 ist die sehr stark während des Beschichtungsprozesses durch Chlorionen angeätzte Substratoberfläche und Titancarbidschicht dargestellt. Das Titancarbid füllt jede Unebenheit und Mikrolöcher so aus, (siehe Bild 17), daß eine gute Verzahnung zwischen der Hartstoffschicht und dem Substrat vorhanden ist. In dieser Schicht wurde Titancarbid mit einem Gitterparameter von a = 4,360 Å und einer Vorzugsorientierung (220) (siehe Bild 23), und einer Härte von 4600 HV gemessen. Mittels Sin^2-verfahren wurden Eigenspannungen in den TiC-Schichten parallel und quer zur Strömungslinien(siehe Bild 24) untersucht. Als Meßebene wurde die Ebene

Titancarbid

Bild 17

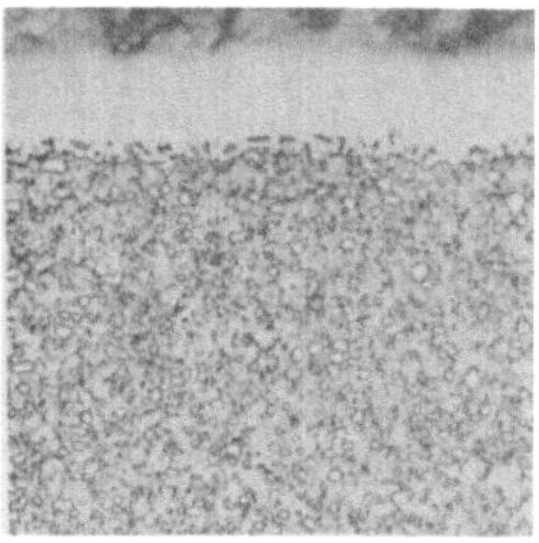

Schichtdicke : 6 - 8 µm
Schichthärte : 3200 HV
Schichtträger: 1.3343
Ätzmittel: HNO_3
V : 500 : 1

Titancarbid + Titancarbonitrid

Bild 18

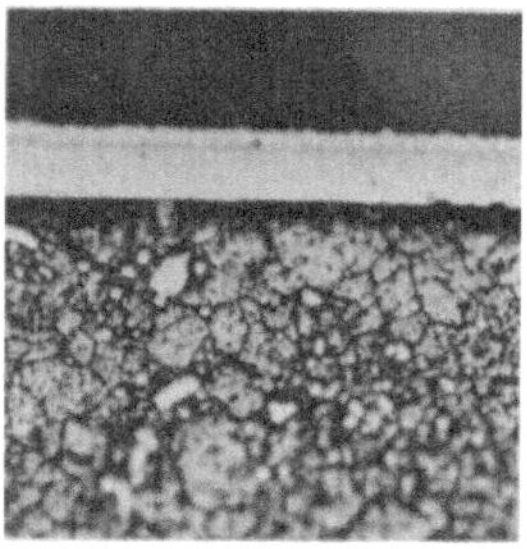

Schichtdicke: 8 - 11 µm
Schichthärte: 2700 HV
Schichtträger: 1.3343
Ätzmittel: $FeCl_2$
V : 500 : 1

Bild 19

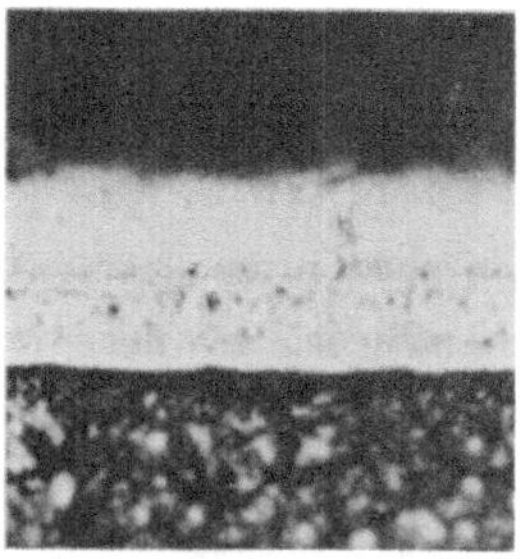

Schichtdicke : 18 µm
Schichthärte : 2450 HV
Schichtträger: 1.3343
Ätzmittel: $FeCl_2$
V : 1000 : 1

Bild 20

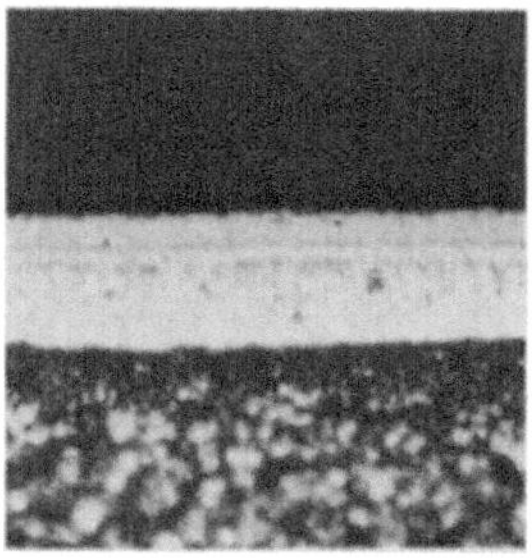

Schichtdicke : 8 - 12 µm
Schichthärte : 2700 HV
Schichtträger : 1.3343
Ätzmittel: $FeCl_2$
V : 1000 : 1

Titancarbid

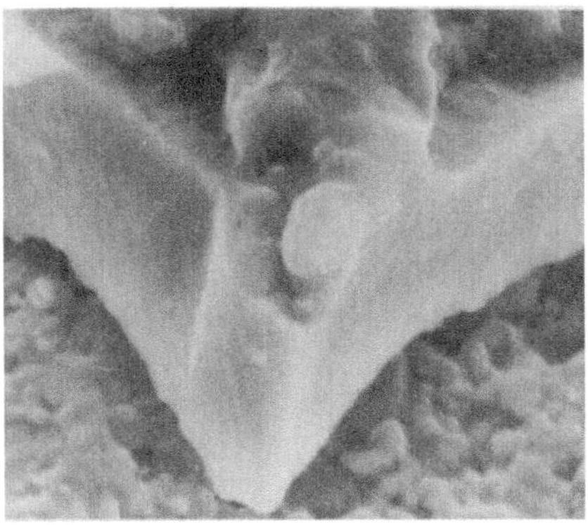

Bild 21

REM-Bruchfläche
U = 25 kV
Kippwinkel: 40°
V : 5000:1
Sekundärelektronen

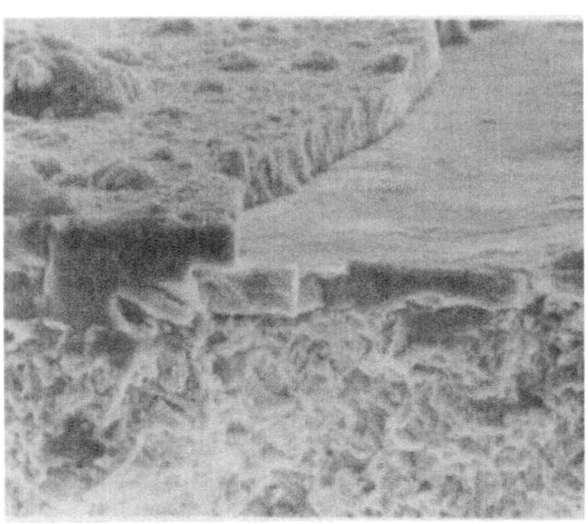

Bild 22

REM-Bruchfläche
U = 25 kV
Kippwinkel: 0°
V : 1000:1
Sekundärelektronen

(220) gewählt und die Messung mittels Co-K_α-Strahlung durchgeführt. Der röntgenographisch beleuchtete Fleck war elliptisch mit einer Höhe von 1,75 mm bei ψ = 0 und einer Breite von 2,9 mm bei 2Θ = 67°. Die Messungen wurden mit ψ-Pendelungen durchgeführt ($\Delta\psi = \pm$ 3°,5 Pendelungen). Es wurden folgende Einspannungswerte gemessen:

Meßstelle	Druckspannungen (N/mm²) parallel zur Strömungsrichtung	Druckspannungen (N/mm²) quer	Substrat
1	-2134,83 ± 80,63	-1866,21 ± 77,52	S 10-4-3-10
2	-2172,04 ± 86,97	-1941,04 ± 63,75	S 10-4-3-10
3	-2213,59 ± 75,01	-2074,45 ± 34,44	S 10-4-3-10
4	-2281,17 ± 26,41	-2095,61 ± 75,86	S 10-4-3-10
5	-2595,06 ± 121,34	-2448,11 ± 133,51	SC 6-5-2

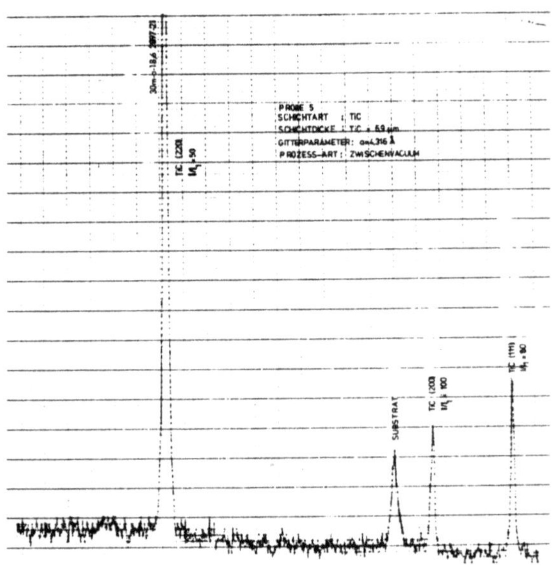

Bild 23 Röntgenographische Phasenanalyse

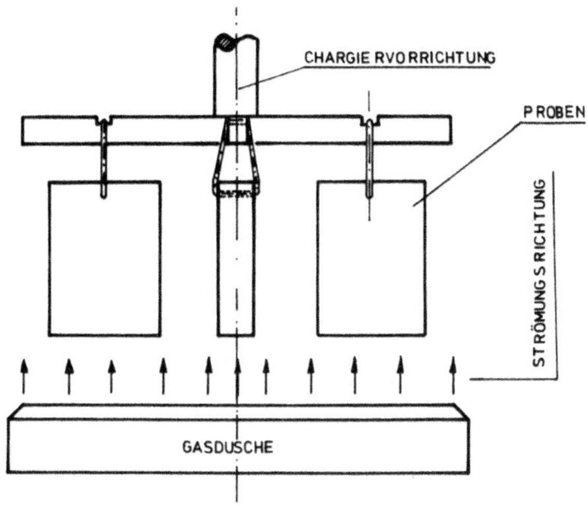

Bild 24 Strömungsrichtung in einer CVD-Anlage

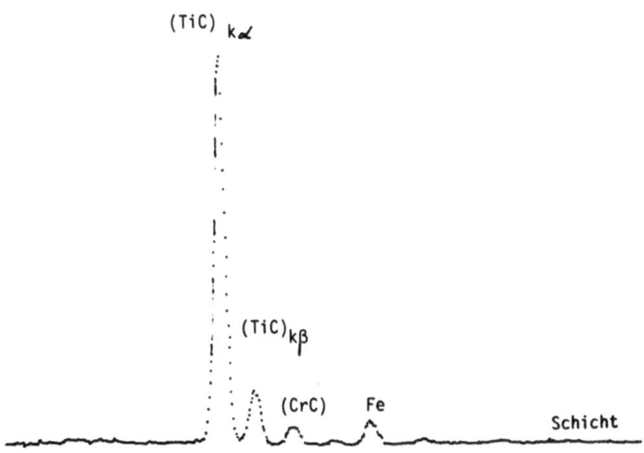

Bild 25 Energiedispersive Röntgenanalyse einer TiC-Schicht

Ein Drittel aller beschichteten Proben wurden einer energiedispersiven Röntgenanalyse unterzogen. Die Ergebnisse erlaubten eine ständige Optimierung der Beschichtungsparameter in der Zweikammer-CVD-Anlage (siehe Bild 25)

b. Titancarbonitrid

b.a. Eigenschaften

Die Farbe der Schicht ist von der Relation von Kohlenstoff zu Stickstoff im Titancarbonitrid abhängig und kann von grau bis zu goldgelb sein. Der thermische Ausdehnungskoeffizient kann Werte von 7,5 bis 8,5 x 10^{-6} je nach Stickstoffsättigung annehmen. Die Oxydationsfähigkeit ist von der TiCN-Phase abhängig, z.B. $TiC_{0,28} N_{0,63}$ fängt bei 600°C, $TiC_{0,4} N_{0,5}$ bei 480°C und $TiC_{0,8} N_{0,12}$ bei 450°C zu oxydieren an. Das Titancarbonitrid ist in Salpetersäure, Schwefelsäure und Salzsäure nicht löslich, die Beständigkeit gegen Alkalien und Königswasser ist sehr gering. Im Vakuum zersetzt es sich oberhalb 1000°C.

b.b. Ausgangsmaterialien

Die Grundmaterialien für die Abscheidung von Titancarbonitrid sind:

1. Titantetrachlorid	($TiCl_4$)	99,953%
2. Methan	(CH_4)	99,950%
3. Wasserstoff	(H_2)	99,993%
4. Stickstoff	(N_2)	99,999%
5. Argon	(Ar)	99,998%

b.c. Abscheidebedingungen

Abscheidedruck:	Grobvakuum
Abscheidetemperatur:	von 85°C bis 1000°C
Abscheidezeit:	2 bis 8 Stunden
Gaszusammensetzung:	von ($TiCl_4$: CH_4 : H_2 : N_2 = A : 0,30: 2,5 :40) bis ($TiCl_4$: CH_4 : CH_4: H_2 : N_2= A: 1: 8: 120)

b.d. Titantetrachlorid-Verflüchtigung

siehe a.d.

b.e. Probenmaterial

siehe a.e.

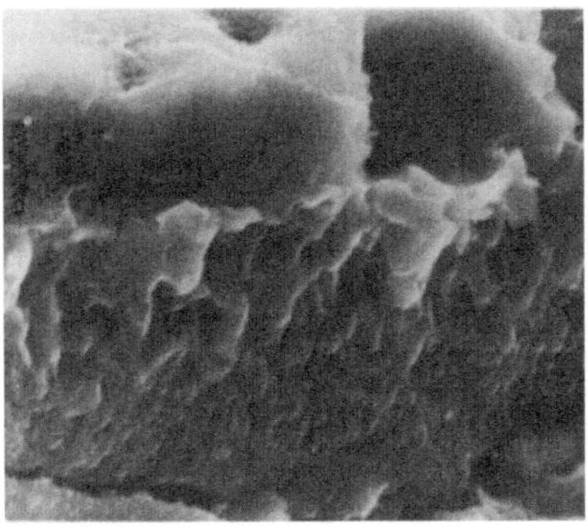

Bild 26

REM-Bruchfläche
U = 25 kV
Kippwinkel: 60°
V : 1000:1
Sekundärelektronen

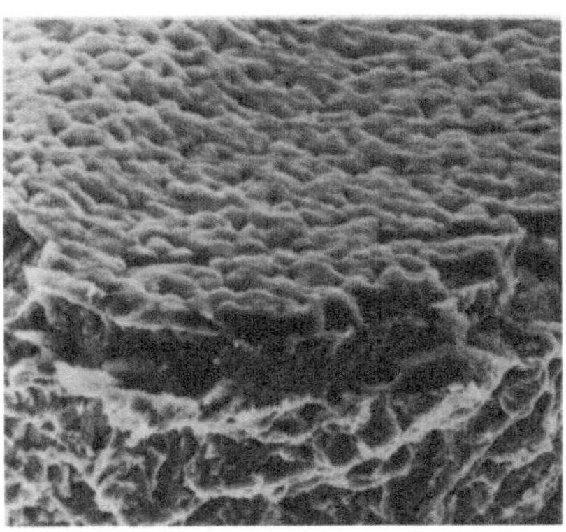

Bild 27

REM-Bruchfläche
U = 25 kV
Kippwinkel: 60°
V : 1000:1
Sekundärelektronen

b.f. Versuchsdurchführung

Das Titancarbonitrid wird meistens als Verbundschicht zwischen Titancarbid und Titannitrid benutzt, seltener als echte Verschleißschicht. Titancarbonitrid wird immer auf Titancarbid als 2. Schicht abgeschieden und führt zur Erweiterung des in Punkt a.e. beschriebenen Titancarbidabscheideprozesses und folgt den Schritten:
Chargenzusammenbau - Chargeneinsatz - Anwärmen (450°C bis 600°C) - Vorwärmen (850°C) - Titancarbidbeschichten (1000°C bis 1050°C) - Titancarbonitridbeschichten (1050°C bis 850°C) - Abschrecken - Anlassen - Chargenauseinanderbau.

b.g. Beschichtungsergebnisse

Die Schichtdicken lagen zwischen 3 und 22 µm. Die Schichten (siehe Bilder 18, 20, 26) sind homogen und gut mit dem Titancarbid zusammengewachsen. Die Oberfläche (siehe Bild 27) ist im Vergleich zu Titancarbid (siehe in Bild 22 die Oberfläche der Zwischenschicht) um 1 - 3 µm rauher und läßt sich mittels Diamantpaste sehr leicht polieren. Die im Zwischenvakuum abgeschiedene Carbonitridschicht wurde mittels Röntgen-Feinstruktur-Difraktometer untersucht und man stellte fest, daß die Gitterparameter Werte (je nach Prozeßführung) von a = 2,244 Å und a = 2.248 Å annähmen. Die Vorzugsorientierung war die (220)-Ebene (siehe Bild 28).

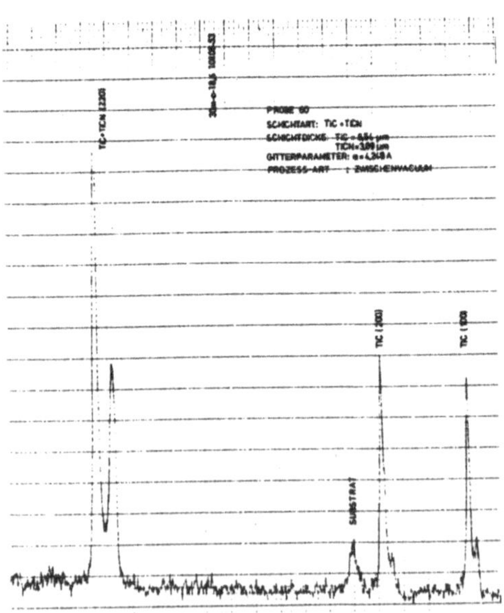

Bild 28
Röntgenographische Phasenanalyse einer TiC-TiCN Schicht

c. Titannitrid

c.a. Eigenschaften

Titannitrid wird mittels CVD-Verfahren als bronzebraune bis zur gelbgoldenen Schicht abgeschieden. Der Homogenitätsbereich liegt zwischen $TiN_{0,25}$ bis $TiN_{1,16}$ und besitzt entsprechende Gitterparameter von 4,213 Å bis 4,234 Å. Das Titannitrid besitzt ein kubisch-flächenzentriertes Gitter, Typ B I, eine Dichte von 5,432 g/cm³, weist eine Härte von 2450 HV auf, besitzt einen Wärmeausdehnungskoeffizient von 9,35 x 10^{-6}, Schmelzpunkt von 2950 ± 50°C und ist im Vakuum bei 1000°C zersetzlich.

c.b. Ausgangsmaterialien

Die Abscheidung von Titannitrid (TiN) auf die Titancarbid oder Stählen mit einem niedrigen Kohlenstoffgehalt ist nur in Gegenwart folgender Reagenzien möglich:

1. Titantetrachlorid	(flüssig)	99,953%
2. Stickstoff	(N_2)	99,999%
3. Wasserstoff	(H_2)	99,993%

c.c. Abscheidebedingungen

Abscheidedruck:	Zwischenvakuum
Abscheidetemperatur:	von 850°C bis 1000°C
Abscheidezeit:	2 bis 8 Stunden
Gaszusammensetzung:	von ($TiCl_4 : H_2 : N_2$ = A : 2,0 : 40)
	bis ($TiCl_4 : H_2 : N_2$ = A : 8 : 150)

c.d. Titantetrachlorid-Verflüchtigung

Wurde mittels Glasverdampfer und Direkteinspritzung durchgeführt.

c.e. Probenmaterial

Siehe a.d.

c.f. Versuchsdurchführung

Wie in Punkt b.d., erweitert durch den Titannitrid-Abscheidprozeß.

c.g. Beschichtungsergebnisse

Es sind insgesamt 15 Versuchschargen gefahren worden, die Schichtdicken lagen (je nach Prozeßlänge) zwischen 1,5 µm und 12 µm. Die maximale Schichthärte betrug 2520 HV und der Gitterparameter war a = 4.238 Å. Die Schichten besaßen eine goldgelbe Farbe, waren homogen und gut mit Titancarbonitrid verwachsen (siehe Bild 19 und 20). Die in Bild 19 und 20 präsentierten Schichten bestehen aus Titancarbid ca. 8 µm, Titancarbonitrid ca. 1,5 µm und Titannitrid ca. 3 µm. Diese Schichten wurden einer energiedispersiven Röntgenanalyse unterzogen (siehe Bild 29).

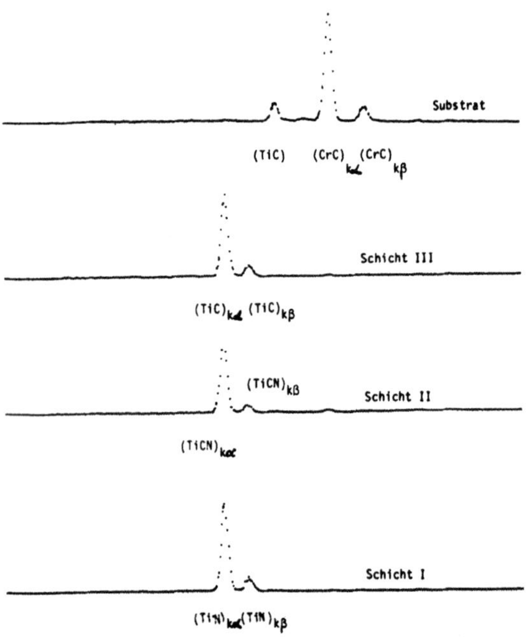

Bild 29
Energiedispersive Röntgenanalyse einer TiC-TiCN-TiN-Schicht

Wie aus den Analysen hervorgeht, besitzen die Schichten III und II Chromverunreinigungen, die in der 1. Schicht nicht mehr zu finden waren. Bild 30 dokumentiert die hohe Homogenität der Titannitridschicht und ihre Verbindung mit dem Titancarbonitrid. In Bild 31 wurde die Topographie der Titannitrid-Oberfläche dargestellt.

d. Vanadincarbid

d.a. Eigenschaften

Vanadincarbidschicht wird im CVD-Prozeß als graumetallische Schicht, die aus hexagonalem V_2C und kubischem VC besteht, abgeschieden. Die Dichte des V_2C beträgt 5,62 g/cm² und des VC 5,71 g/cm³. Die Härte von V_2C beträgt 2150 HV und des VC ca. 2944 HV, der Schmelzpunkt liegt für V_2C bei 1850°C und für VC bei 2800°C. Es ist sehr widerstandsfähig gegen alle Säuren, außer HNO_3 und bei Rotglut gegen Gase wie H_2, H_2O, H_2S und HCl. Chlorionen führen bei 500°C zu totaler Zerstörung der VC-Schicht.

Titannitrid

Bild 30

REM-Analyse
Bruchfläche
U = 25 kV
Kippwinkel: 60°
V : 5000:1
Sekundärelektronen

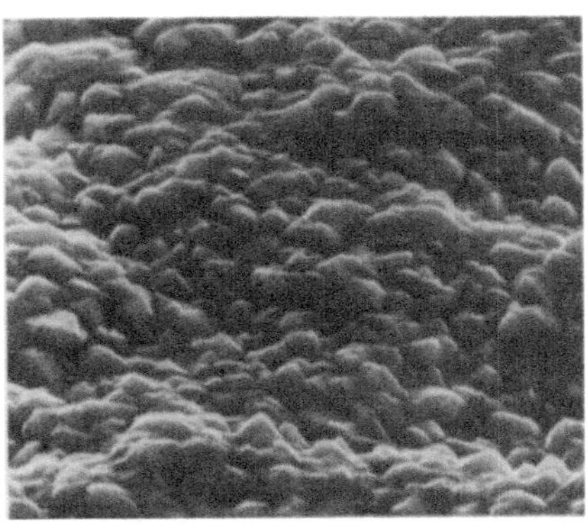

Bild 31

REM-Analyse
Oberfläche
U = 25 kV
Kippwinkel: 60°
V : 3000:1
Sekundärelektronen

d.b. Ausgangsmaterialien

1.	Ferro-Vanadium	(FeV 90)	90,000%
2.	Chlorwasserstoff	(HCl)	99,800%
3.	Wasserstoff	(H_2)	99,993%
4.	Argon	(Ar)	99,998%
5.	Methan	(CH_4)	99,950%

d.c. Abscheidebedingungen

Abscheidedruck:	
	1. Zwischenvakuum
	2. Normaldruck
Abscheidetemperatur:	von 850°C bis 1100°C
Abscheidezeit:	2 bis 3,5 Stunden
Gaszusammensetzung:	von (CH_4 : HCl : H_2 : Ar = 0,12 : 0,65 : 1,9 : 6)
	bis (CH_4 : HCl : H_2 : Ar = 0,9 : 5 : 10 : 16)

d.d. Ferro-Vanadium-Verflüchtigung

Ferro-Vanadium wurde mittels direkter Verflüchtigung außerhalb des Rezipienten, siehe Punkt X. h.e. durchgeführt.

d.e. Probenmaterial

Siehe Punkt a.d.

d.f. Versuchsdurchführung

Siehe Punkt a.e.

d.g. Versuchsergebnisse

Vanadincarbid wurde in der Zweikammer-Vakuum-CVD-Anlage unter Vakuum- und unter Normaldruckbedingungen abgeschieden. Vakuumschichten waren um ca. 27% bis zu 34% dünner als Normaldruckschichten (unabhängig vom Substrat) (siehe Bilder 32 bis 43). Vakuumschichten weisen höhere Homogenität, bessere Haftfestigkeit auf. Beide Schichten werden gegenwärtig sehr intensiv auf ihr tribologisches Verhalten untersucht.

Vanadincarbid

Vakuum

Bild 32

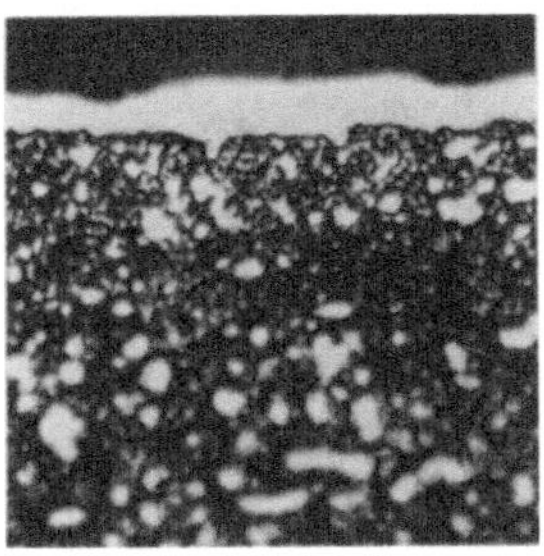

Schichtdicke: 4-4,5 µm
Schichthärte: 2944 HV
Schichtträger: 1.3207
Ätzmittel: HNO_3
V : 1000:1

Normaldruck

Bild 33

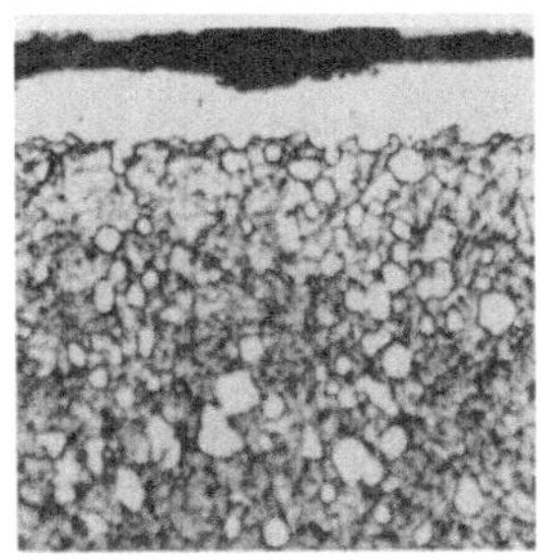

Schichtdicke: 6 µm
Schichthärte: 2340 HV
Schichtträger: 1.3207
Ätzmittel: $FeCl_2$
V : 1000:1

Bild 34

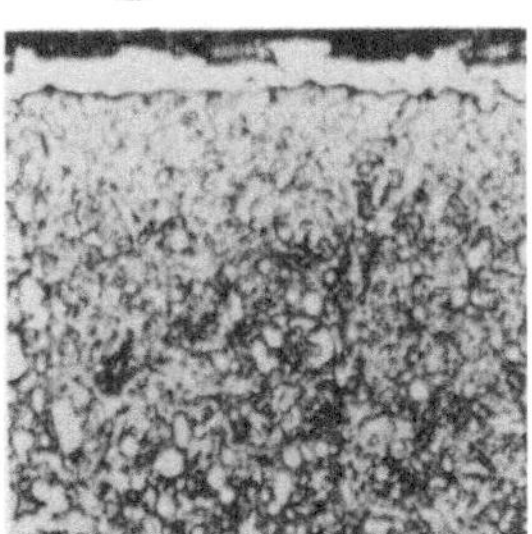

Schichtdicke: 4 µm
Schichthärte: 2944 HV
Schichtträger: 1.3343
Ätzmittel: HNO_3
V : 1000:1

Bild 35

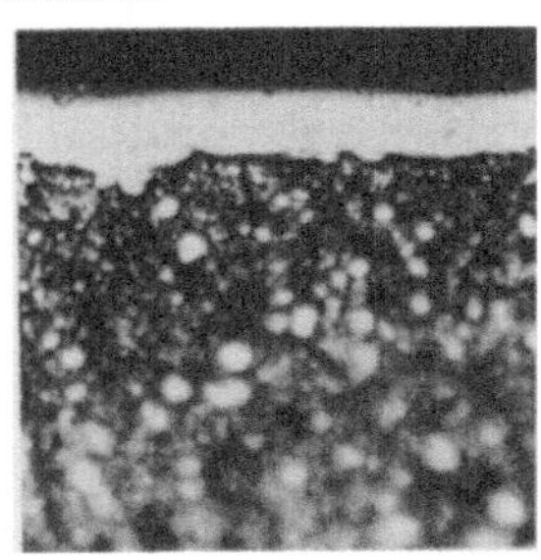

Schichtdicke: 6 µm
Schichthärte: 2340 HV
Schichtträger: 1.3343
Ätzmittel: $FeCl_2$
V : 1000:1

Vanadincarbid

Vakuum

Normaldruck

Bild 36

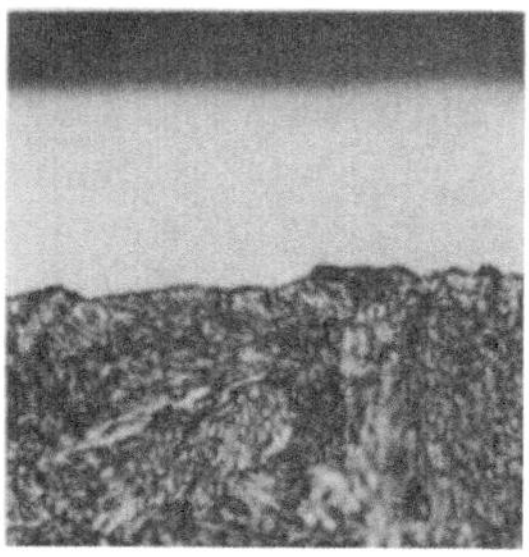

Schichtdicke: 14 µm
Schichthärte: 2944 HV
Schichtträger: 1.2210
Ätzmittel: HNO_3
V : 1000:1

Bild 37

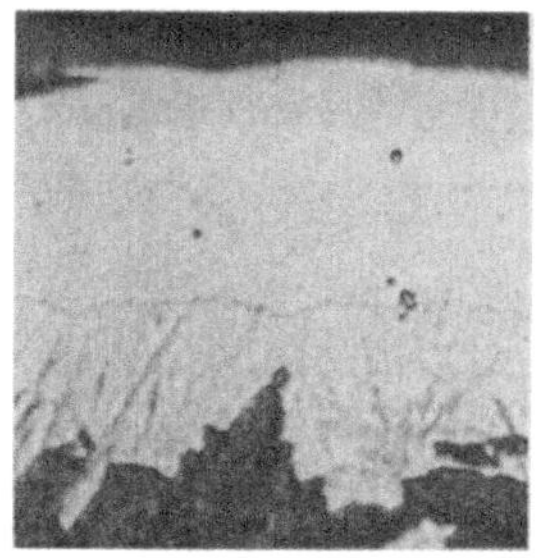

Schichtdicke: 20 µm
Schichthärte: 2340 HV
Schichtträger: 1.2210
Ätzmittel: HNO_3
V : 1000:1

Bild 38

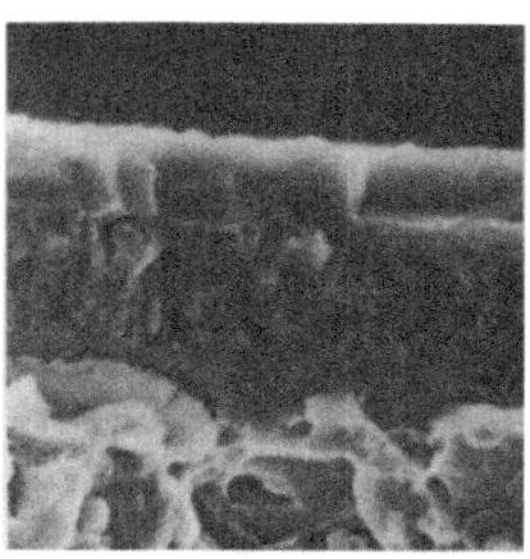

REM-Bruchfläche
U = 25 kV
Kippwinkel: 0°
V : 2000:1
Sekundärelektronen

Bild 39

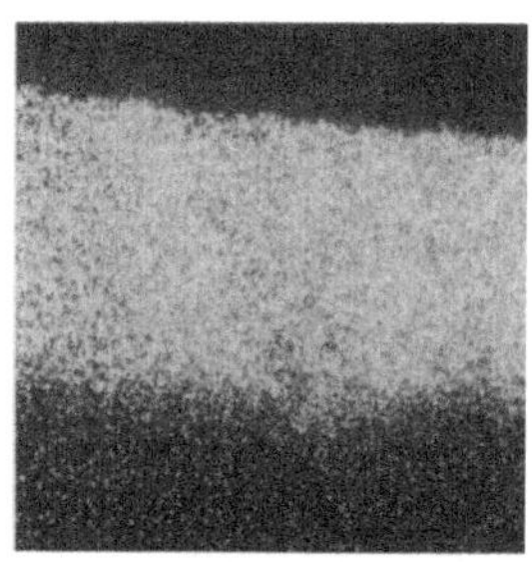

Vanadincarbid-Verteilung
U = 25 kV
Kippwinkel: 0°
V : 2000:1
Roentgen

Vanadincarbid

Vakuum

Bild 40

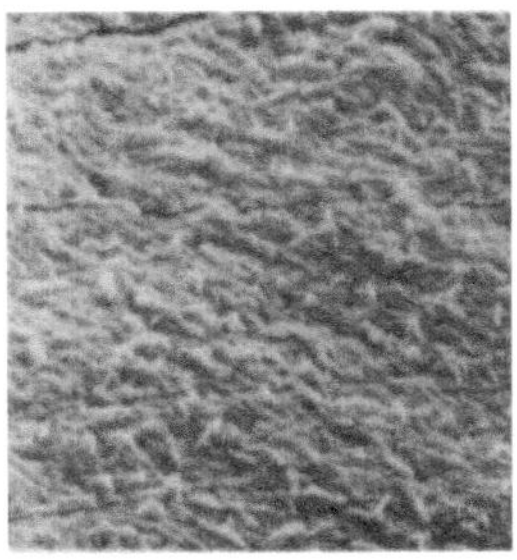

REM-Oberfläche
U = 25 kV
Kippwinkel: 0°
V : 1000:1
Sekundärelektronen

Normaldruck

Bild 41

REM-Oberfläche
U = 25 kV
Kippwinkel: 0°
V : 2000:1
Sekundärelektronen

Bild 42

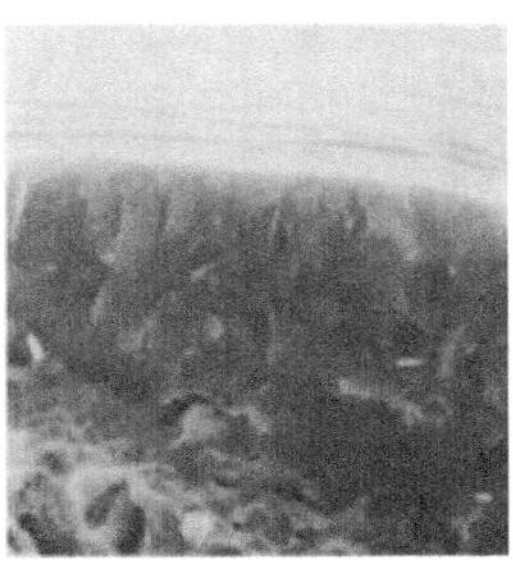

REM-Bruchfläche
U = 25 kV
Kippwinkel: 0°
V : 1000:1
Sekundärelektronen

Bild 43

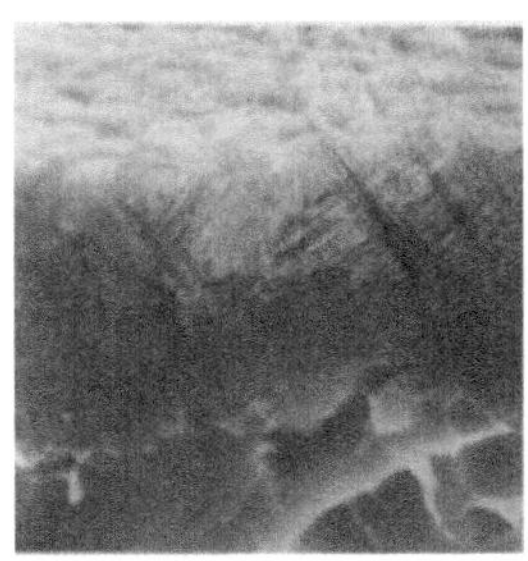

REM-Bruchfläche
U = 25 kV
Kippwinkel: 0°
V : 5000:1
Sekundärelektronen

Es sind folgende Schichtdicken nach drei Beschichtungsstunden erreicht worden:

Werkstoff		Vakuum-Schichten	N.-Druck-Schichten
1.3207	S 10-4-3-10	3 - 7 µm	4 - 8 µm
1.3247	S 2-10-1-8	5 - 7 µm	6,5 - 8,5 µm
1.3265	S 18-1-2-10	1 - 2,5 µm	4 - 3,5 µm
1.3343	S 6-5-2	3 - 4 µm	4 - 5,5 µm
ASP 23	Pulver-metallurgie	4 - 5 µm	4,5-5,5 µm

e. Niobcarbid

e.a. Eigenschaften

Niobcarbidschichten besitzen ein graubraunes metallisches Aussehen, technische Bedeutung haben hexagonales Nb_2C und kubisches flächenzentriertes NbC erlangt. Die Gitterkonstante von NbC kann Werte von 4,432 bis 4,472 Å je nach Prozeßführung annehmen. Die Härte des Nb_2C beträgt 2123 HV und des NbC 2400 HV, der Elastizitätsmodul beträgt 34500 kg/mm² und der Wärmeausdehnungskoeffizient $6{,}65 \cdot 10^{-6}$. Das NbC ist sehr wiederstandsfähig gegen alle Säuren, verbrennt an der Luft schon bei 700°C, es ist beim Erhitzen unter Wasserstoff bis zu höchsten Temperaturen beständig.

e.b. Ausgangsmaterialien

1. Ferro-Niob	(FeNbR)	(67%- 70%) Nb
2. Methan	(CH_4)	99,950%
3. Chlorwasserstoff	(HCl)	99,800%
4. Wasserstoff	(H_2)	99,993%
5. Argon	(Ar)	99,998%

e.c. Abscheidebedingungen

Abscheidedruck:	1. Zwischenvakuum
	2. Normaldruck
Abscheidetemperatur:	von 850°C bis 1100°C
Abscheidezeit:	von 2 bis 4 Stunden
Gaszusammensetzung:	von (CH_4 : HCl : H_2: Ar = 0,105 : 0,65 : 2,1 : 6)

bis (CH_4 : HCl : H_2 : Ar = 0,8 : 2,5 : 16 : 50)

e.d. Ferro-Niob-Verflüchtigung

Ferro-Niob wurde mittels direkter Verflüchtigung im und außerhalb des Rezipienten durchgeführt. Der Wirkungsgrad der Verflüchtigung im Rezipienten liegt im Bereich von 25 - 32%.

e.f. Probenmaterial

Siehe Punkt a.d.

e.g. Versuchsdurchführung

Siehe Punkt a.e.

e.h. Versuchsergebnisse

Unten stehende Stähle wurden unter Vakuum- und Normaldruck beschichtet und ergaben folgende Schichtdicken:

Werkstoffe	Vakuum	Normaldruck
1.3207 S 10-4-3-10	2 - 6 µm	3 - 6 µm
1.3247 S 2-10-1-8	5 - 7 µm	6,5 - 8,5 µm
1.3265 S 18-1-2-10	1 - 2,5 µm	1,5 - 3,0 µm

Niobcarbid-Vakuum- und Normaldruckschichten sind homogen (siehe Bild 44 bis 55), besitzen eine hohe Haftfestigkeit, weisen eine Härte von 2400 HV auf.

Es wurde mittels Phasenanalyse festgestellt, daß die Gitterkonstanten von Vakuum- und Normalsdruckschichten unter der definierten Prozeßführung gleich sind. Der Wert beträgt für:

1. $NbC_{(Vakuum)}$ NbC(Fm3m a = 4,470 Å)

2. $NbC_{(Normaldruck)}$ NbC(Fm3m a = 4,470 Å)

Die Vorzugsorientierung der Karbide ist (222) und (111). Es werden gegenwärtig weitere Untersuchungen auf diesem Gebiet geführt.

Niobcarbid

Vakuum

Bild 44

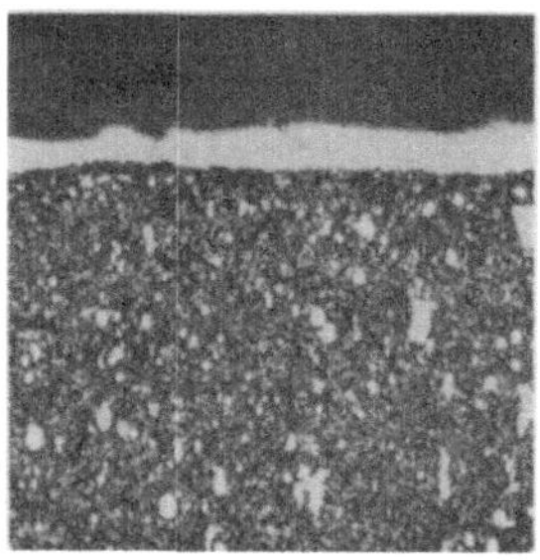

Schichtdicke : 6 µm
Schichthärte : 2250 HV
Schichtträger: 1.3222
Ätzmittel: $FeCl_2$
V : 500 : 1

Normaldruck

Bild 45

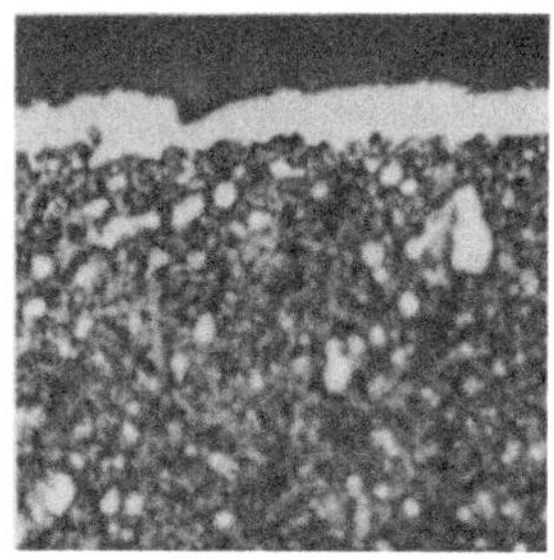

Schichtdicke : 6 µm
Schichthärte : 2260 HV
Schichtträger: 1.3202
Ätzmittel: $FeCl_2$
V : 500 : 1

Vakuum

Bild 46

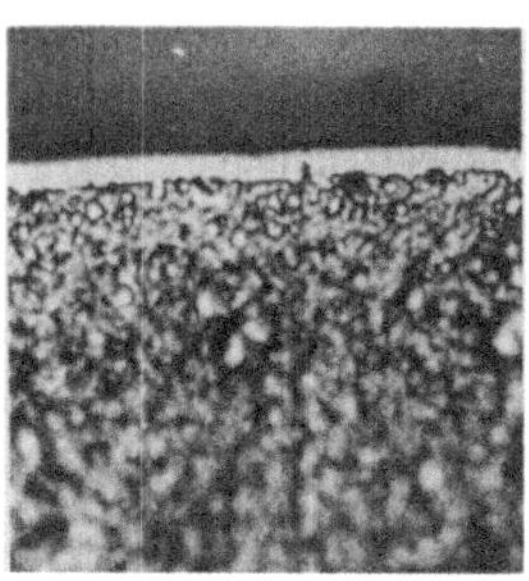

Schichtdicke : 6 µm
Schichthärte : 2240 HV
Schichtträger: 1.3343
Ätzmittel: $FeCl_2$
V = 500 : 1

Normaldruck

Bild 47

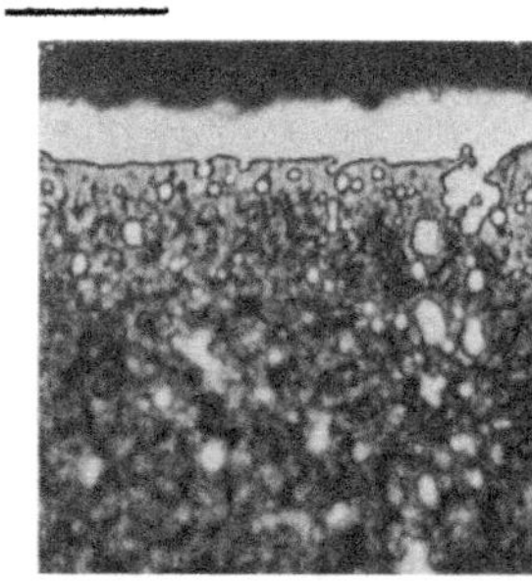

Schichtdicke : 6 µm
Schichthärte : 2190 HV
Schichtträger: 1.3343
Ätzmittel: $FeCl_2$
V = 1000 : 1

Niobcarbid

Vakuum	Normaldruck

Bild 48

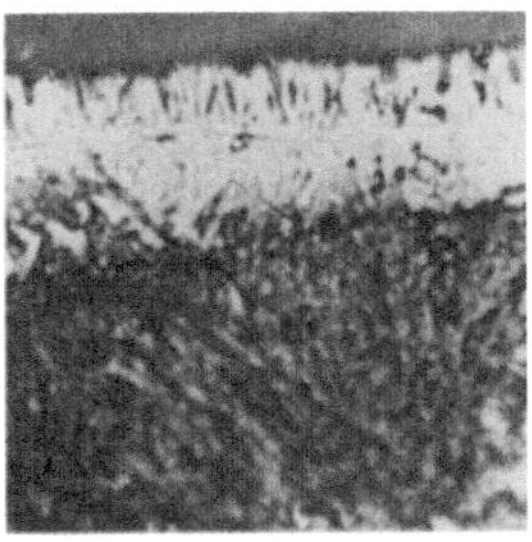

Schichtdicke : 6 µm
Schichthärte : 2400 HV
Schichtträger: 1.0603
Ätzmittel: HNO_3

Bild 49

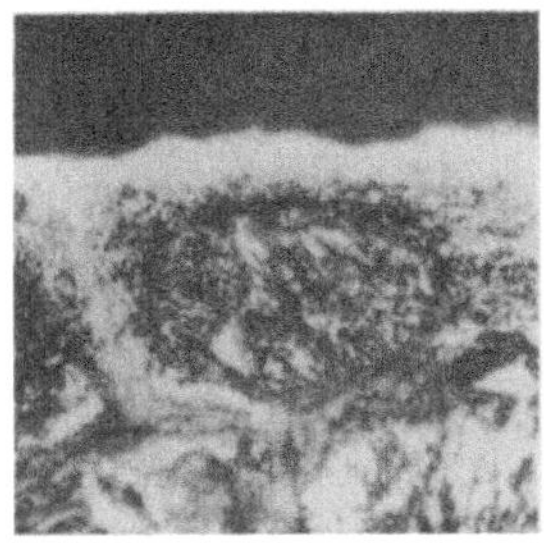

Schichtdicke : 8 µm
Schichthärte : 2160 HV
Schichtträger: 1.0603
Ätzmittel: HNO_3

Bild 50

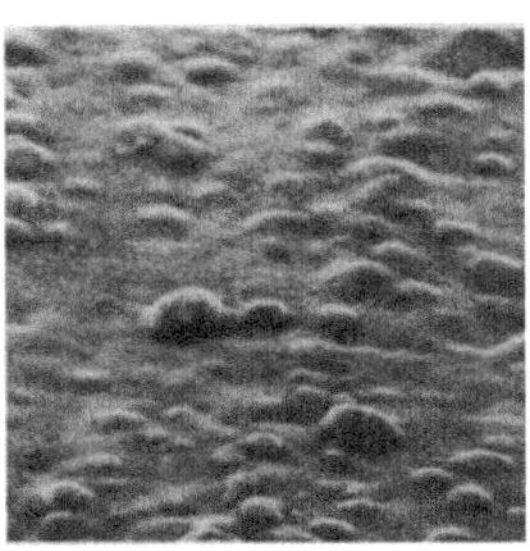

REM-Oberfläche
V : 1000 : 1
U = 25 kV
Kippwinkel: 15°
Sekundärelektronen

Bild 51

REM-Oberfläche
V : 1000 : 1
U = 25 kV
Kippwinkel : 15°
Sekundärelektronen

Niobcarbid

Vakuum

Normaldruck

Bild 52

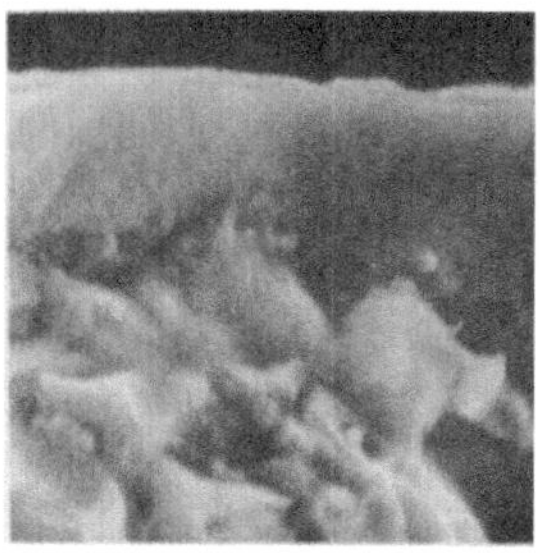

REM-Bruchfläche
V : 5000:1
U = 25 kV
Kippwinkel: 0°
Sekundärelektronen

Bild 53

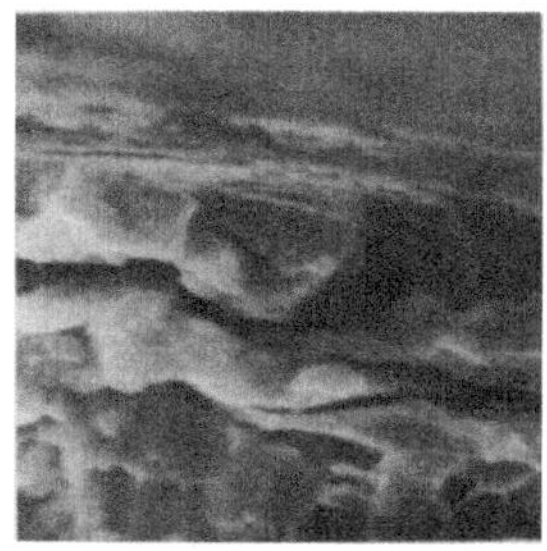

REM-Bruchfläche
V : 3000:1
U = 25 kV
Kippwinkel: 0°
Sekundärelektronen

Vakuum

Bild 54

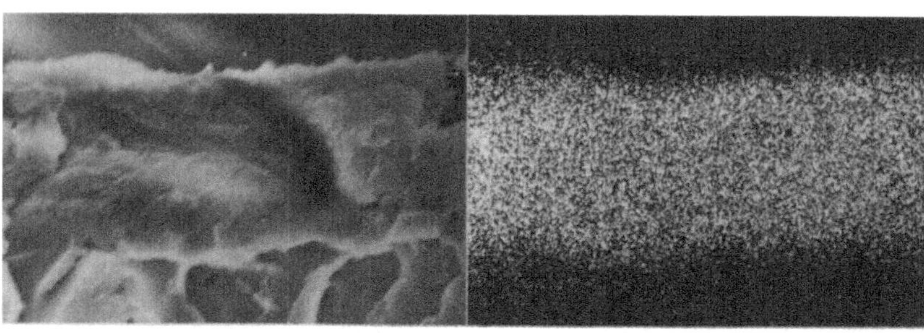

REM-Bruchfläche mit NbC-Verteilung
V : 3000:1
U = 25 kV
Kippwinkel: 0°
Sekundärelektronen

XII. Maßänderungen

Für das Abscheiden von 5 bis 10 µm Hartstoffschichten ist eine Reaktionszeit bis zu 5 Stunden erforderlich. Der größte Teil aller Werkzeuge wird hinsichtlich der Austenitisierung überzeitet. Das führt zu radialen und axialen Formabweichungen wie Ovalität, Rautenförmigkeit (siehe Bild 55), Mehreckigkeit (siehe Bild 56), negativer und positiver Balligkeit, Kugelförmigkeit.

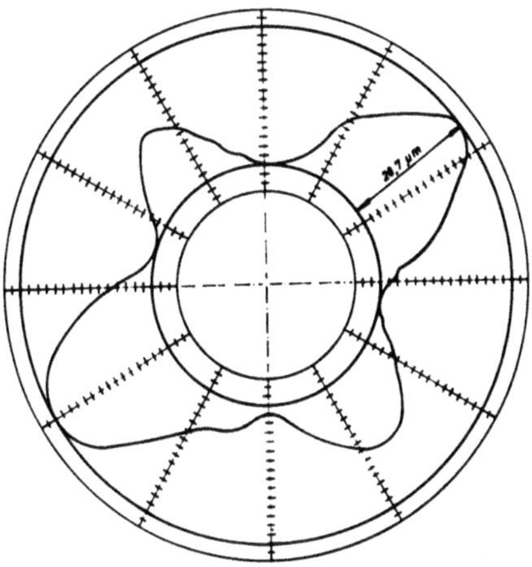

Bild 55

Formabweichung von Rundstäben aus gewalztem S 6-5-2

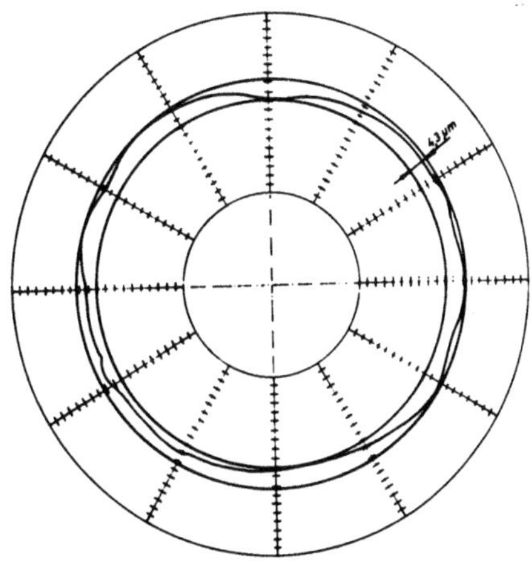

Bild 56

Formabweichung von Rundstäben aus pulvermetallurgisch gesintertem S 6-5-2

Es wurden außer Wendeschneidplatten 25 Werkzeuge 130 mm Durchmesser und 10 Werkzeuge mit einem Außendurchmesser von 300 mm und Innendurchmesser von 100 mm aus den konventionellen und pulvermetallurgisch hergestellten Schnellstählen gefertigt, beschichtet, gehärtet und angelassen. Alle Proben wurden vor und nach dem Beschichten, nach dem Härten und nach dem Anlassen genau vermessen. Die gemessenen Maßänderungen nach dem Vergüten in der Längs- und Querrichtung lagen bei:

1. unbeschichteten Stählen: (%-Angabe in Bezug auf Fertigmaß)

Substrat	Längenänderung	Breitenänderung	Unrundheit
12% Chromstahl	± 0,10%	± 0,02%	-
S 6-5-2	+ 0,20%	± 0,05%	0,04%
ASP-23 (1.3343)	+ 0,15%	+ 0,08%	0,015%

Die Abweichungen stimmen mit den Herstellerangaben überein.

2. beschichteten Stählen: (Beschichtungszeit: 3 Stunden)

Substrat	Längenänderung	Breitenänderung	Unrundheit
12% Chromstahl	± 0,10%	± 0,02%	± 0,01%
S 6-5-2	+ 0,22%	± 0,22%	± 0,055%
ASP-23 (1.3344)	+ 0,16%	± 0,08%	+ 0,016

In einigen Fällen lassen sich sehr geringe Maßabweichungen durch gesteuertes Anlassen korrigieren (siehe Bild 57 und 58).

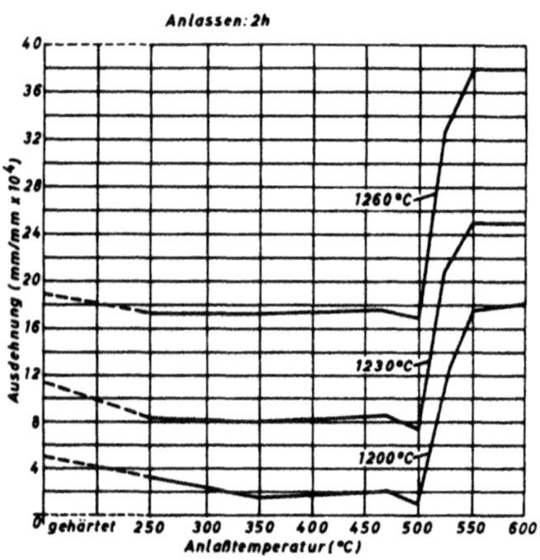

Bild 57

Ausdehnung als Funktion der Anlaßtemperatur von konventionellen S 6-5-2

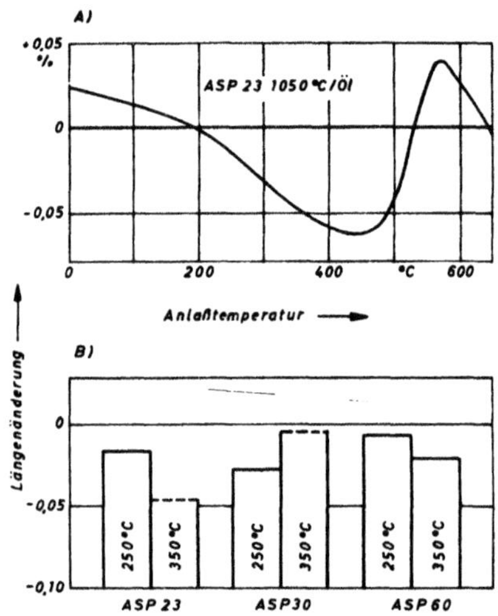

Bild 58

Ausdehnung als Funktion der Anlaßtemperatur von pulvermetallurgisch hergestellten Schnellstählen

XIII. Standzeitversuche

1. Verschleiß an Werkzeugen aus Hochleistungsschnellstählen allgemein

Es ist bekannt, daß jedes Zerspanungswerkzeug nach einer gewissen Einsatzzeit seine Schneideigenschaft verliert. Das führt zur Unterbrechung des Formgebungsprozesses. Die Ursache für den Verlust ist der Verschleiß. Er kann verschiedene Gründe haben, ist aber immer durch das Abtragen von Teilchen an den Kontaktzonen gekennzeichnet. In der Literatur werden sechs Verschleißmechanismen als besonders wichtig angesehen:

- Tribo-Oxydation:	chemische Reaktion von Grund- und/oder Gegensubstrat mit Bestandteilen des Zwischenstoffes oder Umgebungsmediums als Folge einer reibbedingten Aktivierung
- Abrasion:	Anritzung und Zerspanung im Mikrobereich des Werkzeuges durch Rauhigkeitshügel des Werkstoffes oder Partikel der Zwischenzone
- Adhäsion:	Bildung und Trennung von atomaren Bindungen zwischen Grund- und Gegenkörper (Press-Schweißungen)
- Oberflächenzerrüttung:	Rißbildung und Rißwachstum bis zur Abtrennung von Verschleißpartikeln infolge wechselnder Beanspruchungen
- Tribo-Sublimation:	
- Diffusion:	

Tribo-Oxydation wird durch das Wachstum einer Eisenoxydschicht durch Reiboxydation gekennzeichnet. In der Praxis ist die Tribo-Oxydation besonders in den Bereichen gefährlich, in denen die Reaktionsprodukte Lagerspiele verkleinern oder vollkommen zusetzen könnten, so daß die für die Funktion notwendigen Relativbewegungen von Gleitpartnern verhindert werden.

Abrasion Man spricht von einem Abrasionsvorgang nur im Falle, in dem die Reibungshügel des Gegenkörpers oder Werkstoffpartikel, die als Zwischenwerkstoff vorhanden sind, in die Oberfläche (in unserem Fall der Wendeschneidplatte) eindringen und einer Tangentialbewegung unterzogen werden, so daß Risse und Mikrorisse entstehen. Abrasion ist die Verschleißart, die zu sehr hohem Verschleißbetrag von Schneidwerkzeugen führt.

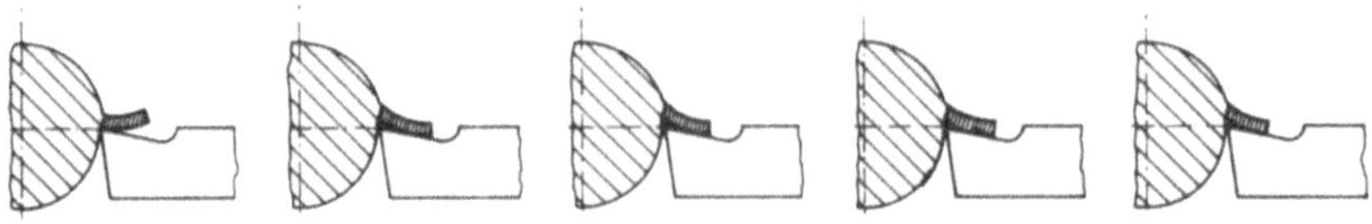

Bild 59 Preßschweißen

Adhäsion kommt nur dann vor, wenn sich zwei Körper unmittelbar berühren, der eigentliche Kontakt erfolgt nur im Mikroflächenbereich. Die Gesamtheit der Mikrokontaktflächen bildet die sogenannte reelle Kontaktfläche, deren Größe nur einen Bruchteil der durch die Abmessung der Kontaktpartner gegebenen Oberfläche ausmacht. In Mikroflächenbereichen herrschen erhebliche mechanische Spannungen, die durch tangentiale Relativbewegungen der Verschleißpartner noch verstärkt werden, so daß die belasteten Rauhigkeitshügel elastisch oder elasto-plastisch verformt werden. Dadurch entstehen auf den Oberflächen haftende Adsorptions- und Reaktionsschichten von großer Festigkeit (siehe Bild 59). Sie müssen bei einer Relativbewegung der Verschleißpartner gelöst werden. Die dazu benötigten Kräfte sind hauptsächlich für die Reibung verantwortlich. Die Höhe des Verschleißes ist von der Abscherung der Schichten von den ursprünglichen Mikrokontaktflächen und von dem Herausreißen der angrenzenden Oberflächenbereiche des Werkzeuges oder auch Gegenmaterials determiniert. In diesem Fall werden die herausgerissenen Oberflächenteile mit dem Span abgeführt. Das Bild solcher Abnutzung sind starke Aufrauhungen mit Riefen und Materialübertrag.

Oberflächenzerrüttung ist die Folge wechselnder mechanischer Spannungen in der Werkzeugoberfläche. Sie äußert sich im allmählichen Entstehen und Verbreiten von Rissen, die schließlich zur Abtrennung von Partikeln aus der beanspruchten Oberfläche führen, so daß Grübchen oder Löcher zurückbleiben, die mit bloßem Auge zu erkennen sind. Die Oberflächenzerrüttung ist die Hauptursache für die Begrenzung der Standzeit z.B. von Wälzlagern.
Tribo-Sublimation und Diffusion Die zwei Verschleißmechanismen treten nebeneinander auf. Beide Mechanismen sind wirksam, wenn in Folge der Reibungswärme die Temperatur der Oberflächenbereiche erheblich ansteigt. Dies führt bei der Tribo-Sublimation zur Ablösung von Atomen oder Molekülen, die mit dem Span von der angrenzenden Oberfläche abgeführt werden. Die Tribo-Subli-

mation ist ein endothermer Prozeß. Die Diffusion von Atomen aus den tribologisch beanspruchten Oberflächenbereichen des Werkzeuges in den Span führt zum unmittelbaren Materialverlust, der als Verschleiß gemessen wird. Viel schädlicher ist dabei die Herabsetzung der mechanischen Festigkeit. So kann man bei allen Schneidwerkzeugen aus Metall-Kohlenstoff-Verbindungen, die bei hohen Schnittgeschwindigkeiten eingesetzt waren, einen erhöhten Verschleiß auf eine Diffusion des Kohlenstoffes in den zerspanenden Werkstoff beobachten, wodurch das Gefüge geschwächt wird.
In der Praxis treten die oben beschriebenen Verschleißmechanismen selten separat auf, man beobachtet sehr oft Überlagerungen oder gegenseitiges Beeinflussen. Für eine unmittelbare Entstehung von Verschleißpartikeln sind Adhäsion und Oberflächenzerrüttung verantwortlich. Durch die Adhäsion erfolgt ein Werkstoffübertrag von Werkzeug auf Werkstoff und umgekehrt. Die übertragenen Partikel können durch Oberflächenzerrüttung und Abrasion abgetragen werden. Sie können aber auch selbst abrasiv wirken. Die abgeführten Partikel werden tribochemisch oxydiert.

2. Gebildeänderung der Schneide durch Erwärmung

Da der Span beim Zerspanungsprozeß seine Form bleibend ändert, und an den Kontaktstellen zwischen Werkstück und Werkzeug hohe Reibkräfte auftreten, wird Wärme erzeugt, die zu einer Temperatursteigerung im Werkzeug führen kann (siehe Bild 60). Obwohl die größere Wärmemenge im Span abgeführt wird, entsteht im Schneidkeil ein Temperaturfeld, das sich in kalottenförmige Isothermen in Richtung Werkzeugschaft ausbreitet. Durch die Temperaturerhöhung wird der Formänderungswiderstand des Schneidstoffes herabgesetzt, und die Schneide durch Einwirkung der Schnittkräfte plastisch verformt. Die Schneide verliert ihre geometrisch ideale Keilform und damit ihre Schneidfähigkeit.

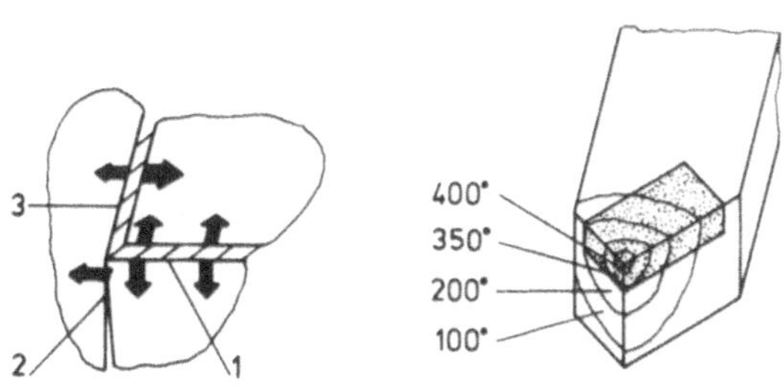

Bild 60 Temperaturverteilung an der Schneidkante

3. Typische Verschleißformen am Schneidwinkel von unbeschichteten HSS- Wendeschneidplatten

Die vorher erläuterten Verschleißmechanismen ergeben folgende auswertbare Verschleißformen am Schneidkeil (siehe Bild 61):

- Spanflächenverschleiß
- Kolkverschleiß
- Freiflächenverschleiß
- Schneidkantenrundung

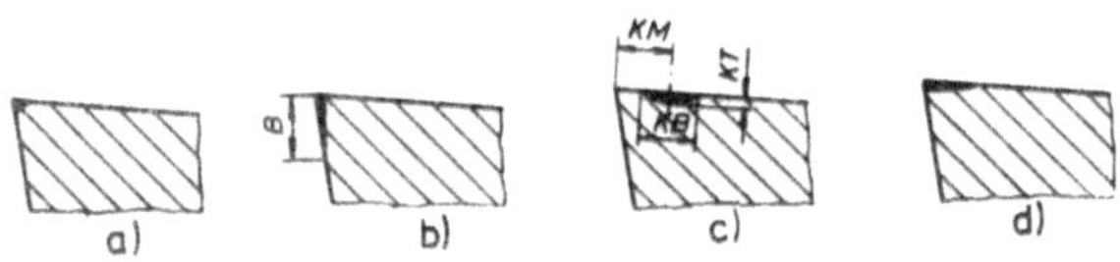

Bild 61 Typische Verschleißformen vom Schneidkeil von unbeschichteten Wendeschneidplatten

Diese Verschleißformen treten kombiniert auf und sind im Gegensatz zu den Schneideausbrüchen normale Abnutzungserscheinungen. Zur zahlenmäßigen Feststellung des Verschleißzustandes an der Freifläche wird die Verschleißmarkenbreite B festgelegt. Sie ist der Abstand der Verschleißmarken-Begrenzung auf der Freifläche von der ursprünglichen Schneide. Mit zunehmender Verschleißmarkenbreite wird der Wirkfreiwinkel bis auf 0° verändert, und dadurch mit wachsender Kontaktfläche die Reibung erhöht. Um die Verschleißmarkenbreite messen zu können, wird vor Beginn der Versuche eine Bezugslinie auf der Freifläche parallel zur Hauptschneide angegeben. Oft wird die Verschleißmarkenbreite B in Abhängigkeit von der Schnittzeit b dargestellt. Es ergibt sich eine degressiv ansteigende Kurve (siehe Bild 62). Der Kolkverschleiß tritt auf der Spanfläche des Schneidkeils auf. Die Größe des Verschleißes wird durch die Kolktiefe KT und den Kolkmittenabstand KM hinreichend definiert (siehe Bild 63). Während KM von der Schnittzeit unabhängig ist, steigt die KT mit zunehmender Schnittzeit in Abhängigkeit des Schneidstoffes linear oder schwach degressiv an (siehe Bild 64).

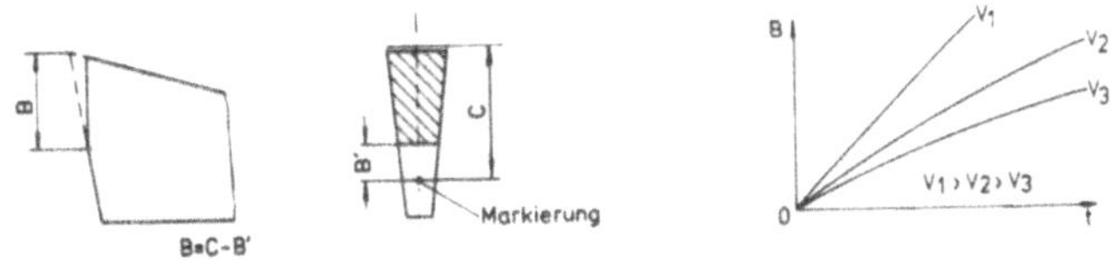

Bild 62 Verschleißmarkierung und B = f (t)

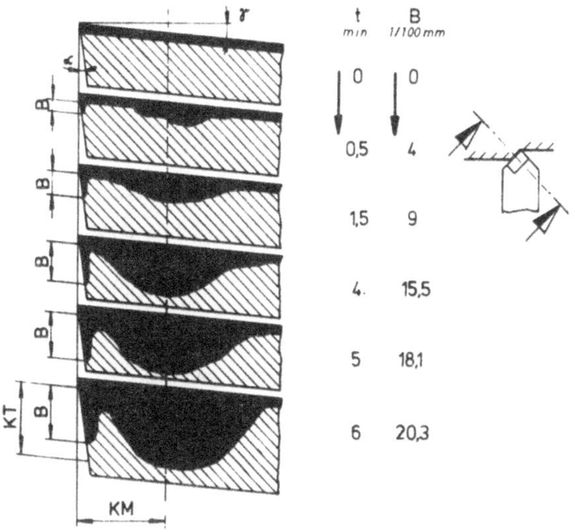

Bild 63

Verschleiß eines unbeschichteten Werkzeuges

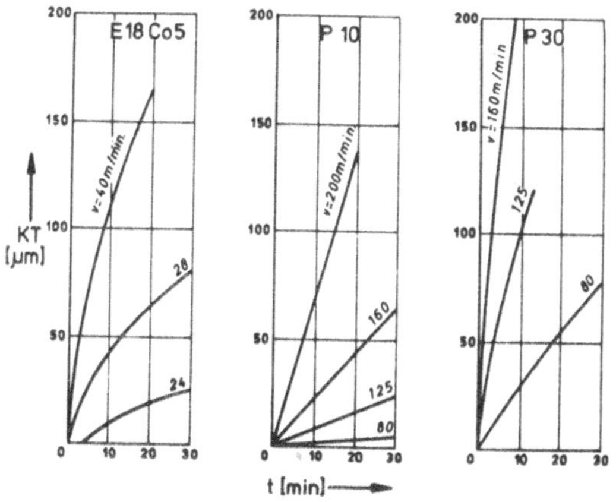

Bild 64
Kolktiefe als Funktion der Drehgeschwindigkeit in Bezug auf Schnellstahl- und Hartmetallwerkzeuge

4. Werkzeugstandzeit

Die Standzeit ist die Schnittzeit des Werkzeuges bis zum Eintreten des Erliegungskriteriums, d.h., die Schnittzeit zwischen 2 aufeinanderfolgenden Werkzeuginstandsetzungen.

5. Standzeitversuche - Durchführungsmethodik

a. Wendeschneidplatten (Proben)

Für die erste Phase der Standzeitversuche wurden 350 Wendeschneidplatten, Typ I aus dem Werkstoff 1.3247 und 350 Wendeschneidplatten, Typ II konstruiert, Material gekauft und in eigener Regie gefertigt.

b. Beschichtung

Die gesamte Gruppe von 500 Wendeschneidplatten wurde gekennzeichnet, auf 8 Untergruppen je 60 Wendeschneidplatten unterteilt und mittels folgender Schichtkombinationen belegt:

60 Stck. mit TiC/ Schichtdicke h = (8 - 13) µm

60 Stck. mit TiC + TiCN/ Schichtdicke h = (13 - 17) µm

60 Stck. mit TiC + TiCN + TiN/ Schichtdicke h = (9 - 24) µm

60 Stck. mit TiC + TiN/ Schichtdicke h = (8 - 12) µm

60 Stck. mit NbC_{Vakuum}/ Schichtstärke h = (8 - 12) µm

60 Stck. mit $NbC_{Normaldruck}$/ Schichtdicke h = (8 - 15) µm

60 Stck. mit VC_{Vakuum}/ Schichtdicke h = (8 - 12) µm

60 Stck. mit $VC_{Normaldruck}$/ Schichtdicke h = (8 - 12) µm

c. Vakuum-Härten

Das Vakuum-Härten wurde mit zweimaligem Anlassen in der halbtechnischen Zweikammer-Vakuum-CVD-Anlage durchgeführt. Die Anlaßhärte entsprach dem angesetzten Kriterium.

d. Festlegung des Drehmaterials

Für die Standzeitversuche sind folgende Stähle in der Form eines Rohlings 100 mm Durchmesser gewählt worden.

1.0711 (9 S 20) Automatenstahl

(Chemische Zusammensetzung in % - $C < 0,13/Si < 0,05, Mn \leq 0,60-1.20$ P =0,1

S =(0,18 - 0,25)
1.2341 (X 6 Cr Mo 4) Kaltarbeitsstahl
(Chemische Zusammensetzung in % - C $\leq$0,07; Si $\leq$ 0,20; Mn $\leq$0,20; P$\leq$ 0,03; S $\leq$0,03; Cr =3,5/4,0; Mo= 0,3/0,6)

e. Festlegung der Drehparameter

Im Vorversuchsfeld wurde die maximale mögliche Schnittgeschwindigkeit und die Schnittiefe für unbeschichtete HSS-Plättchen bei einer Erliegungszeit von 60 Sekunden auf dem empirischen Weg ermittelt.

Stahl 1.0711	Stahl 1.2341
V_{max} = 90 m/min	V_{max} = 60 m/min
t = 1,25 mm	t = 1,25 mm
Vorschub s = 0,63 mm/U	s = 0,63 mm/U

f. Standzeitdrehen

Das Drehen wurde in der gesamten Drehzeit von 20 Minuten viermal unterbrochen. Die Unterbrechung war nach je 3 Minuten und zweimal nach je 4 Minuten. In der weiteren Untersuchungszeit wurde der Drehprozeß nach je 5 Minuten unterbrochen.

g. Einfetten

Nach dem Drehen wurden alle Proben einem Reinigungsprozeß im Perbad unterworfen. Die Verfälschung des Abtrages für ungereinigte Platten lag bis zu 40% höher als für gereinigte.

h. Auswertung

Als Abnutzungskriterium wurde die Verschleißgröße mittels Analysenwaage, Typ Sartorius 2462 mit einer Genauigkeit von 0,1 mg nach jedem Drehversuch und Entfetten festgehalten. Alle Ergebnisse wurden graphisch ausgewertet. In Bild 65 wurden Verschleißeinzelwerte von beschichteten Wendeschneidplatten mittels Titancarbonitrid (Diagramm 1) und Titancarbid (Diagramm 2) dargestellt. Wendeschneidplatten aus dem Diagramm 1 hatten 6 bis 8 µm dicke Titancarbidschicht und 4 bis 6 µm dicke Titancarbonitridschicht. In den ersten 3 Minuten wurden fest an der glatten Titancarbonitridoberfläche klebende Partikelchen abgetragen. Die Kontaktfläche blieb in den nächsten 5 Minuten rein und unbeschädigt. Erst nach 8 Minuten traten die ersten Rißbildungs- und Zerrüttungserscheinungen in der TiCN-Schicht auf. In der 24. Minute wurde die Titancarbonitridschicht zerstört und es entstand eine Kolkfläche von ca. 0,2 mm² (siehe Bild 66 und 67). In der weiteren Untersuchungszeit sind positive Gewichtsveränderungen an der Wendeschneidplatte aufgetreten (siehe Aufbauschneidelinie). In Diagramm 2 ist in den ersten 3 Minuten eine Aufbauschneide von 0,6 mg entstanden, die in der 6. Minute durch den reibenden Span abgetragen wurde. In den nächsten Minuten sind weitere positive Gewichtsveränderungen notiert worden. Der Versuch wurde nach 70 Minuten abgebrochen. Die Schneidkante (siehe Bild 68 bis 72) blieb bis zur 60. Minute unverändert. In Bild 73 sind Verschleißdiagramme 3 und 4 für beschichtete und unbeschichtete Hartmetallwendeschneidplatten K 20 dargestellt worden.

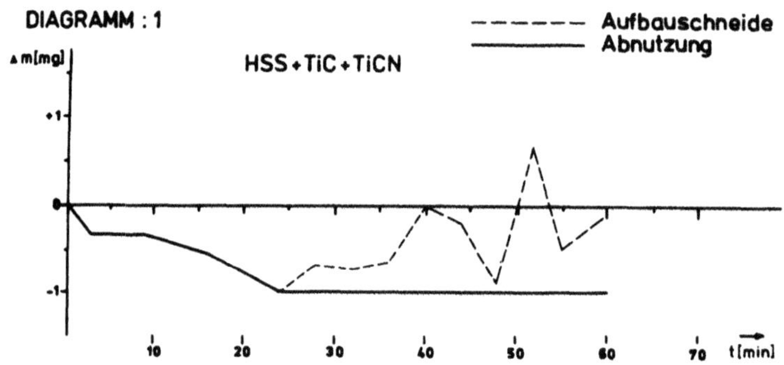

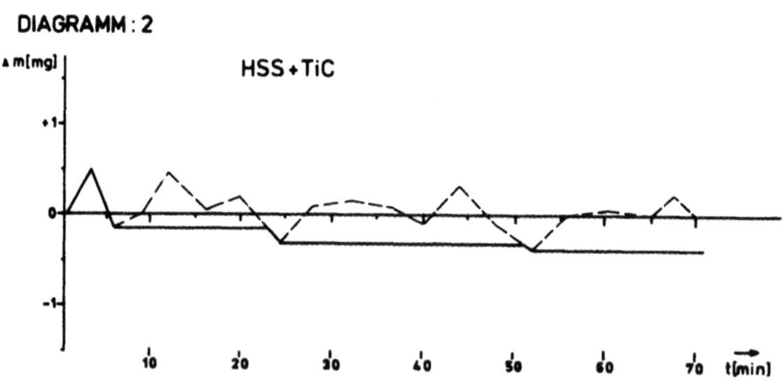

Bild 65 Gewichtsänderung in Abhängigkeit der Drehzeit für HSS-Wendeschneidplatten

Bild 66 Kolkbildung in mit TiC beschichteten Wendeschneidplatten

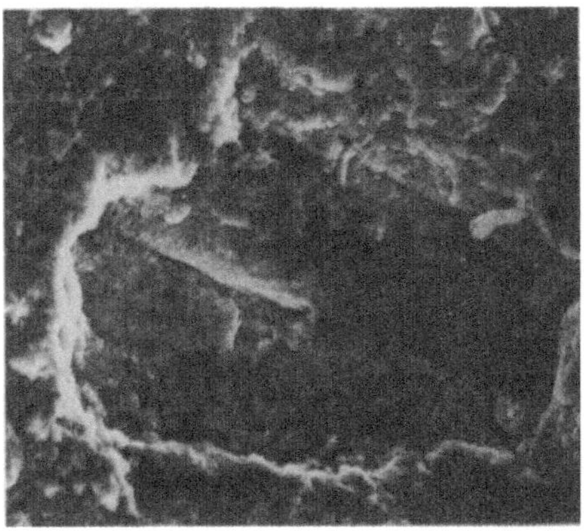

REM
Draufsicht
U = 25 kV
Kippwinkel = 0°
Sekundärelektronen
V = 1000:1

Bild 67 Titanverteilung in der Schicht an der Wendeschneidplattenoberfläche

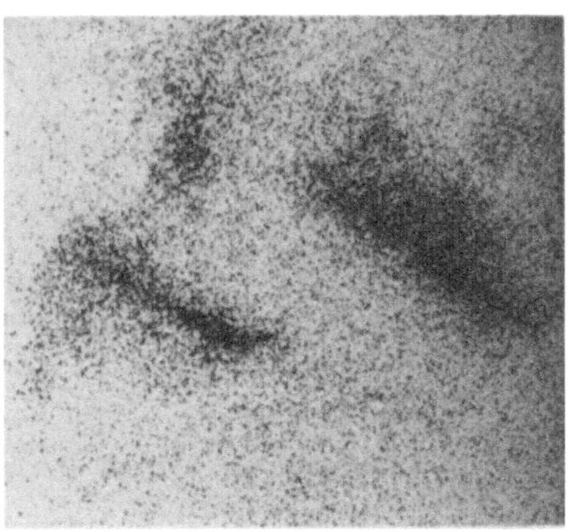

REM
Draufsicht
U = 25 kV
Kippwinkel = 0°
Röntgen: Ti-Verteilung
V = 1000:1

Bild 68

Aussehen der Schneidkante einer beschichteten HSS-Wendeschneidplatte nach 25 Minuten (TiC + TiCN)

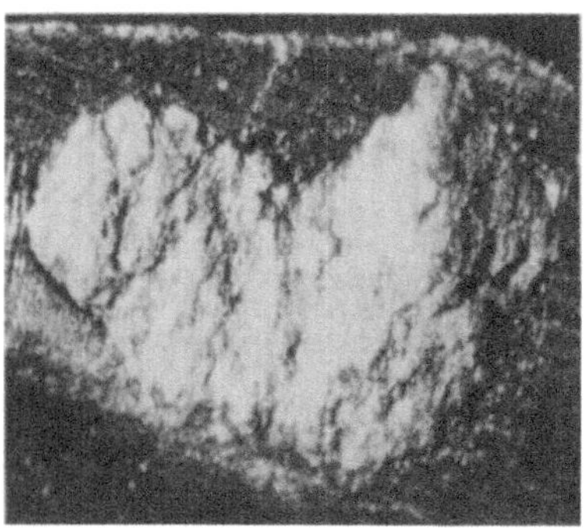

Bild 69

Aussehen nach 30 Minuten

Bild 70

Aussehen nach 35 Minuten

Bild 71

Aussehen nach 40 Minuten

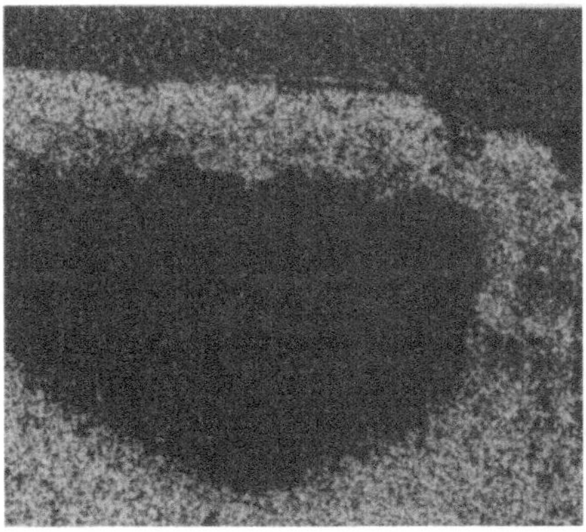

Bild 72 Titanverteilung

Aussehen nach 40 Minuten

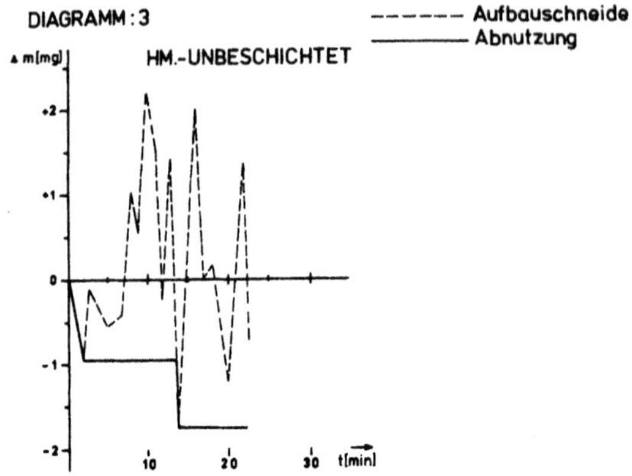

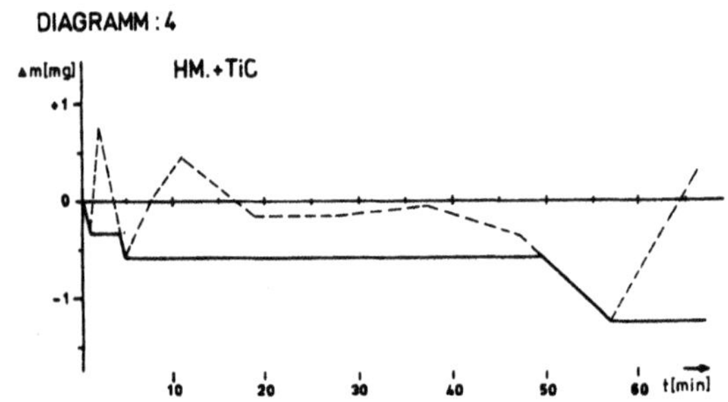

Bild 73

Gewichtsänderung in Abhängigkeit der Drehzeit für HM-Wendeschneidplatten

In Bild 74, Diagramm 5 sind Standzeitcharakteristiken von unbeschichteten und beschichteten HSS- und HM-Wendeschneidplatten dargestellt. Die unbeschichteten HSS-Wendeschneidplatten sind nach 52 Sekunden verschlissen gewesen. Die besten Ergebnisse brachten Hartmetall und Schnellstahl mit Titancarbid beschichtet.

In Bild 75 sind Standzeitkurven von unbeschichteten und beschichteten Wendeschneidplatten dargestellt worden. Alle Kurven präsentieren Mittelwerte von mindestens 10 Drehversuchen. Das interessanteste Ergebnis brachte HSS + TiC-Kombination, welche erst nach 25 Minuten in die Abnutzungsphase eintrat. Die höchste Standzeit weist HM + TiC, HSS + TiC, HSS + TiC + TiCN, HM auf. Diese Untersuchungen sind als 0-Stufe für weitere Untersuchungen anzusehen und es werden gegenwärtig weitere Hartstoffschichten wie Vanadincarbid und Niobcarbid auf Wendeschneidplatten geprüft. In Bild 76 wurde die Kolkbildung an mit Titancarbid und Titancarbonitrid beschichteten Schnellstahl-Wendeschneidplatten dargestellt.

DIAGRAMM: 5

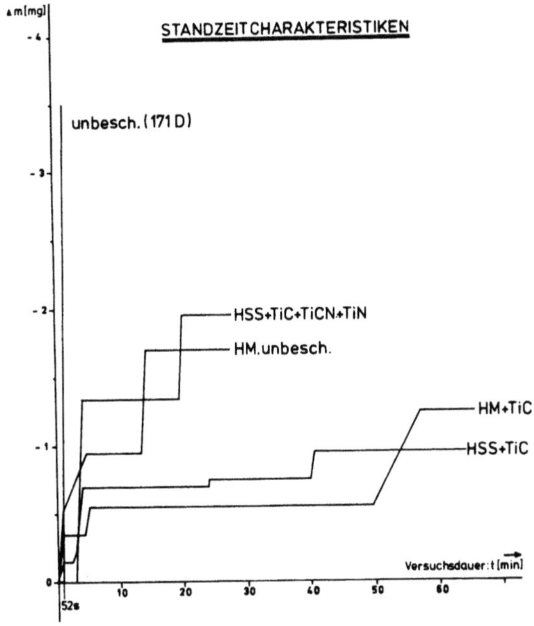

Bild 74

Standzeitcharakteristiken von Wendeschneidplatten, gedrehter Stahl 1.2341

DIAGRAMM: 6 ABNUTZUNG DER SCHNEIDE

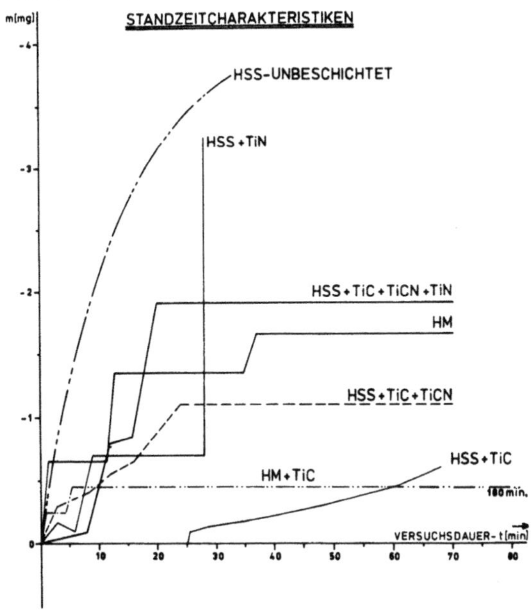

Bild 75

Standzeitcharakteristiken von Wendeschneidplatten, gedrehter Stahl 1.0711

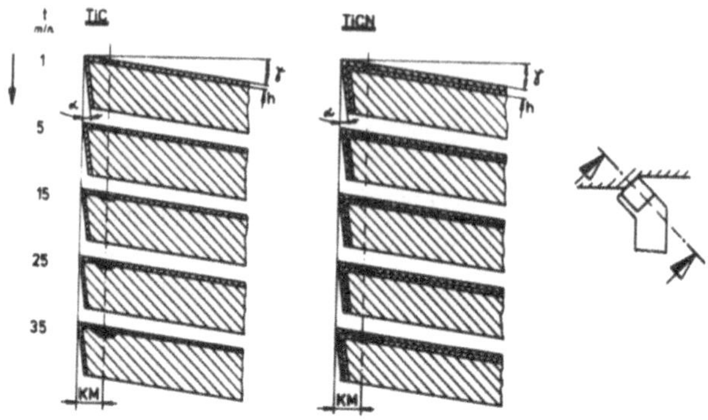

Bild 76 Kolkverschleiß an beschichteten HSS-Wendeschneidplatten

XIV. Untersuchung des Verschleißverhaltens mittels Wälzreibungsprüfmaschine

a) Schichtträger und Hartstoffe

Als Schichtträger wurden Prüfkörper mit einem Außendurchmesser von 50 $\pm$ 0,01 mm, einem Innendurchmesser von 19,95 - 0,05 mm und einer Dicke von 8 $\pm$ 0,01 mm aus dem Stahl S 6-5-2 gewählt. Die Versuche wurden auf einer Wälzreibungsprüfmaschine Typ AMSLER durchgeführt. Die Hartstoffe sind unter Vacuum (V) und Normaldruck (N) in einer Dicke von 8 - 10 µm abgeschieden worden.

b) Festlegung der Reibpartner

1. Partner gegen Titancarbid
   TiC_N/TiC_N, TiC_N/1.2080, TiC_N/CrC, TiC_V/1.4571, TiC_N/1.4571, TiC_V/TiC_V

2. Partner gegen Vanadincarbid
   VC_V/VC_V, VC_N/VC_N, VCV/TiC_V, VC_N/TiC_V, VC_V/1.2080, VC_N/1.2080, VC_N/VC_V, VC_V/1.4571, VC_N/1.4571

3. Partner gegen Niobcarbid

NbC_V/NbC_V, NbC_N/NbC_N, NbC_V/NbC_N, NbC_V/TiC_N, NbC_N/TiC_V, $NbC_V/1.2080$, $NbC_N/1.2080$, $NbC_V/1.4571$, $NbC_N/1.4571$

c) Versuchsbedingungen

Aufgrund von Voruntersuchungen wurden folgende Bedingungen zugrunde gelegt:

Normalkraft	F_N	= 1000 N
Schlupf	s	= 39 %
Luftfeuchtigkeit	p_D	= 80 Pa
Drehzahl	n	= 200 min^{-1}
Versuchszeit	T	= 3 h
Reibungsart		= Trockenreibung

Die Versuchsbedingungen wurden so ausgewählt, daß nach Ablauf der Versuchszeit bereits ein meßbarer Verschleiß bzw. eine Zerstörung oder Ablösung der Hartstoffschichten festgestellt werden konnte.

d) Allgemeines über Wälzreibungsprüfmaschinen

Es sind gegenwärtig 2 Ausführungen von Prüfmaschinen bekannt, die nach 2 verschiedenen Körperantriebsarten aufgebaut sind. Der erste Antrieb erfolgt über einen geschlossenen Leistungskreislauf, d.h. der Antrieb beider Prüfkörper erfolgt durch einen einzigen Motor und Zahnradgetriebe. Der zweite charakterisiert sich durch einen offenen Leistungskreislauf, d.h. jeder Prüfkörper wird mit eigenem Motor angetrieben.

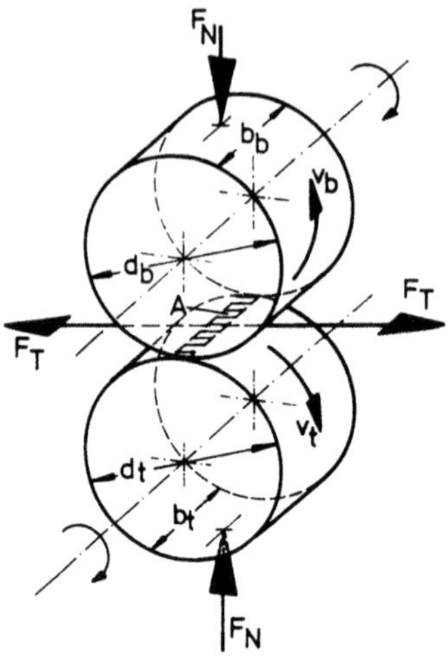

Bild 77

Kraft- und Bewegungsverhältnis von Wälzreibungssystemen

Beanspruchungsparameter:

F_N Normalkraft

F_R Reibungskraft

$F_r = \mu \cdot F_N$

M_R Reibungsmoment

$M_R = F_R \cdot \frac{d}{2}$

v: Umfangsgeschwindigkeit

0: Temperatur der Körper

Geometrische Größen:

d: Durchmesser der Körper

b: Breite der Körper

A: gemeinsame Berührungsfläche der Körper

a: halbe Berührungslänge (Berechnung nach Hertz)

$A = 2\ a\ b$

Indices:

t: Treibender Körper

b: Bewegender Körper

Wälzreibung:

v_g: Gleitgeschwindigkeit

$v_g = v_t - v_b$

$v_t > v_b$

s: Schlupf

$$s = \frac{v_t - v_b}{v_t} \times 100\%$$

e) Beschreibung der Wälzreibungsprüfmaschine - Typ AMSLER

Auf die Wellen der in Bild 78 dargestellten AMSLER-Maschine werden 2 Prüfkörper befestigt (7). Die beiden Wellen werden von einem zweistufigen Elektromotor über die Zahnradsätze (1/2) und (3/4) sowie das Wechselradpaar (8/9) angetrieben. Die untere Welle läuft mit einer konstanten Drehzahl von 200 min^{-1}, bzw. von 400 min^{-1}. Mittels des Zahnradpaares (8/9) wird die Drehzahl der oberen Welle geregelt und somit der Schlupf.

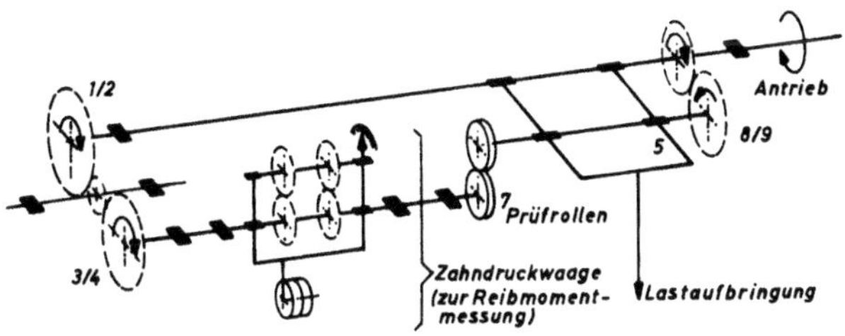

Bild 78

Die obere Welle ist in einem um 90° nach oben schwenkbaren Arm gelagert, der die Verwendung von Körpern mit unterschiedlichem Durchmesser ermöglicht und gleichzeitig den Einbau der Prüfkörper erleichtert. Die Entstehung von Riefen wird dadurch verhindert, daß der obere Prüfkörper zusammen mit der Schwinge axial hin und her bewegt werden kann. Durch einen Exzenter, dessen Hub zwischen 0 und 8 mm einstellbar ist und der mit der Vorlegewelle gekoppelt ist, wird diese Bewegung hervorgerufen und gesteuert. Hier beträgt die Auslenkung der bremsenden Rolle 0,2 mm. An der Schwinge befinden sich 2 Hebel, die die Fixierung der verschiedenen Belastungsgewichte lösen. Die Drehmomentmessung zur Bestimmung der Reibungszahl µ erfolgt mechanisch über eine Zahnradwaage. Zur Einstellung eines definierten Umgebungsmediums befinden sich die Prüfkörper in einer Klimakammer.

f) Versuchsdurchführung

Alle Verschleißversuche wurden nach folgendem Plan durchgeführt:

1. Reinigung der Prüfkörper im Ultraschallbad
2. Gewichtsmessung mittels Feinmeßwaage
3. Messung der Laufflächenrauhheit an 4 um 90° gegeneinander versetzten Stellen der Lauffläche
4. Messung der Oberflächenhärte an 4 Stellen quer zur Lauffläche
5. Verschleißversuch
6. Reinigung im Ultraschallbad
7. Gewichtsmessung mit Feinmeßwaage
8. Messung der Rauhheit, wie unter Punkt 3.
9. Messung der Oberflächenhärte wie unter Punkt 4.

g) Versuchsergebnisse

Die gesamten Meßwerte wurden in Tabellen festgehalten, nach denen die folgenden Verschleißdiagramme (s. Bild 79, 80, 81) erstellt wurden.

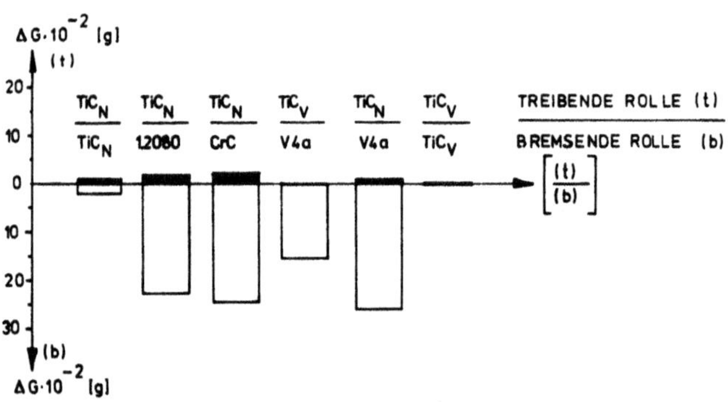

Bild 79 Reibverschleiß von Titancarbid gegen verschiedene Reibpartner (1.4571 = V4a)

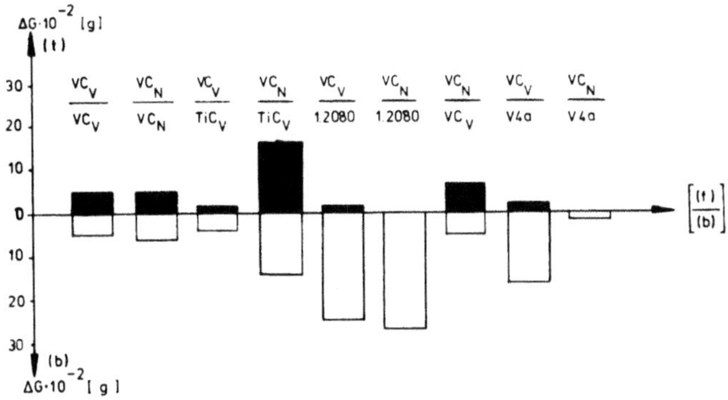

Bild 80 Reibverschleiß von Vanadincarbid gegen verschiedene Reibpartner (1.4571 = V4a)

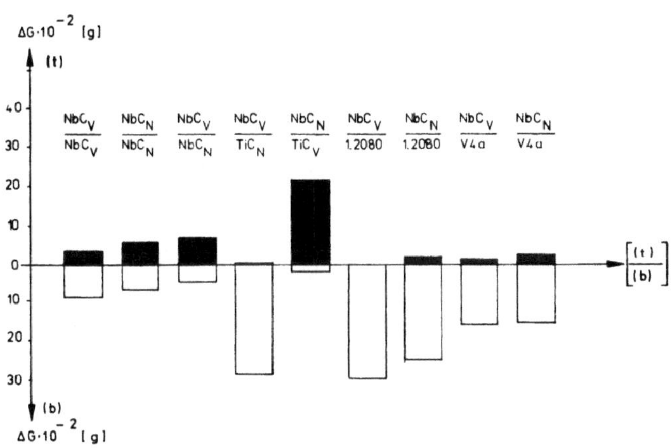

Bild 81 Reibverschleiß von Niobcarbid gegen verschiedene Reibpartner (1.4571 = V4a)

XV. Schrifttum

E. Horvath, H. Zimmermann — CVD-Beschichtungstechnik im Werkzeugbau, Sonderdruck November 1979

E. Broszeit, H.M. Gabriel — Beschichten nach den CVD-Verfahren, Z. Werkstofftechnik 11, 31-40 (1980) Verlag Chemie GmbH, D-6940 Weinheim 1980

G. Ebersbach, E. Mey, G. Ulrich — Herstellung und Eigenschaften von Metallcarbidschichten durch chemisch-thermische Behandlung in der Gasphase, Die Technik. 25. Jg., Heft 7, Juli 1970

H. Feld — Verschleißprüfung und Verschleißverhalten von Hartmetallen. Z Werkstofftechnik, 9. 172-176 (1978), Verlag Chemie GmbH, D-6940 Weinheim 1978

E. Horvath, A.J. Perry — Die Beschichtung von Stählen mit Titancarbid durch CVD, Material und Technik, Separatdruck aus Jg. 7 (1979), Nr. 2, S. 63-69

D. Hacman — Oberflächenbeschichten nach den CVD-Verfahren, VDI-Berichte Nr. 242, 1975

H. Jonsson — Die Verwendung von Hartmetall-Schneidplatten beim unterbrochenen Schnitt. Sonderdruck "TZ für praktische Metallbearbeitung" 68, Jg. 1974, Heft 4, S. 139-142

H. Jäger — Herstellung epitaxialer Halbleiterschichten durch Abscheidung aus der Gasphase, Vortrag anläßl. des Industriekoloquiums am 24.2.1972 im Battelle-Institut

R. Kieffer, P. Ettmayer — Entwicklung auf dem Gebiet der Nitridchemie der Übergangsmetalle. Inst. f. Chemie Technologie anorg. Stoffe der Techn. Hochschule Wien

H. Krause, H.H. Jühe — Röntgenographische Eigenspannungsmessungen an Oberflächen wälzbeanspruchter Kohlenstoffstähle, Härterei techn. Mitteilungen 31 (1976), S. 168-170

H. Krause, H.H. Jühe — Röntgenographische Eigenspannungsmessungen an aus der Gasphase abgeschiedenen Titancarbid-Schichten, Härterei techn. Mitteilungen 32 (1977), 6, S. 302-307

H. Krause, T. Senuma — Die Beanspruchung in Wälzkörpern und ihre Auswirkung auf den Werkstoffzustand, Fortschrittbericht der VDI-Zeitschrift, Reihe 5, Nr. 35, 1978

H. Krause, A. Halim Demirci — Untersuchungen über bevorzugte Kristallorientierungen in den Grenzschichten metallischer Wälzflächen, Forschungsbericht des Landes NRW, Nr. 2667, Opladen 1977

H. Krause, H.-H. Jühe — Automation eines -Difraktometers mit Hilfe von Tischrechnern, Vortrag: Fachausschuß "Spannungsmeßtechnik in der Arbeitsgem. f. Wärmebehandlung und Werkstofftechnik (AVT)" Neuhausen/Schweiz 12. 10. 1979, Veröffentl.: Berichtsband der Tagung, Vortrag (Kurzfass.) Herausg.: V. Hauch

T. Karpinski, T. Wierzhon, J. Bogacki — Aussichten auf die Erzeugung von Titandiffusionsüberzügen in der Glimmentladung, HTM 34 (1979) 4

B. Lux — Oberflächenschutzschichten für metallische Hochtemperatur-Werkstoffe, Teil 2: Schutzschichten aus Superlegierungen und hochschmelzenden Metallen. Z.f. Werkstofftechn. 4. Jg. 1973/ Nr. 8

K. Langhammer — Aufbau und Verschleißverhalten von beschichteten Hartmetallen, Industrieanzeiger 94, Jg. Nr. 68 vom 18. 8. 1972

A. Münster — Eigenschaften und Anwendungen von Titannitrid und Titancarbid, Angewandte Chemie/69, Jg. 1957/ Nr. 9

A. Münster, W. Ruppert — Zur Thermodynamik einiger Reaktionen des Titantetrachlorids, Sonderdruck aus der Zeitschrift für Elektrochemie, Berichte der BG für physikalische Chemie, Bd. 57, Heft 7, 1953 (S. 558-564)

Kuznecov, Babad-Zachrjapin — Die Abscheidung von Carbidüberzügen in einer Glimmentladung, Izvestija VUZ, Cernaja Metallurgija (1978), H. 11, S. 136-138

F. Reif — Physikalische Statistik und Physik der Wärme Walter de Gruyter, Berlin-New York - 1976

Ch. Weißmantel, Z. Hamann — Grundlagen der Festkörperphysik, VEB Deutscher Verlag der Wissenschaften 1979

H. Goelltz, H.G. Schneider, H. Roeßler — Vakuumelektronik, Akademie-Verlag Berlin 1978

H.G. Schneider, A. Häßner — Werkstoffe und Technologie elektronischer Bauelemente, VEB Deutscher Verlag für Grundstoffindustrie Leipzig 1975

W.J. Moore, D.O. Hummel — Physikalische Chemie, de Gruyter, Berlin-New York - 1976

K. Winnacker, L. Küchler — Chemische Technologie, Karl Hanser Verlag, München 1973

W.R.A. Vauk, H.A. Müller — Grundoperationen chemischer Verfahrenstechn. VEB Deutscher Verlag für Grundstoffindustrie Leipzig 1977

V. Boldyrev, K. Meyer — Festkörperchemie, VEB Deutscher Verlag für Grundstoffindustrie Leipzig 1973

Ch. Edelmann — Vakuumphysik und -technik, Akad. Verlagsges. GEEST PORTIG K-G, Leipzig 1978

M. von Ardenne	Tabellen zur angewandten Physik, Bd. 1,2,3 VEB Deutscher Verlag der Wissenschaften, Berlin 1979
Autorenkollektiv	Intermetallische Phasen, VEB Deutscher Verlag für Grundstoffindustrie Leipzig 1976
A. Walz	Strömungs- und Temperatur-Grenzschichten, Verlag G. Braun, Karlsruhe
F. Keune, K. Burg	Singularitätenverfahren der Strömungslehre, G. Braun, Karlsruhe
G. Fleischer, H.G. Wolmerstatt, H. Thum	VEB Verlag Technik, Berlin, 1980
I.W. Kragelski	Reibung und Verschleiß, Carl Hanser Verlag, München 1971
K.H. Habig	Verschleiß und Härte von Werkstoffen, Carl Hanser Verlag, München - Wien 1980
U. Putzier	CVD-Verfahren, Haus der Technik - Vortragsveröffentlichungen 374
W. Ruppert	Nitrier-, Chromcarbid- und Titancarbidschichten bei Werkzeugen, Sonderdruck aus Bänderbleche, Rohre 11 (1970) Nr. 4, S. 216-222
W. Ruppert	Dissertation über Abscheidung-Eigenschaften und Verhalten von Titancarbidschichten auf Stahloberflächen 1959
H.D. Schubert	Eigenschaften von Wendeplatten, titancarbidbeschichtet und konventionell gefertigt, Sonderdruck aus Maschinenmarkt, Vogel Verlag 79, Jg., Heft 18 vom 23. 2. 1973
W. Schintelmeister, O. Pacher	Herstellung und Eigenschaften von Hartstoffschichten für die Metallzerspanung und für Schmuckzwecke, Plan C Berichte für Pulvermetallurgie, Bd. 23, 1975
J. Schlichting	Chemical Vapour Deposition of Siliconcarbyd, Howde Metallurgy international Vol. 12, No. 3, 1980
H. Zickler	Leistungssteigerung an Standwerkzeugen durch CVD-Hartstoffbeschichtungen, VDI-Berichte Nr. 272, 1976

Die Vorgänge beim Schwingungsverschleiß in Abhängigkeit von Beanspruchung und Werkstoff

von

Dr.-Ing. R. Heinz

Dr.-Ing. G. Heinke

Robert Bosch GmbH
Stuttgart

Förderkennzeichen: 01 ZT 037

Laufzeit: 1.1.1978 - 31.12.1980

Zusammenfassung

Aufgabe

Ziel dieser Arbeiten war es, die Vorgänge beim Schwingungsverschleiß unter den verschiedensten Beanspruchungsbedingungen und mit unterschiedlichen Werkstoffen zu untersuchen, um die Frage der Korrelation von Verschleißversuchen bei "Modell", Bauteil- und Erzeugnis zu klären und Hinweise für die gezielte Durchführung von Verschleißversuchen mit einfachen Prüfkörpern zu geben. Weiterhin waren allgemeine Hinweise zur Minderung von Schwingungsverschleiß zu erarbeiten.

Ergebnisse

Es wird begrifflich unterschieden zwischen Schwingungsverschleiß, reversierendem Gleitverschleiß und reinem Gleitverschleiß. Kennzeichnend für die Verschleißart Schwingungsverschleiß ist die oszillierende Gleitbewegung mit Schwingungsweiten, welche kleiner sind als eine charakteristische Länge des Kontaktes in Schwingungsrichtung, z.B. Gesamtlänge der Kontaktfläche, der Abstand zwischen zwei Nuten, oder der mittlere Abstand zwischen Schmierstoffdepots in einer Kontaktfläche. Wesentlich ist die Änderung des Stofftransportes und dessen Auswirkung innerhalb der Kontaktflächen beim Übergang zum reversierenden Gleitverschleiß.

Schwingungsverschleiß beeinträchtigt die Funktion von Maschinen nicht wegen besonders hoher Verschleißraten, sondern durch örtlich begrenzte geometrische Strukturänderungen infolge Verschleiß und deren Folgen.

Nach Art des Stoff-Transportes unterscheidet man:

- leichter Abtransport der Verschleißteilchen
  Folgen: Spielvergrößerung senkrecht zur Schwingungsrichtung
  Formschluß in Schwingungsrichtung

- erschwerter Abtransport der Verschleißteilchen
  Folgen: Verklemmen von Spielpassungen
  Schwergängigkeit von Gleitführungen

Bei der Variation der Beanspruchungsparameter hat es sich gezeigt, daß im Gegensatz zu vielen Gleitverschleißsystemen, bei Schwingungsverschleiß die Stoff-Transportvorgänge einen starken Einfluß auf den Verschleißvorgang haben.

Beispielsweise ist bei Stahl-Stahl-Paarungen mit kleiner werdender Schwingungsweite der Verschleißabtransport stark erschwert, die sich bildende Zwischenschicht wirkt verschleißmindernd und die spezifische Verschleißrate K ($\mu m/(km \cdot N/mm^2)$) sinkt stark ab.

Bei PA66-Stahl-Paarungen ist der Verschleißabfluß mit kleiner werdender Schwingungsweite nicht in gleichem Maße erschwert, die spezifische Verschleißrate ist unabhänig von der Schwingungsweite.

Die Verschleißmenge und die Art ihres Transportes aus der Kontaktfläche ändern die geometrische Struktur der Kontaktflächen, was wiederum Auswirkungen auf die Verschleißrate selbst hat. Beeinflussende Parameter sind die Belastung, Flächenform, Schwingungsweite und Schwingungsform, insbesondere überlagerte Schwingungen in zwei unterschiedlichen Richtungen. Daher ist eine Übertragung der Ergebnisse mit vereinfachten Prüfkörpern auf das Erzeugnis umso eher möglich, je genauer man die Kontaktgeometrie, Bewegungsform und die Belastung den Verhältnissen im Erzeugnis nachbildet.

Die Untersuchungen haben aber auch gezeigt, daß diese strenge Forderung nach betriebsnaher Simulation nicht für alle Schwingungsverschleißsysteme gilt. Es werden drei Systemklassen mit steigender Forderung nach Betriebsnähe des Beanspruchungskollektivs und der Systemstruktur unterschieden.

Für die verschiedensten Werkstoff-Paarungen werden die Verschleiß-Versuchsergebnisse in Form von Bewährungsfolgen zusammengestellt. Besonders übersichtlich ist hierbei die Benutzung der spezifischen Verschleißrate K.

Eine optimale Werkstoffpaarung, welche den Schwingungsverschleiß auf ein für den Anwendungsfall erträgliches Minimum reduziert und gleichzeitig eine kostengünstige Lösung darstellt, muß für den jeweiligen Einzelfall stets neu gesucht werden. Die Arbeit zeigt den Nutzen der Prüftechnik mit einfachen Probekörpern bei der Lösung dieser Aufgabe.

Gliederung

1. Einleitung

Schwingungsverschleiß (SV) entsteht bei Kontakt zweier Festkörper, welche relativ zueinander oszillierende Gleitbewegungen kleiner Schwingungsweite Δx (Hub) mit meist unter 1 mm durchführen. Die Bewegungsform kann dabei gleitend, bohrend und/oder wälzend erfolgen. Die Normalkraft verläuft z.B. infolge Erschütterungen oder Unwuchten oft schwellend oder stoßend.

Bei sehr großen Schwingungsweiten ist der Begriff reversierender Gleitverschleiß [1) bezeichnender, da sich in solchen Fällen der Verschleißvorgang von dem bei Schwingungsverschleiß infolge geänderter Stofftransportbedingungen deutlich unterscheidet. Eine Grenze zwischen beiden Begriffen kann aufgrund kinematischer Betrachtungen definiert werden, wenn man die Schwingungsweite Δx auf eine charakteristische Länge l der Kontaktfläche in Schwingungsrichtung bezieht. Ist die bezogene Schwingungsweite $\Delta x/l < 1$, so liegt man im Gebiet des Schwingungsverschleißes (Bild 1).

Schwingungsverschleiß ist eine im gesamten Maschinenbau sehr häufige Verschleißart. Die Verschleißschäden werden oft in Form von Reibrost oder Passungsrost an Lagersitzen, Passungen, Führungen, Anschlägen mechanischer Verstelleinrichtungen, Einspannungen, Kupplungen usw. beobachtet.
Die Verschleißerscheinungsform Reibrost ist allerdings auf die spezielle, wenn auch große Systemgruppe mit mindestens einem metallischen Körper und kleinen Schwingungsweiten begrenzt (Bild 1). Der Schwingungsverschleiß umfaßt eine größere Klasse von Systemen, wobei alle Verschleißmechanismen, Adhäsion, Abrasion, Oberflächenzerrüttung und tribochemische Reaktionen beteiligt sein können. In manchen Systemen (z.B. Kunststoff-Stahl oder Kunststoff-Kunststoff) sind die tribochemischen Reaktionen sogar vernachlässigbar klein, Reibrostbildung tritt nicht auf. Reibrostbildung durch tribochemische Reaktionen ist daher nicht kennzeichnend und keine Voraussetzung für Schwingungsverschleiß.

---

1) Häufig wird die Art der Bewegung beim reversierenden Gleitverschleiß auch mit hin- und hergehender Bewegung bezeichnet.

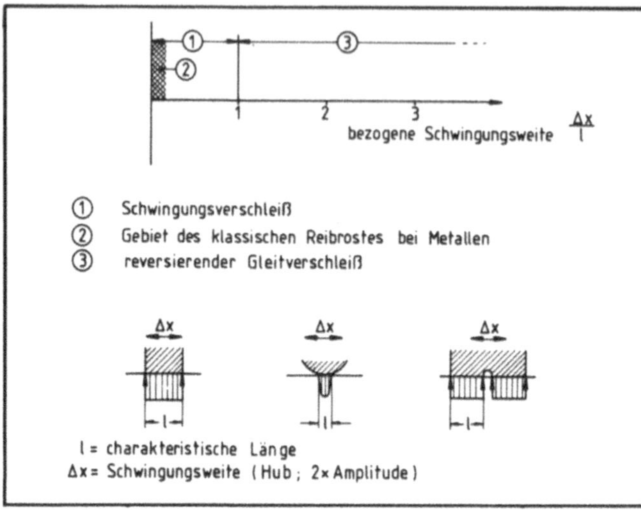

Bild 1: Unterscheidung der Begriffe Schwingungsverschleiß (Gebiet 1) und reversierender Gleitverschleiß (Gebiet 3) aufgrund der bezogenen Schwingungsweite $\Delta x/l$. Bei sehr kleiner bezogener Schwingungsweite (Gebiet 2) ist der Verschleißtransport stark erschwert. Die Verschleißpartikel bilden hier bei Stahl-Stahl-Paarungen auf den Kontaktflächen häufig den bekannten Reibrost.

2. Zielsetzung

In vielen Fällen können bei relativ niedrigen Belastungen optimale Materialpaarungen zwischen Kunststoff und Metall gefunden werden, bei denen die Reibrostbildung vermieden werden kann und bedeutend höhere Lebensdauern erreicht werden, als mit reinen Metall-Paarungen. Ein Großteil der hier durchgeführten Arbeiten, wenn auch nicht alle, befaßt sich daher mit der Untersuchung von solchen Kunststoff-Stahl Paarungen.

Die Arbeiten haben zum Ziel,

- die Auswirkungen und Wechselwirkungen einzelner Beanspruchungs- und Strukturparameter auf den Verschleißvorgang für die verschiedensten Werkstoff-Paarungen in einem möglichst weiten Wertebereich festzustellen.

- Weiterhin sind die tribologischen Vorgänge, d.h. die Verschleißmechanismen und die zeitabhängigen Vorgänge auf Reibung und Verschleiß infolge Veränderung der Systemstruktur während der Versuchszeit in Abhängigkeit vom gewählten Beanspruchungskollektiv und der Ausgangsstruktur zu erfassen.

- Es wird vermutet, daß der Stoff-Transport innerhalb der bei SV ständig belasteten Kontaktflächen besondere Auswirkungen auf die Tribostruktur und den Verschleißvorgang hat. Hierauf ist bei den Versuchen besonders zu achten.
  Zusammen mit den zeitlichen Strukturänderungen soll dies einerseits eine verallgemeinerte Beschreibung des Schwingungsverschleißes ermöglichen und andererseits Unterschiede zum reversierenden Gleitverschleiß bzw. zum reinen Gleitverschleiß aufzeigen.

- Aufgrund dieser Erkenntnisse sind für die Prüfung der ausgewählten Werkstoffpaarungen Standard-Versuchsprogramme mit Modellkörpern aufzustellen und Versuche durchzuführen.

- Die Versuchsergebnisse sind in Form von Bewährungsfolgen zusammenzustellen und so zu quantifizieren, daß sie auch für andere Anwendungsfälle aussagekräftig und nutzbar sind.

- In diesem Zusammenhang sind Kriterien für eine sichere Übertragung der Versuchsergebnisse auf praktische Anwendungsfälle zu finden, d.h. das Problem der Korrelation zwischen Modell-, Bauteil- und Erzeugnisversuch ist neu anzugehen und auf Grund der durchgeführten Versuche Möglichkeiten und Grenzen der Korrelation bei SV darzustellen.

- Auf allgemeine Abhilfemaßnahmen bei SV ist hinzuweisen.

Im Rahmen dieser Arbeiten ist es notwendig, die Modell- und Bauteilprüfstände so auszubauen, daß die unterschiedlichsten Beanspruchungsbedingungen, Bewegungsformen und Systemstrukturen realisiert werden können. Die Vielseitigkeit der Prüfstände hat zwei Vorteile:

1. Es können Parameterstudien auf breiter Basis durchgeführt werden, deren Ergebnisse wieder für die systematische Auswahl und Verbesserung von Werkstoffpaarungen genutzt werden können.

2. Bei Bedarf kann ein Versuch praxisnah simuliert werden, um eine direkte Übertragung der Ergebnisse auf das Erzeugnis sicherzustellen.

3. Stand der Erkenntnisse

Die Verschleißerscheinungsformen sind bei SV entsprechend den unterschiedlichen Systemstrukturen, Bewegungsvorgängen und Beanspruchungsbedingungen sehr vielseitig. Erscheinungsformen bei Metallpaarungen sind

- Reibrost (Tribooxidation) in Form von
  - Fleckigkeit (WATERHOUSE[36], BARTEL[2])
  - Narbenbildung (ALLIANZ[1], BARTEL[2])
  - Riffelbildung (BARTEL[2], ENGEL, KLINGELE[9], MÜLLER[26], HEINZ[19], ALLIANZ[1])
  - Wurmspuren (ALLIANZ[1], REED, BATTER[29], HEINZ[19])

Über die einzelnen Ursachen gibt es im Schrifttum noch keine klaren Aussagen.

- Politurbildung (ohne Tribooxidation) (BARTEL[2], ALLIANZ[1]) insbesondere bei Anwesenheit eines Schmiermittels und guter Spülwirkung.

Die Erscheinungsformen Fleckigkeit, Narbenbildung, Riffelbildung und Wurmspuren können auch bei Kunststoff-Stahl-Paarungen auftreten, ohne daß es dabei zu Tribooxidation kommt. Die Tribooxidation kann bei SV beteiligt sein, ist aber nicht Voraussetzung. Daher spricht man im englischen Schrifttum oft einfach von fretting und nicht von fretting corrosion (HURRICKS[20]).

Zur gezielten Schadensminderung ist es vorteilhaft, den Einfluß der verschleißrelevanten Beanspruchungsparameter und Struktureigenschaften zu kennen. Aus den Literaturangaben darf man keine eindeutigen und allgemein gültigen Daten erwarten, da je nach untersuchtem System andere Verhältnisse vorliegen und die einzelnen Parameter sich noch gegenseitig beeinflussen können. Die Angaben stellen daher lediglich Tendenzen dar.

Schwingungsweite Δx

Die Schwingungsweite Δx (doppelter Wert der Amplitude) ist eine der wichtigsten Parameter beim SV. Nach WATERHOUSE[36],[37] wird heute für Metall-Metall Paarungen allgemein anerkannt, daß bei Schwingungsweiten unter 50 bis 150 µm der Verschleiß stark abfällt. In Bild 2 ist nach WATERHOUSE[37] die spezifische Verschleißrate (s. weiter unten) über der Amplitude aufgetragen. Man erkennt klar die Änderung im Verschleißverhalten bei Schwingungsweiten um 50 bis 150 µm. Im Gegensatz hierzu beobachteten FENG, UHLIG[12] und STOWERS, RABINOWICZ[31] diese Verschleißabnahme auch bei Schwingungsweiten von 10 µm nicht.
FENG, UHLIG[12] prüften mit 9 Hz, hoher Flächenpressung p = 36,5 N/mm^2 aber sehr kurzer Prüfzeit, 2 h (entsprechend einem Gleitweg von 1,4 m ! bei einer Schwingungsweite von 10 µm).
STOWERS, RABINOWICZ[31] prüften mit hoher Frequenz von 150 Hz, geringer Flächenpressung p = 0,72 N/mm^2 und vermutlich ebenfalls mit einer recht

kurzen Prüfzeit. Die Verschleißrate wurde durch Wiegen ermittelt.
In beiden Fällen ist die sehr kurze Prüfzeit die Ursache für die nicht beobachtete Verschleißreduzierung bei Amplituden kleiner 50 bis 150 µm, da der hohe Einlaufverschleiß (Adhäsionsverschleiß) zu stark in das Ergebnis eingeht. Bei FENG, UHLIG[12] kann man einen Fall bei $\Delta x = 100$ µm für die stationäre Verschleißphase umrechnen und erhält dann eine spezifische Verschleißrate $K = 3{,}3 \frac{\mu m}{km \quad N/mm^2}$, einen Wert, der mit den allgemein beobachteten kleinen Verschleißraten bei kleinen Schwingungsweiten übereinstimmt (vgl. Bild 2). STOWERS, RABINOWICZ[31] begründen ihre Ergebnisse mit einer genaueren Messung der am Prüfkörper auftretenden Schwingungsweiten, was aber nach den obigen Betrachtungen nicht die Hauptursache sein kann. Bei sehr kleinen Schwingungsweiten ($<$ 10 µm) wird oft kein Verschleiß mehr beobachtet (elastische Deformation ohne Schlupf), weshalb einige Autoren von einer unteren Grenzamplitude sprechen (MÜLLER[26]). Diese untere Grenzamplitude ist aber stark systemabhängig, z.B. beobachtete JOHNSON[21] noch bei 50 nm Verschleiß und SÄMANN[30] mit Hochfrequenzschwingungen (20 kHz) Verschleiß bei 1 µm.

Die Flächenpressung geht etwa linear in die Verschleißrate ein. Je höher die Flächenpressung ist, desto eher besteht die Gefahr der Verfälschen der Schwingungsbewegung, infolge elastischer Deformation der Probenhalterung und der Proben. Weiterhin können die gebildeten Oxidschichten zerstört werden und der Verschleiß plötzlich ansteigen(MC DOWELL[25]).

Der Schwingfrequenz-Einfluß auf den Verschleiß wird nur sehr wenig beschrieben. Bei Frequenzen unter 17 Hz (WATERHOUSE[37]) steigt der Verschleiß, bei vorgegebener Zahl von Schwingungen, infolge größerer Zeit für die Reaktion Metall-Sauerstoff an. Über 17 Hz ist der Einfluß vernachlässigbar. Frenquenzen über 50 Hz wurden mit wenigen Ausnahmen (150 Hz bei STOWERS, RABINOWICZ[31]) selten gefahren.
Keine Hinweise waren zu finden über den Einfluß von

- Schwingungsform (z.B. beide Richtungen; stochastisch)
- Probengeometrie
- Schlagbelastung.

Das Umgebungsmedium , insbesondere die Luftfeuchte spielt eine große Rolle. Je größer die Luftfeuchte, desto geringer ist der Verschleiß, vermutlich wegen der geringeren Härte der sich bildenden Hydrooxide (FENG, UHLIG[12]) bei Stahl-Stahl Paarungen.

Viele Untersuchungen befassen sich mit der Analyse der tribologischen Vorgänge d.h.

A) den Verschleißmechanismen (den ablaufenden physikalischen und chemischen Prozessen)

B) dem zeitlichen Reibungs- und Verschleißverhalten (den Vorgängen infolge Veränderung des Tribosystems während der Betriebszeit)

C) dem Vergleich mit anderen Verschleißarten (den typischen Unterschieden des SV gegenüber reversierendem Gleitverschleiß bzw. Gleitverschleiß).

Zu A (Verschleißmechanismen):

Nach HURRICKS[20] werden beim SV von Stahl-Stahl nacheinander folgende Verschleißmechanismen durchlaufen:

1. Adhäsion mit Materialübertragung

2. Produktion von Verschleißpartikeln durch tribochemische Reaktionen

3. Stationärer Vorgang der vorzugsweisen Produktion von Verschleißpartikeln durch Oberflächenzerrüttung und nicht so sehr infolge Abrasion .

Für Edelmetalle und Kunststoff-Stahl bzw. Kunststoff-Kunststoff Paarungen oder in nichtoxidierendem Umgebungsmedium spielt Punkt 2, die Tribooxidation, keine wesentliche Rolle.

Für Stoffpaarungen, deren Oxide bzw. Verschleißpartikel härter sind als der Grundwerkstoff (Al-Al; Kunststoff+Glasfaser-Stahl) ist in Stadium 3 die Abrasion von besonderer Bedeutung.

Die Oberflächenzerrüttung unter Punkt 3 kann in Form von Schichtermüdung (Delamination) erfolgen (WATERHOUSE [37]).

Zu B (zeitliches Verschleißverhalten):

Nach WATERHOUSE[37] können beim Verschleißvorgang 4 typische Verläufe über der Zeit unterschieden werden (Bild 3).

Anfänglich ist der Verschleiß groß und verläuft degressiv (Stadium 1 der Adhäsion und Materialübertragung).
Bilden sich oxidierte Verschleißpartikel, so wird der Verschleißvorgang stationär oder degressiv (B, D), typisch für ungehärteten Stahl, wobei kleine Amplituden D und große Amplitude B entsprechen. Progressiven Verschleiß (A, C) zeigen Materialien, welche Verschleißpartikel mit höherer Härte als dem Grundmetall erzeugen (abrasiver Verschleißmechanismus). Wird nach zu kurzer Versuchszeit (auch nach Erreichen des stationären Zustandes) der Verschleiß durch wiegen festgestellt und die spezifische Verschleißrate (s.unten) bestimmt, so erhält man zu hohe Werte, da der Anfangsverschleiß zu stark eingeht. Hierdurch können Unstimmigkeiten beim Vergleich der Ergebnisse verschiedener Autoren entstehen (vergl. oben: Schwingungsweite). Besser ist die Auswertung der mittleren Steigung der Verschleißkurve nach Beendigung der Einlaufphase.

Zu C (Vergleich mit anderen Verschleißarten):

Für die Quantifizierung der Versuchsergebnisse hat sich als Verschleißmeßgröße die spezifische Verschleißrate K im Schrifttum als besonders vorteilhaft für vergleichende Betrachtungen herausgestellt. In der DIN 50321[6] "Verschleißmeßgrößen" ist diese Größe noch nicht enthalten. Eine sehr übersichtliche Betrachtung der Verschleißbereiche, insbesondere für Stähle, führte CHILDS[3] mit Hilfe der spezifischen Verschleißrate K durch (Bild 4).

$$\text{Spezifische Verschleißrate } K = \frac{\text{linearer Verschleißbetrag}}{\text{Gleitweg} \cdot \text{Flächenpressung}}$$

Die Zahlenwerte sind am handlichsten, wenn man die folgenden Dimensionen verwendet

$$K \text{ in } \frac{\mu m}{km \cdot N/mm^2}$$

Andere, in Veröffentlichungen häufig behutzte Dimensionen sind

$1\ \frac{mm^3}{mm \cdot N}$	=	$10^9\ \frac{\mu m}{km \cdot N/mm^2}$
$1\ \frac{m^2}{N}$	=	10^{15} "
$1\ \frac{mm^3}{m \cdot N}$	=	10^6 "
$1\ \frac{g}{m \cdot N}$	=	$\frac{1}{7,85} \cdot 10^9\ \frac{\mu m}{km \cdot N/mm^2}$ 1)

1) für Stahl mit $\rho = 7,85 \cdot 10^{-3}\ \frac{g}{mm^3}$

Ähnliche aus Energiebetrachtungen hergeleitete bezogene Verschleiß-Meßgrößen benutzt FLEISCHER[13] .
Die dort verwendete scheinbare Reibenergiedichte $e_R{}^*$ und die Verschleißintensität I_h ist wie folgt mit K verknüpft

scheinbare Reibenergiedichte $e_R{}^* = \frac{\text{Reibenergie}}{\text{Verschleißvolumen}} = \frac{\mu}{K}$

Verschleißintensität $I_h = \frac{\text{Linearer Verschleiß}}{\text{Gleitweg}} = K \cdot p$

mit p = Flächenpressung, μ = Reibungszahl

Die meisten Autoren stellen für SV bei kleinen Schwingungsweiten (< 100 µm) für Stahl-Stahl Paarungen bis zu 2 Zehnerpotenzen kleinere Verschleißraten gegenüber Gleitverschleiß unter sonst gleichen Bedingungen fest. Bei großen Schwingungsweiten ($\Delta x >$ 50 bis 150 µm) ist der Schwingungsverschleiß mit dem milden Gleitverschleiß ($K \approx 10\ \frac{\mu m}{km \cdot N/mm^2}$) vergleichbar (Bild 2 und Bild 4), bzw. er liegt etwas darüber (WARD[35]) infolge größerem Zerrüttungsanteil, verstärkt durch die Verschleißpartikel.

Für Kunststoff-Stahl-Paarungen fehlt ein direkter Vergleich zwischen SV und Gleitverschleiß. Die Unterschiede sind vermutlich nicht sehr groß, sofern es sich um homogene Kunststoffe handelt. Bei den faserverstärkten Kunststoffen ist ein hoher Anteil Oberflächenzerrüttung zwischen Faser und Grundwerkstoff festzustellen (CLERICO [4], ELIEZER, KHANNA[8]).
Für die Kunststoff-Vorauswahl wird auf das sehr umfangreiche Schrifttum bei Gleitverschleiß zurückgegriffen:
LANCASTER[23] , ESDU-Datenblätter[11] , VDI-Richtlinien 2541[34] ,
LNP-KATALOG[24].

Bezüglich der Korrelation von Versuchsergebnissen mit Erzeugnissen, Bauteilen und "Modellen" (Ersatzsysteme) gibt es viele Veröffentlichungen, die sich allerdings überwiegend verbal mit diesem Thema beschäftigen (UETZ [32], HABIG[14], HEINKE[17],[18], de GEE[5]). Dabei zeichnen sich jedoch einige Regeln ab, welche praktisch nutzbar sind, ohne allerdings z.B. in Form von Ähnlichkeitsgesetzen faßbar zu sein.

UETZ beschreibt in [32] 6 Prüfkategorien (Bild 5). Hieraus und ergänzt durch die Hinweise der anderen Autoren sind für die vorliegende Arbeit folgende Erkenntnisse wichtig:

- Je nach Zielsetzung ist der Einsatz bestimmter Prüfkategorien vorteilhafter gegenüber anderen.

- Günstiger ist nach HEINKE[17] eine Prüfkette mehrerer Prüfkategorien, bei der ständig eine Rückkopplung der Ergebnisse aus z.B. "Modell" -, Bauteil-, Erzeugnis- und Feldversuche angestellt wird (Bild 6).

- Zur Werkstoffvorauswahl genügen allerdings meist der "Modell"- oder Bauteilversuch.

- Die Werkstoffvorauswahl und die Korrelation zum Erzeugnis mit "Modell"-Versuchen kann durch genauere Simulation des Beanspruchungskollektivs und der Systemstruktur verbessert werden. Die Prüfstände müssen dazu entsprechend vielseitig sein.

- Zur Lebensdauerermittlung muß der Feld-, Prüfstands- oder bei ausreichender Erfahrung, der Aggregatversuch verwendet werden, der "Modell"-Versuch ist ungeeignet.

- Je komplexer ein System ist, desto schwieriger ist eine Korrelation zwischen "Modell"- und Erzeugnisversuchen zu erreichen.

- Bei Schwingungsverschleiß sind allerdings die meisten Systeme relativ einfach (keine Wärmeleitprobleme, keine hydrodynamischen Effekte) und eine Übertragbarkeit der Versuchsergebnisse aus "Modell"- oder Bauteilversuchen auf das Erzeugnis ist hier leichter möglich (HEINKE, HEINZ, THIEL[18]).

Die in der Literatur genannten Abhilfen gegen SV sind sehr zahlreich, wobei jeder Einzelfall bezüglich Verschleißminderung neu zu prüfen ist. Es lassen sich nach WATERHOUSE [37] 3 Hauptrichtungen für Gegenmaßnahmen feststellen.

1. Schwingungen durch konstruktive Maßnahmen verhindern.
2. Oberflächen beschichten oder Materialpaarungen so auswählen, daß Adhäsion und Zerrüttung vermieden bzw. gemindert werden.
3. Reibungszahl durch Schmierung vermindern.

Die Wege 2 und 3 insbesondere Weg 2, werden in der vorliegenden Arbeit vorzugsweise zur Verschleißabwehr genutzt. Allgemein bekannt ist, daß SV durch Ersatz der Metall-Metall-Paarung durch Kunststoff-Metall bzw. Kunststoff-Kunststoff vermindert werden kann.

Über die Bewährung solcher Kunststoff-Stahl-Paarungen ist bei Gleitverschleiß viel untersucht worden, aber es ist offen, ob sich das gleiche Verhalten auch bei SV zeigt.

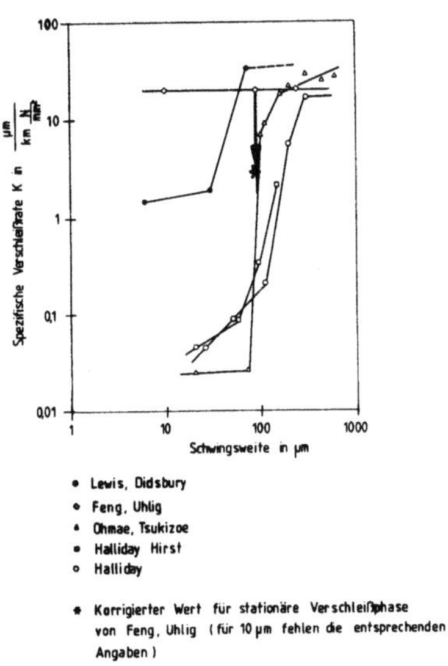

Bild 2: Vergleich der spez. Verschleißrate von Stahl-Paarungen für Schwingungsverschleiß in Abhängigkeit von der Schwingungsweite nach WATERHOUSE [37]

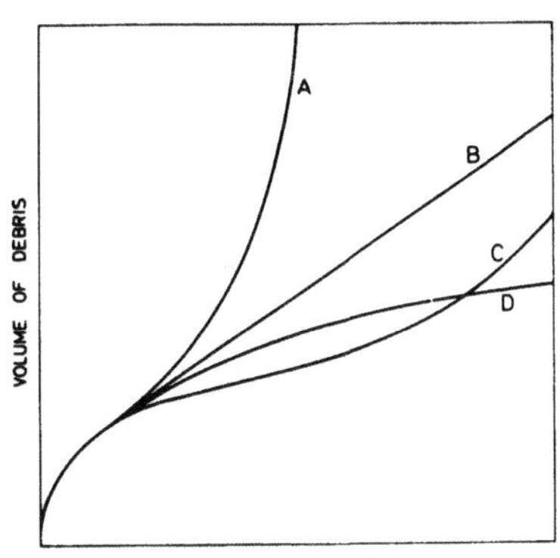

Bild 3: Schematische Darstellung von 4 typischen Verschleißverläufen mit zunehmender Schwingungszahl nach WATERHOUSE [37]

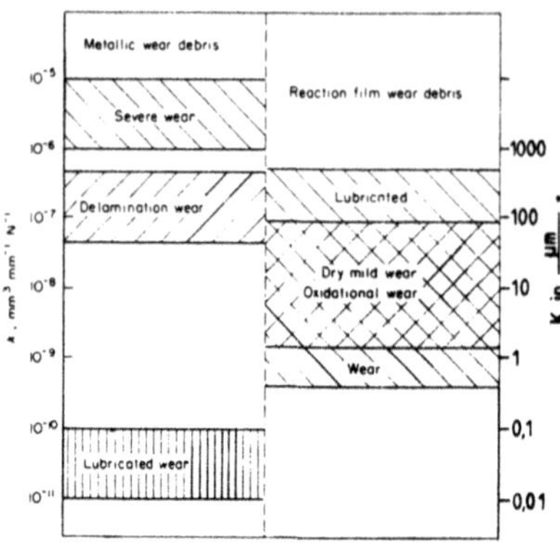

Bild 4: Bereiche der spez. Verschleißrate von Stahlpaarungen unter verschiedenen Bedingungen nach CHILDS [3]

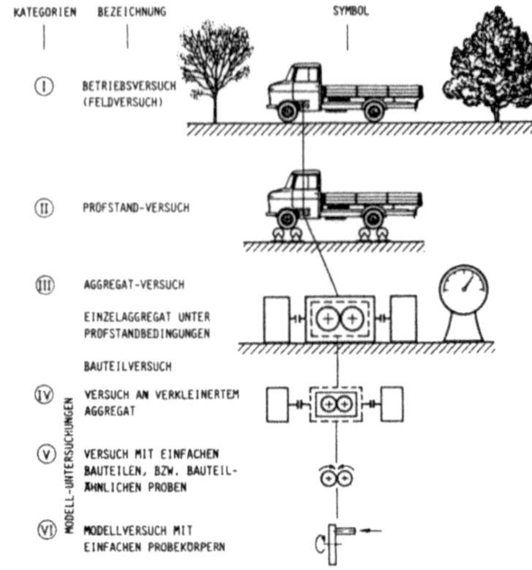

Bild 5: Die sechs Prüfkategorien nach UETZ [32]

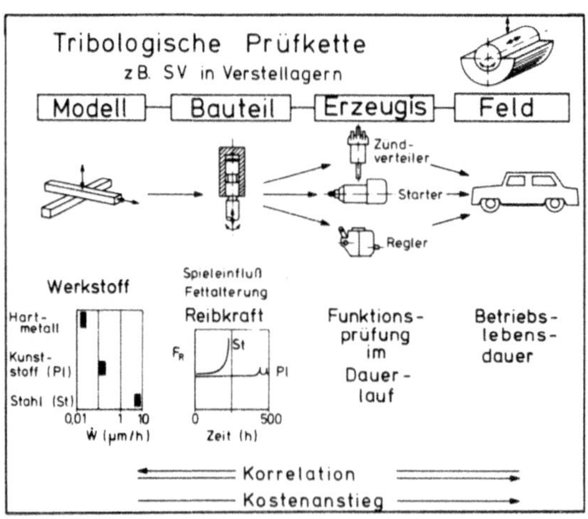

Bild 6: Prüfkette nach HEINKE [17] die bei Modellprüfungen in Form einer Rückkopplung der Ergebnisse angestrebt werden soll

4. Versuchsbeschreibung

4.1 Vorbemerkungen zu den Prüfeinrichtungen

Allgemein gilt für tribologische Prüfeinrichtungen und für SV-Prüfeinrichtungen im besonderen, daß sie eine ausreichende Steifigkeit besitzen müssen, um die Bewegungsform unabhängig von unterschiedlichen Reibkräften und Belastungen exakt übertragen zu können. Weiterhin sollten diese Prüfeinrichtungen eine große Variationsbreite des Beanspruchungskollektivs gestatten. Gerade dieser Punkt ist besonders hervorzuheben, da bei Prüfmaschinen mit sehr eng begrenzten Parameterbereichen die Beanspruchungseinflüsse nicht genügend herausgearbeitet werden können, und für die Korrelation der Meßergebnisse mit dem Erzeugnis wichtige Vorgänge nicht erkannt werden.

4.2 "Modell"-Prüfstände für Schwingungsverschleiß

Es werden 5 verschiedene "Modell"-Prüfstände für Schwingungsverschleiß benutzt und nachfolgend beschrieben:

Bezeichnung	einige kennzeichnende Merkmale 2)
SVM1	- Schwingungen wahlweise in x und/oder y-Richtung, - elektromagnetischer Antrieb - mechanisch konstante <u>und</u> hydraulisch schlagende Belastung
SVM2 SVM3	- hydraulischer Antrieb - Belastung mechanisch, konstant
SVM4 SVM5	- hydrostatische Tischlagerung - hydraulischer Antrieb wie SVM2,3 - Belastung hydraulisch konst. <u>oder</u> schlagend

2) nähere Daten sind in Tabelle 1 angegeben

Bei diesen Prüfständen werden gekreuzte Stäbchen als Standard-Probekörper benutzt (Bild 7). Bei Bedarf können seitlich an der Kontaktfläche Absätze eingearbeitet werden, so daß keine Kantenpressungen entstehen. Außerdem sind 6 mm-breite Stäbchen einspannbar oder durch kurzen Umbau Sonderprobenformen benutzbar.

Der schematische Aufbau der SV-"Modell"-Prüfstände SVM1 bis 3 ist in Bild 8 ersichtlich.
Der Tisch trägt den unteren Halter und ist über flache Stäbe oben am Gestell befestigt, so daß der Probentisch horizontale Bewegungen ausführen kann. Ein schwenkbarer Arm trägt den oberen Probenhalter. Im zusammengeklappten Zustand wird bei SVM 2,3 über den Arm mechanisch durch Gewichte die Normalkraft aufgebracht (10 N bis 2000 N), bei SVM1 kann zusätzlich hydraulisch eine schlagende Belastung erzeugt werden, die in Frequenz, Lastspitze und Stoßform einstellbar ist.
Die Schwingbewegung bei SVM1 wird seitlich über Elektromagnete mit einer Frequenz von 50 Hz erzeugt oder mit variabler Frequenz durch einen Hydraulikzylinder, der durch ein Servoventil und Frequenzgenerator angesteuert wird (0,01 Hz ... 100 Hz) (SVM2,3). Der Schwingweg wird über eine Regelung auf dem jeweiligen Sollwert gehalten (50 µm ... 3000 µm).

Der untere Probenhalter sitzt in einem Topf, so daß auch Paarungen unter flüssigem Zwischenstoff gefahren werden können. Durch jeweils zwei Heizpatronen im unteren Stäbchenhalter können die Proben bis auf 200°C aufgeheizt werden. Bei SVM1 ist für Kraftstoff-Versuche eine Luftheizung eingebaut.
Bild 9 zeigt einen Überblick der Prüfstände SVM1 (rechts) und SVM2,3 (links). Rechts neben den 2 Meßschränken ist der elektromagnetische Pulser für die zusätzlichen Querschwingungen für SVM1 zu erkennen.
Im Prinzip ähnlich aufgebaut ist die "Modell"-Prüfeinrichtung SVM4, SVM5 (Bild 10). Der Probentisch ist hier hydrostatisch gelagert und gestattet bei absolut linearer Führung, Schwingweiten bis 5 mm. Der Antrieb (rechts in Bild 11) erfolgt hydraulisch, durch ein Servoventil gesteuert, genau wie bei SVM2,3. Für sehr kleine Schwingungsweiten ist links ein mechanischer Antrieb über Exzenter und Scheibenläufermotor anschließbar. Alle Probenteller sind mit Heizpatronen bis 200°C heizbar.

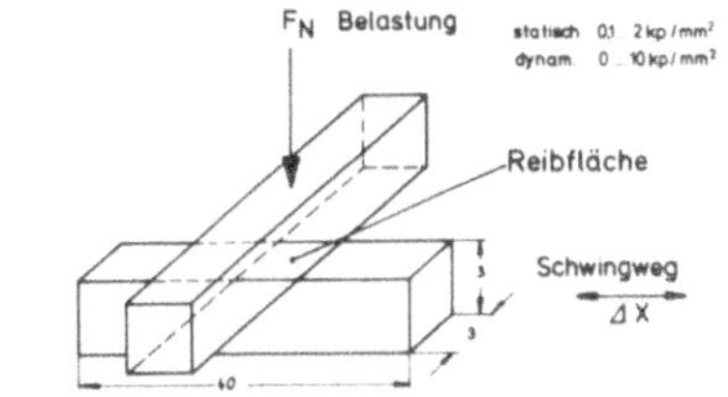

Bild 7: Prüfkörper (Stäbchen) für Schwingungsverschleiß-Versuche

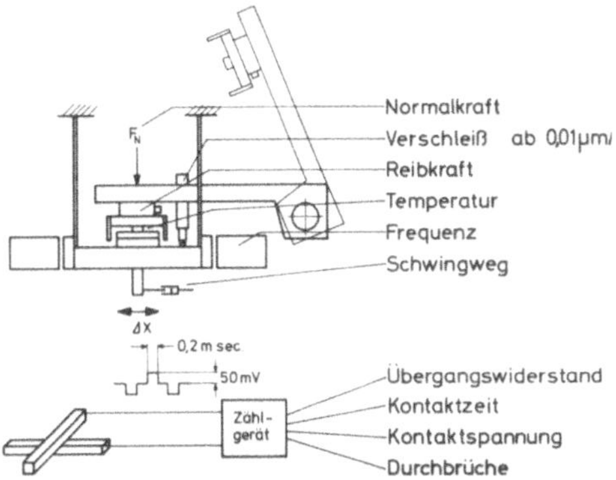

Bild 8: Prinzip der Schwingungsverschleiß-Prüfstände SVM 1,2 und 3 mit mechanischer Lagerung des unteren Probetisches durch Pendelstäbe und mechanischer Belastung des schwenkbaren oberen Probenhalters

Bild 9: Schwingungsverschleiß- "Modell"-Prüfstände SVM1 (rechts) mit elektromagnetischen Antrieben gleichzeitig in zwei Richtungen und SVM 2,3 (links) mit hydraulischem Antrieb

Bild 10: Schwingungsverschleiß "Modell"-Prüfstände SVM 4 und 5 mit hydraulischem Antrieb und hydraulischer Belastung.
Links im Bild das Hydraulikaggregat, rechts die Steuer- und Registriergeräte

Bild 11: Schwingungsverschleiß-"Modell"-Prüfstände SVM 4, 5, Belastungssystem aufgeklappt, obere und untere Probenhalter sichtbar, rechts der hydraulische Tischantrieb. Die unteren Probehalter sind hydrostatisch gelagert.

		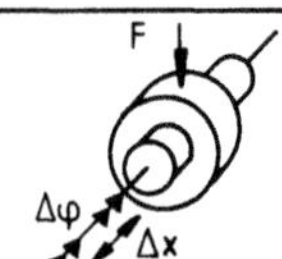		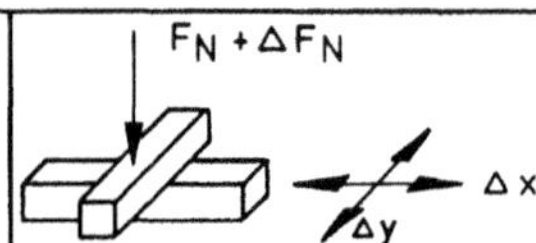		
		SVB 1 SVB 2	SVB 3	SVM 1	SVM 2 SVM 3	SVM 4 SVM 5
Last	F_N (N)	10...500	10...500	10...2000	10...2000	10...2000
	$\Delta F_N^{(1)}$ (N)			0...2000		0...2000
	f_F (Hz)			0...100		0...100
Bewegung	Δx (µm)	10...3000	0...10° ($\Delta\varphi$)	50...1000	50...3000	10...5000
	fx (Hz)	0...500	0...50	50	0...200	0...200
	Δy (µm) fy (Hz)			50...1000 0...300		
Fläche	a×b [mm]	d= 2...20	d= 2...20	3×3 6×6	3×3 6×6	3×3 6×6
Temp.	ϑ (°C)	200	200	20	20	200
Umgeb.	flüssig			×	×	×
	gasdicht			×	×	

(1) Last stoßend, schwellend o. schwingend

Tabelle 1: Parameterbereich der Schwingungs-Verschleiß-Prüfstände

Es sind verschiedene Schwingungsformen realisierbar, z.B. sinusförmig, dreieckförmig, stochastisch und Überlagerungen verschiedener Schwingungsformen. Die Variationsbreite der "Modell"-Prüfstände für SV (SVM1 bis 5) ist in Tabelle 1 zusammengefaßt.
Gemessen werden kontinuierlich der Verschleiß (≥ 0,01 µm/h) mit einem induktiven Weggeber zwischen unterem Probentisch und Schwenkarm, und mit einer auf Schub ansprechenden piezoelektrischen Meßplattform die Reibkraft. Das rechteckförmige Reibkraftsignal wird gleichgerichtet und auf einem Punktdrucker gemeinsam mit dem Verschleiß angezeigt.

4.3 Bauteil-Prüfstände für Schwingungsverschleiß

Für die Untersuchung schwingungsbeanspruchter Bauteile werden 3 verschiedene Bauteil-Prüfstände benutzt.

Bezeichnung	einige kennzeichnende Merkmale 3)
SVB1 SVB2	- elektromagnetischer Antrieb (Axial- bzw. Torsionsbewegungen)
SVB3	- mechanischer Antrieb (Torsionsbewegungen)

3) nähere Daten sind in Tabelle 1 angegeben

Zur Untersuchung von Bauteileigenschaften (z.B. der Einfluß von Lagerspielen) müssen Bauteilprüfstände verwendet werden.
Bild 12 zeigt den prinzipiellen Aufbau eines solchen Bauteilprüfstandes (SVB1,2). Ein elektromagnetischer Pulser bewegt axial die in einer Spannzange befestigte Welle. Die über die Welle geschobene Hülse wird mit einer Körnerspitze an der Bewegung in y-Richtung gehindert. Ein kugelgelagerter Schlitten trägt den Haltekörper und die piezoelektrische Reibkraft-Meßplattform. Mittels einer Abdeckhaube und einem Heißluftgebläse wird die gesamte Prüfeinrichtung bis 120°C beheizt.
Der Aufbau dieser Prüfstände wird immer den jeweiligen Erfordernissen angepaßt.
Bild 13 zeigt eine Teilansicht des umgebauten Bauteilprüfstands SVB1.
In Bild 14 ist die Gesamtanlage für SVB1 und 2 dargestellt.

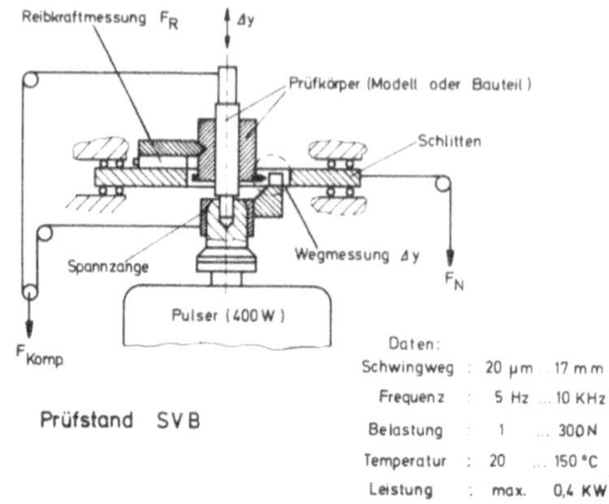

Bild 12: Prinzip der Schwingungsverschleiß-Bauteilprüfstände SVB 1 und SVB 2. Die Prüfstände werden dem jeweiligen Anwendungsfall angepaßt.

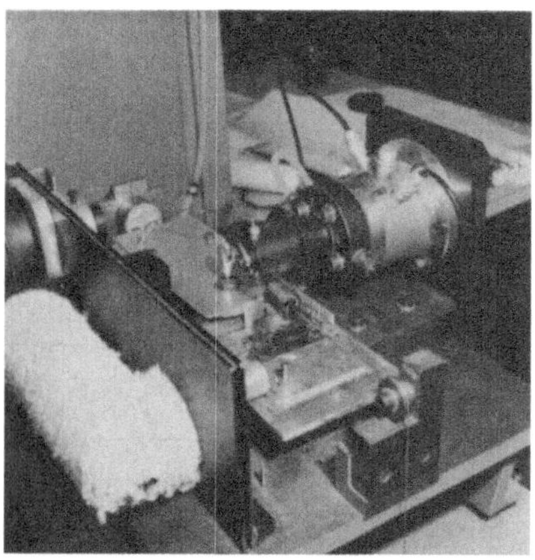

Bild 13: Schwingungsverschleiß-Bauteilprüfstand SVB 1 mit eingespanntem Fliehgewichtshebel zur Lagerprüfung. Erfassung des Reibmomentes mittels Kistler-Meßnabe (rechts). Belastung mechanisch über Seilzug und Gewicht.

Bild 14: Schwingungsverschleißprüfstände SVB 1 und SVB 2. Variabler Prüfaufbau, Schwingungsbewegung elektromagnetisch erzeugt.

Treten in einer Lagerung Torsionsschwingungen auf, (Beispiel einer Probenform s. Bild 15), so können diese Schwingungsverschleißvorgänge mit der im Bild 16 schematisch dargestellten Prüfeinrichtung SVB3 getestet werden. Eine Spannzange wird mechanisch über eine Schwinge angetrieben. Die Schwinge ist an einen verstellbaren Exzenter angekoppelt, der über Zahnriemen von einem Gleichstrommotor angetrieben wird. Die Spannzange ist in einem Wälzlager gelagert, wobei die Wälzlager durch eine Hülse und einen Servomotor langsam gedreht werden, um Schwingungsverschleiß in den sonst nur oszillierenden Wälzlagern zu vermeiden. Auch diese Prüfeinrichtung ist mit einer Abdeckhaube ausgerüstet, so daß die gesamte Versuchseinrichtung über heizbare Luft bis auf 120°C aufgeheizt werden kann. Bild 17 zeigt eine Teilansicht dieses Prüfstandes.

4.3 Gleitverschleiß "Modell"-Prüfstand

Für die Untersuchung von Kunststoff-Stahl-Paarungen werden außer den Schwingungsverschleiß-Versuchen auch Gleitverschleiß-Versuche mit einer Stift-Scheibe-Prüfmaschine (GV1) durchgeführt, um unter anderem festzustellen, ob sich unter Gleitverschleiß die gleichen Bewährungsfolgen ergeben wie unter SV. Wäre dies der Fall, so könnten die vielen Literaturangaben über Kunststoffverschleiß auch für SV-Probleme genutzt werden.
Der konstruktive Aufbau ist in Bild 18 dargestellt. Ein regelbarer Scheibenläufermotor (n = 50 ... 1000 U/min) treibt die Prüfscheibe von 20 mm Durchmesser an. Als Stift werden 3x3 mm Proben von den Stäbchen für SV-Versuche benutzt und in eine Halterung eingespannt. Die Belastung erfolgt mechanisch über Zustellung einer Feder. Belastung und Reibkraft werden über eine piezoelektrische Meßplatte erfaßt, der Verschleiß durch einen induktiven Taster gemessen.
Bild 19 zeigt einen Überblick über den Gleitverschleiß-"Modell"-Prüfstand GV1.

4.4 Versuchsdurchführung

Alle Proben werden vor dem Versuch im Ultraschallbad mit Aceton gereinigt. Kunststoffe, die gegen Aceton unbeständig sind, werden mit FAM-Benzin gereinigt. Nach dem Einbau und Ausrichten werden die Proben noch einmal mit Aceton (Benzin) abgewischt und mit einem Föhn getrocknet.

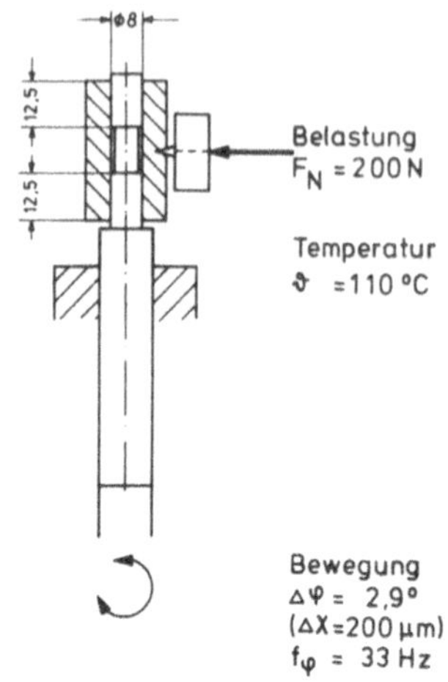

Bild 15: Beispiel der Bauteil-Versuchsbedingungen des Prüfstandes SVB 3 für Torsionsschwingungen

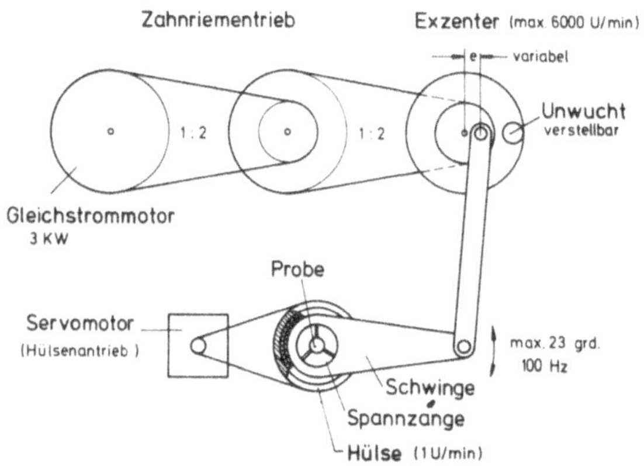

Bild 16: Prinzip des Schwingungsverschleiß-Bauteilprüfstandes SVB 3

Bild 17: Schwingungsverschleiß-Bauteilprüfstand für Torsionsschwingungen, links im Bild die Heißluftzufuhr

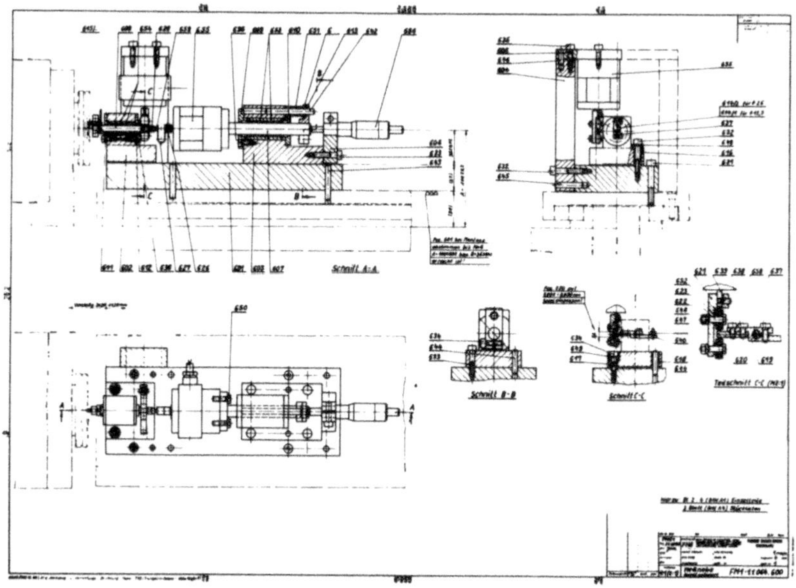

Bild 18: Gleitverschleiß-Prüfstand GV 1
nach dem Stift-Scheibe Prinzip

Bild 19: Gleitverschleiß-Prüfstand GV 1
mit Scheibenläufermotor (links),
mechanischer Belastung über Stell-
spindel (rechts) und Feder

In Vorversuchen wird der Einfluß einzelner, seither noch nicht ausreichend untersuchter Beanspruchungsparameter und Strukturparameter getestet. Diesen Versuchen sind keine starren Versuchsprogramme zugeordnet, sondern eine vielseitige und variable, den Umständen angepaßte Versuchsführung ist hier Voraussetzung für das Erkennen von Einflußgrößen, Wechselwirkungen und Prozessen.

In den Hauptversuchen wird das Verschleißverhalten verschiedener Werkstoffpaarungen mit und ohne flüssige Zwischenmedien festgestellt. Insbesondere für die Kunststoff-Stahl-Paarungen wird mit Standard-Versuchsprogrammen gefahren, um eine einheitliche Basis für die Bewährungsfolgen zu haben. Die Standard-Versuchsprogramme basieren auf den Erfahrungen der Vorversuche.

Standard-Versuchsprogramme:

Programm 1	Bewährungsfolge von Kunststoff-Metall-Paarungen [3]	
	Bestimmung von Reibung und Verschleiß bei einer Standardkombination der Beanspruchungsarten	
	Bewegung	
	- Schwingweite	Δx = 2000 µm
	- Schwingungsform	= sinus
	- Schwingungsfrequenz	f_x = 2,5 Hz
	Last	
	- Belastung konst. [4]	F = 100 N (Paarung geschmiert)
		= 25 N (Paarung ungeschmiert)
	Temperatur	= 30°C
	Kontaktfläche	A = 3 x 3 = 9 mm^2
	Zwischenmedium	= frei wählbar
	Umgebungsbedingungen	= Luft, relat.Feuchte 40%
	Laufzeit	t = 20 ... 70 h

3) Bei Metall-Metall-Paarungen werden etwas geänderte Beanspruchungsbedingungen festgelegt: Δx = 100 µm, Sinusschwingung, Frequenz f_x = 100 Hz.

4) Die angegebenen Kräfte stellen Maximalwerte dar. Durch Vorversuche mit 2 niedrigeren Kräften ist bei jedem Kunststoff zu prüfen, ob der Verschleiß auch bei der Maximalkraft noch proportional zur Kraft ist.

Programm 2: Verschleißverhalten bei verschiedenen Temperaturen (Thermisches Verschleißkennfeld)

Es gelten die gleichen Werte wie für Programm 1, nur wird hier bei 3 verschiedenen Temperaturen geprüft (30°C, 80°C und 120°C). Im Bedarfsfall wird mit noch höherer Temperatur gefahren, um die obere thermische Einsatzgrenze zu erfassen.

Programm 3: Verschleißverhalten bei verschiedener Schwingungsweite (Schwingungsverschleiß-Kennfeld)

Die Prüfdaten sind wieder die gleichen wie in Programm 1, bis auf die Schwingungsweite, welche in folgenden Stufen variiert wird:
Δx in µm = 100, 200, 500, 1000, 2000 und 3000. Ergänzend wird noch ein Gleitverschleißversuch gefahren.

5. Versuchsergebnisse

Der Umfang der Meßaufgabe gestattete es zeitlich nicht, z.B. alle denkbaren und meßtechnisch möglichen Beanspruchungs- und Strukturvarianten zu fahren. Es mußte eine Auswahl getroffen werden. Trotzdem ergeben sich aus der Vielzahl und der Vielseitigkeit der Ergebnisse für den Schwingungsverschleiß allgemein gültige Erkenntnisse, welche in den folgenden 3 Kapiteln zusammengefaßt sind.

Zunächst wird in Kapitel 5.1 die Auswirkung der wesentlichsten Einzelparameter auf den Schwingungsverschleiß dargestellt.

5.1 Der Einfluß einzelner Beanspruchungs- und Strukturparameter auf den Schwingungsverschleiß

5.1.1 Schwingungsweite Δx

Der Einfluß der Schwingungsweite Δx auf den Schwingungsverschleiß wird an Hand von drei in der Praxis häufig eingesetzten aber strukturell verschiedenen Werkstoffpaarungen beschrieben.

1. Ungefüllter Thermoplast gegen Stahl

PA66 - 100 Cr6H	Ultramid A3W Stahl geschliffen in Bewegungsrichtung R_{3Z} = 1....2,5 µm Härte HV = 825 (für 100Cr6H)

2. Fasergefüllter Thermoplast gegen Stahl

PA66 + 20 % CF - 100 Cr6H	Thermocomp RC 1004 HS Kohlefaser Typ HS kurzgeschnitten regellos in PA66 verteilt

3. Stahl gegen Stahl

100 Cr6H - 100 Cr6H und 100 Cr6H - C15	Härte von C15: HB_{30} = 146

Bei der Paarung PA66 - 100 Cr6H ist, wie Bild 20 zeigt, die spezifische Verschleißrate K unabhängig von der Schwingungsweite Δx, sowohl bei sehr kleinen Schwingungsweiten (< 100 µm) als auch bei großen Schwingungsweiten bis hin zum reversierenden Gleitverschleiß (rev. GV) und reinen Gleitverschleiß (GV). Der hier ermittelte Wert von $K = 6{,}5 \frac{\mu m}{km \cdot \frac{N}{mm^2}}$ stimmt gut mit Werten aus der Literatur überein.

rev. GV:

DOWSON [7]	$K = 5{,}5 \ldots 9{,}3\ \mu m/(km \cdot N/mm^2)$	$\Delta x = 19$ mm Proben Ø 3 mm F = 25 N .... 67 N

GV:

VDI-Richtlinie 2541 [34]	K = 4	µm/(km · N/mm²)
	(Gegenprobe $R_z = 1$ µm)	
LNP Katalog [24]	K = 4	"
ESDU Datenblatt [11]	K = 4,4	"

Auf Grund dieser bemerkenswerten Tatsache, daß die spez. Verschleißrate bei PA 66 - 100 Cr6H unabhängig von der Schwingungsweite ist und auch mit der bei Gleitverschleiß ermittelten Verschleißrate übereinstimmt, können die in der Literatur vorhandenen Daten für Gleitverschleiß auf den Schwingungsverschleiß übertragen werden. Wie weiter unten gezeigt wird, gilt dies nur für ungefüllte Kunststoffe.[5] Das verwendete PA66 hat im Vergleich zu anderen Thermoplasten den besonderen Vorteil, bei Gleitverschleiß von der Rauhigkeit des Gegenwerkstoffes wenig abhängig zu sein. In der Literatur wird berichtet, daß die Paarung PA66 gegen Stahl zu Reibrostbildung führt. Bei Schwingungsweiten über 100 µm wurde bei den hier durchgeführten Versuchen (Stahlhärte HV = 825) keine Reibrostbildung festgestellt. Die Gefahr der Tribooxidation besteht offenbar nur bei sehr kleinen Schwingungsweiten unter 100 µm und ist zudem von der Härte des Stahles abhängig.

5) Auch bei den ungefüllten Kunststoffen gibt es Typen mit großen Unterschieden SV zu GV.

Zusammenfassend kann festgestellt werden, daß Paarungen aus ungefüllten Thermoplasten gegen gehärteten Stahl im Trockenlauf zu den relativ einfachen Tribosystemen gehören, unter anderem deshalb, weil der Gegenpartner während des Verschleißvorganges nicht angegriffen wird. Demzufolge sind auch "Modell"-Versuchsergebnisse gut auf Erzeugnisse übertragbar.

Bei der Paarung PA66 + 20 % CF gegen 100 Cr6H liegt, wie Bild 21 zeigt, ein völlig anderes Verschleißverhalten in Abhängigkeit von der Schwingungsweite Δx vor. Die spezifische Verschleißrate K ist abhängig von der Verschleißart.

Beim Schwingungsverschleiß (SV) können eine Verschleißtieflage und eine Verschleißhochlage festgestellt werden.

In der Verschleißtieflage tritt der Verschleiß nur auf der Kunststoffoberfläche auf. Nach längerer Laufzeit (die Zahlen an den Meßpunkten der Hochlage von Bild 21 geben die Laufzeit in Stunden an) wirkt die Kohlefaser abrasiv auf die Stahlgegenfläche, die Fläche wird aufgerauht und verschleißt wiederum verstärkt die Kunststoffprobe, die Verschleißhochlage ist erreicht.

Infolge erschwertem Verschleißabtransport geht die spezifische Verschleißrate bei kleiner werdenden Schwingungsweiten zurück.

Bei reversierendem Gleitverschleiß (rev. GV) ist nur im Übergangsbereich zum SV ($\Delta x \approx$ Probenlänge) eine Tief- und Hochlage des Verschleißes vorhanden, im restlichen Bereich existiert nur die Tieflage.

In diesem Übergangsbereich geht die spezifische Verschleißrate mit steigender Schwingungsweite stark zurück und die Laufzeiten bis zum Sprung in die Hochlage werden größer, bis nur noch die Tieflage auftritt. Hier wirkt die Kohlefaser zwar auch abrasiv auf den Stahl, aber die Stahloberfläche wird nicht so stark wie beim Schwingungsverschleiß aufgerauht.

Bei reinem Gleitverschleiß (GV) existiert nur noch die Verschleißtieflage. Die Kohlefasern wirken polierend auf die Stahlgegenfläche mit entsprechend geringer Verschleißrate.

Bei solchen Paarungen mit abrasiver Wirkung der Füllstoffe (ähnliche Verhältnisse, aber mit noch stärkerem Abrasivverschleiß erhält man auch mit Glasfaser) muß die Schwingungsweite genau beachtet werden, wenn man "Modell"-Versuche auf das Erzeugnis übertragen will.

Der Einfluß der Schwingungsweite auf die Paarung 100 Cr6H gegen 100 Cr6H bzw. zum Vergleich 100 Cr6H gegen C15 weich zeigt Bild22 .

Auffällig ist der Verschleißrückgang beim Übergang vom reversierenden Gleitverschleiß zum Schwingungsverschleiß, der im Bereich $\Delta x < 100$ µm (Reibrostbildung) noch stärker abfällt. Ursache ist die sich bildende Zwischenschicht aus vorwiegend oxidischen Verschleißpartikeln, die verschleißmindernd wirkt und deren Dicke bei kleiner werdender Schwingungsweite wächst, da die Verschleißpartikel nur noch schwer abtransportiert werden.

Bei reinem Gleitverschleiß hat die Paarung 100 Cr6H - 100 Cr6H eine kleinere spez. Verschleißrate ($K = 18$ µm/(km • N/mm^2)) gegenüber 100 Cr6H - C15 weich ($K = 40$ µm/(km • N/mm^2)). Im Schwingungsverschleißbereich $\Delta x < 200$ µm kehren sich die Verhältnisse um, da sich die harten Verschleißpartikel am weicheren Werkstoff festsetzen und ihn damit schützen.

Abschließend soll zur Auswirkung der Schwingungsweite auf die spez. Verschleißrate noch auf zwei Richtungen von Maßnahmen zur Verschleißreduzierung bei kohlefasergefüllten Thermoplasten bei Trockenlauf hingewiesen werden.

Geht man von PA66 + 20 CF aus (Bild 21) so kann man durch Zusatz von Teflon (PTFE) die abrasive Wirkung der Kohlefaser auf den Stahl vermindern, ohne sie allerdings völlig zu vermeiden (Bild 23). Die Verschleißhochlage ist nicht so stark ausgeprägt und die Laufzeit bis zur Verschleißhochlage ist deutlich größer (Zahlen an den Meßpunkten in Stunden).

Statt eines inkorporierten Festschmierstoffes kann man als zweiten Weg zur Reduzierung des Abrasivverschleißes die Härte der Gegenfläche erhöhen.
In Bild 23 ist Hartmetall gewählt worden. Es genügen aber auch schon harte Oberflächenschichten um 1000 HV. Die verschleißmindernde Wirkung der Härte bleibt auch bei PA66 + 20 CF ohne Teflonzusatz erhalten. Im dargestellten Fall geht der Verschleiß fast um den Faktor 10 zurück.

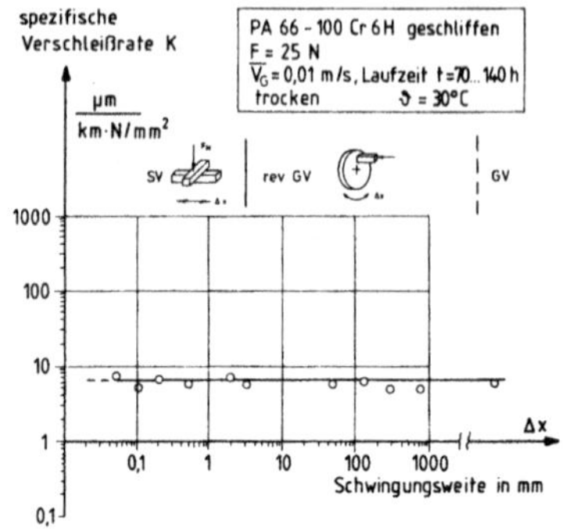

Bild 20: Spez. Verschleißrate K über der Schwingungsweite Δx für PA 66. Die spez. Verschleißrate ist unabhängig von der Schwingungsweite.

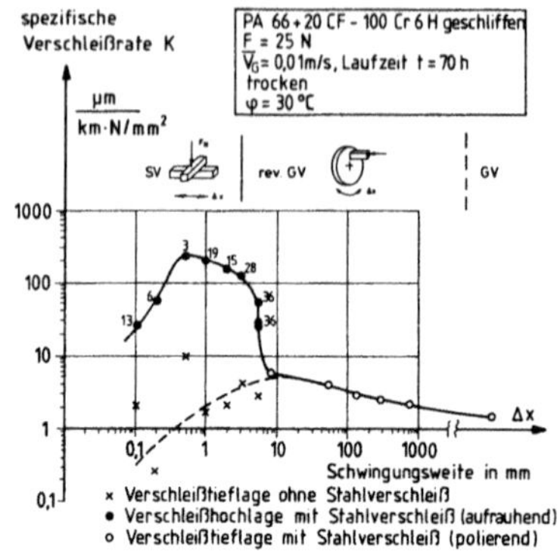

Bild 21: Spez. Verschleißrate K über der Schwingungsweite Δx für PA 66 + 20 CF. Im SV-Gebiet nach kurzer Laufzeit (Zahlenwert in h) infolge Abrasion Verschleißhochlage. Bei rev. Gleitverschleiß geht Hochlage zurück infolge Polierwirkung der Fasern.

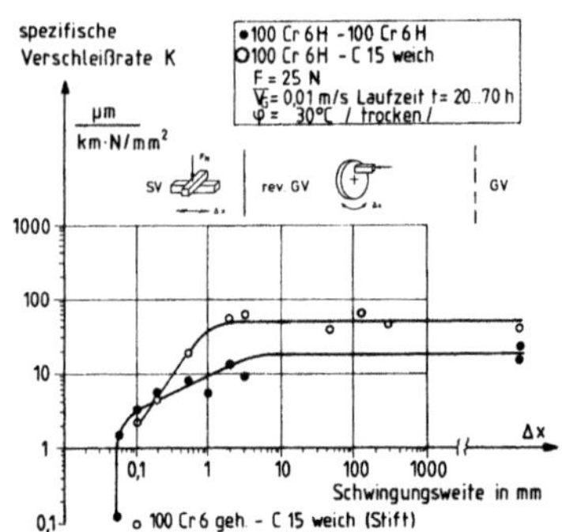

Bild 22: `Spez. Verschleißrate K über der Schwingungsweite Δx für Stahl-Stahl-Paarungen. Infolge erschwertem Verschleißtransport und gleichzeitiger Schutzwirkung der Zwischenschicht geht K im SV-Gebiet mit kleiner werdender Schwingungsweite zurück.

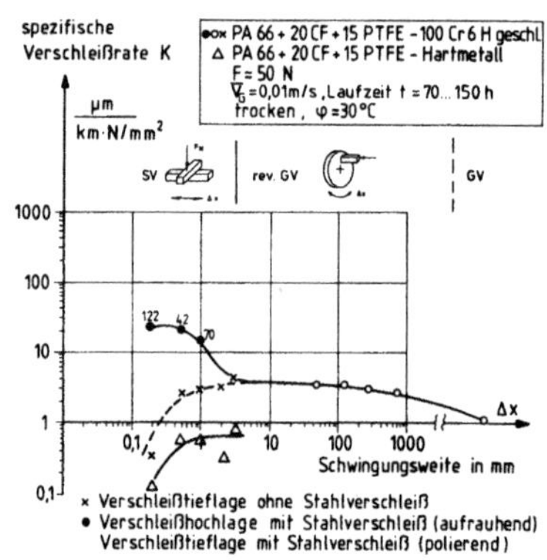

Bild 23: Spez. Verschleißrate K über der Schwingungsweite Δx für PA 66 + 20 CF + 15 PTFE gegen 100Cr6H bzw. Hartmetall. Man erkennt die verschleißmindernde Wirkung des PTFE und die Vermeidung der Abrasion (keine Hochlage) bei Hartmetall.

5.1.2 Frequenz

Innerhalb des untersuchten Frequenzbereiches von 1...150 Hz und einer Schwingungsweite Δx = 3000 .... 50 µm entsprechend einer mittleren Gleitgeschwindigkeit von 0,006.... 0,15 m/s wurde kein Frequenzeinfluß auf die spezifische Schwingungsweite festgestellt.

5.1.3 Schwingungsform

Die Schwingungsform wurde durch Überlagerung zweier Sinusschwingungen sowohl in der gleichen Richtung (Δx) als auch in zwei senkrecht aufeinanderstehenden Richtungen ($\Delta x + \Delta y$) variiert.

In beiden Fällen kann die Schwingungsform die Verschleißrate stark beeinflussen. Ein Beispiel für die Auswirkung auf den Verschleißverlauf bei Schwingungsüberlagerung in einer Richtung wird in Kapitel 5.2 (Bild 32) beschrieben.

Hier werden einige Hinweise zum Verschleißverhalten bei Überlagerung zweier Schwingungen in x und y-Richtung gegeben.

1. System: Stahl-Stahl trockenlaufend | 100 Cr6H-100 Cr6H, F_N = 10 N, VNr. 613/SVM1

Tabelle 2 enthält die spezifische Verschleißrate K für drei verschiedene Frequenzverhältnisse der Schwingungen in x und y-Richtung. Je nach Frequenzverhältnis erhält man unterschiedliche Verschleißraten. Ursache ist hauptsächlich der geänderte Stofftransport in der Kontaktfläche, worauf in Kapitel 5.3 näher eingegangen wird. Das gleiche Verhalten zeigt auch PA66 gegen 100Cr6H.

Varianten Nr.	Δx µm	fx Hz	Δy µm	fy Hz	K $\frac{\mu m}{km \cdot N/mm^2}$
1	-	-	40	150	1
2	40	50	40	51	0,7
3	40	50	40	150	5,4

Tabelle 2: Schwingungsformeinfluß bei Stahl-Stahl trocken.

2. System: Stahl-Stahl geschmiert | 100 Cr 6 H - 100 Cr6H in Dieselkraftstoff zusätzliche Schlagbeanspruchung

Bild 24 zeigt den Verschleißverlauf über der Laufzeit bei verschiedenen Frequenzverhältnissen in den beiden Schwingungsrichtungen x und y.

Wegen der überlagerten Schlagbelastung konnte die spezifische Verschleißrate K nicht exakt berechnet werden. Deshalb ist hier der Verschleißverlauf W_1 angegeben.

Die Verschleißrate ist bei Schwingungen in einer Richtung klein (V.Nr. 479).

Bei zusätzlichen Schwingungen in y-Richtung steigt der Verschleiß stark an (V.Nr. 491, 495). Es genügt bereits eine Schwingung in y-Richtung von 2 Hz, um den Verschleiß zu erhöhen. Die Verschleißrate steigt weiter, wenn die Frequenz z.B. in y-Richtung größer als in x-Richtung ist (V.Nr. 477).

Während im 1. Beispiel bei Trockenlauf (Tabelle 2) der Stofftransport maßgebend für die sich einstellende Verschleißrate ist, wirken sich z.B. unter Kraftstoffschmierung zusätzlich makroskopische Geometrieänderungen der Verschleißfläche auf die Verschleißrate aus.

Durch die zusätzliche y-Schwingungen entstehen aus den planparallelen Kontaktflächen leicht gewölbte Sattelflächen, sodaß auf Grund der stoßartigen Belastung der Schmierfilm stärker verdrängt wird, was eine Erhöhung der Verschleißrate zur Folge hat.

Zusammenfassend kann über den Einfluß der Schwingungsform festgehalten werden: Obwohl über den Einfluß von Frequenz und Schwingungsweite in x und y-Richtung keine systematischen Versuche durchgeführt werden konnten, wird aufgrund der vorliegenden Ergebnisse deutlich, daß in praktischen Schwingungsverschleiß-Systemen (geschmiert oder trocken) auf Schwingung in unterschiedlicher Richtung geachtet werden muß, da derartige Schwingungsformen die Verschleißrate wesentlich beeinflussen können.

5.1.4 Rauhigkeit

Bei metallischen Werkstoffpaarungen unter Trockenlauf spielt die Rauhigkeit beim Schwingungsverschleiß eine untergeordnete Rolle, da die Ausgangsrauhigkeiten während des Einlaufes stark verändert werden.

Für Paarungen Kunststoff gegen Metall liegen in der Literatur umfangreiche Ergebnisse bezüglich des Rauhigkeitseinflusses auf den Verschleiß vor. Danach hängt der Einfluß der Rauhigkeit stark vom Kunststofftyp ab.

VDI-Richtlinie 2541 [34]:

Mit steigender Rauhigkeit nimmt der Verschleiß zu. Bei PA66 ist nach diesen Untersuchungen der Einfluß gering.

Zur Überprüfung des Rauhigkeitsverhaltens bei Schwingungsverschleiß wurde eine PA66- Stahl Paarung gewählt.

Bild 25 zeigt, daß auch unter Schwingungsverschleißbedingungen bei PA66 die Rauhigkeit keinen Einfluß auf den Verschleiß hat.

Dies gilt auch für unterschiedliche Bearbeitungsrichtungen.

5.1.5 Geometrie

Die Geometrie der Kontaktflächen ist beim Schwingungsverschleiß eine wichtige Größe.

1. Die Größe der Fläche beeinflußt den Stofftransport (Verschleißabfluß). Die Zahl der durchgeführten Versuche reicht noch nicht aus, um allgemeine Schlußfolgerungen zu ziehen.
2. Bei geschmierten Schwingungsverschleiß-Systemen, insbesondere bei überlagerter Schlagbeanspruchung, werden zusätzlich hydrodynamische Effekte wirksam, die geometrieabhängig sind und damit die Verschleißrate beeinflussen.

Zu 2. gibt Bild 26 einige Hinweise auf den Geometrieeinfluß. Die Paarungen bestehen aus 100 Cr6H gegen 100 Cr6H. In allen Versuchen ist eine schlagende Belastung überlagert, was zu einer Quetschströmung mit hydrodynamischer Tragwirkung führt.

Im Versuch 491 sind glatte Stäbchen eingesetzt. Nach kurzer Laufzeit erfolgt ein Verschleißanstieg infolge leichter Wölbung der Kontaktflächen, die bereits in Abschnitt 5.1.3 beschrieben wurden.

Prüft man Stäbchen mit Randnuten (V.Nr. 487), um Kanteneffekte zu vermeiden, steigt der Verschleiß an, der Effekt ist aber nicht bedeutsam.

Viel deutlicher wirkt sich eine zusätzliche Mittelbohrung auf den Verschleiß aus (V.Nr. 486).

Der verschleißerhöhende Einfluß ist nicht auf die durch die Bohrung erhöhte Flächenpressung von nur 2 % zurückzuführen, sondern auf die verminderte hydrodynamische Tragwirkung.

Die Verschleißrate erhöht sich weiter (V.Nr. 485), wenn in eine Kontaktfläche 5 Mikronuten (20 µm breit, 50 µm tief) eingearbeitet werden.
Die Ursache der Verschleißerhöhung ist die Minderung des hydrodynamischen Traganteils, da die Kontaktfläche in 6 Einzelflächen unterteilt wird.

Bild 27 zeigt, daß sich die Verschleißrate nur bei zusätzlicher Schlagbeanspruchung erhöht. Bei konstanter Belastung wirken die Nuten unter Schwingungsverschleißbedingungen sich nicht auf die Verschleißrate aus, auch wenn die konstante Belastung höher ist als die gemittelte Belastung bei Schlagüberlagerung.

Zusammenfassend haben die Versuche gezeigt, daß die Verschleißrate von Schwingungsverschleißsystemen unter flüssigen Umgebungsmedien und Schlagbelastung wesentlich von der Geometrie der Kontaktflächen beeinflußt wird.

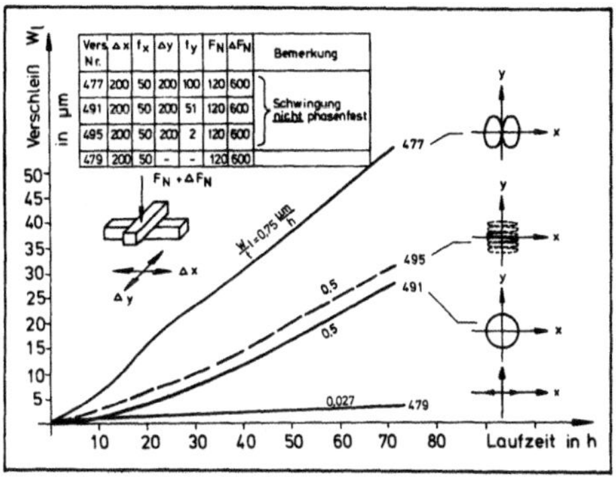

Bild 24: Einfluß der Schwingungsform auf den Verschleiß von 100 Cr6H-100Cr6H, kraftstoffgeschmiert und schlagend belastet

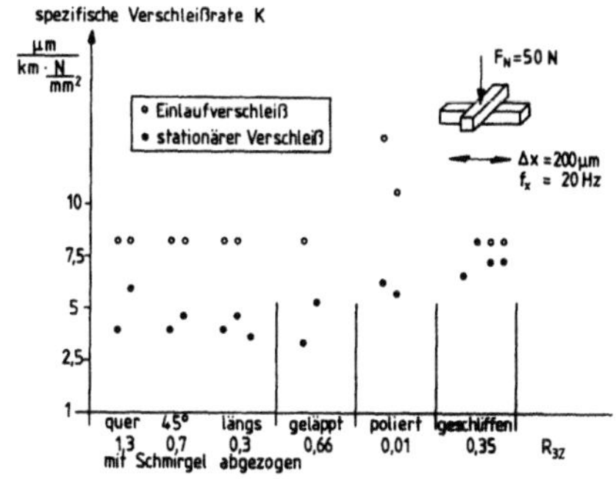

Bild 25: Rauhigkeitseinfluß auf die spezifische Verschleißrate bei PA 66 gegen 100Cr6H. Es wurde die Rauhigkeit des Stahls variiert.

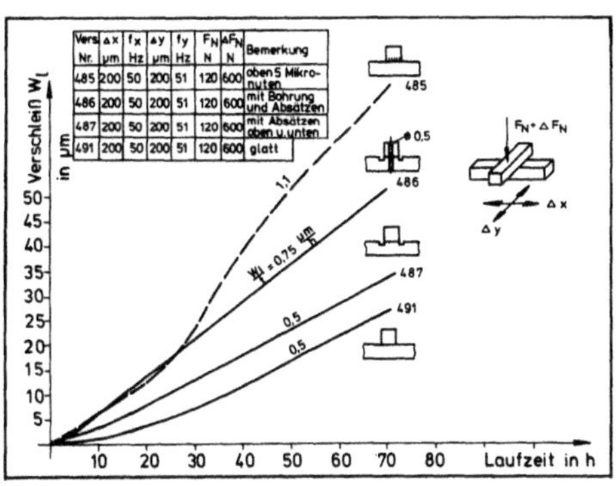

Bild 26: Geometrieeinfluß auf den Verschleiß bei schlagend belasteten Systemen 100Cr6H-100Cr6H kraftstoffgeschmiert. Den stärksten Verschleiß gibt es bei Mikronuten in einem Probestäbchen (V.Nr.485).

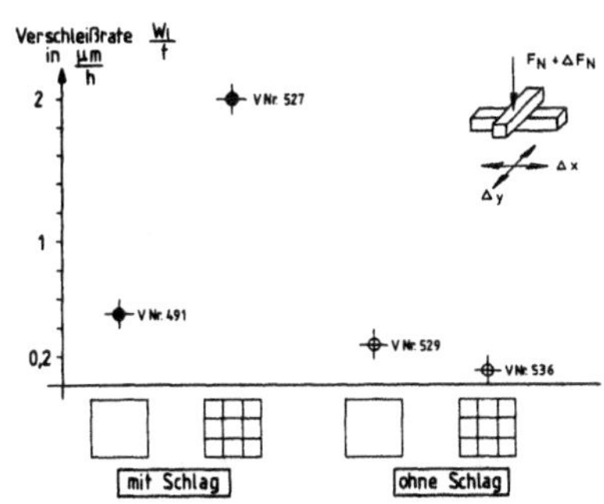

Bild 27: Einfluß von Mikronuten in der Kontaktfläche auf den Schwingungsverschleiß bei schlagend belasteten Systemen 100Cr6H-100Cr6H, kraftstoffgeschmiert

5.2. Tribologische Vorgänge

In diesem Abschnitt werden an einigen ausgewählten Systemen die tribologischen Vorgänge beim Schwingungsverschleiß näher beschrieben, d.h. die zeitlichen Verläufe von Reibung und Verschleiß, die Strukturveränderungen und die beteiligten Mechanismen. Die Vorgänge werden wieder wie bei Kapitel 5.1 an Hand der 3 folgenden Systemgruppen beschrieben:

1. Ungefüllter Thermoplast gegen Stahl

   PA66 gegen 100 Cr6H | gefettet mit Klüber, Topas NB 52 synthetischer Kohlenwasserstoff mit Bariumkomplexseife, Oxidations- und Korrosionsinhibitor, EP Zusatz

2. Faserverstärkter Thermoplast gegen Stahl

   PA66 + 20 % Kohlefaser gegen 100 Cr6H | gefettet wie bei 1

3.1 Stahl gegen Stahl trockenlaufend

100 Cr6H gegen 100 Cr6H

3.2 Stahl gegen Stahl geschmiert mit Diesel-Kraftstoff

100 Cr6H gegen 100 Cr6H

Die Beanspruchungsbedingungen sind bei den nachfolgenden Bildern jeweils angegeben.

System 1 (Ungefüllter Thermoplast - Stahl) ist relativ einfach, da sich der Verschleißvorgang auf den Kunststoff beschränkt. Im Beispiel ist die Paarung gefettet, was aber gegenüber einem Trockenlauf am Charakter des Vorganges nichts ändert. Im Bild 28 sind Reibung und Verschleiß über der Versuchszeit aufgezeichnet. Man kann 2 Phasen unterscheiden:

Phase I Einlaufphase mit ansteigender Reibungszahl und degressivem Verschleißverlauf. Die Verschleißmechanismen sind adhäsiv und abrasiv. Dieses Verhalten findet man auch bei trockenlaufenden Schwingungsverschleiß-Systemen und beim Gleitverschleiß. Im Einlauf bildet sich eine Zwischenschicht, die je nach Struktur und Eigenschaft das Reibungs- und Verschleißverhalten in der folgenden Phase II mit beeinflußt. Durch den Einlauf erhöht sich meist die reale Kontaktfläche und damit die Adhäsion und die Reibungszahl.

Phase II Stationäre Phase mit konstanter Reibungszahl und konstanter Verschleißrate. Solange im vorliegenden Beispiel das Fett wirksam bleibt und der Kunststoff seine Eigenschaften z.B. durch Alterung nicht ändert, verbleibt das System in diesem stationären Zustand. Die Oberflächen bleiben relativ glatt (Bild 29), sofern bei erhöhtem Verschleiß der Stofftransport die Oberflächenstruktur nicht beeinflußt (Kapitel 5.3). Der Verschleiß erfolgt überwiegend durch Adhäsion.

System 2 (Faserverstärkter Thermoplast - Stahl)
Die hier eingesetzten Faserwerkstoffe (Glasfaser, Kohlefaser) sind härter als der Gegenwerkstoff Stahl und wirken deshalb abrasiv auf diesen, womit das System 2 im Vergleich zu System 1 komplizierter wird (Bild 30).

Phase I Einlaufphase. Der Verlauf der Reibungszahl durchläuft in der Einlaufphase ein Maximum. Der anfängliche Reibkraftanstieg steht im Zusammenhang mit der wachsenden realen Kontaktfläche und der damit verbundenen größeren Adhäsion. Im Reibungsmaximum sind beide Oberflächen sehr glatt. Gleichzeitig entstehen Verschleißpartikel. Die Verschleißpartikel trennen stellenweise die Oberflächen, wodurch die Reibkraft wieder sinkt. Kurz vor der stationären Phase können die ersten Oberflächenschäden am Kunststoff in Form von Vertiefungen infolge behinderter Verschleißabfuhr beobachtet werden.

Phase II Der stationäre Zustand ist gekennzeichnet durch eine sehr konstante Reibungszahl. Neben dem adhäsiven Verschleiß wird eine steigende Oberflächenzerrüttung am Kunststoff beobachtet. Diese Ermüdung, Risse und Brüche zwischen Faser und Matrixwerkstoff sind bei Schwingungsverschleiß besonders intensiv, werden aber auch bei Gleitverschleiß beobachtet (CLERICO[4], ELIEZER, KHANNA[8]). Diese intensive Oberflächenzerrüttung gilt bei Schwingungsverschleiß allgemein für alle gefüllten Kunststoffe, auch wenn die Zusätze in Pulverform zugegeben werden, wie z.B. bei PA66 + Graphit. Die Schwingungsverschleißrate ist hier deutlich größer als bei reinem PA 66. Die Laufzeit der stationären Phase hängt vom Verschleißzustand des Kunststoffes, der Härte des Gegenwerkstoffes und bei geschmierten Proben vom Öl- oder Fettzustand ab.

Phase III Die Phase III, die Verschleißhochlage kündigt sich durch einen Anstieg der Reibungszahl an. Das Schmiermittel ist nicht mehr wirksam, so daß das System trocken läuft. Die in der Zwischenschicht liegenden Faserreste (Bild 31)wirken jetzt stark abrasiv auf die Stahlgegenfläche, es kommt zu Stahlabrieb, der sich in der Kunststoffoberfläche festsetzt. Bei Schwingungsverschleiß rauht sich nun die Stahlfläche auf, es entsteht Tribooxidation in Form von Reibrost auf den Kontaktflächen. Der Verschleiß von Stahl und Kunststoff steigen stark an.

System 3 (Stahl - Stahl) wird untergliedert in trockenlaufend und geschmiert, da hier im Gegensatz zu Kunststoff-Stahl-Systemen deutliche Unterschiede im tribologischen Verhalten zu beobachten sind.

System 3.1 (Stahl - Stahl trockenlaufend) Diese Systeme sind komplexer als die beiden vorhergehend betrachteten, da von Anfang an beide Körper verschleißen und damit auch die geometrische Struktur makroskopisch verändert wird, was sich wiederum auf das Verschleißverhalten auswirkt.

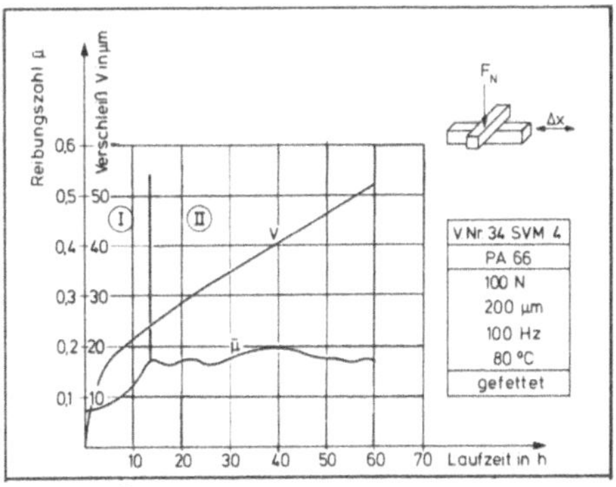

Bild 28: Reibungs- und Verschleißverlauf der Paarung PA66 - 100Cr6H.
Gebiet I: Einlaufphase
Gebiet II: stationäre Phase

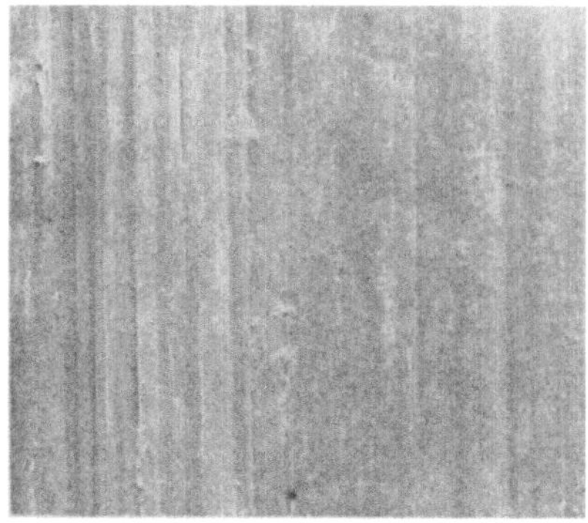

Bild 29: Oberfläche des PA 66 nach 60 h Laufzeit (VNr. 34/4). Die Oberfläche ist relativ glatt.
|——| 25 µm

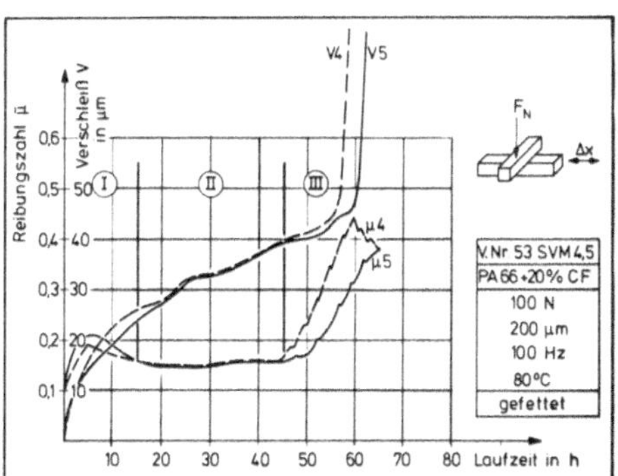

Bild 30: Reibungs- und Verschleißverlauf der Paarung PA66 + 20 CF-100Cr6H
Gebiet I: Einlaufphase
Gebiet II: stationäre Phase
Gebiet III: Verschleißhochlage

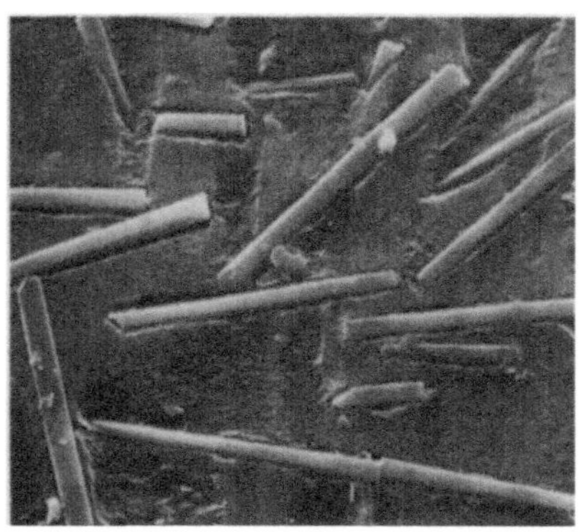

Bild 31: Oberfläche des PA66 + 20 CF nach 65 h Laufzeit (VNr. 53/4). Die abrasive Wirkung der Kohlefaser auf den Stahl bewirkt die Verschleißhochlage in Bild 30.
|——| 25 µm

Phase I Die relativ kurze Einlaufphase ist durch überwiegend adhäsiven Verschleiß gekennzeichnet und wurde in der Literatur eingehend beschrieben (WATERHOUSE [37]). Bemerkenswert ist in dieser Phase, daß infolge behindertem Verschleißabtransport sich eine Zwischenschicht aufbaut, die infolge Oxidation ihr Volumen vergrößert. Dadurch entsteht die Gefahr des Verklemmens bei Passungen und Führungen. Bei kontinuierlicher Verschleißmessung ist ein "negativer" Verschleiß festzustellen.

Phase II Die relativ kurze Phase I geht kontinuierlich in Phase II über. Bild 32 zeigt daher nur Phase II, der tribochemischen Oxidation und Oberflächenzerrüttung. Infolge des Verschleißes ändert sich im dargestellten Beispiel die Kontaktgeometrie, was zu einem degressiven Verschleißverlauf führt (Kurve 1). Durch Überlagerung einer langsamen Schwingung größerer Schwingungsweite bleibt die anfänglich hohe Verschleißrate erhalten (Kurve 2). Die Oberflächen bleiben dabei eben. Bild 32 zeigt, wie verschiedene Bewegungsformen den Verschleißverlauf entscheidend beeinflußen können.

System 3.2 (Stahl - Stahl, kraftstoffgeschmiert)

Bei diesem System ist der Verschleißverlauf von 100 Cr6H gegen 100 Cr6H progressiv (Kurve 1, Bild33) im Gegensatz zum Trockenlauf in Bild 32. Die Ursache für den progressiven Verlauf ist die Formänderung der Kontaktfläche, die allmählich ballig wird, sodaß der hydrodynamische Traganteil (Quetschströmung) während der Schlagbeanspruchung ständig kleiner wird.

Wählt man als Gegenpartner den verschleißfesteren Werkzeugstahl X82WMO Cr V 654, so bleiben die Kontaktflächen eben, der hydrodynamische Traganteil bleibt konstant und der Verschleißverlauf ist linear (Kurve 2).

Die Kontaktflächen bleiben für beide Werkstoffpaarungen metallisch blank.

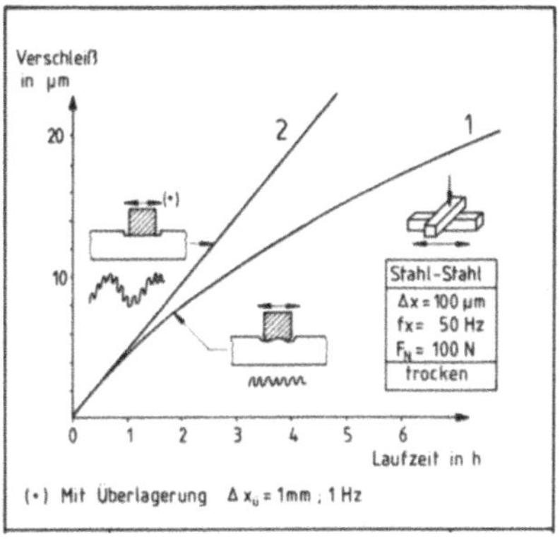

Bild 32: Verschleißverlauf der Paarung 100Cr6H - 100 Cr6H bei Trockenlauf. Ersichtlich ist die Wechselwirkung von Schwingungsform, Oberflächengeometrie-Änderung und Verschleißverlauf.

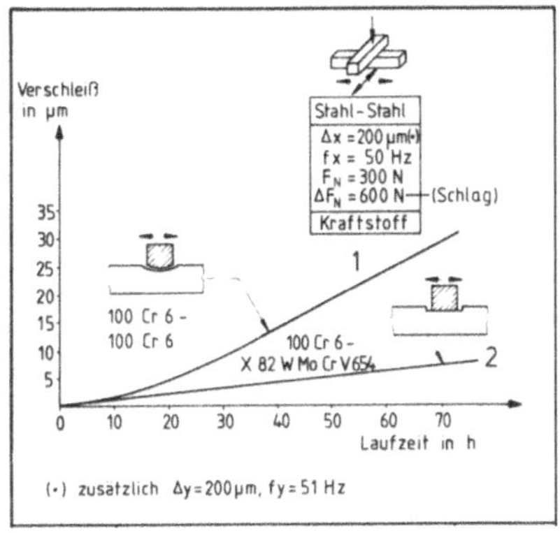

Bild 33: Verschleißverlauf der Paarungen 100Cr6H - 100Cr6H bzw. X82 WMoCrV 654 (leg. Werkzeugstahl), bei Kraftstoffschmierung. Wechselwirkung von Stoffpaarung, Oberflächengeometrie-Änderung und Verschleißverlauf.

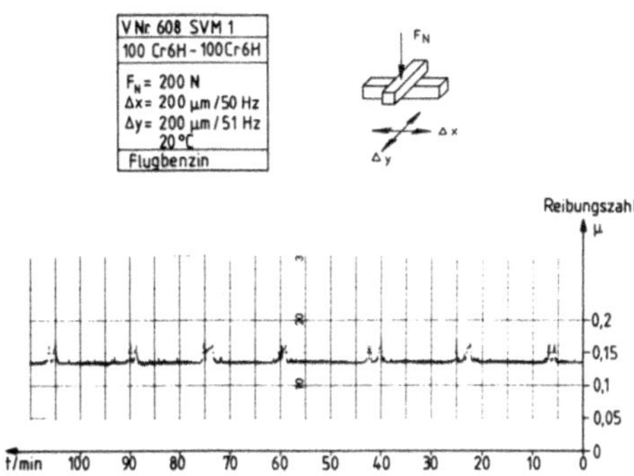

Bild 34: Zyklische Pulsation der Reibungszahl ("tribologische Uhr") einer kraftstoffgeschmierten Stahl-Stahl-Paarung unter Schwingungsverschleißbedingungen. Solche zyklischen Zustandsänderungen von Systemen werden neuerdings in der Thermodynamik der irreversiblen Prozesse unter allgemeinen Gesichtspunkten behandelt. Ursache ist in diesem Beispiel die Wechselwirkungen von Schmierstoffzufluß, Verschleißabfluß und Oberflächenrauhigkeit.

Bei den Schwingungsverschleißversuchen mit konstanter Belastung (ohne Schlag) wird eine zyklische Pulsation der Reibkraft beobachtet (Bild 34). Nach jeweils etwa 15 Minuten steigt die Reibkraft um 30 % an und fällt dann innerhalb kurzer Zeit (3 Minuten) wieder auf den alten Wert zurück. Während des Reibkraftanstieges wird die Oberfläche aufgerauht. Durch ungehinderten Schmierstoffzufluß verbessert sich der Schmierzustand und die Reibkraft fällt ab. Die Flächen glätten sich allmählich wieder durch Verschleiß und die Kontaktflächen nähern sich soweit, bis Mangelschmierung eintritt. Danach rauhen sich die Flächen auf und der Vorgang setzt erneut mit einem Reibkraftanstieg ein.

Die im System 3.2 beschriebenen Beispiele haben gezeigt, daß eine enge Wechselbeziehung zwischen der Kontaktgeometrie, den Transportvorgängen von Schmierstoff und Verschleißpartikel und den Reibungs- und Verschleißvorgängen besteht.

Bei der Simulation mit Modellsystemen muß deshalb Struktur und Beanspruchungskollektiv des entsprechenden Bauteils sehr genau nachgebildet werden.

5.3 Der Stofftransport beim Schwingungsverschleiß

Reibungs- und Verschleißvorgänge sind stets gekoppelt mit Transportvorgängen energetischer und stofflicher Natur.

Die Analyse von Tribosystemen nach energetischen Gesichtspunkten, d.h. eine Energiebilanzierung wurde besonders von FLEISCHER[13] in ausführlicher Form vorgenommen. Auch UETZ, FÖHL[33] untersuchten den Energietransformationsprozeß bei verschiedenen Verschleißvorgängen und gelangten dadurch zu einem tieferen Verständnis für das unterschiedliche Verhalten einzelner Systeme.

Bei Systemen mit Schwingungsverschleiß spielt nach unseren Beobachtungen die zweite Transportgröße, der Stofftransport die bedeutsamere Rolle und ist für die richtige Beurteilung solcher Systeme nicht zu vernachlässigen. Da die Schwingungsverschleiß-Systeme ständig überlappte Kontaktflächen haben, ist insbesondere der Transport der Verschleißpartikel innerhalb dieser Kontaktflächen und aus ihnen heraus entscheidend für die sich einstellenden Reibungs- und Verschleißvorgänge. Ebenso wichtig sind die Stoff-Zuflüsse von außen zwischen die Kontaktflächen, wie z.B. flüssige Schmierstoffe und Gase.

Die Transportmechanismen können in zwei Formen aufgeteilt werden, in Konvektions-und Leitungsvorgänge. Im Rahmen dieser Arbeit werden Leitungsvorgänge, d.h. Vorgänge im atomaren Bereich (z.B. Diffusion) nicht betrachtet.

Zu den betrachteten Konvektionsvorgängen zählen hier als besondere Gruppe der Transport kleiner, diskreter Teilchen, z.B. Ansammlungen und Transport von Verschleißteilchen.

Aus den geometrischen Veränderungen der Kontaktflächen kann man bei Schwingungsverschleiß Rückschlüsse auf die Mechanismen des Stofftransportes ziehen und die Auswirkungen auf das gesamte Verschleißverhalten besser verstehen. Je vielseitiger hierbei die Systemstrukturen und die Beanspruchungsbedingungen gewählt werden, d.h. je größer die Variationsbreite der untersuchten Systeme ist, desto eher erkennt man die Bedeutung des Stofftransportes bei Tribosystemen und insbesondere bei Systemen mit Schwingungsverschleiß.

Die Wirkung des Stofftransportes wird wieder an den drei unterschiedlichen Stoffpaarungen gezeigt.

1. <u>Ungefüllter Thermoplast gegen Stahl, trockenlaufend</u>

   PA66 - 100 Cr6H

2. <u>Fasergefüllter Thermoplast gegen Stahl, trockenlaufend</u>

   PA66 + 20 % CF - 100 Cr6H

3. <u>Stahl gegen Stahl</u>

3.1 <u>Stahl gegen Stahl, trockenlaufend</u>

   100 Cr6H - C 15

3.2 <u>Stahl gegen Stahl, geschmiert</u>

   100 Cr6H - 100 Cr6H

System 1 (PA66 - 100 Cr6H), obwohl zu den tribologisch einfachen Systemen gehörend (Gegenfläche ändert sich nicht), zeigt doch sehr deutlich die Auswirkung unterschiedlicher Beanspruchungsbedingungen auf das Verschleißbild und damit auch auf den Stofftransport der Verschleißpartikel.

Kleine Schwingungsweite in einer Richtung

Bild 35 zeigt die PA66-Oberfläche einer 3 x 3 mm-Probe nach einem Trockenlauf bei kleiner Schwingungsweite (Δx = 100 µm).

Δx	=	100 µm / 50 Hz
F_N	=	25 N
Laufzeit t	=	20 h

Die Verschleißmenge, genauer die Verschleißflußdichte, ist hier schon zu groß, um in Form feinster Verschleißpartikel abtransportiert zu werden ohne die Fläche zu schädigen. Infolge der kleinen Schwingungsweite sammelt sich der Verschleiß an zufälligen Stellen an, wird in Form von "Würstchen" quer zu Schwingungsrichtung gestreckt (Bild 36, Oberfläche vor der Reinigung) und überwiegend seitlich in Form wollförmiger Fäden herausgequetscht (Bild 37). Dabei drücken sich diese Verschleißrollen in die Kunststoffoberfläche hinein und erzeugen sogenannte Wurmspuren, wie man sie auch bei Stahl-Stahl-Paarungen beobachten kann, z.B. bei Zahnkupplungen).

Große Schwingungsweite in einer Richtung

Im nächsten Beispiel wurde die Schwingungsweite auf Δx = 2 mm erhöht.

Δx	=	2 mm / 2,5 Hz
F_N	=	25 N
Laufzeit t	=	20 h

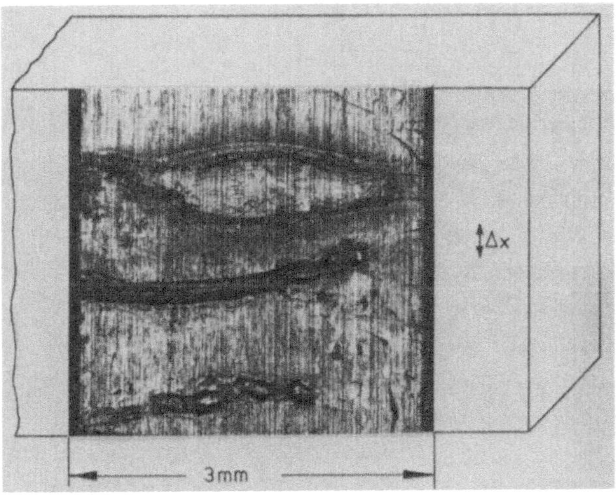

Bild 35: Wurmspurbildung auf PA66 bei kleiner Schwingungsweite Δx = 100 µm. Die Verschleißpartikel sammeln sich fadenförmig quer zur Schwingungsrichtung, drücken sich in die Oberfläche ein und werden seitlich herausgepreßt.

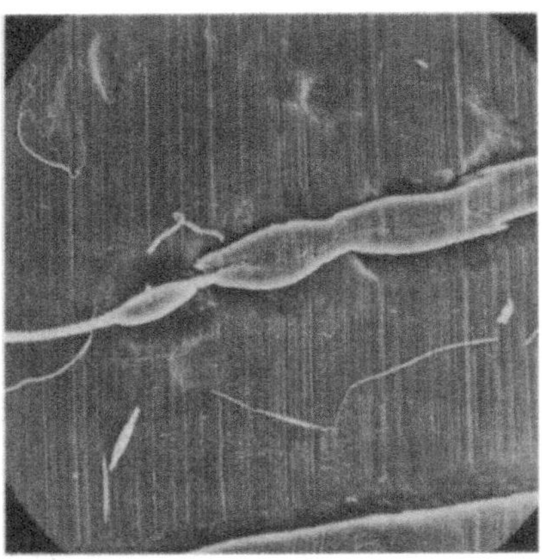

Bild 36: Detailansicht der oberen Probe im ungereinigten Zustand. Deutlich ist die seitliche Extrudierung eines fadenförmigen Verschleißpartikels zu erkennen.

20 µm

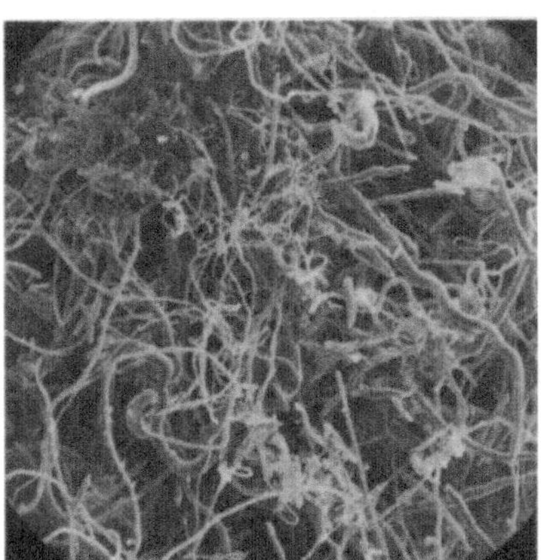

Bild 37: Typische PA66-Verschleißpartikel in Form von Fäden bei Schwingungsverschleiß kleiner Schwingungsweite

20 µm

Der Verschleiß ordnet sich in diesem Fall nicht mehr quer zur Laufrichtung an, sondern wird infolge der großen Schwingungsweite (groß bezogen auf die Kontaktflächenlänge von 3 mm) in Schwingungsrichtung abtransportiert und hinterläßt Verschleißspuren in dieser Richtung (Bild 38). Die Verschleißpartikel fallen an den Stirnseiten in Schwingungsrichtung schuppenförmig an (Bild 39), seitlich an den Rändern ist kein Verschleißabfluß zu beobachten. Die Verschleißtransport-Richtung hat sich also gegenüber dem vorhergehenden Beispiel mit kleiner Schwingungsweite genau umgekehrt.

Das dritte Beispiel soll abschließend zeigen, wie stark die Schwingungsform, d.h. hier die Überlagerung von zwei Schwingungen in zwei senkrecht aufeinander stehenden Richtungen ($\Delta x + \Delta y$) den Verschleiß-Transport und auch die Verschleißrate beeinflussen kann.

Kleine Schwingungsweite in zwei Richtungen

Bild 40 zeigt die Oberfläche eine PA66-Probe nach dem Lauf bei folgenden Versuchsbedingungen:

Δx	=	100 µm / 50 Hz
Δy	=	100 µm / 51 Hz
F_N	=	25 N
Laufzeit t	=	10 h

Der Verschleiß bleibt in der Kontaktfläche liegen, da die treibende Kraft für den Stoff-Transport keine Vorzugsrichtung hat. Die spezifische Verschleißrate geht in diesem Fall von K = 5 µm/(km·N/mm^2) bei Schwingungen in einer Richtung auf K = 0,5 µm/(km·N/mm^2) zurück.

Diese Ansammlung von Verschleißpartikeln bei ungerichteter Schwingung in zwei Richtungen führt zur Grübchenbildung,wie Bild 41 deutlich an einer PI-Probe unter den gleichen Bedingungen wie PA66 in Bild 38 gelaufen zeigt. Die gleichen Grübchen können auch bei Stahl-Stahl-Paarungen gefunden werden, wie z.B. an äußeren Halteverzahnungen von Planetengetrieben.

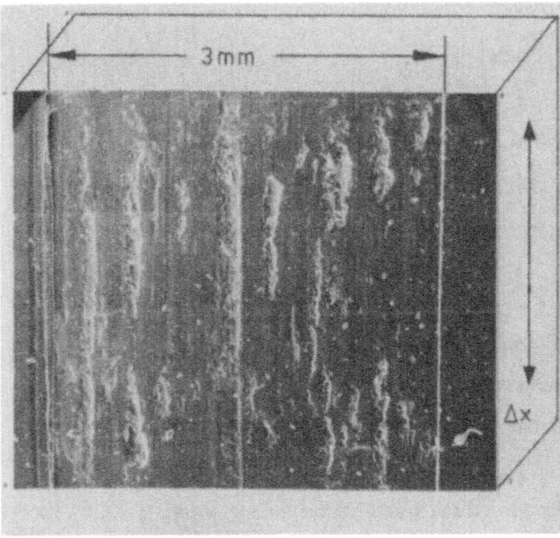

Bild 38: **Riefenbildung auf** PA66 bei großer Schwingungsweite Δx = 2 mm. Transport der schuppenförmigen Verschleißpartikel in Schwingungsrichtung.

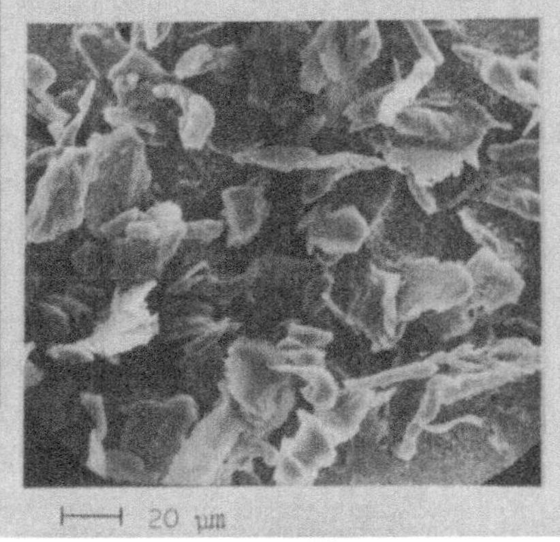

Bild 39: Typische PA66-Verschleißpartikel in Form von Schuppen bei großer Schwingungsweite.

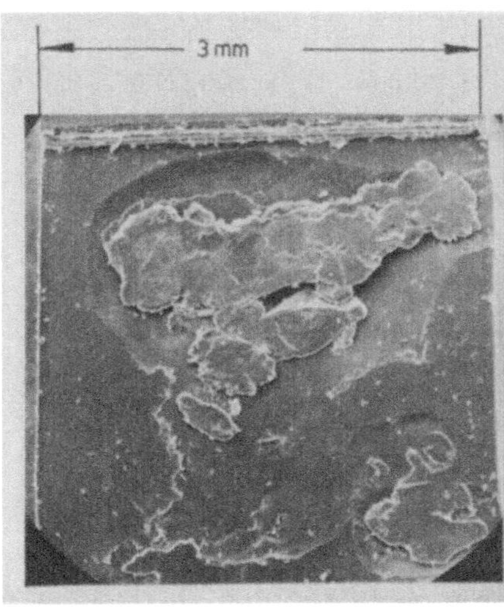

Bild 40: Grübchenbildung auf PA66 bei kleinen Schwingungsweiten $\Delta x = \Delta y$ = 100 µm in zwei Richtungen. Die Verschleißpartikel sammeln sich innerhalb der Kontaktfläche an und bilden die Grübchen.

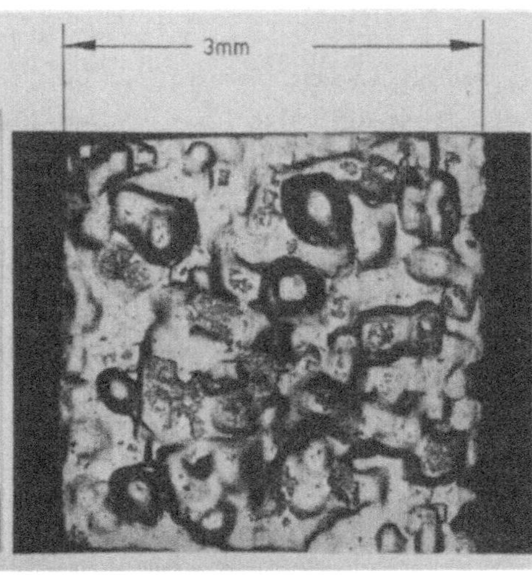

Bild 41: Grübchenbildung auf PI gegen Stahl gelaufen. Versuchsbedingungen wie in Bild 40. Hier ist die Grübchenbildung noch deutlicher zu erkennen.

Am System 2 (PA66 + 20 % CF - 100 Cr6H) können ähnliche Erscheinungen wie bei reinem PA66 beobachtet werden, allerdings nicht in der gleichen deutlichen Form, da der heterogene Aufbau des Werkstoffes den Stoff-Fluß behindert und damit auch das Schadensbild der Kontaktfläche verändert. Zusätzlich spielen Ermüdungseffekte zwischen den Kohlefasern und dem Grundwerkstoff eine große Rolle auf den Verschleißvorgang.

Bild 42 zeigt eine Probenoberfläche mit Furchen quer zur Schwingungsrichtung. Dies sind nun keine Wurmspuren, sondern Spalten an den Rändern von großen Grübchen. Die Spalten sind meist paarweise angeordnet, wie Bild 43 in stärkerer Vergrößerung zeigt. Zwischen den Spalten haben sich Verschleißpartikel angesammelt und in die Oberfläche eingepreßt. Dort wo man noch deutlich die Kohlefaser erkennt, ist noch die unzerrüttete Oberfläche vorhanden.

System 3.1 (100 Cr6H - C 15,trockenlaufend) wurde bereits in Kapitel 5.1 im Abschnitt Einfluß der Schwingungsweite auf den Schwingungsverschleiß erwähnt. Hier soll an Hand von 3 Bildern der Kontaktfläche des C 15 gezeigt werden, wie die Verschleißpartikel die Oberflächenstruktur in Abhängigkeit von der Schwingungsweite verändern und damit auch die Verschleißrate beeinflussen.

Bild 44 zeigt die Oberfläche des C 15 bei $\Delta x = 3$ mm, dem Übergang zum reversierenden Gleitverschleiß. Sie ist metallisch blank und zeigt lange Riefen infolge Adhäsion und Abrasion in Schwingungsrichtung.

Bild 45 zeigt die Oberfläche bei $\Delta x = 1$ mm. Der innere Teil der Fläche bleibt auf einer Länge von 2 mm ständig überlappt, mit entsprechend behindertem Stoff-Transport. Die teilweise oxidierten Verschleißpartikeln, insbesondere die des härteren Gegenwerkstoffes, bleiben in der Oberfläche haften und wirken verschleißmindernd.

Bild 46 zeigt nun die Oberfläche bei kleiner Schwingungsweite $\Delta x = 100$ µm. Fast die gesamte Oberfläche ist ständig überlappt und die Verschleißpartikel, teilweise oxidiert, bleiben in der Oberfläche und bilden eine Zwischenschicht, die stark verschleißmindernd

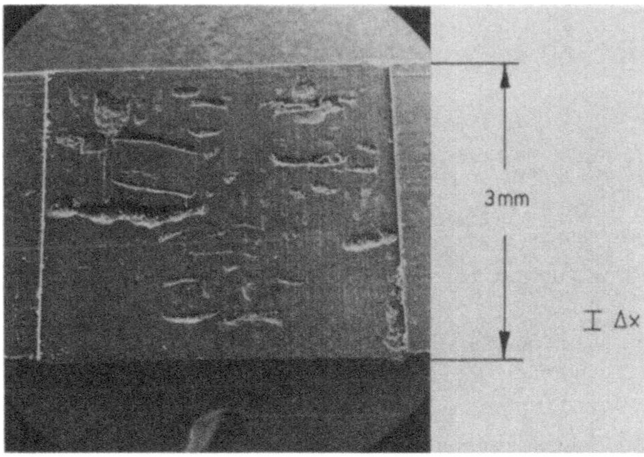

Bild 42: Verschleißbild einer PA66 + 20 % CF-Probe (VNr. 61, 50 N, 100 Hz, 200 µm, 100°C, 31 h, trocken). Die meist paarweisen Furchen quer zur Schwingungsrichtung sind keine Wurmspuren, sondern die Ränder von mit Verschleißpartikeln aufgefüllten Grübchen. Der Verschleißabtransport wird durch die Fasern erschwert.

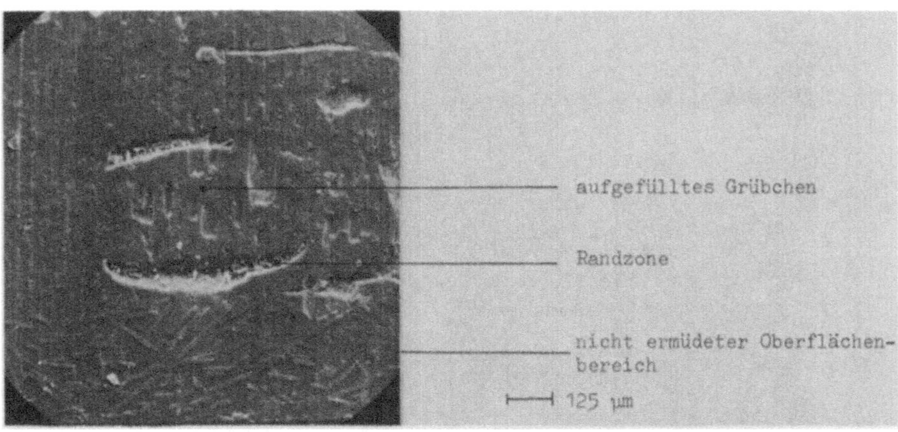

Bild 43: Detailansicht der Fläche von Bild 42 mit einem Furchenpaar. Dazwischen liegt ein mit Verschleißpartikeln aufgefülltes Grübchen.

Bild 44: Verschleißbild von C15 bei großer Schwingungsweite im Übergangsbereich Schwingungsverschleiß - rev. Gleitverschleiß. Oberfläche riefig aber metallisch blank. Große spezifische Verschleißrate K.

Δx	=	3000 µm
$\Delta x/l$	=	1
K	=	55 µm/(km·N/mm^2)

Bild 45: Schwingungsweite Δx = 1 mm. Entsprechend der bezogenen Schwingungsweite $\Delta x/l$ = 0,33 sind etwa 70 % der Fläche mit Oxidprodukten (Reibrost) belegt. Die spez. Verschleißrate K ist kleiner als bei Bild 44.

Δx	=	100 µm
$\Delta x/l$	=	0,33
K	=	30 µm/(km·N/mm^2)

Bild 46: Schwingungsweite Δx = 0,1 mm. Nahezu die gesamte Fläche ist mit Oxidprodukten belegt. Die spez. Verschleißrate K ist stark zurückgegangen.

Δx	=	100 µm
$\Delta x/l$	=	0,033
K	=	2 µm/(km·N/mm^2)

wirkt (vgl. Bild 22). Der Verschleißmechanismus ist jetzt vorwiegend oberflächenzerrüttend und tribooxidativ.

Mit System 3.2 (100 CrH - 100 Cr6H geschmiert) wurden umfangreiche Messungen durchgeführt, deren Ergebnisse zum Teil in den vorhergehenden Kapiteln beschrieben wurden. Unter dem Aspekt Stofftransport werden nachfolgend der enge Zusammenhang zwischen Beanspruchungsbedingungen, Oberflächenstruktur (Schadensbild) und Verschleiß an drei Beispielen gezeigt. Alle drei Versuche liefen in Petroleum.

1. Beispiel

VNr. 480/1	Δx	=	200 µm bei 50 Hz
	F_N	=	185 N konstant
	Laufzeit t	=	70 h

Bild 47 zeigt die Oberflächenstruktur nach Schwingungen in einer Richtung ohne Schlagbelastung. Typisch für eine Schwingung bei konstanter Schwingungsweite über längere Zeit, konstanter Last und kleiner Verschleißrate, ist die Ausbildung eines Riffelmusters. Bei größeren Verschleißraten oder behindertem Verschleißabfluß kann das Riffelmuster nicht entstehen oder wird zerstört. Die im Bild 47 abgebildeten Riffel haben eine Höhe von 1 µm und einen Abstand zwischen 200 bis 300 µm. Durch die Verschleißpartikel entsteht eine rauhe Oberfläche mit Riefen in Schwingungsrichtung. Die Oberfläche ist metallisch blank. Die Verschleißrate beträgt 0,03 µm/h.

2. Beispiel

VNr. 482/1	Δx	=	200 µm bei 50 Hz
	Δy	=	200 µm bei 50 Hz
	F_N	=	200 N konstant
	Laufzeit t	=	70 h

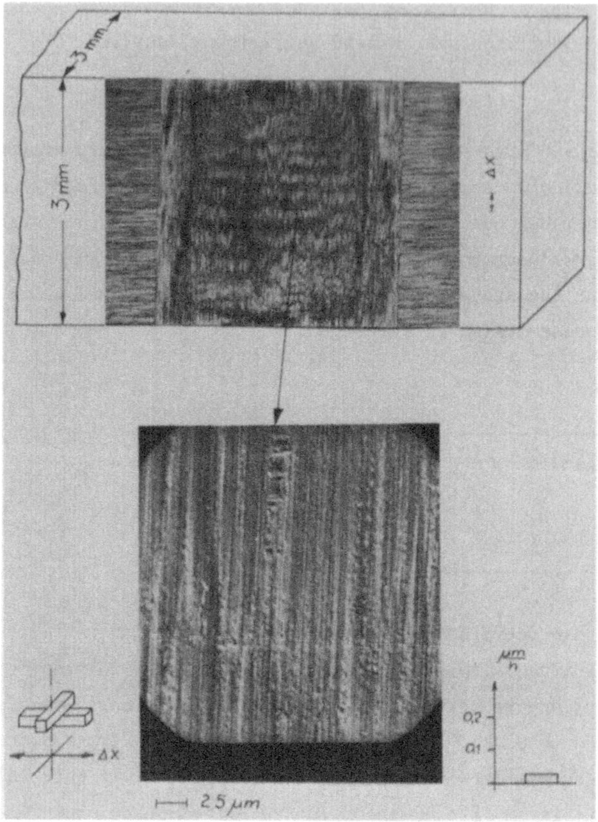

Bild 47: Riffelbildung bei einer 100Cr6H-100Cr6H-Paarung, kraftstoffgeschmiert. Die Strukturbildung ist typisch für konstant belastete Systeme, konstanter Schwingungsweite über längere Zeit und kleiner Verschleißrate. Bei größeren Verschleißraten (oder genauer:Verschleißflußdichten) kann es infolge stärker behindertem Stofftransport zu unregelmäßigeren Riffelstrukturen oder zu Wurmspuren kommen.

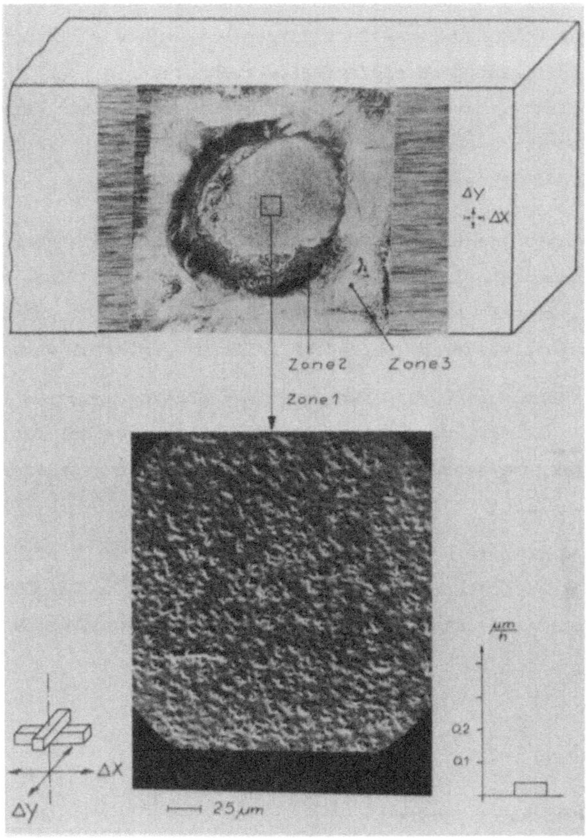

Bild 48: Verschleißbild einer 100Cr6H - 100Cr6H-Paarung, kraftstoffgeschmiert nach einer Schwingungsbewegung in zwei Richtungen (kreisförmig, $\Delta x = \Delta y$ = 100 µm) und konstanter Belastung.

Zone 1: Gleichmäßige leicht löchrige Struktur, ähnlich einer geläppten Oberfläche, infolge größerer Ansammlung von Verschleißpartikeln.

Zone 2: Überganszone, kreisförmige Riefen entsprechend der Schwingungsbewegung durch einzelne Verschleißpartikel.

Zone 3: Glatt polierte Oberfläche infolge intensivem Stofftransport von Schmierstoff und Verschleißpartikeln.

In Bild 48 ist die sich ausbildende Oberflächenstruktur bei einer kreisförmigen Schwingungsbewegung dargestellt. In der Flächenmitte (Zone 1) bildet sich eine riefenfreie,leicht löchrige aber metallisch blanke Oberfläche, hervorgerufen durch die angesammelten Verschleißpartikel. Der Abtransport dieser Verschleißteilchen ist wegen der Kreisbewegung stark erschwert.

In der sich anschließenden Übergangszone sind kreisförmige Riefen zu erkennen. Schmierstoffzufluß und Verschleißabfluß sind hier bereits erleichtert, weshalb einzelne Verschleißpartikel Riefen erzeugen können, die nicht sofort wieder durch andere Teilchen geglättet werden.

Die äußere, dritte Zone von etwa doppelter Schwingungsweite ist nahezu völlig glatt poliert. Hier werden die Verschleißpartikel aus der Verschleißfläche herausgespült. Die Verschleißrate ist wegen der schützenden Zwischenschicht in Zone 1 klein (0,045 µm/h).

Gerade dieses Beispiel zeigt deutlich, wie stark die Oberflächenstruktur durch das Wechselspiel von Schmierstoffzufluß und Verschleißabfluß beeinflußt wird und sich letztlich auch auf die Verschleißrate auswirkt.

3. Beispiel

VNr. 477/1	Δx	=	200 µm bei 50 Hz
	Δy	=	200 µm bei 100 Hz
	F_N	=	120 N
	F_N	=	600 N (Schlag)[6]
	Laufzeit t	=	70 h

6) 10 % der Zeit wirkend, Schlagfrequenz f_N = 30 Hz.

Die Oberfläche ist nahezu völlig glatt (Bild 49) und metallisch blank. An einzelnen Stellen sind Pittings entstanden, welche auf beiden Oberflächen sich gleich ausgebildet haben. Bei Schwingungen in einer Richtung können auf diese Weise Wurmspuren entstehen. Die Verschleißrate ist, verglichen mit den beiden ersten Beispielen deutlich größer (W_1 = 0,75 µm/h), trotz metallisch blanker und glatter Oberfläche.

Diese hohe Verschleißrate erhält man nur, wenn man Schlagbelastung und Schwingungsbewegungen in zwei unterschiedlichen Richtungen kombiniert.

Zusammenfassend kann festgestellt werden, daß die Höhe des Schwingungsverschleißes unter flüssigem Zwischenmedium eng an die Stoffflüsse, d.h. Schmierstoffzufluß (Spülwirkung, Schmierwirkung) und Verschleißabfluß gekoppelt ist. Diese Stoffflüsse werden von der Probengeometrie, den Beanspruchungsbedingungen und der Bewegungsform beeinflußt.

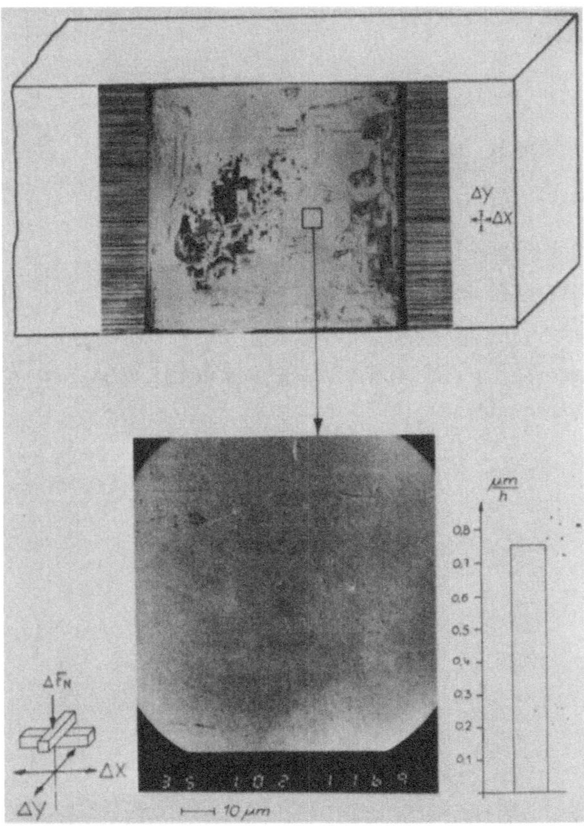

Bild 49: Gleiche Paarung wie Bild 47 und 48. Schwingbewegung in zwei Richtungen, jetzt aber schlagend 600 N belastet (≙ 200 N mittlere Last wie in Bild 47 und 48). Oberfläche völlig glatt poliert. Durch die Schläge werden die Verschleißpartikel aus dem Spalt herausgepumpt. Die Kontaktfläche wird leicht ballig, der Verschleiß steigt stark an.

5.4 Bewährungsfolgen

Nach den Angaben in den Versuchsprogrammen 1 und 2 in Kap. 4.4 wurden eine Reihe von Werkstoffpaarungen geprüft. Die Ergebnisse sind in den folgenden Listen in Form von Bewährungsfolgen dargestellt. Bei der Beurteilung der Ergebnisse und deren Nutzung für andere Anwendungsfälle sind insbesondere die in Kapitel 5.1 gemachten Hinweise zu den einzelnen Parametereinflüssen zu beachten.

Werkstoffgruppen:

1. Metall-Metall, trocken	SV (Schwingungsverschleiß)	Tabelle 3
2. Metall-Kunststoff, trocken	SV	Tabelle 4 ...7
3. Metall-Kunststoff, gefettet	SV	Tabelle 8
4. Kunststoff-Kunststoff, trocken	SV	Tabelle 9
5. Metall-Kunststoff, trocken	GV (Gleitverschleiß)	Tabelle 10

Tabelle 10 (Gleitverschleiß) ist zu vergleichen mit Tabelle 4 (Schwingungsverschleiß).

Gruppe 1 Metall-Metall, trocken

Schwingungsverschleiß			
lfd. Nr.	Werkstoffpaarung	spezifische Verschleißrate K in $\frac{\mu m}{km \cdot N/mm^2}$	Reibungszahl μ
	C 15 weich / C 15 weich	5,7	0,57
	100Cr6H / 100Cr6H	9,4	0,67
	leg. Werkzeugstahl / leg. Werkzeugstahl	1,2	0,65
	C10 E / C10 E	7,8	0,56
	C60 G weich / C60 G weich	13,1	0,58
	16 Mn CrS5 H / 16 Mn CrS5 H	9	0,51
	X210Cr12H / X210Cr12H	8,9	0,62
	C10 / C10	8,5	0,56
	Hartmetall / Hartmetall	0,08	0,6
	Rotguß [1] / Rotguß	111	0,5

1) GZ - Cu Sn 7 Zn Pb

Versuchsbedingungen
100 µm/50 Hz/100 N
trocken/20°C

Tabelle 3

Es wurden jeweils gleiche Werkstoffe gepaart. Auffällig ist der relativ geringe Unterschied in der Verschleißrate zwischen den ungehärteten und gehärteten Stählen. Dies ist eine Folge der sich bildenden Zwischenschicht aus Oxidprodukten. Bei größeren Schwingungsweiten wird der Verschleiß der ungehärteten Stähle größer (Bild 22). Auch bei Schmierung mit flüssigen Medien und verstärktem Abtransport der Verschleißpartikel haben die ungehärteten Werkstoffe eine größere Verschleißrate. Geeignete Werkzeugstähle und Hartmetalle haben kleinere Verschleißraten. Bei überlagerter Schlagbeanspruchung haben sich die legierten Werkzeugstähle gut bewährt, da sie sowohl zäh sind als auch durch die harten Karbide einen günstigen Verschleißwiderstand besitzen.

Gruppe 2 Metall-Kunststoff, trocken

Schwingungsverschleiß

lfd. Nr.	Werkstoffpaarung [1]	spezifische Verschleißrate K in $\frac{\mu m}{km \cdot N/mm^2}$	Reibungszahl µ
	PA 66 (Ultramid A 3 W)	5,5	0,7
	PA 12 (Vestamid L 1941)	2,2	0,4
	PA 66 + 20 CF (Thermocomp RC 1004 Hs)	4 [2]	0,7
	PA 66 + 20 CF + 15 PTFE (Thermocomp RCL 4034)	3,3	0,4
	PA 66 + 20 CF + 13 PTFE + 2 Silikonöl (Thermocomp RCL 4536)	2,5	0,3
	POM-C (Hostaform C9021)	1	0,4
	POM + chem. Schmiermittel (Delrin 500 CL)	0,83	0,22
	PBTP (Ultradur B 4500)	1,5	0,54
	PI (Vespel SP1)	1,5	0,44

1) Jeweils gegen 100Cr6H längs geschliffen
2) Verschleißrate bevor Stahlverschleiß einsetzt

Versuchsbedingungen
Δx = 2 mm/2,5 Hz/10 N
trocken/30°C

Tabelle 4

Die gleichen Werkstoffe wie in Tabelle 4 wurden auch unter Gleitverschleiß-Bedingungen gefahren. Die Ergebnisse sind in Tabelle 10 zusammengestellt. Die geringe Verschleißrate von PA 12 und PBTP gegenüber PA 66 ist vermutlich auf den Aufbau einer verschleißmindernden Zwischenschicht zurückzuführen. Der Zusatz von Kohlefaser zu PA 66 vermindert die Verschleißrate, aber es besteht wie auch bei Glasfaserzusatz die Gefahr des Stahlverschleißes durch Abrasion und damit ein Sprung in die Verschleißhochlage. Eine Verbesserung durch Zusatz von PTFE und Silikonöl mindert die Tendenz zur Verschleißhochlage, wobei das Verschleißverhalten noch stark von der Amplitude abhängig ist (Bild 23).

Die folgenden Tabellen 5 bis 7 zeigen die spezifische Verschleißrate K von Kunststoffen gegen gehärteten Stahl bei den Temperaturen 30°C, 80°C und 120°C.

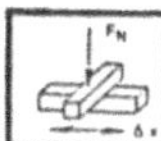

Schwingungsverschleiß

lfd. Nr.	Werkstoffpaarung [2)]	spezifische Verschleißrate K in $\frac{\mu m}{km \cdot N/mm^2}$	Anmerkung	Reibungszahl μ
	PA 66 (Ultramid A 3 W)	10 (30°C)		0,7
		7 (80°C)		0,7
		3 (120°C)		0,7
	PA 66 (Technyl A216)	7,5		0,72
		3		0,64
		5		0,64
	PA 66 + 20 CF (Thermocomp RC 1004 HS)	4	1)	0,7
		3	1)	0,4
		5,6	1)	0,5
	PA 66 + 20 CF + 15 PTFE (Thermocomp RCL 4034)	3,3	1)	0,4
		2,1	1)	0,4
		3,4	1)	0,4
	PA 66 + 20 CF + 13 PTFE + 2 Silikonöl (Thermocomp RCL 4534)	2,5	3)	0,3
		2	3)	0,36
		3,5	3)	0,25
	PA 66 + 10 CF (Mischung 1:1 aus Thermocomp 1004 HM + Ultramid A3W)	6	3)	0,64
		6,3	3)	0,6
		10		0,4
	PA 66 + 30 CF + 13 PTFE + 2 Silikon (Thermocomp RCL 4536)	0,8		0,5
		10	4)	0,34
		3,75	4)	0,36
	PA 66 + 20 CF + 2 Silikonöl (Thermocomp RCL 4414 HS)	7,5	1)	0,4
		3,75	3)	0,2
		5,25	3)	0,14

1) Bevor Stahlverschleiß eintritt
Die spez. Verschleißrate ist abhängig von Schwingungsweite
2) Jeweils gegen 100Cr6 H, geschliffen
3) Tendenz zur Verschleißhochlage durch Stahlverschleiß
4) Mit Stahlverschleiß

Versuchsbedingung: $\Delta x = 2$ mm/10Hz; F_N = 25...50 N; 30°C/trocken

Tabelle 5

Die PA 66-Proben zeigen eine Minderung der spezifischen Verschleißrate mit steigender Temperatur bzw. bei den kohlefasergefüllten Proben ein Minimum bei 80°C. Vermutlich spielen hier die Eigenschaften der sich bildenden Zwischenschicht und das Erweichungsverhalten (Glaszustand-weichelastischer Zustand der amorphen Anteile) des Kunststoffes eine große Rolle.

Schwingungsverschleiß				
lfd. Nr.	Werkstoffpaarung [1]	spezifische Verschleißrate K in $\frac{\mu m}{km \cdot N/mm^2}$ K	(0 1 2 3 4 5 6 7 8 9 10)	Reibungszahl μ
	PA 12 + 20 CF	6,9	30°C	0,36
	(Thermocomp	1,9	80°C 2)	0,36
	SC 1004 HS)	6,3	120°C	0,24
	PA 12 + 15 PTFE	2		0,24
	(Thermocomp	6,3		0,36
	SL 4030)	22		0,24
	PA 12 + 30 GF +	0,06		0,14
	15 PTFE	1,8	4)	0,46
	(Thermocomp S FL 4036)	8,8	4)	0,34
	PI	1,5		0,44
	(Vespel SP1)	3		0,42
		4,4		0,42
	PI + 15 C	0,8		0,38
	(Vespel SP 21)	0,25		0,28
		0,25		0,32
	PI + 12 C + 3PTFE	1		0,46
	(Polyamidimid)	0,94		0,32
	(Torlon 4301)	2,5		0,36

1) Jeweils gegen 100Cr6H geschliffen
2) Bevor Stahlverschleiß eintritt
3) Verschleißhochlage mit Stahlverschleiß
4) Tendenz zur Verschleißhochlage durch Stahlverschleiß

Versuchsbedingung
Δx = 2mm/10Hz
F_N = 25...50 N
30°C/trocken

Tabelle 6

PA 12 sollte bei 120°C und trockener Reibung auch kurzzeitig nicht eingesetzt werden. Eine deutliche Verbesserung des Verschleißverhaltens bei PI erhält man durch Zusatz von Graphit und PTFE, gleichzeitig wird auch die hohe Reibungszahl verringert.

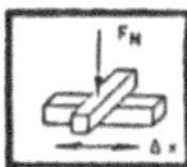

Schwingungsverschleiß

lfd. Nr.	Werkstoffpaarung 1)	spezifische Verschleißrate K in $\frac{\mu m}{km \cdot N/mm^2}$		Reibungszahl µ
	POM-C (Hostaform C 9021)	1	30°C	0,4
		0,7	80°C	0,32
		0,63	120°C	0,32
	POM-C + 30 GF (Hostaform C 9021 GV1/30)	22,5	2)	0,6
		81	2)	0,66
		>100	2)	0,7
	POM-Hom.pol. + 2 Silikonöl (Fulton 441 D + 2 Silikonöl)	0,63		0,4
		0,41		0,36
		3,1		0,44
	POM-C + PTFE-Pulver (Hostaform C 9021 TF)	0,25		0,18
		0,25		0,08
		0,25		0,06
	POM + 5 % Mineralöl (Railko PV 80)	0,55		0,28
		0,38		0,22
		0,8		0,28
	POM + chem. Schmiermittel (Delrin 500 CL)	0,83		0,22
		0,5		0,16
		1,9		0,22
	PBTP (Ultradur B 4500)	1,7		0,72
		2,5		0,6
		3,5		0,6
	PBTP + 30 GF (Ultradur B 4300 G6)	2,5	3)	0,44
		5	3)	0,6
		12,5	3)	0,56
	PBTP + 15 CF 1 : 1 Mischung (Ultradur B 4500 + Thermocomp WC 1006)	2,5	3)	0,44
		3,1	3)	0,3
		3,8	3)	0,38

1) Jeweils gegen 100Cr6H geschliffen
2) Verschleißhochlage mit Stahlverschleiß
3) Tendenz zur Verschleißhochlage durch Stahlverschleiß

Versuchsbedingung
Δx = 2 mm/10Hz
F_N = 25...50 N
30°C/trocken

Tabelle 7

Die spezifische Verschleißrate von POM, insbesondere mit Zusätzen welche die Schmierung verbessern, sind unter Schwingungsverschleiß-Bedingungen bemerkenswert klein.

Gruppe 3 Metall-Kunststoff, gefettet

Wie man im Vergleich mit den Tabellen 4 bis 7 der Gruppe 2 erkennen kann, geht bei Fettschmierung die spezifische Verschleißrate um rund den Faktor 10 zurück und die Eigenschaften der Kunststoffarten machen sich auf den Verschleiß weniger bemerkbar.

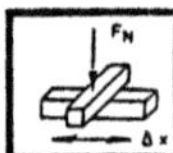

Schwingungsverschleiß

lfd. Nr.	Werkstoffpaarung 1)	spezifische Verschleißrate K in $\frac{\mu m}{km \cdot N/mm^2}$ K (Skala 0 1 2 3 4 5 6 7 8 9 10)	Reibungszahl µ
	PA 66 (Ultramid A 3 K)	0,34	2)
	PA 66 (Ultramid A 3 W)	0,34	
	PA 66 + 20 CF (Thermocomp RC 1004 HS)	0,22	
	PA 66 + 20 CF + 15 PTFE (Thermocomp RCL 4034)	0,16	
	PA 12 (Vestamid L 1941)	1	
	PA 12 + 20 CF (Thermocomp SC 1004 HS)	0,21	
	PA 12 + 20 C Mischung 1 : 1 (Vestamid X 1933 Vestamid L 1901)	0,4	
	PA 12 + 40 C (Vestamid X 1933)	0,6	
	POM-C (Hostaform C 9021)	0,3	
	PBTP (Ultradur B 4500)	0,5	

1) Jeweils gegen 100Cr6H geschliffen
2) Die Reibungszahlen liegen zwischen 0,1 und 0,15

Versuchsbedingung
Δx = 100 µm/100 Hz
F_N = 100 N/80°C
gefettet

Tabelle 8

Gruppe 4 Kunststoff-Kunststoff, trocken

Die Versuche in dieser Gruppe wurden mit dem Ziel durchgeführt, Stoffpaarungen zu finden, welche eine besonders niedrige Reibungszahl bei Schwingungsverschleiß ergeben. Die Reibungszahlen liegen meist tiefer als die gleichen Kunststoffe gegen Stahl.

Schwingungsverschleiß

lfd. Nr.	Werkstoffpaarung	spezifische Verschleißrate K in $\frac{\mu m}{km \cdot N/mm^2}$		Reibungszahl µ
	POM-C-PBTP	8,33		0,48
	POM-C-PBTP + 10 GF	1,4	1)	0,38
	POM-C-PBTP + 30 GF	2,6	2)	0,28
	POM-C + chem. Schmiermittel (Delrin 500 CL) - PBTP + 30 GF	1	1)	0,4
	POM-C-PBTP + 30 GK (Vestadur B GK 30)	2,5	1)	0,46
	POM-C-PA 66	0,1	3)	0,36
	POM-C-PA 66 + 20GF	0,1	3)	0,22
	PA 12 + 15 PTFE (Thermocomp SL 4030) PBTP + 30 GF	1,67		0,16
	PTFE-PBTB + 30 GF	0,7		0,116
	POM-C-PBTP + 30 GF	0,24	4)	0,036

1) Tendenz zur Verschleißhochlage
2) Nach 43 h Verschleißhochlage, K=50
3) F_N = 50 N, bei 25 N nach 80 h noch kein Verschleiß > 1µm
4) gefettet

Versuchsbedingung
Δx = 2 mm/2,5 Hz
F_N = 25 N/30°C
trocken

Tabelle 9

Die Reibkraft setzt sich allgemein aus zwei Anteilen zusammen, dem Adhäsions- und dem Deformationsanteil. Der Adhäsionsanteil wird umso größer, je größer die Oberflächenenergie des Kunststoffes ist (ERHARD [10]). PTFE hat eine kleine Oberflächenenergie, PA66 infolge "polarer" Moleküle hat eine große Oberflächenenergie. Bei glasfasergefüllten Kunststoffen steigt mit dem Verschleiß der Deformationsanteil durch die herausgetretenen Glasfasern, so daß solche Paarungen eine hohe Reibungszahl ergeben können.

Gruppe 5 Metall-Kunststoff, trocken. Gleitverschleiß

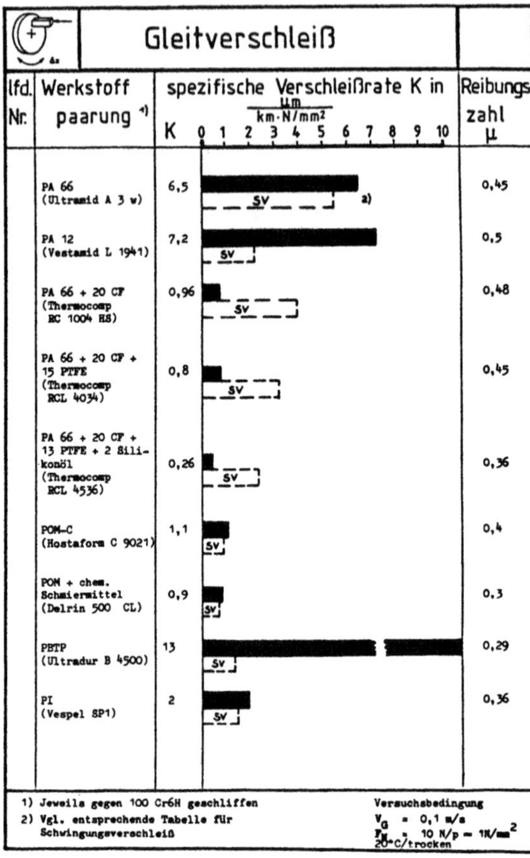

Tabelle 10

Diese Versuche wurden durchgeführt, um festzustellen, ob die spezifischen Verschleißraten K bei Kunststoff gegen Stahl unter Gleitverschleiß auch für Schwingungsverschleiß übernommen werden können. Wie Tabelle 10 zeigt, ist bei den ungefüllten Kunststoffen eine Korrelation bei PA 66, POM-C und PI vorhanden. Bei PA 12 und PBTP ist allerdings unter Gleitverschleiß die Verschleißrate deutlich größer gegenüber Schwingungsverschleiß. Vermutlich hängt dies mit der Struktur der sich ausbildenden Zwischenschicht zusammen. Es wurde kein gleichmäßiger Film, sondern schuppenförmige Ablagerungen beobachtet. Die mit Kohlefaser gefüllten PA 66-Werkstoffe haben unter Gleitverschleiß kleinere Verschleißraten, da die abrasive Wirkung der Fasern bei Gleitverschleiß die Stahlgegenfläche glättet, bei Schwingungsverschleiß aber aufrauht (vgl. Kap. 5.1.1 und Bild 21).

Die Oberflächenrauhigkeit der Gegenfläche (Stahl) und deren Struktur (z.B. quer geschliffen, in Umfangsrichtung geschliffen) hat einen wesentlich größeren Einfluß auf den Gleitverschleiß als bei Schwingungsverschleiß. Ursache sind die unterschiedlichen Transportvorgänge der Verschleißpartikel bei Gleitverschleiß und gegenüber Schwingungsverschleiß. Vermutlich gibt es aus dem gleichen Grund bei einzelnen Werkstoffen sogar deutliche Unterschiede bei Gleitverschleiß, wenn man von Stift-Scheibe-Versuchen auf Ring-Scheibe-Versuche übergeht.

Abschließend zeigt Bild 50 einen Überblick über das Verschleißverhalten einiger Werkstoffpaarungen. Multipliziert man die angegebenen Werte der Verschleißgeschwindigkeit mit 2,5, so erhält man die spezifische Verschleißrate K in µm/(km · N/mm^2).

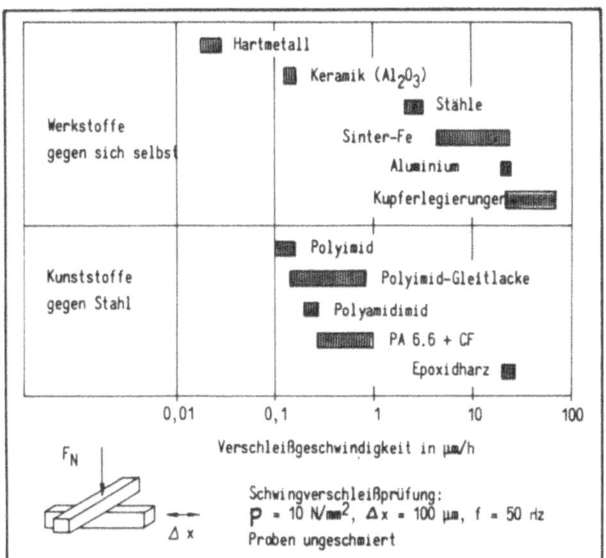

Bild 50: Bewährungsfolge verschiedener Werkstoffpaarungen bei Schwingungsverschleiß

5.5 Korrelation von "Modell"-Bauteil- und Erzeugnisversuchen

Zielsetzung ist, Werkstoff-Vorauswahlen über Bewährungsfolgen zu treffen bzw. Bauteile zu optimieren. Besteht dabei die Aufgabe, den Verschleiß im Erzeugnis mittels vereinfachter Prüfkörper ("Modell") oder Bauteilen zu simulieren, so stellt sich die Frage nach der Korrelation der Ergebnisse von "Modell"-, Bauteil- und Erzeugnisversuch.

Eine vollstände Korrelation erreicht man nur, wenn Identität aller Prozesse in den Tribosystem herrscht. Nur dann sind die Bewährungsfolgen mit Sicherheit identisch, sowohl in der Reihenfolge (qualitative Korrelation) als auch im Betrag (quantitative Korrelation). Meist genügt aber eine qualitative Korrelation.

Will man eine solche qualitative Korrelation erreichen, so kann man aufgrund der neueren Literatur nach HABIG[14] zusammenfassend sagen, daß nur die folgenden Größen bei der Simulation mit vereinfachten Prüfkörpern von Erzeugnissystemen verschieden sein dürfen:

- Belastung
- Geschwindigkeit
- Zeit (Beanspruchungszeit)
- Geometrie der Verschleißpartner (Formeigenschaften)

Diese Parameter sind allerdings nicht völlig frei variierbar, sondern es sind einschränkende Gesichtspunkte zu beachten. Aufgrund der in dieser Arbeit durchgeführten Versuche sind zur Auswahl der Parameter bei der Simulation von Schwingungsverschleiß-Vorgängen folgende Hinweise zu beachten:

1. Der Stofftransport beeinfluß über die sich ausbildende Zwischenschicht und Veränderungen der Kontaktgeometrie den Schwingungsverschleiß. Belastung, Bewegung und Flächenform sind so zu wählen, daß sich zwischen den vereinfachten Prüfkörpern analoge Stoff-Transportvorgänge wie im Erzeugnis einstellen.
   Die beiden folgenden Bilder verdeutlichen den Stofftransport im Zusammenhang mit der Frage nach gleicher Last oder gleicher Flächenpressung. In Bild 51 ist zunächst für Gleitverschleiß die Verschleiß-Flußdichte in Abhängigkeit von der Flächengröße bei gleicher

Flächenpressung dargestellt. Die Verschleiß-Flußdichte ist bei der neunmal größeren Fläche um den Faktor 3 größer. Dieser vergrößerte Stofffluß zur Austrittskante hin kann sowohl verschleißerniedrigend als auch verschleißerhöhend wirken, was in der Literatur nur vereinzelt Beachtung findet. LANCASTER[22] schlägt zum Beispiel bei Gleitverschleißversuchen Kunststoff gegen Stahl vor, bei Modellversuchen mit verkleinerten Prüfflächen im Vergleich zum Bauteil die Last und nicht die Flächenpressung konstant zu halten, um eine Korrelation der Bewährungsfolgen zu erreichen. Dies unterstützt die Forderung nach vergleichbarem Stofftransport. Daneben sind die thermischen Auswirkungen bei geänderter Reibleistungsdichte ($\mu \cdot p \cdot v$) stets zu beachten.

Für Schwingungsverschleiß verdeutlicht Bild 52 die Auswirkung veränderter Verschleiß-Flußdichte auf das Schadensbild und die spezifische Verschleißrate.

Versuchsbedingungen:

PA66 - 100 Cr6H trockenlaufend	
Δx = 1 mm / 5 Hz	
$A_1 = 3 \times 3 = 9\ mm^2$	$A_2 = 6 \times 6 = 36\ mm^2$
$F_1 = 25$ N	$F_2 = 100$ N
$p_1 = p_2 = 2{,}8\ N/mm^2$	

Die 3 x 3 mm PA66-Probe zeigt eine Oberfläche ohne Wurmspuren. Die spezifische Verschleißrate K = 10 N/(km • N/mm^2).

Bei der 6 x 6 mm PA66-Probe mit der höheren Verschleiß-Flußdichte ist eine deutliche Wurmspur entstanden. Die spezifische Verschleißrate beträgt K = 15 N/(km • N/mm^2). Der Unterschied in den spezifischen Verschleißraten ist bei der hier vorliegenden Kunststoff-Stahl-Paarung gering, kann aber bei geschmierten metallischen Paarungen beträchtlich sein (REED, BATTER [29]).

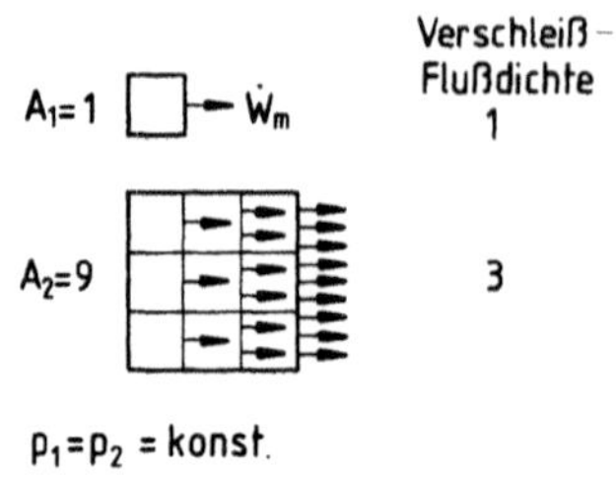

Bild 51: Schematische Darstellung des unterschiedlichen Stoff-Transportes (Verschleißflußdichten) bei geänderter Flächengröße aber konstanter Flächenpressung.

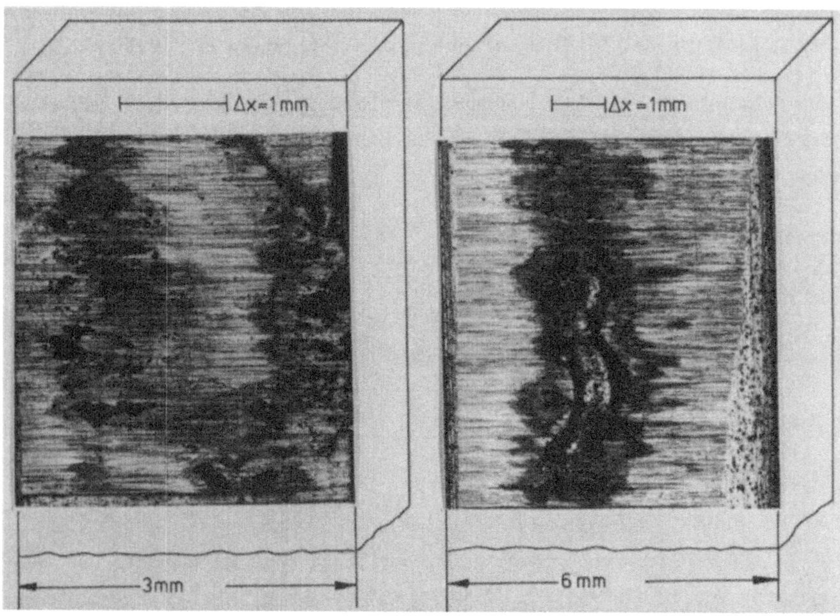

Bild 52: Verschleißbilder von PA66 unterschiedlicher Probengröße bei gleicher Flächenpressung gelaufen (F = 25 N bzw. 100 N, p = 2,8 N/mm^2, Δx = 1 mm, f = 5 Hz, trocken, 30°C.

Die kleinere Probe zeigt Riefen in Schwingungsrichtung, der Verschleißabtransport erfolgt in Schwingungsrichtung. Spezifische Verschleißrate K = 10 µm/(km·N/mm^2).
Die große Probe zeigt in der Flächenmitte eine Warmspur infolge erschwertem Verschleißabtransport, der Verschleiß wird sowohl in Schwingungsrichtung als auch quer dazu abtransportiert. Spezifische Verschleißrate K = 15 µm/(km·N/mm^2).

Analoge Stoff-Transportvorgänge erreicht man durch:

- erzeugnisnahe Verschleißraten
- erzeugnisnahe Reibleistungsdichten
- erzeugnisnahe geometrische Abmessungen der Kontaktflächen
- erzeugnisnahe Schwingungsform

Diese Forderungen bedeuten einen engen Spielraum für "Modell"-Versuche. Die Untersuchungen haben aber gezeigt, daß dies nicht für alle Systeme gilt.

2. Es gibt verschiedene Systemklassen, bei welchen die Betriebsnähe des Beanspruchungskollektivs und der Systemstruktur bei Simulation unterschiedlich streng eingehalten werden muß (Bild 53). Es werden drei Systemklassen aufgestellt, die sich nach der Zahl ihrer wesentlichsten Wechselwirkungen unterscheiden. Je größer die Zahl der Wechselwirkungen, umso genauer müssen Beanspruchung und Struktur des Betriebssystems bei der Simulation eingehalten werden.

Die Beachtung dieser Zusammenhänge erleichtert die Versuchsplanung und erhöht die Korrelation der Bewährungsfolgen zwischen Simulation und Erzeugnis.

Es ist noch darauf hinzuweisen, daß bei breitangelegten, systematischen Stoffuntersuchungen eine Variation des Beanspruchungskollektivs in weiten Grenzen vorausgehen sollte, um insbesondere Sprungstellen (Hoch-Tieflage) im Systemverhalten zu erkennen.

Zwei praktische Beispiele sollen die obigen Ausführungen zur Korrelation ergänzend veranschaulichen.

1. Beispiel (einfache Systemklasse I)

Die Verschleißrate ist bei fettgeschmierten Kunststoffen gegen Stahl in weiten Bereichen linear abhängig von der Belastung, der Frequenz und der Schwingungsweite (Bild 54).

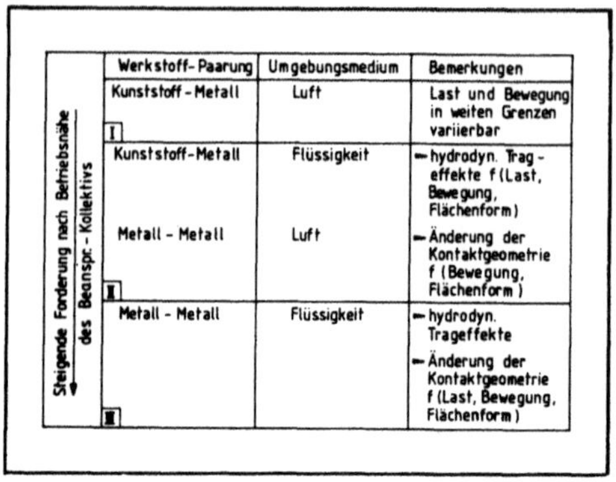

	Werkstoff-Paarung	Umgebungsmedium	Bemerkungen
I	Kunststoff-Metall	Luft	Last und Bewegung in weiten Grenzen variierbar
II	Kunststoff-Metall	Flüssigkeit	– hydrodyn. Trag-effekte f (Last, Bewegung, Flächenform)
	Metall-Metall	Luft	– Änderung der Kontaktgeometrie f (Bewegung, Flächenform)
III	Metall-Metall	Flüssigkeit	– hydrodyn. Trageffekte – Änderung der Kontaktgeometrie f (Last, Bewegung, Flächenform)

Bild 53: Forderung nach steigender Betriebsnähe des Beanspruchungs-Kollektivs und der Systemstruktur in Abhängigkeit einzelner Systemklassen bei der Simulation mit vereinfachten Prüfkörpern.

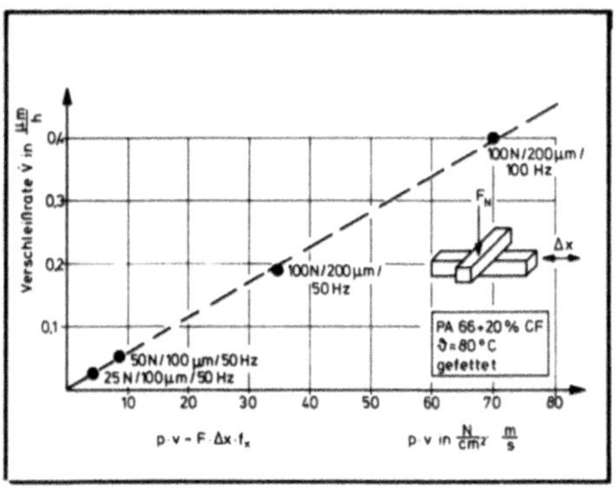

Bild 54: Linearer Zusammenhang zwischen Belastung, Frequenz und Verschleißrate für Kunststoff-Stahl, gefettet.
Dieses System zählt zu der relativ einfach zu simulierenden Systemklasse I (Bild 53).

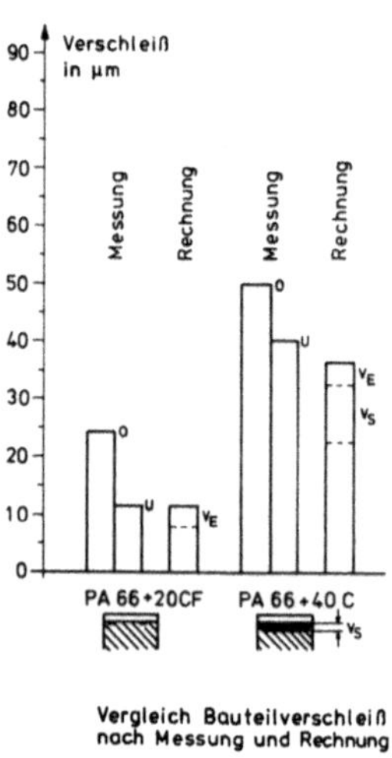

Bild 55: Vergleich des Bauteilverschleißes nach Messung im Bauteilversuch und Rechnung nach der in "Modell"-Versuchen ermittelten spez. Verschleißrate K für zwei Werkstoffe.

O = Oberes Lager (vgl. Bild 15)
U = Unteres Lager

Bild 55 zeigt gute Übereinstimmung von Messungen im Bauteil verglichen mit der Rechnung. Dabei wurde mit "Modell"-Versuchen gemäß den Daten der Tabelle 11 die spezifische Verschleißrate K experimentell ermittelt.

Beanspruchungs-Bedingungen		Modell	Bauteil
P_N	$\frac{N}{mm^2}$	10	1
P_N	$\frac{N}{mm^2}$	-	-
f_p	Hz	-	-
Δx	µm	200	200
fx	Hz	100	33
Δy	µm	-	-
fy	Hz	-	-
ϑ	°C	100	110
t [7]	h	50	500

Tabelle 11

7) t = Versuchsdauer

In der gleichen Tabelle sind die Beanspruchungsbedingungen enthalten, mit denen die Bauteilversuche durchgeführt wurden. Der Bauteilverschleiß W_l wurde nach Versuchsende für das obere (o) und untere (u) Lager messen.

Gemäß der Beziehung

$$K = \frac{W_l}{s_G \cdot p} = \frac{W_l}{2 \cdot \Delta x \cdot f_x \cdot \Delta t \cdot p} \quad \text{in} \quad \frac{\mu m}{km \cdot N/mm^2}$$

wurde nun der Verschleiß W_l des Bauteils aus der spezifischen Verschleißrate und den Beanspruchungsbedingungen des Bauteils berechnet.

2. Beispiel (komplexe Systemklasse III)

Die Paarung besteht aus Stahl-Stahl in Kraftstoff laufend. Bild 56 zeigt eine gute Übereinstimmung des Verschleißverlaufes zwischen Erzeugnis (Produkt) und "Modell". Diese Übereinstimmung wurde nur erreicht, nachdem das Beanspruchungskollektiv in der Simulation (Bild 57) dem Beanspruchungskollektiv im Produkt angepaßt wurde. Zusätzlich wurde die Schwingung in y-Richtung zwischen 50 Hz und 150 Hz ständig variiert, damit der Beanspruchungszustand instationär blieb wie im Erzeugnis.

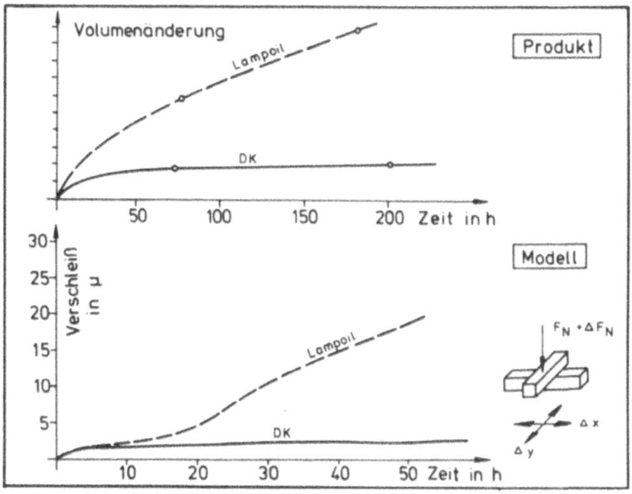

Bild 56 : Vergleich der Verschleißverläufe eines Bauteils mit zwei Kraftstoffen im Erzeugnis (Produkt) und nach der Simulation mit vereinfachten Probekörpern (Modell).

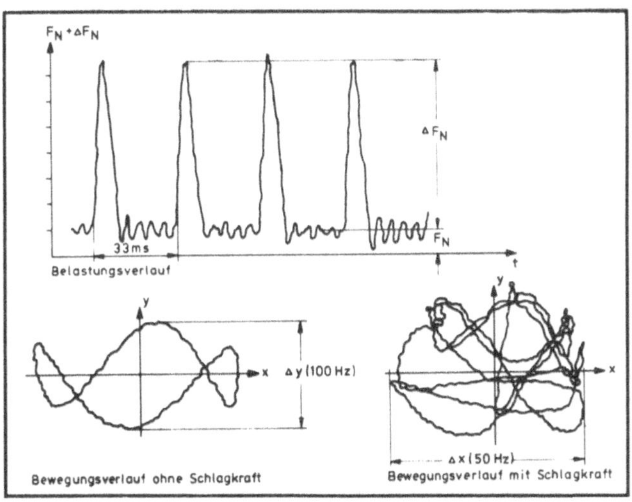

Bild 57 : Die in Bild 55 dargestellten Verschleißverläufe konnten unter Simulationsbedingungen erst erreicht werden mit schlagender Belastung und Schwingungsbewegungen in zwei Richtungen, wie es auch dem Original im Erzeugnis entsprach.

5.6 Ergänzende Hinweise zu Abhilfemaßnahmen bei Schwingungsverschleiß

Es ist sehr schwierig, konkrete Abhilfemaßnahmen gegen Schwingungsverschleiß unabhängig vom jeweiligen System zu nennen. In diesem Abschnitt werden daher nur allgemeine Gesichtspunkten zur Schwingungsverschleißminderung aus der Literatur und aus den hier durchgeführten Versuchen genannt.

Da der Schwingungsverschleiß zunächst mit einem überwiegend adhäsiven Verschleißmechanismus beginnt, und in einem späteren Stadium die Oberflächenzerrüttung eine maßgebende Rolle spielt, sind alle Werkstoffmaßnahmen günstig, welche die Adhäsion und Oberflächenzerrüttung mindern. Dies wird erreicht nach WATERHOUSE[37] durch:

- harte Werkstoffoberflächen (z.B. Härten, Einsatzhärten, Nitrieren, legierte Werkzeugstähle, Hartmetalle).
- Beschichtungen hoher Härte (z.B. Hartchrom, PVD- bzw. CVD-Beschichtungen u.a. Titannitrid, Titancarbid, Plasmaspritzschichten).
- Erzeugung von Druckeigenspannungen (z.B. Rollieren, Kugelstrahlen, Sandstrahlen).
- Nichtmetallische Gegenflächen (z.B. Kunststoffe, MoS_2-Gleitlacke)
- Schmieren der Oberflächen mit flüssigen Schmierstoffen.

Hinweise zu einigen besonderen Maßnahmen:

Die Wirkung von Nuten und Depots zur Verbesserung der Schmierung ist bei Schwingungsverschleiß stark systemabhängig, wie teilweise in Kapitel 5.1.5 bereits beschrieben wurde. Das Maß der Bewährung steht im engen Zusammenhang mit den Stofftransportvorgängen.

Allgemein haben Nuten zwei Vorteile:

1. Nuten sorgen für ausreichende Schmierstoffzufuhr, sofern die Schwingungsweiten groß genug sind ($\Delta x >$ Nutabstand). Bei Schwingungsweiten unter etwa 1 mm wird es sehr schwierig sein, die Schmierwirkung durch Nuten zu verbessern.

2. Nuten erleichtern den Verschleiß-Abtransport und können daher die Gefahr des Verklemmens von Teilen in engen Passungen vermindern.

Da Nuten bei kleineren Schwingungsweiten nicht mehr verschleißmindernd wirken, können durch Aufrauhen, Ätzen, Schaben, Rändeln oder ähnlichen Maßnahmen Depots auf der Oberfläche geschaffen werden, welche Festschmierstoffe, Fette oder Öle aufnehmen können und die Schmierung verbessern. Auch hier ist die Wirkung ähnlich wie bei Nuten abhänig vom Verhältnis Schwingungsweite zum mittleren Abstand der einzelnen Depots. Bei Schwingungsweiten $\Delta x < 100$ µm wird es zunehmend schwierig, eine ausreichende Schmierwirkung aufrecht zu erhalten. Besser verhalten sich dann geschlossene Festschmierstoffschichten.

Ein anderer Weg zur Verminderung des Schwingungsverschleißes und Vermeidung von Reiboxidation mit seinen nachteiligen Folgen durch Aufrauhung und Verklemmen von Passungen ist bei entsprechender Wahl der Werkstoffpartner der Einsatz von Kunststoff. Verwendung von Kunststoff bietet weitere Vorteile:

1..Dämpfende Wirkung bei Schlagbelastungen

2. Minderung der Reibkraft

Durch Fettschmierung erreicht man bei Kunststoff-Stahl-Paarungen eine Verminderung der Verschleißrate um etwa den Faktor 10. Für die Optimierung solcher Paarungen werden abschließend einige dabei zu beachtende Hinweise zu Einzelparametern gegeben.

<u>1. Kunststoff</u>	Art	Temperatureinsatzgrenze Quellverhalten
	Faserfüllstoffe	Mögliche abrasive Wirkung auf Gegenkörper beachten
	Pulverfüllstoffe	Tendenz zur Erhöhung der Verschleißrate infolge Zerrüttungsprozesse zwischen Pulverteilchen und Matrix, gilt auch für Faserfüllstoffe

2. Gegenwerkstoff	Rauhigkeit	Auch bei Schwingungsverschleiß sollte die Rauhigkeit klein sein (z.B. $R_{3z} < 3\mu m$)
	Härte	Um Verschleiß am metallischen Partner zu vermeiden, ist die Härte auf den Kunststoff abzustimmen. Allgemein gilt HRC > 50.

Schrifttum

1. Allianz: Handbuch der Schadenverhütung. München und Berlin: Allianz Versicherungs-AG 1976.

2. Bartel, A.A.: Reibkorrosion. VDI-Berichte Nr. 243 (1975), S. 157 - 170.

3. Childs, T.H.C.: The sliding wear mechanisms of metals, mainly steels. Tribology international (1980) Heft 12, S. 285 - 293.

4. Clerico, M.; Patierno, V.: Sliding wear of polymeric composites. Wear 53 (1979), S. 279 - 301.

5. De Gee, A.W.J.: Selection of materials for tribomechanical applications - the role of tribometry. Tribology international (1978) Heft 10, S. 233 - 239.

6. DIN 50321: Verschleiß-Meßgrößen. (1979).

7. Dowson, D.; Furber, J.R.; Atkinson, J.R.: Wear Mechanisms for Nylon 66. In: The wear of non-metallic materials. Dowson, D.; Godet, M.; Taylor, C.M. London: Mechanical engineering publications 1978, S. 25 - 31.

8. Eliezer, Z. ; Khanna, V.D.: Wear mechanism in composites: A qualitative model. Wear 51 (1978) S. 169 - 179.

9. Engel; Klingele: Rasterelektronenmikroskopische Untersuchungen von Metallschäden. Köln: Gerling Institut für Schadenforschung und Schadenverhütung GmbH 1974.

10. Erhard, G.: Zum Reibungs- und Verschleißverhalten von Polymerwerkstoffen. Diss. Uni Karlsruhe 1980.

11. ESDU: A guide on the design and selection of dry rubbing bearings. Item Number 76029 (1976). Engineering sciences data unit, Regent Street London W1R 7 AD.

12. Feng, I.M.; Uhlig, H.H.: Fretting Corrosion of Mild Steel in Air and in Nitrogen.. Trans ASME, Journal of appl. Mech. 21 (1954), S. 395 - 400.

13. Fleischer, G: Energiebilanzierung der Festkörperreibung als Grundlage zur energetischen Verschleißberechnung. Schmierungstechnik 7 (1976) 8, S. 225-230, Teil II 7 (1976) 9, S.271-279, Teil III 8 (1977)2, S. 49-58.

14. Habig, K.-H.: Verschleiß und Härte von Werkstoffen. München, Wien: Carl Hanser Verlag 1980.

15. Halliday, J.S.: Experimental investigation of some processes involved in fretting corrosion. Proc. Conf. on lubrication and wear, 1957 London Ber.Nr. 39, S. 640-646, London: Institution of Mechanical Engineers 1958.

16. Halliday, J.S.; Hirtst,W. : The fretting corrosion of mild steel. Proc. Royal Soc. A 236 (1956), S. 411 - 425.

17. Heinke, G.: Verschleiß - eine Systemeigenschaft, Auswirkungen auf die Verschleißprüfung. Zeitschrift für Werkstofftechnik 6 (1975) Heft 5, S. 164 - 169.

18. Heinke, G.; Heinz, R.; Thiel, I.: Betriebserfahrungen mit tribotechnischen Modell- und Bauteilversuchen. Schmiertechnik Tribologie (1981) (in Vorbereitung).

19. Heinz, R.: Schäden an Zahnkupplungen. Antrieb und Steuerung und Regelung 5 (1977) Heft 6, S. 23 - 26.

20. Hurricks, P.L.: The mechanism of fretting-a review. Wear 15 (1970), S. 389-409.

21. Johnson, K.L.: Surface interaction between elastically loaded bodies under **tangential** forces . Proc. Roy. Soc. A 230 (1955) Nr. 1183, S. 531-548.

22. Lancaster, J.K.: Geometrical Effects on the wear of Polymer and Carbons. Journal of Lubrication Technology (1975) April, S. 187-216.

23 Lancaster, J.K.; Evans, D.C.: The wear of Polymers. S. 85 - 139. In Scott, D.: Treatise of materials science and technology Vol 13, Wear. New York, San-Francisco, London: Academic Press 1979.

24. LNP-Bulletin 254-277: LNP Selbstschmierende, verstärkte Thermoplaste, München 50, Franz-Albert-Str. 6 a.

25. Mc Dowell, J.R.: Fretting Corrosion S. 4.21-4.26. In: Standard Handbook of lubrication Engineering. New York, London: Mc Craw-Hill Book Comp. 1968. 4.21 - 4.26

26. Müller, K.: Einfluß von Schmierstoffen auf die Reibkorrosion, dargestellt durch Reibarbeit und Schwingreibverschleiß. VDI-Forschungsheft 564 (1974).

27. Ohmae, N.; Tsukizoe, T.: Wear performance of unidirectionally oriented carbonfiber reinforced plastics. Tribology International 8 (1975) Nr. 4, S. 171 - 175.

28. Ohmae, N.; Tsukizoe, T.: The effect of slip amplitude on fretting. Wear 27 (1974), S. 281 - 294.

29. Reed, F.E.; Batter, J.F.: An Experimental Study of Fretting and Galling in Dental Couplings. ASLE Trans. 2 (1960), S. 159 - 172.

30. Sämann, K.-H.: Reibungszahl und Verschleiß bei Gleitreibung unter der Einwirkung mechanischer Hochfrequenzschwingungen. Schmiertechnik und Tribologie 18 (1971). Heft 3, S. 113 - 116.

31. Stowers, I.F.; Rabinowicz, E: The Mechanism of Fretting wear. Journal of lubrication Technology 1 (1973) S. 65-70

32. Uetz, H.; Sommer, K.; Khosrawi, M.A.: Übertragbarkeit von Versuchs- und Prüfergebnissen bei abrasiver Verschleißbeanspruchung auf Bauteile. In: VDI-Berichte 354, Übertragbarkeit von Versuchs- und Prüfergebnissen auf Bauteile. Düsseldorf: VDI-Verlag GmbH 1979.

33. Uetz, H.; Föhl, J.: Wear as an energy transformation process. Wear 49 (1978), S. 253 - 264.

34. VDI-Richtlinie 2541: Gleitlager aus thermoplastischen Kunststoffen. (1975).

35. Ward, R.: A comparison of reciprocating and continuous sliding wear. Wear 15 (1970), S. 423 - 434.

36. Waterhouse, R.B.: Fretting Corrosion. Oxford, New York, Braunschweig: Pergamon Press 1972.

37. Waterhouse, R.B.: Fretting. In: Treatise of materials science and technology, Vol. 13, Wear. New York, San Francisco, London: Academic Press 1979.

Schwingungsverschleiß an Turbinenwerkstoffen

von

Dr.-Ing. P. Adam
Dr.-Ing. E. Broszeit
und
Prof. Dr.-Ing. K. H. Kloos

MTU-München
und
Institut für Werkstoffkunde
der Technischen Hochschule Darmstadt

Förderkennzeichen: 01 ZT 017 u. 01 ZT 027
Laufzeit: 1. 1. 1978 - 31. 12. 1980 bzw.
1. 11. 1979 - 31. 12. 1981

Zusammenfassung

Die Schwingfestigkeit von Turbinenschaufeln aus Titanlegierungen oder Stahl wird durch Schwingungsverschleiß (Fretting) erheblich abgesenkt. Zur Untersuchung von Abhilfemaßnahmen, die zu einer Steigerung der Schwingfestigkeit führen, wurden im Rahmen dieser Vorhaben zwei Apparaturen entwickelt, mit denen einerseits Reibung und Verschleiß zwischen zwei Reibpartnern untersucht werden können und andererseits die Schwingfestigkeit unter Schwingungsverschleißbeanspruchung geprüft werden kann.

Die bisher untersuchten Schichten ergaben für die Reibungs- und Verschleißapparatur folgende Wertungsfolge:

1. PVD-Hartschichten mit weicher Haftschicht
2. Nitrierschichten
3. Spritzschichten und galvanisch aufgebrachte Schichten

Die Schwingfestigkeitsuntersuchungen mit Schwingungsverschleißbeanspruchung haben ergeben, daß die Veränderung der Oberflächenstruktur und die Einbringung von Druckeigenspannungen durch eine optimale Kugelstrahlbehandlung die höchste Steigerung der Schwingfestigkeit ergibt.

Gliederung

1. Einleitung

Trotz zahlreicher Fortschritte in der Grundlagenforschung und angewandten Forschung auf dem Gesamtgebiet der Tribologie sind die mit Fragen der Werkstoffauswahl, der konstruktiven Gestaltung und dem Betriebsverhalten tribologisch beanspruchter Bauteile befaßten Ingenieure überwiegend auf empirische Methoden angewiesen. Die Gründe für diesen Sachverhalt sind vor allem auf die schwierige mechanische, thermische und chemische Beanspruchungsanalyse tribologisch beanspruchter Bauteilpaarungen und die vielfältigen Wechselwirkungen zwischen den Reibpartnern in Form meist gekoppelt auftretender Verschleißmechanismen (Adhäsion, Abrasion, Tribooxidation, Oberflächenermüdung) zurückzuführen.

Ein Beispiel für eine derartige, zu beträchtlichen Schäden führende Komplexbeanspruchung mit gleichzeitiger Wirkung mehrerer Verschleißmechanismen liegt im Schwingungsverschleiß vor. Schwingungsverschleiß (Reibkorrosion, fretting corrosion) wird ausgelöst durch oszillierende Relativbewegungen mit kleiner Amplitude (einige Angström-Einheiten bis einige Millimeter) bei breitem Frequenzspektrum. Schwingungsverschleiß führt zu Oberflächenveränderungen und Materialverlust, die meist zu einer negativen Veränderung der Funktionsfähigkeit von Bauteilpaarungen z.B. durch Vergrösserung oder Verkleinerung des Spiels in Passungen usw. führen.

Über die Verschleißschädigung (lokale Oberflächenschädigung) hinaus kann bei schwingbeanspruchten Bauteilen -durch Schwingungsverschleiß ausgelöst- ein Bruch des Bauteiles die Folge sein. Aus der Praxis sind Beispiele für durch Schwingungsverschleiß verursachte Dauerbrüche bekannt bei: Wellen-Naben-Verbindungen, Wälzlagern, Pleuelstangen und Turbinenschaufeln.

Bild 1 zeigt am Beispiel einer schwingbeanspruchten Wellen-Nabenverbindung die Zunahme der Parameterzahl in einer Ge-

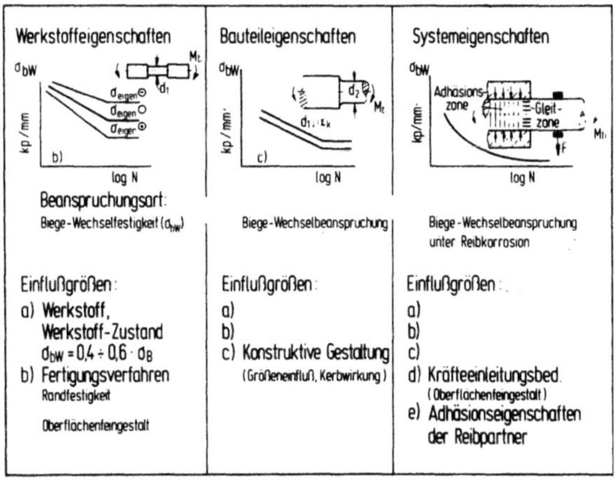

Bild 1. Vergleich zwischen Werkstoff-, Bauteil- und Systemeigenschaften unter Berücksichtigung der wichtigsten Einflußgrößen

genüberstellung von Werkstoff-, Bauteil- und Bauteilpaarungs- oder Systemeigenschaften. Durch Werkstoff und Fertigungsverfahren werden die Biegewechselfestigkeitseigenschaften beeinflußt, zu denen beim Bauteil noch die konstruktiven Gestaltungseinflüsse wie Kerbwirkung und Größeneinfluß hinzukommen. Die Systemeigenschaften der Wellen-Nabenverbindung werden schließlich noch durch die Krafteinleitungsbedingungen und die Adhäsionsneigung der Reibpartner beeinflußt.

In der Regel führt die phasengleiche Überlagerung einer schwingenden Grundwerkstoffbeanspruchung mit einer Schwingverschleißbeanspruchung zu einer beträchtlichen Abnahme der Lebensdauer. Bild 2 zeigt die Ergebnisse [1] systematischer Untersuchungen verschiedener Einflußparameter auf die Reibdauerhaltbarkeit (Zugschwellfestigkeit mit Schwingungsverschleißbeanspruchung).

Für die Untersuchungen wurde die im linken Teilbild dargestellte Flachprobe verwendet. Die Zugschwellfestigkeit

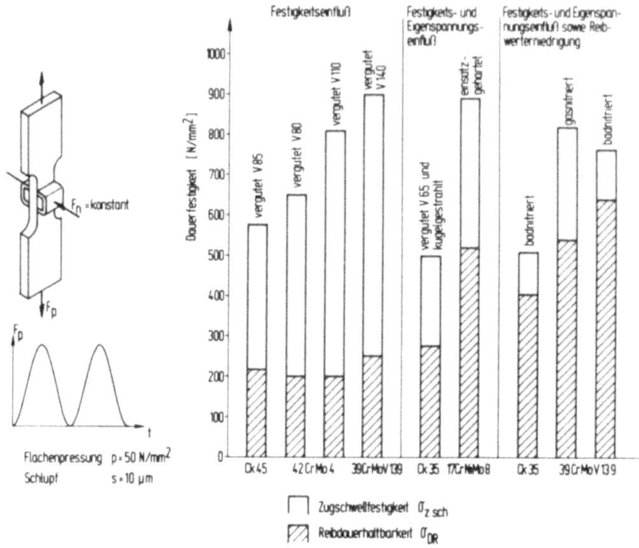

Bild 2. Werkstoff- und Wärmebehandlungseinflüsse auf die Reibdauerhaltbarkeit

wird mit der Probe allein bestimmt, die Reibdauerhaltbarkeit wird mit seitlich angepreßten Reibklötzchen ermittelt.

Aus der Gegenüberstellung der Zugschwellfestigkeitswerte und der Reibdauerhaltbarkeitswerte bei Bruchlastspielzahlen von $N = 4 \cdot 10^7$ läßt sich die beträchtliche Schädigung am Beispiel vergüteter Stahlproben unterschiedlicher Festigkeit unmittelbar nachweisen. Durch partielle Festigkeitssteigerung der Zugproben infolge Kugelstrahlen oder Einsatzhärten kann die Dauerhaltbarkeit vor allem durch die günstige Wirkung von Druckeigenspannungen angehoben werden.

Wird darüber hinaus neben einer partiellen Festigkeitssteigerung und Druckeigenspannungsausbildung im Falle nitrierter Oberflächen zusätzlich der Reibwert durch den günstigen Einfluß der Verbindungszone erniedrigt, so können die höchsten Reibdauerhaltbarkeitswerte erzielt werden.

Bei modernen Flugantrieben werden im Verdichterbereich für die Laufschaufeln und für die Rotorscheiben hochfeste Legierungen eingesetzt. Bei routinemäßigen Überprüfungen werden an den Schaufelfüßen aus Titan und Stahl Oberflächenveränderungen durch Schwingungsverschleiß und Anrisse festgestellt, die zum Auswechseln der Schaufeln zwingen.

Bild 3 zeigt im Fußbereich einer Verdichterrotorschaufel einen Dauerbruchanriß, der von einer Reibschweißstelle seinen Ausgang nahm. Von der anderen Seite des Schaufelfußes ist ein zweiter Riß auf gleiche Weise entstanden. Beide Risse laufen in der seitlichen Fläche zusammen. Der verbleibende tragende Querschnitt des Fußes hat sich dadurch beträchtlich verringert.

Bild 3. Reibdauerbruchanrisse im Schaufelfuß

In Bild 4 ist die durch Schwingungsverschleiß ausgelöste Schädigung in ihrer typischen Form in einem früheren Stadium bei stärkerer Vergrößerung noch einmal dargestellt.

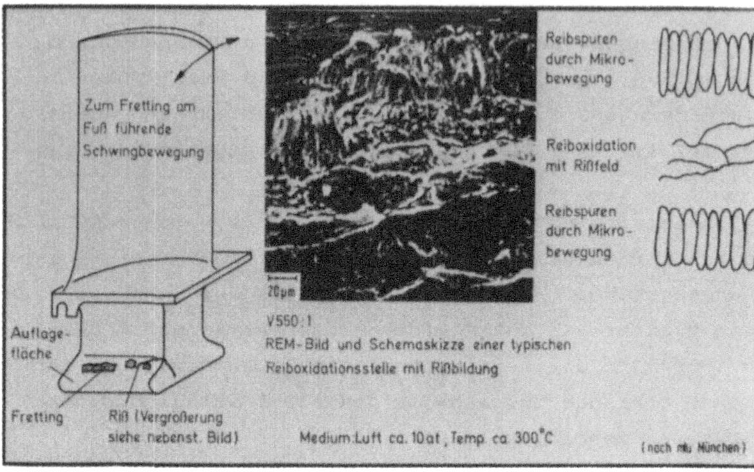

Bild 4. Schäden durch Schwingungsverschleiß an Verdichterschaufeln

Die Schadenshöhe wird von der Größe des örtlich umgesetzten Reibenergiebetrages, d.h. von der örtlichen Flächenpressung, dem Gleitschlupf und dem Reibwert zwischen den eingespannten Teilen maßgeblich beeinflußt.

Die Simulation des Schwingungsverschleißvorganges mit Flachproben und Reibklötzchen aus der Titanlegierung TiAl 6 V 4 mit dem in Bild 2 gezeigten Prüfprinzip gibt Aufschluß über Größe und Veränderung des Reibwertes in Abhängigkeit von der Zahl der Belastungen. Der anfänglich vorliegende Reibwert von $\mu = 0{,}4$ (oxidische Annäherung) steigt allmählich auf $\mu = 0{,}9$ (metallische Annäherung, Bild 5). Die Untersuchung der Oberflächen mit stärkerer Vergrößerung zeigt in unmittelbarer Nähe der kaltverschweißten Oberflächenbereiche zahlreiche Mikrorisse senkrecht zur Bewegungsrichtung, die mögliche Ausgangsstellen für einen makroskopischen Ermüdungsanriß der schwingbeanspruchten Probe sind.

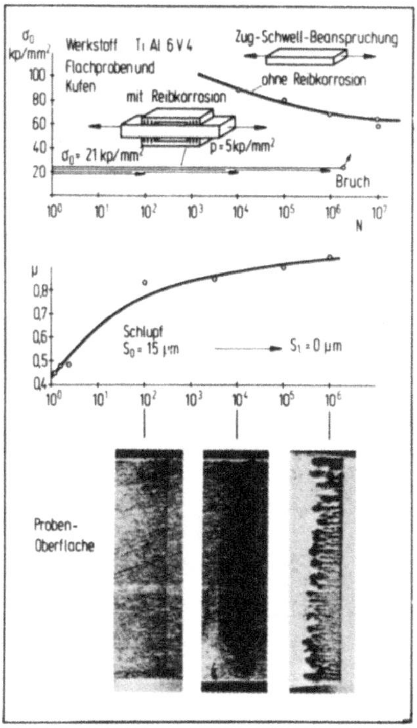

Bild 5.
Reib-Dauerbeanspruchung und Verschleißmechanismen

Die in Bild 5 dargestellten Ergebnisse aus Voruntersuchungen zu diesem Programm zeigen ferner, daß mit zunehmender Reibung zwischen Probe und Kufen der rechnerisch für eine bestimmte Maximalspannung in der Probe ermittelbare Schlupf S_0 abnimmt. Der Probe stab zwischen den Kufen trägt mit zunehmender Reibung nur noch eine geringere Last, da die Reibklötze entsprechend der Reibkraft einen Teil der Belastung übernehmen. Ein Restschlupfbetrag wird jedoch stets vorhanden sein ($S_u > 0$).

2. Experimentelle Untersuchungen zur Reibdauerhaltbarkeit

2.1 Prüfprinzip

Zur Simulation der Beanspruchung, wie sie am Schaufelfuß

auftritt, wurden Zug-Schwell-Beanspruchungen mit sinusförmigem Kraftverlauf auf Flachproben aufgebracht, wobei gemäß Bild 6 gleichzeitig 2 U-förmige Reibklötzchen aus der gleichen Legierung mit konstanter Kraft an die Flachproben angedrückt werden. Die Andrückkraft wurde dabei mit der Spitzenlast in der Flachprobe gekoppelt, so daß die Flächenpressung zwischen der Probe und den Kufen 40 % der maximalen Zugspannung in der Probe betrug. Die Versuchsmethode erfolgte in Anlehnung an das in [1] näher erläuterte Prüfverfahren.

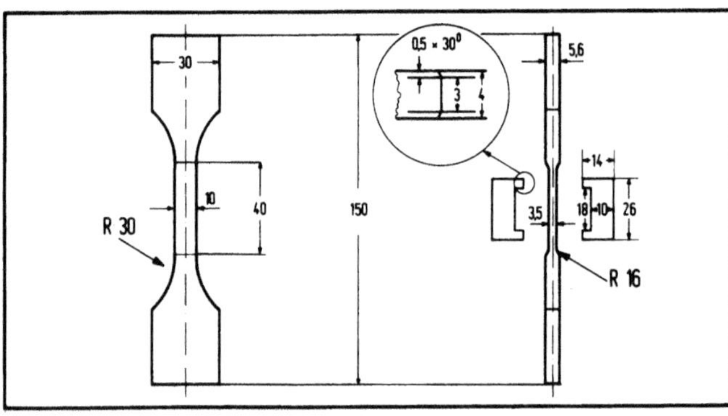

Bild 6. Probenabmessungen

Der Schlupf zwischen Probe und Kufe hängt von der Belastung der Flachprobe, dem Reibwert und den geometrischen Abmessungen der Reibklötzchen ab. Bei einer maximalen Zug-Schwellspannung von z.B. 550 N/mm^2 ergibt sich ein rechnerischer Schlupfbetrag von S_0 = 62,5 μm (bei μ = 0, E_{Ti} = 110 000 N/mm^2 und einem maximalen Kufenabstand von 25 mm). Zwischen S_0 und der maximalen Schwellspannung besteht damit ein linearer Zusammenhang. Wie bereits oben erwähnt, hängt auch die Pressung zwischen Kufe und Probe in linearer Form von der Maximallast in der Probe ab.

2.2 Versuchswerkstoff

Tabelle 1 enthält die chemische Zusammensetzung des Flachproben-Werkstoffes sowie mechanische Kennwerte. Der Werkstoff lag im gewalzten und geglühten Zustand vor.

Element	Fe	C	O	N	H	Al	V	Ti
Gew. %	0,13	0,024	0,16	0,014	0,0026	6,5	4,1	Rest
mech. Kennwerte	E-Modul: 110 000 N/mm^2 Härte HV30 - 320							
(Werkstoff-Nr. 3.7164)	Zugfestigkeit R_m = 1 113 N/mm^2 Streckgrenze $R_{p0,2}$ = 1 019 N/mm^2							

Tabelle 1. Chemische Zusammensetzung und mech. Kennwerte des Werkstoffes TiAl 6 V 4

Die Flachproben wurden aus ca. 7 mm starkem Blech gefertigt; die Probenoberfläche im Bereich der Prüflänge liegt im gefrästen Zustand vor.

2.3 Kugelstrahlbehandlung

Zur Untersuchung des Einflusses einer Kaltverfestigung wurde eine Kugelstrahlbehandlung* mit unterschiedlichen Strahlbedingungen durchgeführt.

Für die Strahlbehandlung wurde eine Maschine nach dem Schleuderradprinzip (Bild 7) verwendet. Als Strahlmittel war Stahlgußschrot mit einem Korndurchmesser d = 0,2 mm

---

* Für die Durchführung der Strahlbehandlung sei auch an dieser Stelle der Fa. Georg Fischer, Scha ffhausen/Schweiz und insbesondere den Herren K. Egli und H.P. Walter gedankt.

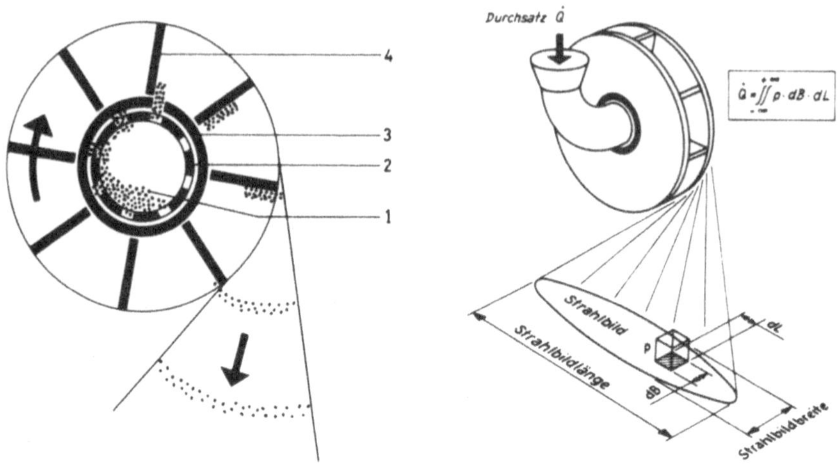

Bild 7. Schleuderrad-Kugelstrahl-Verfahren

verwendet worden. Zur Erzielung unterschiedlicher Strahlintensitäten wurde die Abwurfgeschwindigkeit und die Zahl der Durchgänge variiert. Die Strahlintensität wurde durch Messung des Almen-Wertes A_2 bestimmt. In Tabelle 2 sind die Strahlparameter und die damit erzielten Almen-Intensitäten zusammengestellt.

Abwurfgeschwindigkeit V_{ab} [m/s]	Zahl der Durchgänge	Almenwert A_2 [mm]
34	4	0,13
52	1	0,17
48	4	0,20
78	1	0,24
78	4	0,30

Tabelle 2. Strahlbedingungen

Durch das Kugelstrahlen werden die randnahen Werkstoffbereiche plastisch verformt. Hierdurch kommt es einerseits zu einer veränderten Mikrogeometrie, andererseits aber auch zur Ausbildung von Druckeigenspannungen 1. Art im Werkstoff.

In den Bildern 8 a - c sind rasterelektronenmikroskopische Aufnahmen von den mit A_2 = 0,13 (Bild 8 a), mit A_2 = 0,17 (Bild 8 b) und mit A_2 = 0,3 (Bild 8 c) gestrahlten Probenoberflächen bei jeweils gleicher Vergrößerung wiedergegeben. Mit zunehmender Almen-Intensität tritt eine Vergröberung der Oberflächenstruktur ein. Eine Rißbildung

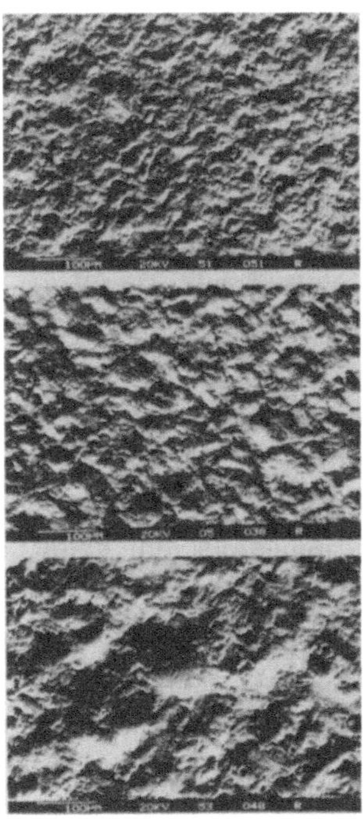

A_2=0,13

A =0,17

A =0,3

Bild 8.
Gestrahlte
Oberflächen

ist bei der A_2 = 0,3-Intensität -die einen bei den Versuchen beobachteten Abfall der Schwingfestigkeit erklären könnten- nicht zu erkennen.

Die Näpfchenstruktur der gestrahlten Oberflächen führt dazu, daß die Kufen der Reibklötzchen nur auf wenigen Punkten aufliegen und sich anfänglich gebildete oxidische Abriebpartikel in den Vertiefungen ansammeln können.

2.4 Eigenspannungen 1. Art

An gefrästen und kugelgestrahlten Proben wurden auf röntgenographischem Wege Eigenspannungen 1.Art in Längs- und Querrichtung gemessen. Hierzu wurden die Proben nach jeder Messung schichtweise elektrolytisch abgetragen und erneut vermessen.

Bild 9 zeigt den Verlauf der Eigenspannungen in Längs- und Querrichtung einer gefrästen und einer nach dem Fräsen kugelgestrahlten Flachprobe. Bei der gefrästen Probe werden Druckeigenspannungen von ca. 260 N/mm^2 erreicht, die in einer Tiefe von ca. 50 µm auf 0 abgeklungen sind. Die mit einem Almen-Wert von A_2 = 0,24 kugelgestrahlte Probe zeigt vor der Beanspruchung (N = 0) ein Druckeigenspannungsmaximum bei ca. 40 µm von ca. 640 N/mm^2. Zur Oberfläche hin und in Richtung größerer Tiefe werden die Eigenspannungen dann wieder kleiner. In einer Tiefe von ca. 200 µm sind die Spannungen dann auf 0 abgesunken. Eine zweite Probe mit einem Almen-Wert von ebenfalls A_2 = 0,24 zeigt nach einer Beanspruchungszahl von $N = 10^7$ ($\sigma_{Sch\ max}$ = 450 N/mm^2) einen in der Tendenz ähnlichen Spannungsverlauf mit etwas geringeren Werten. Ob es sich hierbei um einen Abbau von Eigenspannungen durch die Zugschwellbeanspruchung handelt ist noch nicht geklärt.

---

* Die Messungen wurden von Professor Hauk und Mitarbeitern, TH Aachen durchgeführt. Hierfür sei auch an dieser Stelgedankt.

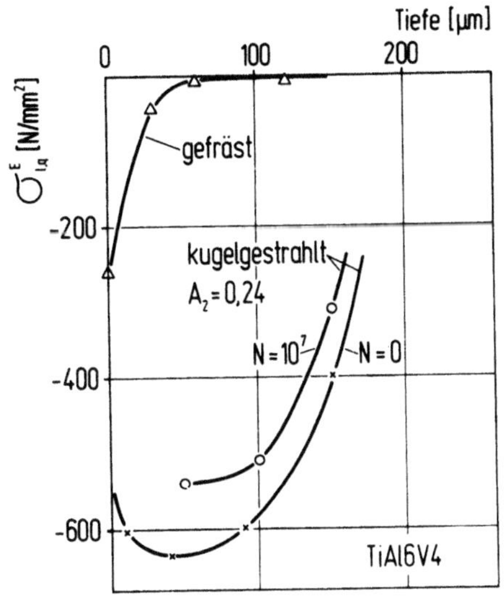

Bild 9. Eigenspannungen 1. Art durch Kugelstrahlen und Fräsen

Bild 10 zeigt den Eigenspannungs-Tiefenverlauf für eine weitere Probe, die mit A_2 = 0,17 kugelgestrahlt und danach einer Zugschwellbeanspruchung von N = 10^7 bei einer maximalen Zugspannung von σ_{Sch} = 450 N/mm^2 ausgesetzt wurde. Die gemessenen Höchstwerte der Eigenspannungen liegen in diesem Fall dichter an bzw. in der Oberfläche und betragen ca. 660 N/mm^2. Diese Eigenspannungen sind in einer Tiefe von ca. 140 µm auf 0 zurückgegangen.

Die Abwurfgeschwindigkeit zum Erreichen dieses Almen-Wertes von A_2 = 0,17 betrug V_0 = 52 m/s während bei der Probe mit A_2 = 0,24 ein V_0 = 78 m/s erforderlich war. Die geringere kinetische Energie des Strahlmittels hat somit bei der Probe mit A_2 = 0,17 eine geringere Tiefenwirkung zur Folge gehabt.

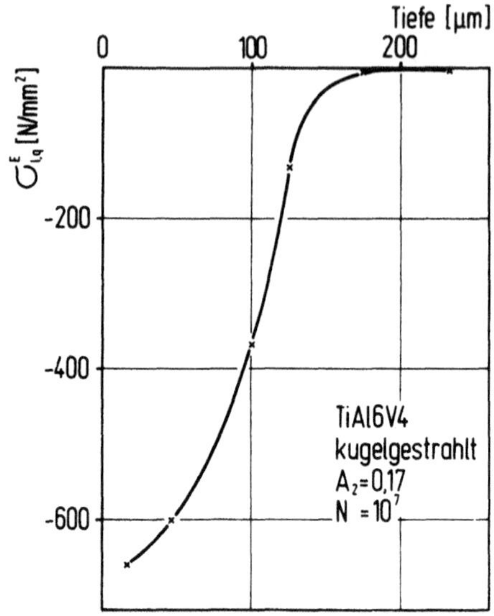

Bild 10. Eigenspannungen 1. Art durch Kugelstrahlen

2.5 Versuchsdurchführung

Zur Bestimmung der Reibdauerhaltbarkeit wurden Wöhlerversuche mit 6 bis 7 Proben durchgeführt; die Zahl der Beanspruchungen wurde auf 10^7 begrenzt, die Prüffrequenz lag bei ca. 160 Hz.

2.6 Ergebnisse

Die Ergebnisse der Schwingfestigkeitsuntersuchungen sind in Tabelle 3 zusammengestellt. In Bild 11 ist die Schwellfestigkeit in Abhängigkeit vom Almen-Wert dargestellt.

Die erzielten Dauerfestigkeitswerte -maximal 480 N/mm^2- liegen um 120 % höher als diejenigen von unbehandelten (gefrästen) Proben.

Almenwert A_2 [mm]	Reibdauerhaltbarkeit σ_{DR} [N/mm^2]	
0	220	gefräst
0,13	410	
0,17	480	kugel-gestrahlt
0,20	-	
0,24	480	
0,30	420	

Tabelle 3. Reibdauerhaltbarkeit von kugelgestrahlten Proben als Funktion des Almen-Wertes

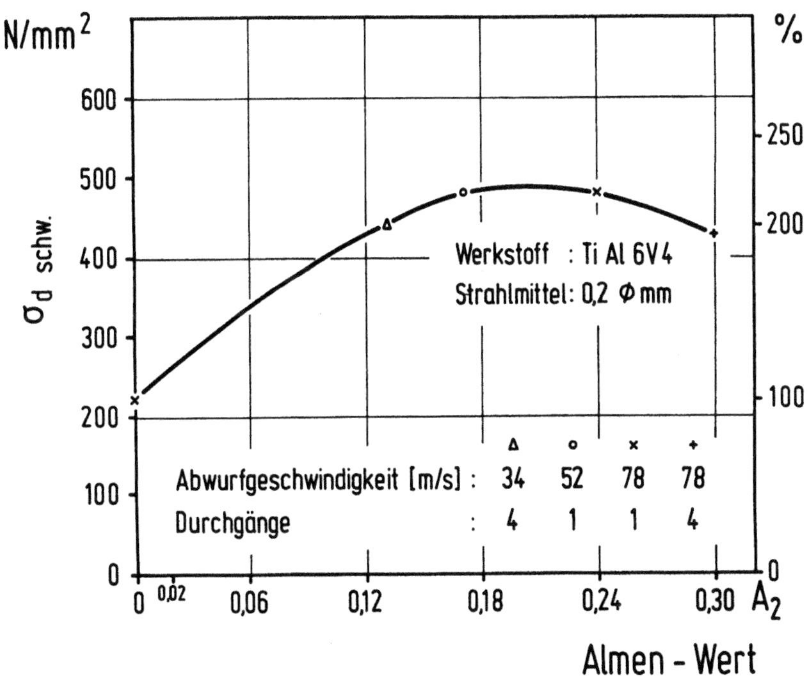

Bild 11. Reibdauerhaltbarkeit kugelgestrahlter Proben aus TiAl 6 V 4

Im Vergleich zu den nicht durch Schwingungsverschleiß beanspruchten Proben (Zugschwellfestigkeit: 600 N/mm^2) wird durch den zusätzlichen Schwingungsverschleiß bei den nicht gestrahlten Proben ein erheblicher Abfall der Schwingfestigkeit beobachtet (Abfall auf 220 N/mm^2). Die Kugelstrahlung führt nun dazu, daß ein wesentlicher Teil der eingebüßten Dauerfestigkeit durch die Strahlbehandlung wieder gewonnen werden kann. Mit zunehmendem Almen-Wert steigt die Reibdauerhaltbarkeit zunächst bis auf ca. 480 N/mm^2 an und fällt danach bei A_2 = 0,3 wieder auf ca. 420 N/mm^2 ab. Offenbar ist bei diesem Werkstoff und unter den hier vorliegenden Prüfbedingungen ein "Überstrahlen" mit hoher Almen-Intensität möglich.

2.7 Plasmagespritzte Proben

Bei MTU wurden gefräste Proben kugelgestrahlt und danach mit WCCo bzw. MoCoSi plasmagespritzt. Die Schichten wurden nach dem Spritzen geschliffen. Bild 12 zeigt die Ober-

Bild 12. Unbeanspruchte Oberfläche

fläche der WCCo-Schicht im unbeanspruchten Zustand. Die beanspruchte Probe zeigt zahlreiche Risse senkrecht zur Richtung der Schlupfbewegung (Bild 13).

Bild 13. Beanspruchte Oberfläche

Eine zusammenfassende Darstellung der Ergebnisse ist in Bild 14 zusammengestellt. Die Kugelstrahlbehandlung erweist sich als sehr wirksame Maßnahme, den stark negativen Einfluß der Schwingungsverschleißbeanspruchung zu reduzieren. Hier läßt sich ein deutliches Maximum erkennen. Die Plasmaspritzschichten bringen keine Vorteile in Bezug auf die Lebensdauer der Proben.

2.8 Möglichkeiten der Kaltverfestigung durch Rollen

Die durch eine Kugelstrahlbehandlung erzielbare Tiefenwirkung ist verhältnismäßig gering. Durch plastische Verformung hervorgerufene Druckeigenspannungen in Längs- und Querrichtung in etwa gleicher Größe sind in einer

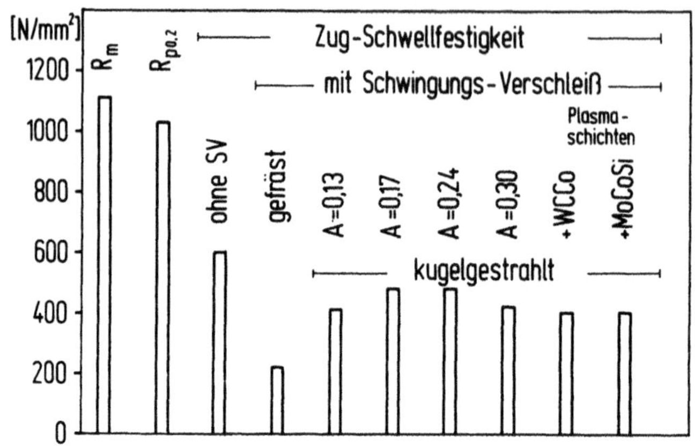

Bild 14. Mechanische Eigenschaften und Reibdauerhaltbarkeit

Tiefe von einigen 100 µm wieder vollständig abgeklungen.

Wesentlich größere Tiefenwirkungen lassen sich durch Rollen mit punktförmiger Berührfläche erreichen. Hierdurch ist es möglich, auch bei Werkstoffen mit hoher (Rand-) Festigkeit noch eine wirkungsvolle Verfestigungswirkung und Druckeigenspannungsausbildung zu erhalten. Bei randschicht- und einsatzgehärteten Werkstoffen läßt sich der Verlauf der Eigenspannungen in die Tiefe durch Variation der Drückkraft und ggf. auch der Geometrie der Drückrolle optimal an die Härte-Tiefen-Kurve anpassen.

Zur Kaltverfestigung von Flachproben durch Rollen mit punkt- bzw. ellipsenförmiger Berührfläche zwischen Werkzeug und Werkstück (Flachprobe) wurde eine Vorrichtung entwickelt und gebaut (Bild 15). Durch Variation der Drückkräfte und der Geometrie der Drückrolle ist es möglich, die Werkstoffanstrengung an die lokalen mechanischen Werkstoffeigenschaften anzupassen. Die nach der Verfesti-

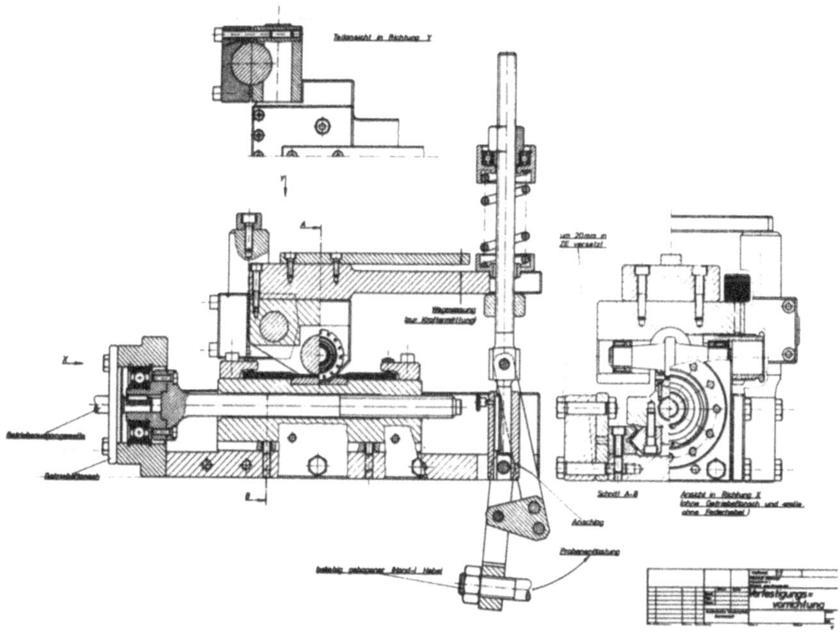

Bild 15. Verfestigungsvorrichtung für Flachproben

gung auf röntgenographischem Wege ermittelten Druckeigenspannungen lassen sich in die Werkstoffanstrengung nach verschiedenen Anstrengungshypothesen einbeziehen.

3. Experimentelle Untersuchungen des Verschleißverhaltens beschichteter Proben

3.1 Verschleiß und Ermüdungsrisse

Zweifellos besteht ein Zusammenhang zwischen den beobachtbaren Verschleißerscheinungen und den Ermüdungsrissen an beanspruchten Maschinenpaarungen. Dieser Zusammenhang ist jedoch quantitativ bisher nicht zugänglich, etwa in Form einer Lebensdauervorhersage aus der Größe oder Morphologie des geschädigten Bereichs. Selbst die Erkennung von Ermüdungsrissen im geschädigten Bereich ist unsicher wegen starker Oberflächenverformung mit begleitendem Verdrücken des Rißeingangs. Um so mehr wird es von Interesse, Oberflächenschutzschichten nicht allein in der Reibklötzchen-Apparatur, sondern unabhängig auf ihr Verschleißverhalten zu untersuchen. Dafür gibt es im wesentlichen folgende Gründe:

- Die Vorauswahl auf ihre Eignung muß möglichst einfach erfolgen, ohne die Reibdauerhaltbarkeit an zahlreichen Proben ermitteln zu müssen,

- die Reibklötzchen-Apparatur ist nicht in allen Punkten ausreichend, um Schutzschichtverhalten zu simulieren, z.B. durch die Tatsache, daß die Normal- und Tangentialkräfte getrennt eingeleitet werden, insbesondere in die Schicht,

- die getrennte Untersuchung des Verschleißverhaltens sowie seiner Korrelation zur Reibdauerhaltbarkeit schafft ein modellmäßiges Verständnis des Schädigungsablaufs, was Aussicht auf Anwendung in der Qualitätssicherung hat.

3.2 Verschleißprüfstand

Für die Simulation des Frettingverschleißes wurde ein Prüfstand gebaut (Bild 16, 17). Während der ersten Phase der Untersuchungen mit diesem Prüfstand wurden die Verschleißproben (Bild 18) mit unveränderlichen Einstellparametern (Tabelle 4) beansprucht und ihre Oberflächen untersucht. Besonderer Wert wurde auf einen gleichmäßig hohen Traganteil der Proben gelegt. Zu diesem Zweck wurden alle Probenoberflächen vor der Verschleißbeanspruchung mit einer mechanischen Vorrichtung (Läppleiste) (Bild 19) im Prüfstand geläppt. Der Traganteil betrug auf diese Weise 100 %. Ohne diese Maßnahme wird der Vergleich der Zeit bis zum Auftreten von Verschleißerscheinungen aufgrund der Unterschiede im Traganteil erschwert.

Bild 16. Gesamtansicht des Fretting-Verschleißprüfstandes (mtu)

Bild 17. Teilansicht des Fretting-Verschleißprüfstandes mit eingebauter Läppleiste

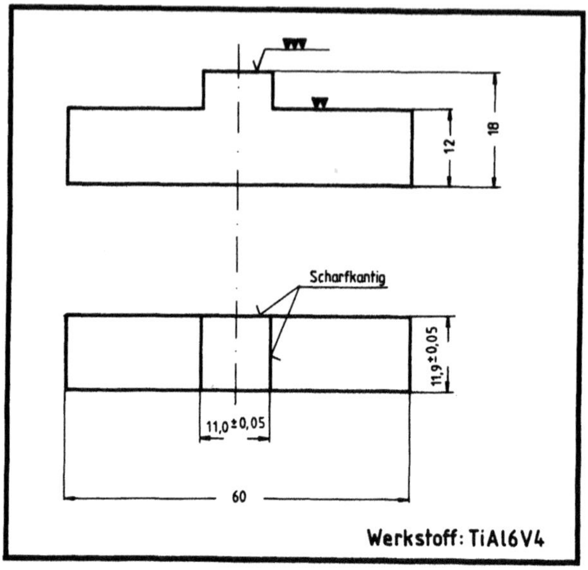

Bild 18. Prüfprobenform

Tabelle 4

- Normalkraft	5,2 N/mm^2
- Frequenz	50 Hz
- Amplitude	0,2 mm
- Schmierung	keine
- Fläche	144 mm^2 (12 x 12)
- Versuchsdauer	3, 6, 9 min mit Zwischenkontrolle

- Entfettung mit Papier und Aceton, anschließend Läppen mit Diamantschicht

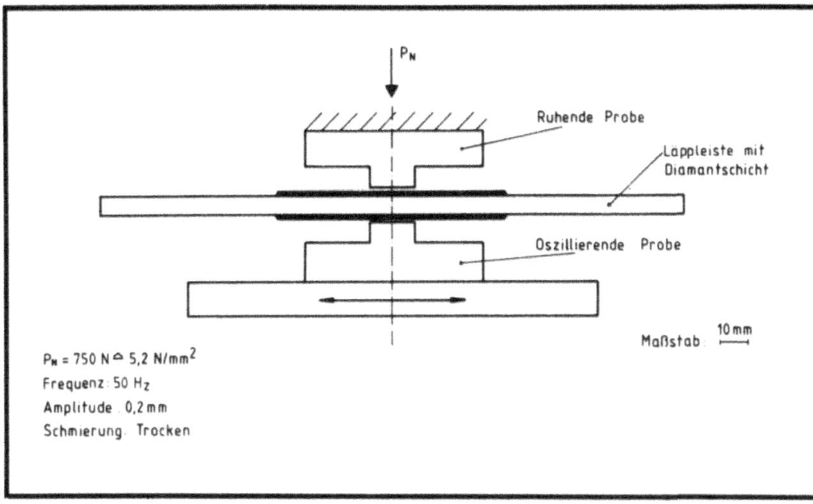

Bild 19. Prinzip der Läppbearbeitung

3.3 Untersuchungsergebnisse

3.3.1 Nitrier-, PVD- und Spritzschichten

In der Tabelle 5 sind die bisher untersuchten Schichtkombinationen zusammengestellt.

Die Oberflächenzustände vor dem Verschleißversuch sind aus Bild 20 a und b zu entnehmen. In Bild 21 sind die nach unterschiedlichen Zeiten beobachteten Verschleißerscheinungen makroskopisch dargestellt. Auf die Wiedergabe der REM-Untersuchungen wird an dieser Stelle aus Platzgründen verzichtet.

Allen Paarungen gemeinsam ist die Ausbildung einer einzelnen Freßstelle und die Entstehung von Verschleißpartikeln, die sich zwischen Reibflächen befinden, wo sie überwiegend an die Ränder der Freßstelle auf beiden Oberflächen geschoben werden, erkenntlich an den Materi-

Tabelle 5 Untersuchte Schichtkombinationen

Schicht	Titan unbe-schichtet,dto	WC/Co WC/Co	Tiduran[1] Tiduran	Tifran[2] Tifran	PVD-TiN PVD-TiN	PVD-TiN WC/Co -
Härte	320 HV 10	750 HV0,3	533 HV0,025	504 HV0,025	525 HV0,025	Titan
Aufbau	Titan 6 Aluminium 4 Vanadium	88 % WC 12 % Co	Badnitrier-schicht Degussa	Oxid-schicht	Titan-Auf-lage unter PVD-TiN	Tiduran, - Tifran
Dicke µm	-	400	40	25	2 + 12	Tiduran, -
Rauhtiefe R_t um	2,67	4,98	3,7	3,09	2,91	wie bei Kombinationen gleicher Schichten

1) Badnitrierschicht der Firma Degussa

2) Oxidschicht der Firma Hydromechanique et Frottement

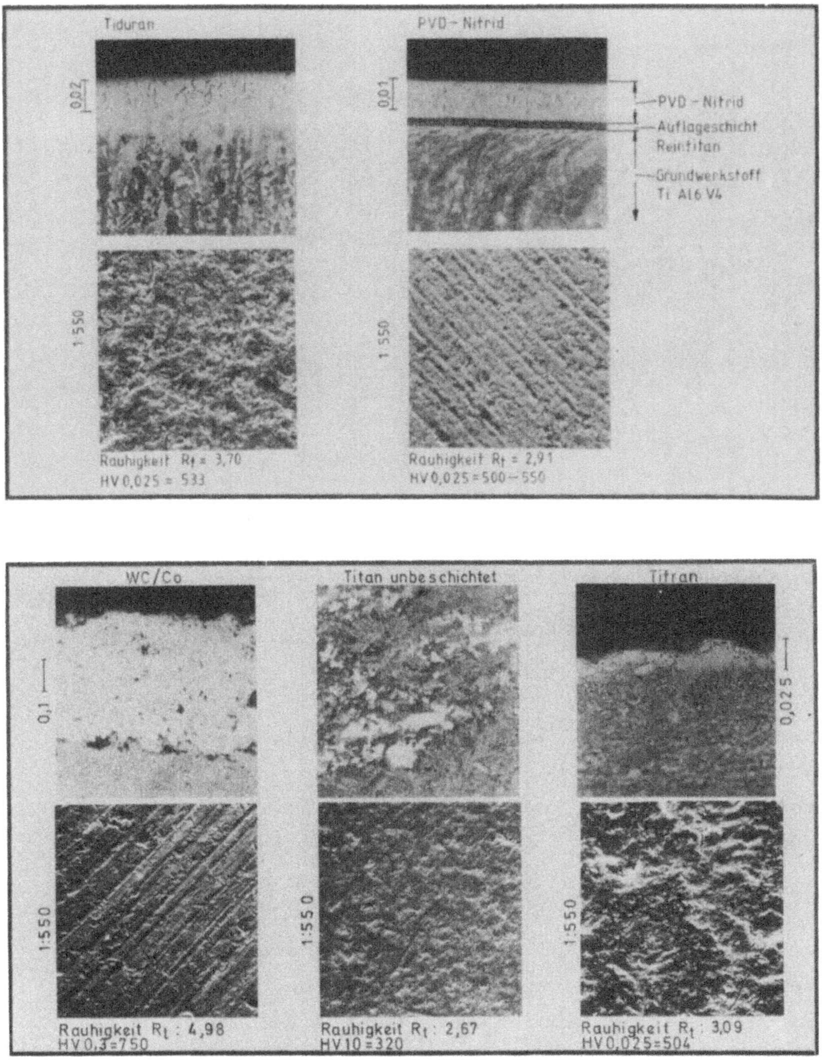

Bild 20. **Oberflächen und Querschliffe untersuchter Schichten**

alanhäufungen an den Rändern der Freßstelle. Aus den REM-Aufnahmen ist deutlich zu erkennen, daß Adhäsion Metallbrücken bildet, die abgeschert werden und die in Analogie zum Aussehen der Tifran (Oxid-) Oberflächen,

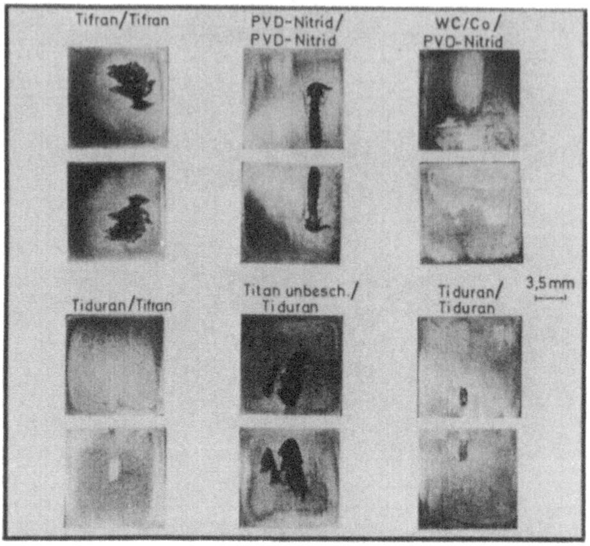

Bild 21. Verschleißerscheinungen makroskopisch

unter Beteiligung von Oxidbildung als Abrieb (Verschleiß) zwischen den Flächen bleiben, sofern der Überdeckungsgrad groß ist. Ein kleiner Teil kann am Rand den Bereich der Überdeckung verlassen.

Im Gegensatz zur Ähnlichkeit der Freßstellen steht die Reibdauer bis zum Auftreten dieser Freßerscheinungen. Waren bei allen anderen Schichtkombinationen bereits nach 3 min (9 000 Zyklen) Freßstellen in der Größenordnung von 3 - 16 % der Reibflächen vorhanden, so zeigte die Kombination Tiduran/Tiduran nach erst 6 min, die Kombination PVD-TiN/PVD-TiN erst nach 9 min derartige Freßstellen, während die Oberflächen bei 3 bzw. 6 min noch keinerlei Freßstellen aufwiesen.

Daraus wird gefolgert, daß die Wirkung von Schutzschichten darauf abzielen muß, entweder die verschleißfreie

Zeitdauer zu verlängern, oder den Grundwerkstoff nicht am Verschleiß zu beteiligen, was alleine vermutlich keine durchgreifende Verbesserung mit sich bringt. Demnach ist eine Erniedrigung des Reibkoeffizienten bei gleichzeitig ausreichender Verformbarkeit der Schicht erforderlich, die relativ zum Grundwerkstoff größer ist, um dort die Werkstoffermüdung herabzusetzen.

Inwieweit die Ergebnisse mit Untersuchungen der Reibdauerhaltbarkeit im Reibklötzchenversuch korrelieren, werden die weiteren Untersuchungen mit beschichteten Flachproben zeigen, die gleichartig wie die Verschleißproben beschichtet und untersucht werden.

3.3.2 Dispersionsschutzschichten

Unabhängig von dem im Abschnitt 3.2 beschriebenen Prüfstand wurden Verschleißuntersuchungen an galvanisch aufgebrachten Dispersionsschutzschichten im Schwingreibverschleißgerät (SRV) nach Lohmeier durchgeführt.

In diesem Gerät besteht eine Reibpaarung zwischen einem liegenden Vollzylinder und einem stehenden Vollzylinder, die aufeinandergepreßt und oszillierend bewegt werden.

In den Tabellen 6 - 8 sind die Ergebnisse dargestellt, aus denen zusammenfassend folgendes festgestellt werden kann:

Bei einseitiger Beschichtung eines Prüfkörpers tritt der höchste Titan-Verschleiß auf bei Ni-Dispersionsschichten, mittlerer Verschleiß bei Ni-, Cu-, Cr- und chemischen Nickelschichten, geringster Titan-Verschleiß bei Ag-, Co- und Co-Dispersionsschichten.

Es ist vorgesehen, die bereits getesteten Schichten im Reibklötzchenprüfstand auf ihre Eignung zur Erhöhung der Reibdauerhaltbarkeit zu prüfen.

Tabelle 6

Verschleiß der Schicht (SRV)

Bedingungen:
- Schicht : auf Stirnseite eines 15 mm ∅ Prüfkörpers
- Gegenkörper: liegender TiAl6V4 Zylinder (11 mm ∅)
- Last = 30 N
- Frequenz = 50 Hz
- Schwingweite = 1 mm
- Temperatur = RT

Schicht gegen Ti	Schichtabriebvolumen (mm^3)
Ag*	n.b.
Co	0,057
Co+Al_2O_3 (4,8 %)	0,057
Co+Cr_3C_2 (3 %)	0,085
Cu	0,28
Cr	0,12
Ni	0,12
Ni chem. (9 % P)	0,12
Ni+SiC+P (6 % SiC, 2 % P)	0,288
Ni galv. + P	0,36
Ni + Graphit (1 %)	0,56
Ni + SiC (50 g/l)	0,56
Ni + Al_2O_3 (3 Gew. %)	0,56
Ni + SiC (150 g/l, 6 Gew. %)	

* zulässige max. Einsatztemperatur bei Ag-Ti z.Zt.: 300 °C

Tabelle 7

Titanverschleiß

Prüfkörper: TiAl6V4 ∅ 11 mm
Schicht : Beschichtung der Stirnseite einer Rolle ∅ 15 mm
Last : 30 N
Frequenz : 50 Hz
Temperatur: RT

Ti gegen Schicht	Ti-Abriebvolumen (mm^3)*
Ag**	<0,020
Co	0,036
Co + Al_2O_3	0,057
Co + Cr_3C_2	0,085
Cu	0,085
Cr	0,12
Ni	0,085
Ni chem.	0,085
Ni + SiC + P	0,22
Ni + P galv.	0,46
Ni + Graphit	0,46
Ni + SiC	0,56
Ni + Al_2O_3	0,56
Ni + SiC	0,56

** zulässige max. Einsatztemperatur bei Ag-Ti z.Zt. 300 °C

* berechnet aus Abriebbreite und Radius des Prüfkörpers

Tabelle 8

Verschleiß der Schicht (SRV)

Bedingungen: Schicht : auf Stirnseite eines 15 mm ∅ Prüfkörpers

Gegenkörper: liegender TiAl6V4 Zylinder (11 mm ∅)

Last = 30 N

Frequenz = 50 Hz

Schwingweite= 1 mm

Temperatur : RT

vor Test WB 400 °C/1h

Schicht	oberer Prüfkörper Ti-Abriebvolumen (mm^3)			unterer Prüfkörper Schichtabriebvolumen(mm^3)		
	RT	400 °C	Δ	RT	400 °C	Δ
Co	0,036	0,085	0,049	0,057	0,085	0,028
Co+Cr_3C_2	0,085	0,12	0,035	0,085	0,12	0,035
Ni	0,085	0,17	0,085	0,12	0,17	0,05
Ni chem.	0,085	0,12	0,035	0,12	0,12	-
Cr	0,085	0,12	0,035	0,12	0,17	0,05
Ni+P galv.	0,46	0,36	0,1	0,36	0,36	-
Ni+SiC	0,56	0,56	-	0,56	0,68	0,12
Ni+SiC	0,56	0,68	0,12	0,68	0,68	-

4. Zusammenfassung

Die Schwingfestigkeit von Turbinenschaufeln aus Titanlegierungen oder Stahl wird durch Schwingungsverschleiß (Fretting) erheblich abgesenkt. Zur Untersuchung von Abhilfemaßnahmen, die zu einer Steigerung der Schwingfestigkeit führen, wurden im Rahmen dieser Vorhaben zwei Apparaturen entwickelt, mit denen einerseits Reibung und Verschleiß zwischen zwei Reibpartnern untersucht werden können und andererseits die Schwingfestigkeit unter Schwingungsverschleißbeanspruchung geprüft werden kann.

Die bisher untersuchten Schichten ergaben für die Reibungs- und Verschleißapparatur folgende Wertungsfolge:

1. PVD-Hartschichten mit weicher Haftschicht
2. Nitrierschichten
3. Spritzschichten und galvanisch aufgebrachte Schichten

Die Schwingfestigkeitsuntersuchungen mit Schwingungsverschleißbeanspruchung haben ergeben, daß die Veränderung der Oberflächenstruktur und die Einbringung von Druckeigenspannungen durch eine optimale Kugelstrahlbehandlung die höchste Steigerung der Schwingfestigkeit ergibt.

Literatur

1. L. Kreitner und H.W. Müller, Die Auswirkung der Reibdauerbeanspruchung auf die Dauerhaltbarkeit von Maschinenteilen.
   Konstruktion 28, 209 - 216 (1976)
   und
   K.H. Kloos, Werkstoffauswahl und Oberflächenbehandlung unter tribotechnischen Gesichtspunkten.
   Z. Werkstofftechnik 10, 456 - 466 (1979)

Reibung und Verschleiß der thermoplastischen Außenschicht bei feinmechanischen Kunststoffbauteilen.

von

Dipl.-Ing. Fritz Dürr

Forschungsinstitut der Forschungsgesellschaft
für Uhren- und Feingerätetechnik e.V.
Stuttgart

Förderzeichen: 01 ZT 048
Laufzeit: 01.09.1978 - 31.08.1980

Kurzfassung:

Es wird über den Einfluß der thermoplastischen Außenschicht, deren Eigenschaften durch eine Variation der Verarbeitungsparameter verändert wurden, auf das Reibungs- und Abriebverhalten von Polymerwerkstoffen berichtet. Die Untersuchungen erstreckten sich auf Acetalhomopolymer und Polyamid 66. Die Reibungs- und Abriebteste wurden mit selbst entwickelten Geräten untersucht. Der Einfluß der verarbeitungsbedingten thermoplastischen Außenschicht wirkt sich auf das Abriebverhalten stärker als auf das Reibungsverhalten aus. Die verminderte Formtemperatur führte bei den Acetalhomopolymer-Gleitpaarungen zu erhöhtem Abrieb. Im Gegensatz zu den Acetalhomopolymer-Scheiben und Platten zeigten sich bei der Verwendung von Polyamid 66-Prüfkörpern keine Unterschiede in ihrem Abriebverhalten.

Abstract:

An report is given of the influence of the thermoplastic surface modified by variation of the moulding conditions on friction and wear of plastic materials. The investigations included acetal homopolymer and nylon 6.6. Friction and wear tests were carried out on test equipment developed by our own. The modified thermoplastic surface affects wear stronger than friction. Reduced mould temperature results in higher wear rate of acetal homopolymer bearings. In contrast to acetal homopolymer plates and sheets no different wear rate was found with nylon 6.6 test specimens.

Inhaltsverzeichnis:

1. Aufgabenstellung:

Die Forderungen nach Wartungsfreiheit und - bei nicht allzustark belasteten Gleitelementen - nach Schmierstofffreiheit sind, neben der Forderung einer optimalen Herstellung von multifunktionellen Bauteilen, die Hauptgründe für den Einsatz von Kunststoffen in der Feinwerktechnik. Im allgemeinen stellen ausgesuchte Stahl/Kunststoff- und Kunststoff/Kunststoff-Gleitpaarungen reibungs- und abriebarme Gleitelemente dar.
Solange Kunststoffe für Gleitelemente eingesetzt werden, steht die Frage im Vordergrund, in welcher Richtung (positiv oder negativ) und in welcher Größenordnung sich die thermoplastische Außenschicht (Spritzhaut) auf das Reibungs- und Abriebverhalten auswirkt.
Die Beantwortung der Frage, welchen Einfluß die Außenschicht auf das tribologische Verhalten ausübt, spielt sicherlich bei Gleitelementen der Maschinentechnik eine wichtige Rolle. In der Feinmechanik und insbesondere in der Mikrotechnik haben jedoch die Bauteile in vielen Fällen dieselbe Dimension wie die Spritzhaut. Sie, und weniger das Grundmaterial, ist also bei diesen Bauteilen weitgehend für das tribologische Verhalten maßgebend.
In bezug auf dieses Verhalten soll geklärt werden:

a) in welchem Maß das Reibungs- und Verschleißverhalten von gespritzten Kunststoffbauteilen durch die Außenschicht (Spritzhaut) beeinflußt wird;
b) ob durch eine geeignete Wahl der Verarbeitungsparameter eine Außenschicht mit optimalen tribologischen Eigenschaften erreicht werden kann;
c) in welchem Maß das tribologische Verhalten der Außenschicht bei feinmechanischen Kunststoffteilen mit demjenigen von Kunststoffbauteilen der Maschinentechnik (kleineres Randzone/Kernzone-Verhältnis) übereinstimmt.

2. Voraussetzungen und Einschränkungen bei den Untersuchungen:

Eine Änderung der Eigenschaften der thermoplastischen Außenschicht muß über eine Variation der Verarbeitungsparameter vorgenommen werden. Eine mechanische Abtragung der Oberflächen zur Entfernung der Außenschicht kommt nicht in Frage.

Der Abrieb an feinmechanischen Bauteilen darf in vielen Fällen aus funktionellen Gründen nicht größer als die Außenschicht selbst sein. Ferner würden sich durch eine feinmechanische Bearbeitung der Prüfkörper andere Oberflächen - insbesondere andere Oberflächenrauhigkeiten - ergeben. Diese würden dann das Reibungs- und Abriebverhalten zumindest in der Einlaufphase, der in der Feinwerktechnik eine wesentlich größere Bedeutung als in der Maschinentechnik zukommt, stark beeinflussen.
Auf eine extrem hohe Prüfkörperqualität wird bewußt verzichtet, da die mit solchen Prüfkörpern erhaltenen Ergebnisse sonst zu praxisfremd würden. Ebenso darf sich die Variation der Verarbeitungsbedingungen von den für die Praxis realistischen Bereichen nicht allzuweit entfernen, obwohl eine Wahl extremerer Verarbeitungsbedingungen sicher zu noch kontrastreicheren Ergebnissen führen würde.
Um die Eigenschaften der thermoplastischen Außenschicht zu beeinflussen, können nur die wichtigsten Verarbeitungsparameter (Formtemperatur, Massetemperatur, Spritzdruck und Druckzeit) variiert werden. Eine darüber hinausgehende Kombination der Spritzbedingungen würde zu umfangreich werden. Ebenso muß die Anzahl der zu untersuchenden Polymerwerkstoffe auf Acetalhomopolymer und Polyamid beschränkt werden.
Als Formen der Prüfkörper werden Kugel und Ebene gewählt (kontraformer Kontakt). Zweifellos ändert sich bei der Verwendung einer Kugel als Prüfkörper mit steigendem Abrieb die Flächenpressung, doch eine definierte Auflage der Prüfkörper ist für die tribologischen Untersuchungen der thermoplastischen Außenschicht unbedingt notwendig.

3. Stand der Technik:

Der Einfluß der Verarbeitungsbedingungen bei der Herstellung von Teilen aus Polymerwerkstoffen wird hauptsächlich in bezug auf die innere Struktur [1] bis [5], die Maßstreuung und Formabweichung [3,6,7] und auf die mechanischen Eigenschaften [3,8] untersucht.
In der Maschinentechnik werden im allgemeinen mechanisch bearbeitete Polymerwerkstoff-Prüfkörper bei Reibungs- und Abriebun-

tersuchungen eingesetzt [9]. An anderen Stellen werden die Prüfkörper durch eine entsprechend lange Einlaufzeit so aneinander angepaßt, daß eine definierte Auflage gewährleistet ist. In vielen Fällen ist deshalb die Außenschicht schon abgerieben, bevor die eigentlichen Untersuchungen beginnen. Den beabsichtigten Arbeiten kommt entgegen, daß die meisten eigenen Reibungs- und Abriebuntersuchungen im Bereich der Polymerwerkstoff-Außenschicht durchgeführt werden. Für diesen Bereich hat sich die Kugel als Prüfkörperform bewährt, da durch sie eine definierte Auflage schon zu Beginn der Versuche gegeben ist.
Eindeutige Aussagen über das tribologische Verhalten der Außenschicht in Abhängigkeit von den Verarbeitungsbedingungen liegen nach unseren derzeitigen Kenntnissen insbesondere für Bauteile der Feinwerktechnik nicht vor.

4. Planung und Ablauf der Arbeiten:

Es ist zweckmäßig - vor der Konzeption des Prüfgerätes - durch die Erprobung von Meßprinzipien und den Bau eines einfachen Prüfgerätes zu klären, ob die geplante Konstruktion die von ihr erwarteten Aufgaben erfüllen kann. Nach der Konstruktion, dem Bau und den Vorversuchen mit den Reibungs- und Abriebprüfgeräten zeigte sich, daß das auf dem Torsionsbandprinzip aufgebaute Reibungsprüfgerät und die Abriebprüfgeräte sich gut bewähren.
Die optimalen Verarbeitungsbedingungen für die einzelnen Prüfkörper, die zu variierenden Parameter und die Größe ihrer Verminderung werden zusammen mit dem Rohstoffhersteller und dem Verarbeiter festgelegt. In Tabelle 1 sind die einzelnen Verarbeitungsbedingungen zusammengefaßt.

Als Stahlprüfkörper werden Wälzlagerkugeln der Güteklasse III (poliert, Ra < 0,03 µm) aus 100 Cr6 und einer Härte von HRC 63 ± 3 verwendet. Die Formabweichung der Kugeln ist < 1 µm.

4.1 Bestimmung der Dimensionen und der Masse der Prüfkörper:

Die Dimensions- und Massebestimmungen der Prüfkörper sollen darüber Auskunft geben, inwieweit sich die Variation der Verarbeitungsbedingungen auf die Füllung der Werkzeugformen auswirkt.

4.1.1 Dimensionen: Bestimmung mit induktivem Tastsystem (Anpreßkraft: 16×10^{-2}N).

Je Polymerwerkstoff werden 1000 Platten und 1000 Scheiben ausgemessen (Platten: Länge an zwei Stellen, Breite und Dicke an je drei Stellen; Scheiben: Durchmesser an zwei Stellen und Dicke an drei Stellen).

4.1.2 Masse: Von 2000 Prüfkörpern je Polymerwerkstoff wird die Masse einzeln bestimmt (10^{-4}g).

4.2 Oberflächenuntersuchungen:

Diese Untersuchungen sollen zeigen, ob durch die Variation der Verarbeitungsbedingungen charakteristische Merkmale auf den Oberflächen zu erkennen sind.

4.2.1 Form und Oberflächenrauhigkeit der Gleitflächen:

Diese Untersuchungen werden mit dem Tastschnittgerät Talysurf 10 durchgeführt. Bei den Platten werden Längs- und Querschnitte von je 5 Platten pro Verarbeitungsparameter ausgemessen. Die Scheiben werden in bezug auf die Form radial - um 90° zum Anspritzpunkt versetzt - gemessen.
Die Oberflächenrauhigkeit (R_z und R_a) wird bei den Platten an drei Stellen längs vom Anspritzpunkt aus und bei den Scheiben in radialer Richtung (um 90° zum Anspritzpunkt versetzt), ebenfalls an drei Stellen, bestimmt.

4.2.2 Mikrotomschnitte:

Die Herstellung der Dünnschnitte erfolgt mit dem Leitz-Serienschnitt - Mikrotom 1515. Die Dicke beträgt 20 µm. Es werden Längs- und Querschnitte bei den Platten und Radialschnitte bei den Scheiben angefertigt. Die Proben werden im Durchlicht betrachtet und fotografiert.

4.2.3 Differential-Interferenzschicht-Kontrast-Mikroskopie:

Die Untersuchungen werden an Platten und Scheiben durchgeführt. Die Prüfkörper werden gereinigt und in der Kontrastrierkammer

Verarbeitungs-Parameter	Masse-temp. °C	Form-temp. °C	Spritz-druck °C	Einspr.-geschw. 1 - 5	Druck-zeit s	Kühl-zeit s	Pause s
Kugel: Delrin 500 NC 10 ø 12,7 mm							
optimale Einstellung	210	120	1.300	3	140	4	3
Delrin 500 NC 10 Platte: 10 x 6 x 2 mm							
optimale Einstellung	215	120	1.300	5	17	4	2
verm. Massetemperatur	190	120	1.300	5	17	4	2
verm. Formtemperatur	215	60	1.300	5	17	4	2
verm. Spritzdruck	215	120	650	5	17.	4	2
verm. Druckzeit	215	120	1.300	5	8,5	4	2
Delrin 500 NC 10 Platte: 10 x 6 x 1 mm							
optimale Einstellung	215	120	1.300	5	10	2	2
verm. Massetemperatur	190	120	1.300	5	10	2	2
verm. Formtemperatur	215	60	1.300	5	10	2	2
verm. Spritzdruck	215	120	650	5	10	2	2
verm. Druckzeit	215	120	1.300	5	5	7	2
Zytel E 101 L NC 10 Platte: 10 x 6 x 2 mm							
optimale Einstellung	285	90	1.300	3	17	4	2
verm. Massetemperatur	260	90	1.300	3	17	4	2
verm. Formtemperatur	285	60	1.300	3	17	4	2
verm. Spritzdruck	285	90	650	3	17	4	2
verm. Druckzeit	285	90	1.300	3	8,5	12,4	2
Zytel E 101 L NC 10 Platte: 10 x 6 x 1,2 mm							
optimale Einstellung	285	90	1.300	5	6	4	2
verm. Massetemperatur	260	90	1.300	5	6	4	2
verm. Formtemperatur	285	60	1.300	5	6	4	2
verm. Spritzdruck	285	90	.650	5	6	4	2
verm. Druckzeit	285	90	1.300	5	3	7	2
Delrin 500 NC 10 Scheibe ø 37 x 2 mm							
optimale Einstellung	215	120	1.300	5	20	28	3
verm. Massetemperatur	190	120	1.300	5	20	28	3
verm. Formtemperatur	215	60	1.300	5	20	28	3
verm. Spritzdruck	215	120	650	5	20	28	3
verm. Druckzeit	215	120	1.300	5	10	28	3
Delrin 500 NC 10 Scheibe ø 37 x 1 mm							
optimale Einstellung	215	120	1.300	5	20	28	3
verm. Massetemperatur	190	120	1.300	5	20	28	3
verm. Formtemperatur	215	60	1.300	5	20	28	3
verm. Spritzdruck	215	120	1.300	5	20	28	3
verm. Druckzeit	215	120	1.300	5	10	28	3
Zytel E 101 L NC 10 Scheibe ø 39 x 2 mm							
optimale Einstellung	285	90	1.300		20		
verm. Massetemperatur	260	90	1.300		20		
verm. Formtemperatur	285	60	1.300		20		
verm. Spritzdruck	285	90	650		20		
verm. Druckzeit	285	90	1.300		10		
Zytel E 101 L NC 10 Scheibe ø 39 x 0,9 mm							
optimale Einstellung	285	90	1.300		10		
verm. Massetemperatur	260	90	1.300		10		
verm. Formtemperatur	285	60	1.300	3	10		
verm. Spritzdruck	285	90	650		10		
verm. Druckzeit	285	90	1.300		5		

Tabelle 1: Verarbeitungsbedingungen der Polymerwerkstoffplatten und -scheiben.

(Leitz) besputtert. Als Mikroskop wird ein Differentialinterferenzschicht-Mikroskop (Orthoplan, Leitz) verwendet.

4.2.4 Elektronenrastermikroskopie:

Die Rasterelektronenmikroskopaufnahmen wurden freundlicherweise vom Institut für Industrielle Fertigung und Fabrikbetrieb der Universität Stuttgart erstellt.
Die Aufnahmen wurden von den 2 mm dicken Acetalhomopolymer- und den Polyamid 66 - Platten unter einem Kippwinkel von 45° gemacht.

4.2.5 Mechanische Eindrückverfahren:

Die Härte der Prüfkörper wird mit drei verschiedenen Geräten (Vickers und Brinell) untersucht, und zwar mit dem Mikrohärteprüfer Durimet (Leitz, Prüflast 0,1471 N, Belastungszeit 30 s), einem von der Fa. Bosch entwickelten Mikrohärteprüfer (Saphirspitzenradius 0,06 mm, Eindringtiefe 3 bis 8 µm, Belastungszeit 30 s), und mit einem selbstentwickelten Kugel-Eindruckgerät (Kugelradius 0,5 mm, Prüflast 9,81 und 4,9 N, Belastungszeit 30 s). Mit einem vierten Härtemeßgerät, Optidur (Göttfert, Vickers-Pyramide, Prüflast 2,943 und 1,962 N, Belastungszeit 30 s), bei dem die Bestimmung der Härte unter Last erfolgt, werden die Serienmessungen, die mit den drei erwähnten Geräten vorgenommen werden, zum Teil stichprobenartig überprüft.

4.2.6 Benetzungsuntersuchungen:

Es werden je 20 Öltropfen (Ätheralkohol, Moebius 9024) auf die vorher sorgfältig gereinigten 2 mm dicken Acetalhomopolymerplatten aufgesetzt und ihre Flächenzunahme in Abhängigkeit von der Zeit bestimmt. Das durchschnittliche Breitlaufen des Öles ist auf den Platten, die bei verminderter Formtemperatur und verminderter Druckzeit hergestellt werden, am stärksten. Weitere Differenzierungen über den Einfluß der Verarbeitungsparameter auf das Breitlaufen des Öles sind - infolge der starken Streuungen der einzelnen Öltropfen - nicht sinnvoll.

4.3 Durchbiegungsuntersuchungen:

Da der Einfluß der Verarbeitungsparameter nur auf die Füllung der Form und die Oberflächen untersucht wird, soll eine zusätzliche Versuchsreihe Auskunft über das mechanische Verhalten der Teile geben. Die 2 mm dicken Acetalhomopolymer-Platten werden in der Plattenmitte konstant mit einer Last von 40 N beansprucht und die Durchbiegung in Abhängigkeit von der Zeit gemessen. Die Platten, die bei verminderter Druckzeit abgespritzt werden, biegen sich am stärksten durch. Eine ebenfalls hohe Durchbiegung zeigen die bei verminderter Formtemperatur hergestellten Platten.

4.4 Prüfgeräte für die Reibungs- und Abriebuntersuchungen:

Reibungs- und Abriebprüfgeräte für hohe Normalkräfte bei oszillierender Gleitbewegung stehen zur Verfügung. Solche für kleine Normalkräfte müssen entwickelt, gebaut und erprobt werden. Zur Erfassung der Reibung ist eine Meßvorrichtung notwendig, die es erlaubt, Reibungsmessungen bei kleinen Normalkräften von 10^{-2} N durchzuführen. Bild 1 zeigt das gebaute Meßgerät. Das durch die Reibungskraft erzeugte Drehmoment wird über ein Torsionsspannbandsystem und die Normalkraft über die Durchbiegung einer am Torsionsband befestigten Blattfeder induktiv aufgenommen. Die Normalkraft läßt sich über die Biegefeder stufenlos einstellen.

Bild 1: Reibungs- und Verschleißprüfgerät (UTI-RV-Gerät) bei oszillierender Gleitbewegung mit Meßkopf für kleine Belastungen.

In Bild 2 ist ein zweites Prüfgerät abgebildet. Die Reibungskraft, die zwischen rotierender Scheibe und der in einem Trägerarm eingespannten Kugel auftritt, biegt zwei Blattfedern durch. Der Trägerarm kann sich in Saphirspitzenlagern in dem zwischen den Blattfedern eingespannten Rahmen leicht drehen. Durch die Auflage von Massen über der Kugel kann die gewünschte Normalkraft aufgebracht werden. Da die durch Spritzgießen hergestellten Scheiben nie exakt flach laufen können, ist es wichtig, daß der drehbare Trägerarm und damit die Kugel den Unebenheiten der rotierenden Scheibe folgen können, ohne daß es dabei zu einer Normalkraftänderung kommen kann.

Bild 2: Reibungs- und Verschleißprüfgerät (Kugel-Scheibe-Prüfgerät) bei gleichförmig drehender Gleitbewegung für kleine Belastungen.

Die Abriebprüfgeräte bei gleichförmig drehender Gleitbewegung funktionieren ähnlich wie das hier beschriebene Reibungsmeßgerät. Bild 3 zeigt einen Ausschnitt der Abriebprüfgeräte. Die Größe des Abriebes wird mit Hilfe von Tastschnittgeräten ermittelt.
Die Reibungsprüfgeräte, bei denen hohe Normalkräfte aufgebracht werden können, wurden bereits ausführlich beschrieben [10, 11]. Beim Reibungspendel handelt es sich um eine im Prinzip von

Bild 3: Verschleißprüfgerät bei gleichförmig drehender Gleitbewegung.

Barker und Mitarbeitern vorgeschlagenen Meßmethode. Eine Prüfkugel, die fest im Pendelrahmen eingespannt ist, gleitet auf zwei unter 45° zur Senkrechten geneigten Prüfplatten. Die Größe der Reibung wird über die Dämpfung des Pendels aus einem freien Ausschwingversuch ermittelt.
Beim UTI-RV-Gerät gleitet eine auf einem oszillierenden Tisch befestigte Prüfplatte unter einer an zwei Blattfedern fest eingespannten Prüfkugel. Die Normalkraft kann über den Rahmen, in dem sich die Blattfedern mit der Kugel befinden, durch Entlastung oder Belastung eingestellt werden. Die Reibungskraft wird über die Durchbiegung der Blattfedern induktiv ermittelt. Die Polymerwerkstoff-Prüfkörper werden mit Frigen 113 TR-T gereinigt und im Klimaschrank bei $23^{o}C$ und etwa 50 % relativer Luftfeuchtigkeit gelagert. Die Stahlkugeln werden, kurz bevor die Untersuchungen angesetzt werden, in einer Reinigungsflüssigkeit (Ultralin) gereinigt, in Toluol gespült und danach in Heißluft getrocknet.

5. Versuchsergebnisse:

5.1 Dimensionsuntersuchungen:

Sämtliche Dimensionen von allen untersuchten Platten und Scheiben (verschiedener Dicke) werden erwartungsgemäß durch die Verminderung des Spritzdruckes von 1300 bar auf 650 bar stark reduziert. Die Einfriertemperatur wird mit abnehmendem Spritzdruck erniedrigt. Die Formschrumpfung nimmt mit fallendem Spritzdruck zu.

Eine Erniedrigung der Druckzeit um die Hälfte wirkt sich nur teilweise und dann nur schwach auf die Dimensionen der Prüfkörper aus. In Bild 4 sind die Plattendickenübermaße dargestellt.

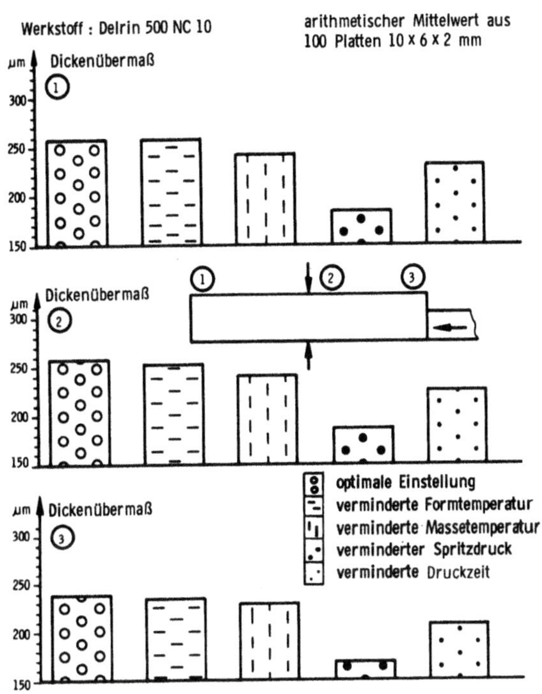

Bild 4: Plattendickenübermaß in Abhängigkeit von den Verarbeitungsbedingungen (Acetalhomopolymer, 2 mm)

5.2 Massebestimmung:

Eine Verminderung des Spritzdruckes von 1300 bar auf 650 bar führt erwartungsgemäß bei allen untersuchten Prüfkörpern zu Teilen geringerer Masse.

Bei den 2 mm dicken Platten und Scheiben aus Acetalhomopolymer ergibt sich die erwartete Reihenfolge. Sie ist in Bild 5 dargestellt. Die größte Masse weisen diejenigen Teile auf, die bei optimalen Verarbeitungsbedingungen hergestellt werden. Die bei verminderter Form- und Massetemperatur hergestellten Platten und Scheiben weichen in bezug auf ihre Masse nur wenig von den optimal verarbeiteten Teilen ab. Besonders stark wirkt sich, wie schon erwähnt, die Verminderung des Spritzdruckes aus. Es wird deutlich, daß die um die Hälfte gegenüber der optimalen Druckzeit reduzierte Druckzeit nicht mehr ausreicht, um die Form voll zu füllen.

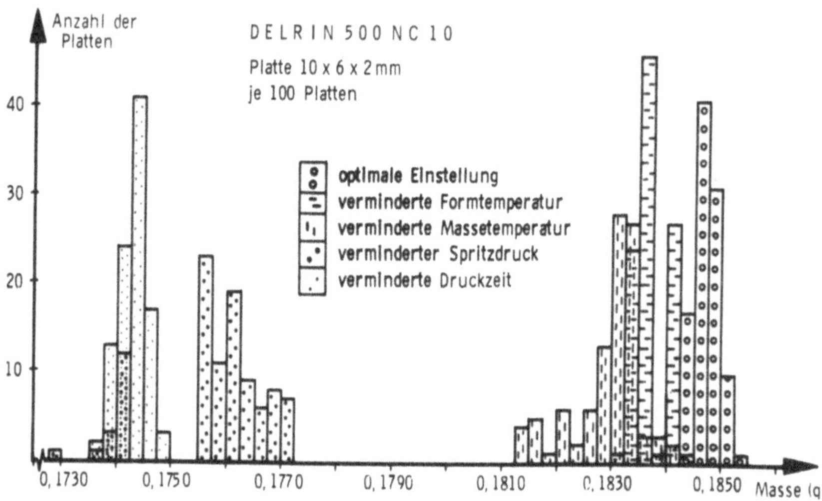

Bild 5: Plattenmasse in Abhängigkeit von den Verarbeitungsbedingungen (Acetalhomopolymer, Plattendicke 2 mm).

Im Gegensatz dazu reicht schon die um die Hälfte reduzierte Druckzeit bei den 2 mm dicken Polyamid 66-Platten und -Scheiben aus, um die Form noch vollständig zu füllen. Bei den 1 mm dicken Platten und Scheiben aus Acetalhomopolymer und Polyamid 66 liegt die Masse der bei verschiedenen Verarbeitungsparametern hergestellten Prüfkörpern - abgesehen von den bei vermindertem Druck abgespritzten Teilen - eng beieinander. Aufgrund der Massebestimmungen kann bei den 1,2 mm dicken Polyamid 66-Platten festgestellt werden, daß allein bei den unter "optimalen" Bedingungen abgespritzten Platten eine um etwa 0,013 Gramm zu geringe Masse auftritt. Dieser Masseverlust ist, wie die Dimensionsbestimmungen ganz klar bestätigten, allein auf eine zu geringe Plattenstärke zurückzuführen. Die um etwa 200 µm verminderte

Plattenstärke - gerade bei den unter "optimalen" Bedingungen hergestellten Platten - kann nicht mehr durch die Verarbeitungsparameter erklärt werden, sondern sie muß von einer Veränderung der Werkzeugform herrühren.

5.3 Formuntersuchungen:

Die bei der Ermittlung der Dimensionen nicht feststellbare Auswirkung der Druckzeitminderung wirkt sich, wie Bild 6 zeigt, bei den 2 mm dicken Acetalhomopolymerplatten ganz deutlich auf die Form aus. Die Druckzeitreduzierung bewirkt bei diesen die größten Einfallstellen in der Plattenmitte (100 µm bis 120 µm). Die dazugehörenden 1 mm dicken Platten zeigen nur kleine Einfallstellen (< 10 µm).

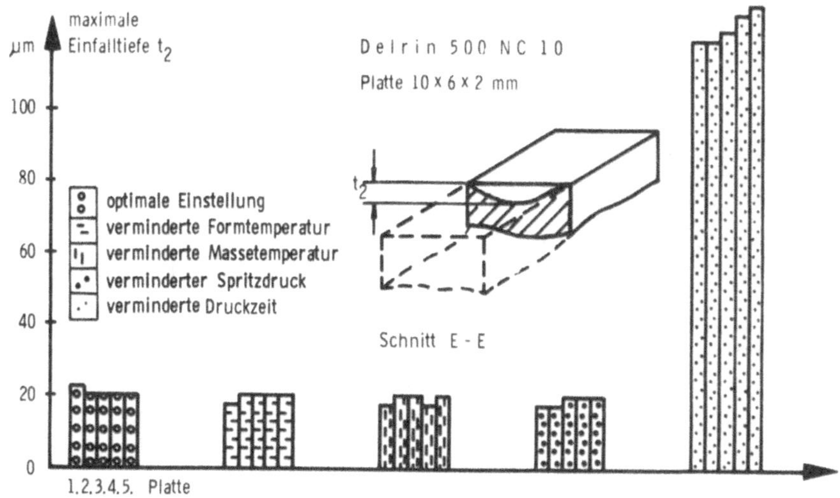

Bild 6: Einfallstellentiefe der 2 mm dicken Acetalhomopolymer-Platten in Abhängigkeit von den Verarbeitungsbedingungen (aus Querschnitt ermittelt).

Bei den 2 mm dicken Acetalhomopolymer-Scheiben wirkt sich die verminderte Druckzeit und der verminderte Spritzdruck nur schwach auf die Größe der Einfallstellen aus. Bei den Polyamid-66-Platten und -Scheiben treten keine Unterschiede in bezug auf den Einfluß der Verarbeitungsbedingungen auf die Größe der Einfallstellen auf. Die auf die Hälfte reduzierte Druckzeit reicht offensichtlich noch aus, um eine nahezu volle Füllung der Form zu erreichen.

5.4 Oberflächenrauhigkeitsuntersuchungen:

Bei den 1 mm- und 2 mm dicken Acetalhomopolymer-Platten und -Scheiben führt die Reduzierung der Formtemperatur von 120°C auf 60°C zu rauheren Oberflächen. Der R_z-Wert verdoppelt sich und der R_a-Wert verdreifacht sich. Lediglich bei den 2 mm dicken Acetalhomopolymer-Scheiben wirken sich die Unterschiede weniger stark aus. In Bild 7 ist der arithmetische Mittenrauhwert der 2 mm dicken Acetalhomopolymer-Platten dargestellt.

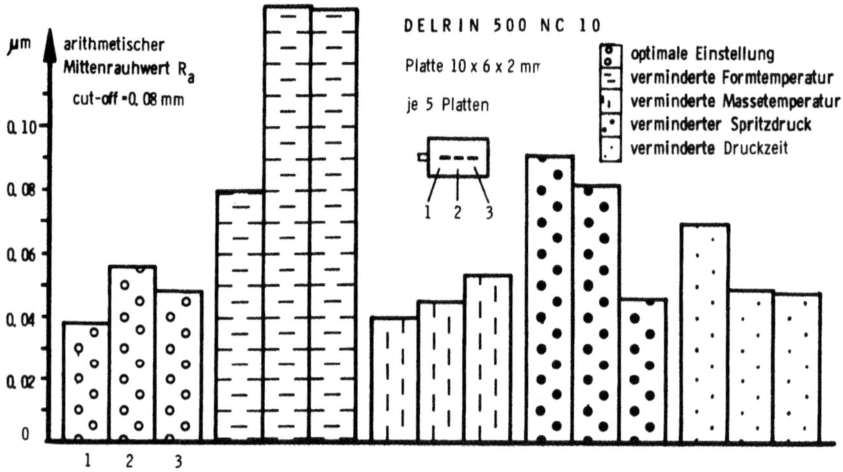

Bild 7: Arithmetischer Mittenrauhwert der 2 mm dicken Acetalhomopolymer-Platten in Abhängigkeit von den Verarbeitungsbedingungen.

Abgesehen von der höheren Oberflächenrauhigkeit bei den unter verminderter Formtemperatur hergestellten 2 mm dicken Acetalhomopolymer-Teilen, und bei den 1 mm dicken Acetalhomopolymer-Platten, die unter verminderter Form- und Massetemperatur abgespritzt werden, sind die Polyamid 66-Prüfkörper um etwa den Faktor 3 glatter. Hervorzuheben sind die besonders glatten Oberflächen der 0,9 mm dicken Polyamid 66-Scheiben (R_z = 0,04 µm, R_a = 0,005 µm). Bei den 2 mm dicken Polyamid 66-Platten ist die Oberflächenrauhigkeit in Angußnähe um ungefähr den Faktor 2 größer als in den angußferneren Stellen.

5.5 Mikrotomschnitte:

Die bekannten drei Kristallisationsbereiche sind bei den 2 mm dicken Acetalhomopolymer-Platten deutlich zu erkennen. Die äußerste Schicht der feinkristallinen und kristallarmen Randzone ist etwa 50 µm dick. In ihr sind die Lamellen quer zur Schnittrichtung - also in Spritzrichtung - orientiert [1, 12]. In der zweiten Zone sind die Lamellen senkrecht zur Oberfläche ausgerichtet. Durch den starken Temperaturgradienten in Richtung zur Formwand bilden sich die richtungsorientierten kristallinen Überstrukturen (Sphärolithe) aus. In der Mitte des Schnittes ist das Kerngefüge nur als schmaler Bereich mit nichtorientierten Sphärolithen ausgebildet. Beim Längsschnitt ist diese Zone deutlicher als im Querschnitt zu sehen. Die zweite Schicht ist bei den unter verminderter Formtemperatur gespritzten Acetalhomopolymer-Platten beträchtlich breiter als dieselbe bei optimaler Einstellung. Die dritte Zone, die durch die runden - nicht orientierten - Sphärolithe gekennzeichnet ist, kann sich infolge der geringen Formtemperatur nicht mehr ausbilden.
Bild 8 zeigt die richtungsorientierten Sphärolithe in den 2 mm dicken Acetalhomopolymerplatten.

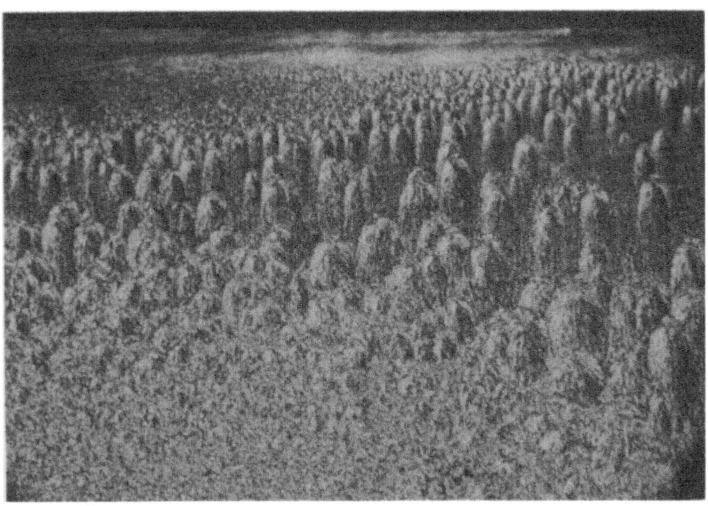

Bild 8: Gefügeausbildung in einer 2 mm dicken Acetalhomopolymerplatte, die bei optimaler Einstellung hergestellt wurde (Vergrößerung 400fach).

Die bei verminderter Massetemperatur abgespritzten Platten zeigen einen recht inhomogenen Gefügeaufbau durch das nicht voll aufgeschmolzene Material.
Eine Verminderung des Spritzdruckes um die Hälfte bewirkt keine starke Veränderung des Gefüges gegenüber dem der bei optimalem Spritzdruck abgespritzten Platten.
Eine nicht genügend lange Druckzeit ist im Dünnschnitt an der Druckzeitlinie zu sehen. Durch den vorzeitigen Abbruch des Druckes kommt es zu einem momentanen Abfall des Innendruckes, der ein Absinken der Einfriertemperatur zur Folge hat.
Bei den unter optimalen Bedingungen abgespritzten 1 mm dicken Acetalhomopolymer-Platten fehlt in der Plattenmitte der Bereich mit nichtorientierten Sphärolithen. Von außen nach innen zeigt sich eine breite Zone mit wenigen großen Sphärolithen. Die 2 mm dicken Acetalhomopolymer-Platten enthalten in derselben Schicht eine größere Anzahl dieser Sphärolithe, die enger beieinander liegen. Erst in der Mitte der 1 mm dicken Acetalhomopolymer-Platten ergibt sich eine Schicht, die derjenigen der äußeren Schicht entspricht, welche bei den 2 mm dicken Acetalhomopolymer-Platten bis nach außen reicht. Die unter den restlichen

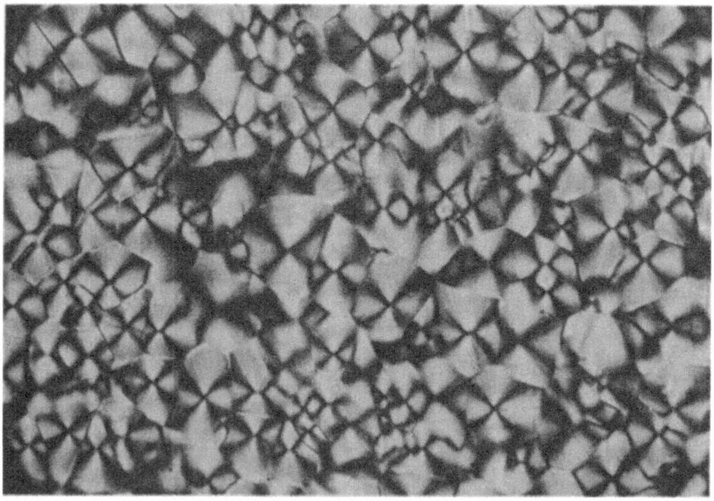

Bild 9: Gefügeausbildung in einer 2 mm dicken Polyamid 66-Platte, die bei optimaler Einstellung hergestellt wurde (Vergrößerung 400fach).

Verarbeitungsparametern abgespritzten 1 mm dicken Platten zeigen - abgesehen von den unter verminderter Druckzeit hergestellten Platten, bei denen die Rückflußzone fehlt - einen entsprechenden Zonenaufbau. Die Gefügeausbildung der 1 mm und 2 mm dicken Acetalhomopolymer-Scheiben ist ähnlich derjenigen der Platten. Bei den Polyamid 66-Prüfkörpern lassen sich aus den Dünnschnitten keine bemerkenswerten Gefügeunterschiede in bezug auf die Verarbeitungsparameter erkennen. Die Sphärolithausbildung ist, wie Bild 9 zeigt, gleichmäßig. Die Größe der Sphärolithe nimmt von außen nach innen zu.

5.6 Differential-Interferenzschicht-Kontrast-Mikroskopie

Mit Hilfe der Differential-Interferenzschicht-Kontrast-Mikroskopie werden die Polymerwerkstoff-Oberflächen untersucht. Im allgemeinen sind die Oberflächen der mit verminderter Formtemperatur gespritzten Acetalhomopolymer-Prüfkörper rauh. Eine eindeutige Bestätigung dieses Ergebnisses liefern die mit dem Tastschnittgerät durchgeführten Untersuchungen. Die 2 mm dicken Prüfkörper zeigen - abgesehen von den bei verminderter Formtemperatur hergestellten Platten - eine homogenere Oberfläche als die 1 mm dicken Prüfkörper. Die 1 mm dicken Platten und Scheiben, die bei verminderter Massetemperatur gespritzt werden, weisen ebenfalls eine relativ hohe Oberflächenrauhigkeit auf. Die schon mit dem Tastschnittgerät nachgewiesenen Formunterschiede bei den mit verminderter Druckzeit hergestellten 2 mm dicken Acetalhomopolymer-Platten werden durch die starken Farbunterschiede in den Differential-Interferenzschicht-Kontrast-Mikroskopaufnahmen bestätigt. Die Polyamid 66-Prüfkörper zeigen unabhängig von Probenart, Probendicke und deren Verarbeitungsbedingungen immer in etwa gleich aussehende Oberflächen. Die Oberflächenstruktur selbst ist gleichmäßig, wesentlich glatter und ebener als die der entsprechenden Acetalhomopolymer-Platten.

5.7 Elektronenmikroskopie:

Die Aufnahmen bestätigen bei den Acetalhomopolymer-Platten die bisherigen Untersuchungsergebnisse. Auffallend ist die rauhe Oberflächenstruktur der unter verminderter Formtemperatur abgespritzten Acetalhomopolymer-Platten. Die Aufnahmen der Polyamid 66-Platten (Dicke 2 mm) zeigen gleichmäßigere Oberflächenstrukturen gegenüber denjenigen der Acetalhomopolymer-Platten. Ein Unterschied in bezug auf die Oberflächenstruktur durch die verschiedenen Verarbeitungsbedingungen läßt sich bei den Polyamid 66-Platten nicht feststellen.

5.8 Mechanische Eindrückverfahren:

An den Härteuntersuchungen zur Charakterisierung von Polymerwerkstoffen wird schon seit Jahren gearbeitet [13,14,15,16]. Bei den Untersuchungen mit den Vickersgeräten kann quantitativ eine gute Übereinstimmung der Meßergebnisse erreicht werden. Qualitativ stimmen auch die Meßergebnisse, die mit den beiden Brinell-Geräten gemessen wurden, überein.
Abgesehen von der Mikrohärtemessung nach Vickers bei den 2 mm dicken Acetalhomopolymer-Platten scheinen bei allen anderen untersuchten Prüfkörpern - auch bei zum Teil verschiedenen Belastungen - die 2 mm dicken Platten und Scheiben härter als die 1 mm dicken Prüfkörper.
Im allgemeinen kann festgestellt werden, daß die bei verminderter Formtemperatur hergestellten Acetalhomopolymer-Platten und

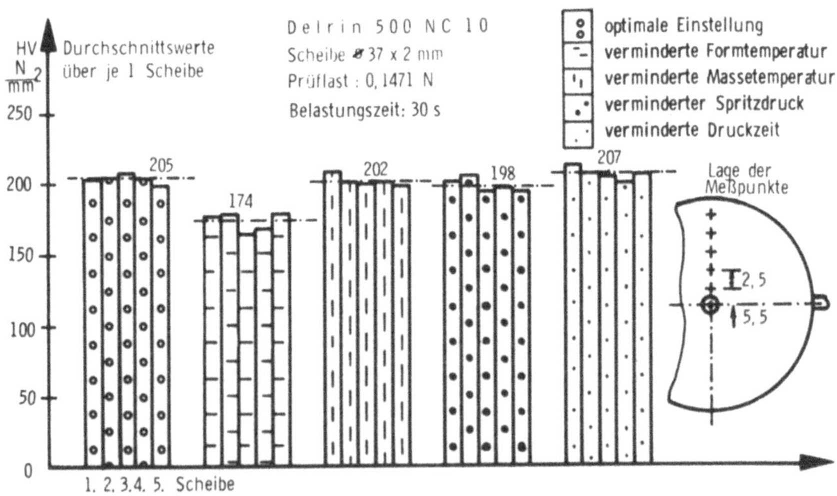

Bild 10: Einfluß der Verarbeitungsbedingungen auf die Vickers-Mikrohärte von 2 mm dicken Acetalhomopolymer-Scheiben (Prüfzeit 30 s).

-Scheiben eine um 7 ... 15 % niederere Härte als die unter optimalen Bedingungen gespritzten Acetalhomopolymer-Prüfkörperoberflächen aufweisen.
Die Acetalhomopolymeroberflächen der Prüfkörper sind um die Faktoren 2 ... 4 härter als die Oberflächen der Polyamid 66-Teile. Bei beiden Werkstoffen sind die Platten am Anguß weicher als an den angußentfernteren Stellen. Am Plattenende stellt sich wiederum ein Härteabfall ein, der jedoch wesentlich geringer als derjenige am Anguß ist. In Bild 10 ist die Vickers-Härte der 2 mm dicken Acetalhomopolymer-Scheiben dargestellt.

5.9 Reibungsuntersuchungen:

5.9.1 Stahl/Polymerwerkstoff-Gleitpaarungen.

Bei gleichförmig drehender Gleitbewegung liegen die Reibungszahlen der Stahl/Polymerwerkstoff-Paarungen in derselben Größenordnung, wie die in früheren Untersuchungen unter hoher Belastung gewonnenen Werte. Überraschend ergeben sich bei oszillierender Gleitbewegung und kleinen Normalkräften (F_N= $1x10^{-2}$N bis $1x10^{-1}$N, Punktberührung) bei den Gleitpaarungen Stahl/Acetalhomopolymer und Stahl/Polyamid 66 hohe Reibungszahlen. Bei den Stahl/Acetalhomopolymer-Gleitpaarungen treten die hohen Reibungszahlen (μ=0,9) nach einer Laufzeit von etwa 2 Stunden (1 mm dicke Platten) und nach etwa zwei Tagen (2 mm dicke Platten) auf. Die Stahl/Polyamid 66-Paarungen zeigen Werte in derselben Größenordnung schon innerhalb kurzer Zeit. Als Ursache für diese hohen Reibungszahlen bei besonders glatten Stahloberflächen könnten die großen Adhäsionsanteile an der Reibungskraft angesehen werden [17]. Der bei diesen Untersuchungen auftretende Abrieb war in keinem einzigen Versuch technisch meßbar. Interessant ist, daß diese hohen Reibungszahlen bei großen Belastungen unter den angegebenen Versuchsbedingungen nicht erreicht werden. Eine eindeutige Erklärung für dieses Verhalten kann zur Zeit noch nicht gegeben werden. Bei den Stahl/Acetalhomopolymer-Gleitpaarungen (1 mm und 2 mm dicken Platten) ließ sich bei hohen Belastungen kein Einfluß der durch die verschiedenen Verarbeitungsbedingungen hergestellten thermoplastischen Außenschicht feststellen. Die Reibungszahlen liegen bei dieser Gleitpaarung (1 mm und 2 mm dicken Platten) zwischen μ = 0,3 und 0,4. Gerätebedingte Unterschiede treten bei oszillierender Bewegung nicht auf. Erwartungsgemäß liegen die Reibungszahlen (μ = 0,4 ... 0,6) bei der Stahl/Polyamid 66-Gleitpaarung höher als bei der Stahl/Acetalhomopolymer-Gleitpaarung. Unterschiede, die auf eine Variation der Verarbeitungsbedingungen zurückzuführen wären, treten bei dieser Gleitpaarung ebenfalls nicht auf.

5.9.2 Polymerwerkstoff/Polymerwerkstoff-Gleitpaarungen.

Die durch verschiedene Verarbeitungsbedingungen hergestellten Acetalhomopolymer- und Polyamid 66-Außenschichten der Platten und Scheiben zeigen in bezug auf die Größe der Reibung bei kleinen Belastungen keine bemerkenswerten Unterschiede. Die ungleichartige Acetalhomopolymer/Polyamid 66-Gleitpaarung weist bei oszillierender Gleitbewegung eine um 20 bis 25 % niederere Reibung und bei gleichförmig drehender Gleitreibung eine um etwa 30 bis 40 % niederere Reibung als die gleichartige Acetalhomopolymer/Acetalhomopolymer-Gleitpaarung auf. Bild 11 zeigt die Reibungszahlen der gleichartigen Acetalhomopolymer/Acetalhomopolymer-Gleitpaarung in Abhängigkeit von der Normalkraft und den Verarbeitungsbedingungen.

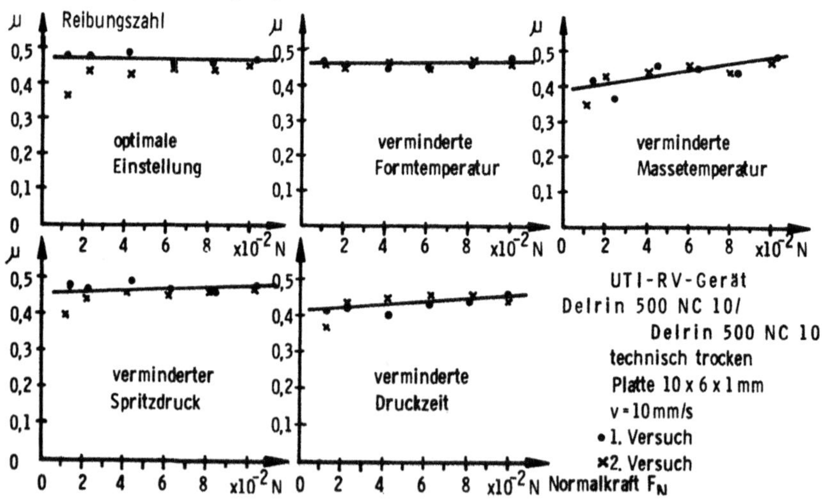

Bild 11: Reibungszahlen in Abhängigkeit von den Verarbeitungsbedingungen und der Normalkraft bei oszillierender Gleitbewegung (UTI-RV-Gerät, Acetalhomopolymer/Acetalhomopolymer, Plattendicke 1 mm).

Die Reibungszahlen der gleichartigen Acetalhomopolymer/Acetalhomopolymer-Gleitpaarungen liegen bei oszillierender und gleichförmig drehender Gleitbewegung in etwa derselben Größenordnung. Ein bemerkenswerter Unterschied in bezug auf die Größe der Reibung tritt bei der Verwendung von 1 mm und 2 mm dicken Platten und Scheiben bei den beiden Polymerwerkstoff/Polymerwerkstoff-Gleitpaarungen nicht auf. Die Veränderung der thermoplastischen Außenschicht durch die Variation der Verarbeitungsbedingungen führt unter den angegebenen Versuchsbedingungen bei hoher Belastung zu keinen bedeutenden Unterschieden in bezug auf die Größe der Reibung. Die Reibung der gleichartigen Acetalhomopolymer-Gleitpaarung bei großer Belastung unterscheidet sich von der bei niederer Belastung hauptsächlich durch den Einfluß der Prüfplattendicke. Die Reibungszahlen beim Gleiten der Acetalhomopolymer-Kugeln auf den 1 mm dicken Acetalhomopolymer-Platten sind bei hoher Belastung größer als beim Gleiten der Acetalhomopolymer-Kugeln auf den 2 mm dicken Acetalhomopolymer-Platten.

5.10 Abriebuntersuchungen:

Die Einflüsse der Verarbeitungsbedingungen zur Erzielung von optimalen tribologischen Eigenschaften der Außenschichten wirken sich auf das Abriebverhalten stärker als auf das Reibungsverhalten aus.

5.10.1 Stahl/Polymerwerkstoffpaarungen.

Die Stahl/Acetalhomopolymer- und die Stahl/Polyamid 66-Gleitpaarungen zeigen bei kleinen Belastungen ($F_N = 2x10^{-1}N$) einen nicht meßbaren Kugelabrieb und auf den Polymerwerkstoff-Platten und -Scheiben nur Schleifspuren. Selbst bei hohen Belastungen (F_N = 6,3 N und 7 N) tritt unter den angegebenen Versuchsbedingungen kein meßtechnisch erfaßbarer Abrieb auf.

5.10.2 Polymerwerkstoff/Polymerwerkstoff-Gleitpaarungen.

Im Gegensatz zu den Stahl/Polymerwerkstoff-Gleitpaarungen tritt bei den gleichartigen Acetalhomopolymer/Acetalhomopolymer-Gleitpaarungen unter den gleichen Versuchsbedingungen Abrieb auf.

5.10.2.1 Oszillierende Gleitbewegung.

Aus den Untersuchungen bei oszillierender Gleitbewegung lassen sich folgende Ergebnisse erkennen:

Prinzipiell ist der Abrieb bei oszillierender Gleitbewegung größer als bei gleichförmig drehender Gleitbewegung.
Die unter verminderter Formtemperatur abgespritzten Platten zeigen beim Gleiten auf den unter optimalen Spritzbedingungen hergestellten Kugeln eindeutig einen großen Abrieb. In Bild 12 ist sowohl der Kugel- als auch der Plattenabrieb nach einem Gleitweg von 8 km dargestellt.

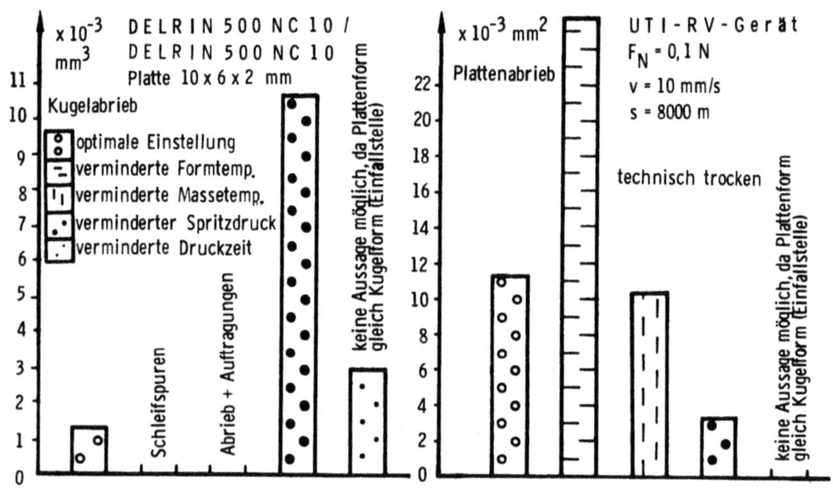

Bild 12: Abrieb in Abhängigkeit von den Verarbeitungsbedingungen bei oszillierender Gleitbewegung (UTI-RV-Gerät, Acetalhomopolymer/Acetalhomopolymer, Plattendicke: 2 mm).

Der Abrieb, der unter verminderter Massetemperatur, vermindertem Spritzdruck, verminderter Druckzeit und optimalen Bedingungen abgespritzten Platten ist demgegenüber kleiner. Eine weitergehende Differenzierung der Einflüsse der Verarbeitungsbedingungen auf die tribologischen Eigenschaften der Außenschicht erscheint aus diesen Untersuchungsergebnissen wenig sinnvoll, da sich zum einen der Abrieb auf Platte und Kugel verteilt und zum anderen die Streuungen der Prüfkörperoberflächeneigenschaften innerhalb eines Verarbeitungsparameters schon relativ groß sind. Bild 13 zeigt, wie weit sich die Kugel nach einem Gleitweg von 4 km in die thermoplastische Außenschicht der Acetalhomopolymerplatte eingearbeitet hat.

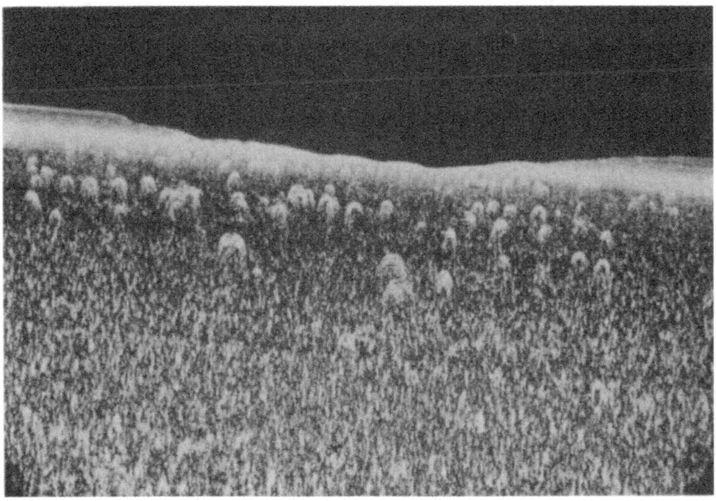

Bild 13: Abriebmulde in einer mit verminderter Formtemperatur hergestellten 2 mm dicken Acetalhomopolymer-Platte (UTI-RV-Gerät, Gleitweg 4 km, Vergrößerung 160fach).

Durch die vorausgegangenen Oberflächenuntersuchungen mit dem Tastschnittgerät, Kontrastinterferenzmikroskop und dem Rasterelektronenmikroskop sowie mit den Mikrohärteprüfgeräten war zu erwarten, daß die rauhere und weichere Außenschicht der unter verminderter Formtemperatur abgespritzten Acetalhomopolymer-Platten sich negativ auf das Abriebverhalten auswirken kann.
Die 1 mm dicken Acetalhomopolymer-Platten weisen im Vergleich zu den 2 mm dicken Acetalhomopolymer-Platten unter denselben Versuchsbedingungen einen etwas geringeren Abrieb auf. Der Abrieb tritt nur auf den 1 mm dicken Acetalhomopolymer-Platten und nicht auf den dazugehörigen Kugeln auf.
Im Gegensatz zu den Acetalhomopolymer-Platten zeigen sich bei den Polyamid 66-Platten keine Unterschiede im Abriebverhalten sowie in ihrer Oberflächenstruktur.
Bei hoher Belastung tritt auf den bei verminderter Formtem-

peratur abgespritzten Platten ebenso wie bei den Untersuchungen mit kleiner Belastung der größte Abrieb sowohl bei den Untersuchungen mit dem Reibungspendel als auch mit dem UTI-RV-Gerät auf.
Bei den mit hoher Belastung und dem UTI-RV-Gerät durchgeführten Versuchen tritt auf den Kugeln großer Abrieb und gleichzeitig Materialauftragungen durch eine hohe Autoadhäsion auf. Der Kugelabrieb läßt sich dadurch nicht mehr von den Materialauftragungen trennen. Eine meßtechnische Erfassung des Kugelabriebes ist deshalb nicht möglich.
Auf den Kugeln, die beim Reibungspendel und unter hoher Belastung eingesetzt werden, ergeben sich unter den entsprechenden Versuchsbedingungen nur Schleifspuren und keine Materialübertragungen von der Platte her. Die Größe des Plattenabriebes ist aus Bild 14 zu entnehmen.

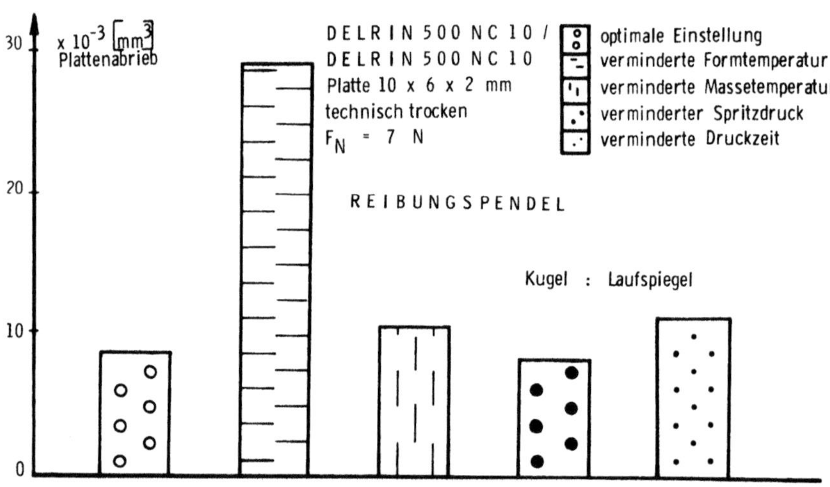

Bild 14: Abrieb in Abhängigkeit von den Verarbeitungsbedingungen bei oszillierender Gleitbewegung (Reibungspendel, Acetalhomopolymer/Acetalhomopolymer, Plattendicke: 2 mm).

Der Plattenabrieb, der auf dem UTI-RV-Gerät bei hoher Belastung untersuchten Acetalhomopolymer-Platten, ist bei den 1 mm dicken Platten um etwa 15 % kleiner als bei den 2 mm dicken Platten.

Qualitativ werden die Ergebnisse, die auch bei kleinen Belastungen gewonnen werden, bestätigt.
Abgesehen von den Untersuchungen der bei verminderter Formtemperatur abgespritzten Acetalhomopolymer-Platten ist der mit dem Reibungspendel hervorgerufene Plattenabrieb auf den 1 mm dicken Platten größer als auf den 2 mm dicken Platten. Auf den Kugeln, die mit den 1 mm dicken Acetalhomopolymer-Platten gepaart werden, sind neben Schleifspuren Materialübertragungen zu erkennen. Das Abriebverhalten in Abhängigkeit von der Plattenstärke (1 mm und 2 mm) ist bei den Untersuchungen mit dem Reibungspendel und dem UTI-RV-Gerät umgekehrt.
Eine Bestätigung dieses umgekehrten Abriebverhaltens liefern die für dieses Vorhaben durchgeführten Vorversuche mit denselben Acetalhomopolymer/Acetalhomopolymer-Gleitpaarungen, die jedoch aus einer anderen Abspritzung stammen.
Die Ursache für dieses reziproke Abriebverhalten läßt sich zur Zeit noch nicht exakt erklären. Hierzu wären noch weitere Versuchsreihen notwendig. Die Untersuchungen mit dem Reibungspendel und dem UTI-RV-Gerät unterscheiden sich vor allem im Bewegungsablauf, der Führung der Gleitpartner zueinander und in der vertauschten höheren spezifischen Anstrengung von Kugeln und Platten.
Trotz hoher Belastung tritt unter den gegebenen Versuchsbedingungen bei den Acetalhomopolymer/Polyamid 66-Gleitpaarungen weder an den Kugeln noch auf den Platten Abrieb auf. Alle weiteren Untersuchungen bestätigen das gute tribologische Verhalten dieser Gleitpaarung.

5.10.2.2 Gleichförmig drehende Gleitbewegung.

Bei der gleichförmig drehenden Gleitbewegung sind bei den Acetalhomopolymer/Acetalhomopolymer-Gleitpaarungen wesentlich längere Gleitwege als bei der oszillierenden Gleitbewegung notwendig, um zu brauchbaren Differenzierungen im Abriebverhalten zwischen den unter verschiedenen Verarbeitungsparametern abgespritzten Prüfkörpern zu kommen.
Jeder einzelne Meßpunkt in den Diagrammen entspricht einem Versuch, der jeweils neu angesetzt wird. Eine größere Streuung der

einzelnen Meßwerte muß in Kauf genommen werden, um in den tribologischen Vorgang nicht eingreifen zu müssen. Eine Entnahme der Kugel aus dem Prüfstand zur Abriebmessung und ihre Neujustierung zur Fortsetzung der Versuche ist praktisch nicht möglich. Eine wünschenswerte kontinuierliche Abriebmessung ist bei den geringen Abriebmengen zu ungenau.

Nachdem Vorversuche bei Normalkräften von $F_N < 10^{-1}$N gezeigt haben, daß die Versuchsdauer - um einen meßtechnisch erfaßbaren Abrieb zu erhalten - innerhalb des Gesamtvorhabens viel zu lange dauern würden, wird die erste große Versuchsreihe bei $F_N = 1x10^{-1}$N angesetzt.

Eine Variation der Gleitgeschwindigkeit von 10 mm/s bis 100 mm/s hat keinen Einfluß auf das Abriebverhalten der Acetalhomopolymer/Acetalhomopolymer-Gleitpaarung. Eine Erwärmung der Prüfkörper ist nicht festzustellen.

Aus den Untersuchungen bei gleichförmig drehender Gleitbewegung lassen sich folgende Ergebnisse erkennen:

Der Kugelabrieb bei der gleichartigen Acetalhomopolymer/Acetalhomopolymer-Gleitpaarung ($F_N = 1x10^{-1}$N) zeigt nach einem Gleitweg von 154 km (Versuchsdauer 89 Tage), daß durch die unter optimalen Bedingungen und unter vermindertem Spritzdruck abgespritzten Acetalhomopolymer-Scheiben die unter optimalen Bedingungen abgespritzten Acetalhomopolymer-Kugeln am wenigsten angegriffen werden.

Die unter verminderter Formtemperatur, verminderter Massetemperatur und unter verminderter Druckzeit hergestellten Scheiben verhalten sich gegenüber der Acetalhomopolymer-Kugel agressiver. Auffallend ist, daß die unter verminderter Formtemperatur abgespritzten Acetalhomopolymer-Scheiben schon bei relativ kleinen Gleitwegen zu erhöhtem Kugelabrieb führen. Dieses Ergebnis wird durch alle mit der Acetalhomopolymer/Acetalhomopolymer-Gleitpaarung bei gleichförmig drehender Gleitbewegung durchgeführten Untersuchungen bestätigt.

Bei Gleitwegen von 4 km, 8 km und 16 km ist der Scheibenabrieb quantitativ nicht zu erfassen, da lediglich Schleifspuren (erhöhte Oberflächenrauhigkeit) auftreten. Beim Gleitweg von 70 km ist der Scheibenabrieb bei der unter verminderter Formtemperatur abgespritzten Scheibe erwartungsgemäß schon groß. Den eben-

falls bei einem Gleitweg von 70 km schon großen Abrieb auf der unter optimalen Bedingungen abgespritzten Scheibe dürfte eine Materialinhomogenität auf der Kugel ausgelöst haben.
Obwohl die Kugeln wesentlich höher belastet sind als die Scheiben, ist die Größe des Abriebes auf den Scheiben um mehr als den Faktor 10 größer als auf den Kugeln. Der Grund für dieses Verhalten liegt darin, daß sich die Kugeln auf dem ganzen Umfang der Scheibe einarbeiten.
Nach der ersten größeren Versuchsreihe mit einer Normalkraft von $F_N = 1x10^{-1}N$ werden die nächsten Versuchsreihen bei der doppelten Normalkraft ($F_N = 2x10^{-1}N$) durchgeführt. Zuerst werden unter den gleichen Versuchsbedingungen bei 5 Acetalhomopolymer/Acetalhomopolymer-Gleitpaarungen der Kugel- und Scheibenabrieb ermittelt. Die Ergebnisse zeigen, daß die Reproduzierbarkeit - unter der Berücksichtigung der kleinen Größe des Abriebes - als ausreichend angesehen werden kann. Die bei verminderter Formtemperatur, verminderter Massetemperatur und verminderter Druckzeit abgespritzten Scheiben von 1 mm Dicke verursachen einen deutlich höheren Abrieb an den Acetalhomopolymer-Kugeln als die unter optimalen Bedingungen und die unter vermindertem Druck abgespritzten Scheiben. In bezug auf die Größe des Abriebes bewirkt die Normalkrafterhöhung von $1x10^{-1}N$ auf $2x10^{-1}N$ keine Steigerung des Abriebes.
In <u>Bild 15</u> und <u>16</u> ist das Abriebverhalten der Acetalhomopolymer/Acetalhomopolymer-Gleitpaarung (Scheibendicke: 2 mm) dargestellt.
Die unter vermindertem Spritzdruck hergestellten 1 mm dicken Acetalhomopolymer-Scheiben verursachen im Vergleich zu den unter optimaler Einstellung gespritzten Scheiben einen etwas grösseren Abrieb an der Kugel. Darüber hinaus verhält sich der Kugelabrieb durch die 1 mm dicken Acetalhomopolymer-Scheiben wie derjenige, der durch die 2 mm dicken Acetalhomopolymer-Scheiben verursacht wird. Die Rangfolge im Abriebverhalten der 1 mm dicken Acetalhomopolymer-Scheiben ist nicht direkt mit derjenigen der 2 mm dicken Acetalhomopolymer-Scheiben zu vergleichen. Dies dürfte auf die relativ schlechtere Oberflächenausbildung bei den 1 mm dicken im Vergleich zu den 2 mm dicken Acetalhomopolymer-Scheiben zurückzuführen sein.

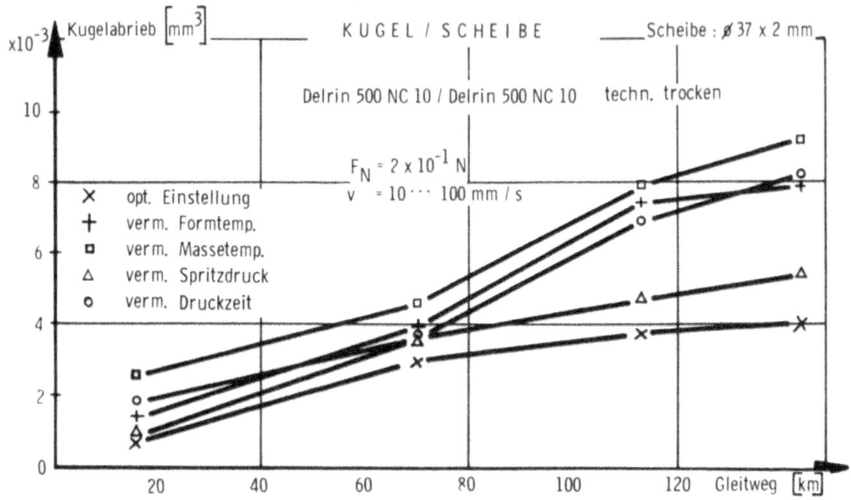

Bild 15: Kugelabrieb in Abhängigkeit von den Verarbeitungsbedingungen und dem Gleitweg bei gleichförmig drehender Gleitbewegung (Kugel-Scheibe-Prüfgerät, Acetalhomopolymer/Acetalhomopolymer, Scheibendicke: 2 mm).

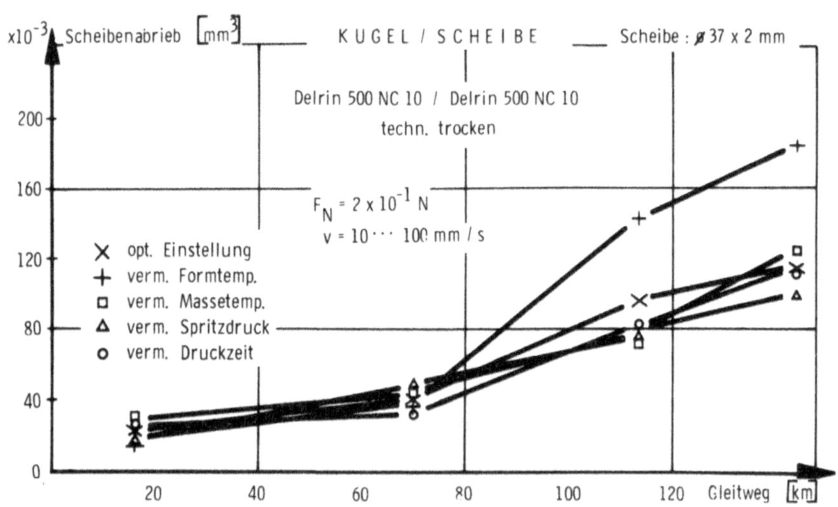

Bild 16: Scheibenabrieb in Abhängigkeit von den Verarbeitungsbedingungen und dem Gleitweg bei gleichförmig drehender Gleitbewegung (Kugel-Scheibe-Prüfgerät, Acetalhomopolymer/Acetalhomopolymer, Scheibendicke: 2 mm).

Das Abriebverhalten der 2 mm dicken Acetalhomopolymer-Scheiben zeigt deutlich, daß die bei verminderter Formtemperatur abgespritzten Scheiben durch die unter optimalen Bedingungen hergestellten Kugeln am stärksten angegriffen werden. Der Abrieb, der unter den anderen Verarbeitungsbedingungen abgespritzten Scheiben, ist gleich groß. Dieses Ergebnis stimmt qualitativ mit denjenigen, die bei oszillierender Gleitbewegung erhalten werden, überein.
Die ungleichartigen Acetalhomopolymer/Polyamid 66-Gleitpaarungen weisen einen um etwa den Faktor 10 kleineren Kugelabrieb gegenüber demjenigen der gleichartigen Acetalhomopolymer/Acetalhomopolymer-Gleitpaarungen auf. Nach einem Gleitweg von 120 km sind auf den Polyamid 66-Scheiben nur Schleifspuren zu erkennen, die meßtechnisch nicht mehr erfaßbar sind.
Einen Einfluß der Verarbeitungsbedingungen der Polyamid 66-Scheiben auf den Abrieb der Gleitpaarung Acetalhomopolymer/Polyamid 66 läßt sich unter den gegebenen Versuchsbedingungen nicht feststellen.
Dieses Ergebnis bestätigt die Oberflächenuntersuchungen bei denen ebenfalls keine Unterschiede zwischen den Polyamid 66-Scheiben auftreten, die unter den 5 verschiedenen Verarbeitungsparametern abgespritzt werden.
Ein Abschleifen der Kunststoffscheiben mit Schleifpapier - ähnlich wie bei der Herstellung von Metallschliffen - ist nicht möglich, da sich einzelne Schleifkörner in die weichen Kunststoffscheiben einlagern können.
Bei zwei Versuchsreihen mit unterschiedlich abgeschliffenen Acetalhomopolymer-Scheiben hat sich der Acetalhomopolymer-Kugelabrieb um die Faktoren 3 bis 7 bzw. 7 bis 12 erhöht. An sämtlichen Acetalhomopolymer-Scheiben sind nur Laufspiegel zu erkennen.

6. Zusammenfassung:

Die Eigenschaften der thermoplastischen Außenschicht werden über eine Variation der Verarbeitungsparameter, die durch den Polymerwerkstoffhersteller und den Verarbeiter festgelegt wurden, erreicht. Eine mechanische Abtragung der Oberflächen zur

Entfernung der Außenschicht kommt nicht in Frage, da in vielen Fällen der Abrieb an den feinmechanischen Bauteilen aus funktionellen Gründen nicht größer sein darf, als die Außenschicht selbst.
Der Einfluß der Verarbeitungsbedingungen auf die verwendeten Prüfkörper wird durch Dimensions- und Massebestimmungen erfaßt. Wertvolle Hinweise auf den Charakter der Polymerwerkstoff-Oberflächen lieferten Form- und Rauhigkeitsuntersuchungen mit dem Tastschnittgerät, Dünnschnitte, Kontrastinterferenzaufnahmen und Mikrohärtemessungen.
Bei den Platten und Scheiben aus Polyoxymethylen, die bei verminderter Formtemperatur abgespritzt wurden, ergeben sich rauhere Oberflächen und niederere Härtewerte gegenüber denjenigen, die unter optimalen Bedingungen hergestellt wurden. Dagegen können bei den Polyamid 66-Platten und -Scheiben keine Unterschiede festgestellt werden.
Das Reibungsverhalten wird durch die thermoplastische Außenschicht, die durch die verschiedenen Verarbeitungsbedingungen der Acetalhomopolymer- und Polyamid 66-Prüfkörper variiert wurde, wenig beeinflußt.
Eine deutlich niedere Reibung weist die Polyoxymethylen/Polyamid 66-Gleitpaarung sowohl bei gleichförmig drehender als auch bei oszillierender Gleitbewegung auf.
Der Einfluß der Verarbeitungsbedingungen wirkt sich über die thermoplastische Außenschicht stärker auf das Abriebverhalten als auf das Reibungsverhalten aus. Die Abriebuntersuchungen erstrecken sich bis zu einer maximalen Oberflächentiefe von 40 µm. Die dazugehörenden Abriebvolumina sind klein und erschweren die Abrieberfassung.
Die bei verminderter Formtemperatur erzeugte thermoplastische Außenschicht führt bei den Acetalhomopolymer/Acetalhomopolymer-Gleitpaarungen gegenüber denjenigen Oberflächen, die durch andere Verarbeitungsparameter (verminderte Massetemperatur, verminderter Spritzdruck, verminderte Druckzeit und optimale Einstellung) hergestellt wurden, zu erhöhtem Abrieb. Die dazugehörenden Oberflächenuntersuchungen bei den mit verminderter Formtemperatur abgespritzten Polyoxymethylen-Prüfkörpern zeigen eindeutig, daß diese eine rauhe und weiche Oberfläche aufweisen.

An der unter optimaler Einstellung hergestellten Kugel (Gleitpartner) rufen die Acetalhomopolymer-Scheibenoberflächen, die unter verminderter Formtemperatur, verminderter Massetemperatur und verminderter Druckzeit abgespritzt wurden, den größten Abrieb hervor.
Daraus läßt sich für die Praxis ableiten, daß bei der Wahl der Verarbeitungsparameter zur Erzielung einer Außenschicht mit optimalen tribologischen Eigenschaften für feinmechanische Bauteile aus Polyoxymethylen der Formtemperatur und einer gleichmäßigen Temperierung des Werkzeuges eine hohe Bedeutung beizumessen ist. Sie darf, um die Zykluszeit kurz zu halten, nicht zu nieder gewählt werden.
Im Gegensatz zu den Acetalhomopolymer-Scheiben und -Platten zeigen sich bei den Polyamid 66-Scheiben und -Platten unter den gegebenen Versuchsbedingungen keine Unterschiede im Abriebverhalten. Dies ist, aufgrund der bei den vorausgegangenen Oberflächenuntersuchungen gefundenen Ergebnissen, auch nicht zu erwarten.
Die Größe des Abriebes ist jedoch bei der Polyoxymethylen/Polyamid 66-Gleitpaarung um ein Vielfaches kleiner als bei der Polyoxymethylen/Polyoxymethylen-Gleitpaarung.
Größere tribologische Beanspruchungen (F_N = 6,3 und 7 N bei Punktberührung) wie sie in der Maschinentechnik auftreten können, führen bei oszillierender Gleitbewegung (POM - 2 mm Platten) ebenfalls zu dem Ergebnis, daß die verminderte Formtemperatur einen hohen Abrieb der Außenschicht bewirkt.

Den Firmen Du Pont de Nemours (Deutschland) und der Firma Dieter Gräßlin, St. Georgen sei für die großzügige Unterstützung dieser Arbeit gedankt. Für die wertvolle Diskussion bei der Deutung der Dünnschnitte möchte ich mich bei Frau E. Böhme, Du Pont de Nemours International S.A. Genf (Schweiz) bedanken. Für die freizügige Benutzung der Härteprüfgeräte bei den Firmen R. Bosch GmbH, Waiblingen, Göttfert, Buchen und dem Institut für Kunststoffprüfung und Kunststoffkunde der Universität Stuttgart sowie für die Erstellung der Rasterelektronenmikroskopaufnahmen durch das Institut für Industrielle Fertigung und Fabrikbetrieb der Universität Stuttgart wird besonders gedankt.

Meinen Mitarbeitern Frau I. Gehrlach, Herrn W. Schulz und Herrn H. J. Grabein möchte ich für die sorgfältige und mühevolle Durchführung der Versuche danken.

7. Schrifttum:

1 Böhme, E.: Strukturuntersuchungen an Acetalhomopolymerisat und deren praktische Anwendung. Kunststoffe 60 (1970), 4, S. 273 ... 276

2 Speil, Th.: Die Mikroskopie, ein Mittel zur Qualitätskontrolle beim Präzisionsspritzgießen. Kunststoffe 63 (1973), 7, S. 1 ... 6

3 Kohlhepp, K.; Petrak, H.: Untersuchung zur Qualitätssicherung von Kleinstformteilen aus Acetalhomopolymerisat, "STZ" (1965), 40/41, S. 915 ... 925

4 Wübken, G.: Einfluß der Verarbeitungsbedingungen auf die innere Struktur thermoplastischer Spritzgußteile unter besonderer Berücksichtigung der Abkühlverhältnisse. Dissertation Technische Hochschule Aachen 1974.

5 Friedrich, K.: Mikroskopische Methoden zur Gefügeanalyse teilkristalliner Thermoplaste. Prakt. Metallographie 16 (1979), S. 321..334.

6 Woebken, W.: Einfluß der Verarbeitung auf die Maßgenauigkeit von thermoplastischen Spritzgußteilen. Kunststoffe 67 (1977), 4, S. 179 ... 183.

7 Geyer, H.: Präzisionsspritzgießen von Acetal-Copolymerisat, Kunststoffe 61 (1971), 1, S. 12 ... 17, S. 81 ... 86.

8 Wolters, E.; Buchberger, H.L.; Kohlhepp, K.G.: Zusammenhang zwischen Gefügeaufbau und mechanischem Verhalten von Spritzgußteilen aus Hostaform. Plastverarbeiter (1973) H9, Hoechst Sonderdruck 6086, S. 1 ... 7.

9 Halach, G.: Gleitreibungsverhalten von Kunststoffen gegen Stahl und seine Deutung mit molekularmechanischen Modellvorstellungen. Dissertation Universität Stuttgart 1975.

10 Barker, G.D.; Mc. Knight, C.E.; Mc. Klveen, J.R.; Hood, D.M.; Alter, G.E.: The Comprehensive Laboratory Testing of Instrument Lubricants. A.S.T.M.-Bulletin (1946), März, S. 25 ... 35.

11 Dürr, F.: Das Reibungs- und Verschleißverhalten des neuentwickelten Delrin 500 CL im Vergleich mit demjenigen der Standardtypen. UT 3 (1976), 1, S. 1 ... 22.

12 Clark, E.S.: Molecular orientation in injection molding acetalhomopolymer. SPE - J. (1967), S. 46 ... 49.

13 Racké, H.H.; Fett, Th.: Grundlagen und Vorteile einer neuen Kugeldruckhärte-Prüfung an Kunststoffen. Materialprüfung 10 (1968), 7, S. 226 ...230.

14 Müller, K.: Anwendung einer neuentwickelten Kleinlast-Härtemeßmethode auf der Basis des Vickers-Verfahrens. Kunststoffe 60 (1970), S. 265 ... 273.

15 Eyerer, P.; Lang, G.: Wichtige Einflußgrößen bei Vickers-Mikrohärtemessungen an Kunststoffen. Kunststoffe 62 (1972), 5, S. 322 ... 328.

16 Racké, H.H.; Fett, Th.: Ermittlung der Härteverteilung unterhalb von Kunststoffoberflächen durch Vickers-Härtemessungen mit mehreren Laststufen. Materialprüfung 17 (1975), 12, Kunststoffe Hoechst Sonderdruck 6152 S.1..5

17 Erhard, G.: Zum Reibungs- und Verschleißverhalten von Polymerwerkstoffen. Dissertation Universität Karlsruhe (TH) 1980.

Als Informationsdienst wurden die Literatur-Recherchen: Tribologie von Kunststoffen, Spritzparameter beim Spritzgießen von Kunststoffen, und Struktur von Kunststoffen der Dokumentationszentrale Feinwerktechnik im Fachinformationszentrum Technik e.V., Ostbahnhofstr. 13, 6000 Frankfurt/M 1 benutzt.

Rheologische Eigenschaften von Ziehseife

für das Drahtziehen

von

Oskar Pawelski, Wolfgang Rasp und Tetsuya Hirouchi

Max-Planck-Institut für Eisenforschung

Düsseldorf

Förderkennzeichen: 01 ZT 347
Laufzeit: 1.1.1978 - 30.6.1980

Gliederung

1. Einführung

Vor einigen Jahren wurde am Max-Planck-Institut in Düsseldorf als Forschungsschwerpunkt das Gebiet des hydrodynamischen Ziehens untersucht [1]. Es konnte bereits theoretisch gezeigt werden, daß der hydrodynamische Effekt mit der sogenannten Doppelziehsteintechnik zu erwarten ist. Dabei ist eine mit Ziehseife gefüllte Druckkammer von zwei Ziehsteinen eingeschlossen, wobei der eine als Einlaß für den Draht und die Seife bestimmt ist, während im anderen die eigentliche Abnahme erfolgt.

Diese theoretischen Überlegungen wurden später von Experimenten [2] unterstützt, die jedoch zeigten, daß die Ziehseife nicht mehr wie bisher als Newtonscher Körper betrachtet werden kann. Die Seife schmilzt offensichtlich bei höheren Temperaturen, und ein fester Pfropfen stellt sich ein.

Diese Beobachtung führte zu der Fragestellung, welche rheologischen Eigenschaften die Trockenziehseifen haben. In dieser Arbeit wurde daher der Versuch unternommen, die Seife als sogenannten Bingham-Körper zu betrachten, der durch einen Pfropf-Fluß charakterisiert ist. Dazu wurden zwei Rheometertypen eingesetzt, und zwar eine selbstentwickelte Rückwärtsfließpreßapparatur und ein gekauftes Extrusiometer.

2. Allgemeine Bemerkungen zum rheologischen Verhalten von Fluiden

2.1 Scherspannung und Schergeschwindigkeitsgefälle

Der Fluß eines Fluids oder eines Granulates wird beschrieben durch seine Geschwindigkeitsverteilung und die Tangentialkraft F, die für die Bewegung verantwortlich ist, Bild 1. Wird nun die auf das Medium wirkende Kraft F auf die zugehörige Berührfläche A bezogen, so erhält man die Scherspannung τ zu

$$\tau = \frac{F}{A} \; . \tag{1}$$

Andererseits kann das Schergeschwindigkeitsgefälle $\dot{\gamma}$ vom Geschwindigkeitsfeld hergeleitet werden. Im einfachen Fall ebener Formänderung gilt

$$\dot{\gamma} = \frac{du}{dy} \; . \tag{2}$$

In dieser Gleichung bedeutet u den in der x-Richtung wirkenden Geschwindigkeitsvektor.

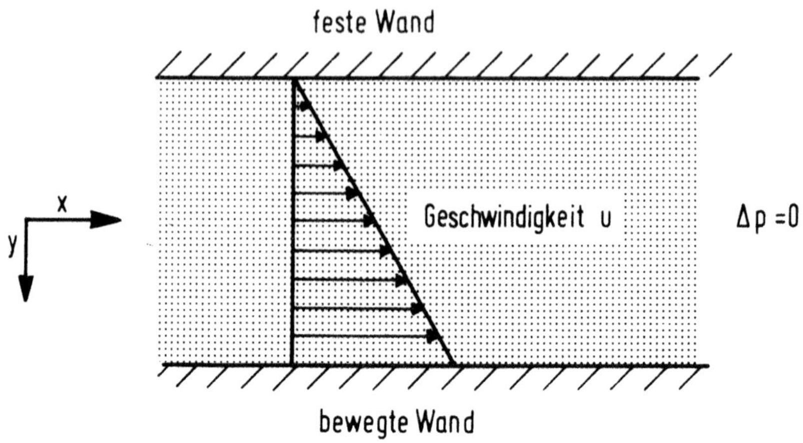

Bild 1: Geschwindigkeitsverteilung zwischen zwei parallelen Platten

2.2 Fließkurve

Bei einem flüssigen Medium zeigt die Fließkurve die Abhängigkeit zwischen der Scherspannung τ und dem Schergeschwindigkeitsgefälle $\dot{\gamma}$

$$\tau = f(\dot{\gamma}) . \tag{3}$$

Diese Funktion ist charakterisch für jedes Fluid. Als Zustandsvariable hängt sie ab vom Druck und von der Temperatur.

Idealisierte Modellvorstellungen werden herangezogen, um manche Fluide zu beschreiben. So folgt der bekannte Newton-Körper der einfachen Fließregel einer proportionalen Abhängigkeit von τ und $\dot{\gamma}$

$$\tau = \eta \dot{\gamma} . \tag{4}$$

Der Proportionalitätsfaktor η wird "dynamische Viskosität" genannt. Bei Raumtemperatur liegt er in der Größenordnung von 10^{-5} Pas für Luft und Wasserstoff und 10^{3} Pas für Asphalt und Schmierseife.

Eine große Zahl nicht-Newtonscher Fließkurven werden zur Beschreibung anderer Fluide benutzt. Bild 2 zeigt die Kurven des Ostwald-Körpers, der einem Potenzgesetz folgt

$$\tau = \eta (\dot{\gamma})^{n} . \tag{5}$$

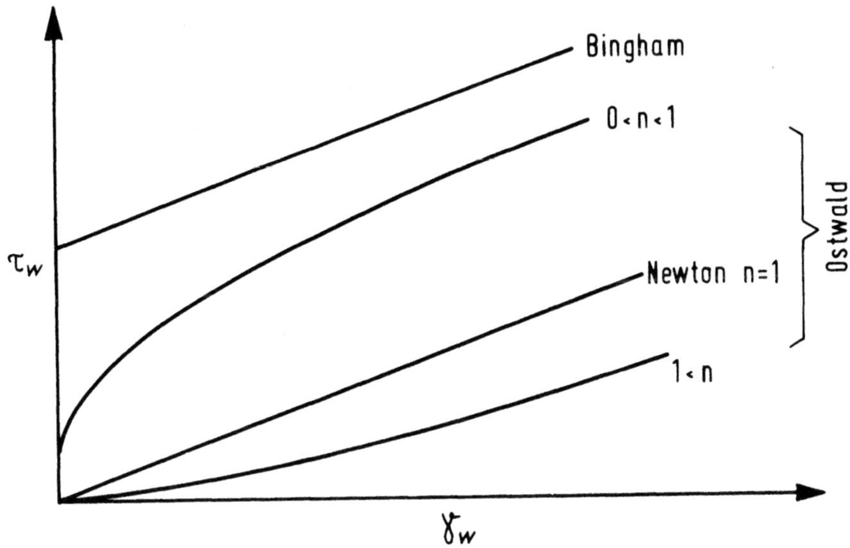

Bild 2: Idealisierte Fließkurven

Hier handelt es sich um einen verallgemeinerten Newton-Körper (n = 1). Der Bingham-Körper verallgemeinert den Newton-Körper, indem der St.-Vernant-Ausdruck für Plastizität hinzugefügt wird in der Form

$$\tau = \tau_0 + \eta_B \dot{\gamma} \quad . \qquad (6)$$

In dieser Gleichung beschreiben zwei Konstanten, die Fließspannung τ_0 und die Bingham-Viskosität η_B, den linearen Zusammenhang zwischen τ und $\dot{\gamma}$. Daneben wird häufig der Begriff scheinbare Viskosität η_a (engl.: apparent) gebraucht, der einen Bingham-Körper in der gleichen Weise beschreibt, wie es das Newton-Gesetz, Gl. (4), vorschreibt

$$\tau = \eta_a \dot{\gamma} \quad . \qquad (7)$$

Vergleicht man Gl. (6) und Gl. (7), so erhält man für die scheinbare Viskosität η_a

$$\eta_a = \frac{1}{\dot{\gamma}} \tau_0 + \eta_B \quad , \qquad (8)$$

und man sieht, daß sich mit wachsendem Schergeschwindigkeitsgefälle die scheinbare Viskosität der Bingham-Viskosität nähert.

2.3 Fließverhalten im Spalt

2.3.1 Kapillare

An der Wand eines Rohres (Radius R, Länge L) führt der Ansatz der Gleichgewichtsbedingungen zu einem Ausdruck für die dort herrschende Scherspannung τ_w

$$\tau_w = \frac{R\Delta p}{2L} , \tag{9}$$

wobei Δp der Druckabfall über der Länge L ist. Der Durchfluß $\dot{Q}$ ist eine Funktion des Geschwindigkeitsfeldes und damit abhängig vom Schergeschwindigkeitsgefälle $\dot{\gamma}$. Für einen Newton-Körper mit parabolischer Geschwindigkeitsverteilung kann $\dot{Q}$ aus dem Hagen-Poiseuille-Gesetz ermittelt werden.

$$\dot{Q} = \frac{\pi \Delta p R^4}{8\eta L} \tag{10}$$

Mit Gleichungen (4), (9) und (10) ergibt sich nun der Ausdruck

$$\dot{\gamma}_w = \frac{4\dot{Q}}{\pi R^3} . \tag{11}$$

Bei gegebenen Geometriewerten R und L erhält man mit den Meßwerten $\Delta\rho$ und $\dot{Q}$ die Lösungen der Gleichungen (9) und (11), so daß die Newtonsche Viskosität aus Gl. (4) zu

$$\eta = \frac{\pi \Delta p R^4}{8 \dot{Q} L} \tag{12}$$

bestimmt werden kann.

Da die Geschwindigkeitsverteilung des Bingham-Körpers von der des Newton-Körpers abweicht, muß das Hagen-Poiseuille-Gesetz, Gl. (10), durch das sogenannte Buckingham-Reiner-Gesetz [3] modifiziert werden

$$\dot{Q} = \frac{\pi \Delta p R^4}{8\eta_B L} \left[1 - \frac{4}{3} \left(\frac{2\tau_o L}{R\Delta\rho}\right) + \frac{1}{3} \left(\frac{2\tau_o L}{R\Delta p}\right)^4 \right]. \tag{13}$$

Diese Gleichung berücksichtigt den Pfropf-Fluß des Bingham-Mediums. Sie geht für $\tau_o = 0$ in Gl. (10) über. Definiert man nun ein modifiziertes Geschwindigkeitsgefälle $\dot{\Gamma}$ zu

$$\dot{\Gamma} = \frac{4 \dot{Q}}{\pi R^3} , \tag{14}$$

so folgt mit Gleichungen (9) und (13)

$$\dot{\Gamma} = \frac{\tau_w}{\eta_B} \left[1 - \frac{4}{3} \left(\frac{\tau_o}{\tau_w}\right) + \frac{1}{3} \left(\frac{\tau_o}{\tau_w}\right)^4 \right] . \tag{15}$$

Diese Gleichung genügt dem nach Rabinowitsch, Mooney und Weissenberg [4][5] benannten Gesetz

$$\dot{\gamma}_w = f(\tau_w) = \frac{3}{4} \dot{\Gamma} + \frac{\tau_w}{4} \left(\frac{d\dot{\Gamma}}{d\tau_w}\right) , \tag{16}$$

das für alle rheologischen Körper gültig ist.

Um nun die rheologischen Kennwerte eines Bingham-Mediums zu ermitteln, ist es sinnvoll, die modifizierte Fließkurve, Gl. (15), zu zeichnen, Bild 3.

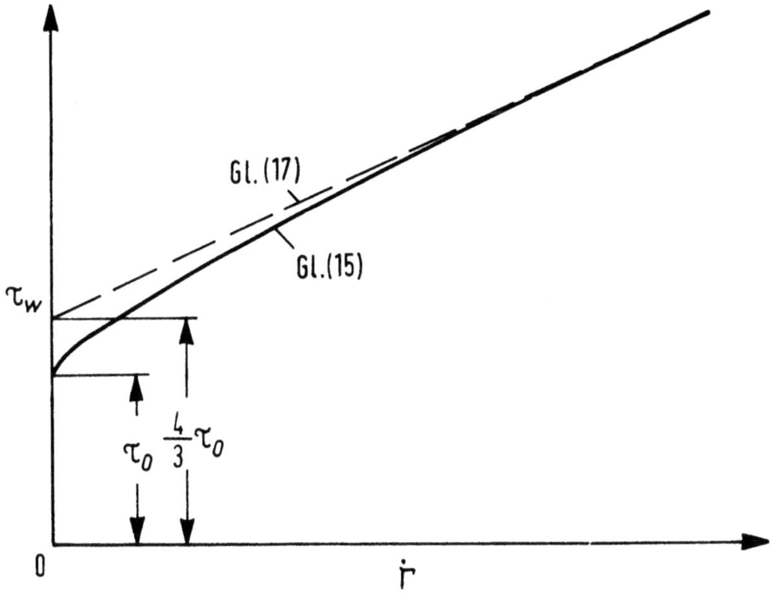

Bild 3: Modifizierte Fließkurve für ein Bingham-Medium

Falls die Scherspannung τ_w groß genug ist, kann Gl. (15) mit genügender Genauigkeit linearisiert werden zu

$$\dot{\Gamma} = \frac{4}{3}\frac{\tau_0}{\eta_B} + \frac{1}{\eta_B}\tau_w \quad \text{oder} \quad \tau_w = \frac{4}{3}\tau_0 + \eta_B\dot{\Gamma} . \tag{17}$$

Hierbei handelt es sich um die Asymptote der Gl. (15). Der Ausdruck $\dot{\Gamma}$ ist formal identisch mit dem Wert $\dot{\gamma}_w$, der für einen Newton-Körper berechnet wurde. Er hat jedoch eine andere Bedeutung, da er nicht das wirkliche Geschwindigkeitsgefälle an der Wand $\dot{\gamma}_w$ beschreibt.

Das Vorgehen, die Bingham-Konstanten τ_0 und η_B zu ermitteln, erfolgt also kurz zusammengefaßt:

a) Messung von Δp und $\dot{Q}$
b) Berechnung von τ_w und $\dot{\Gamma}$ (Gl. (9) und (14))
c) Zeichnen der Funktion $\tau_w = \tau_w(\dot{\Gamma})$

d) Ermittlung der Steigung $\frac{d\Gamma}{d\tau_w}$ aus c)

e) Berechnung von $\dot{\gamma}_w$ (Gl. (16))

f) Zeichnen der Fließkurve $\tau_w = \tau_w(\dot{\gamma}_w)$.

2.3.2 Rückwärtsfließpressen

Sowohl das hydrodynamische Drahtziehen als auch das Rückwärtsfließpressen zeigen, wie in Bild 4 dargestellt, eine ähnliche Geschwindigkeitsverteilung der Seife. Beide Verfahren unterscheiden sich in der Hauptfließrichtung und im örtlichen Geschwindigkeitsgradienten.

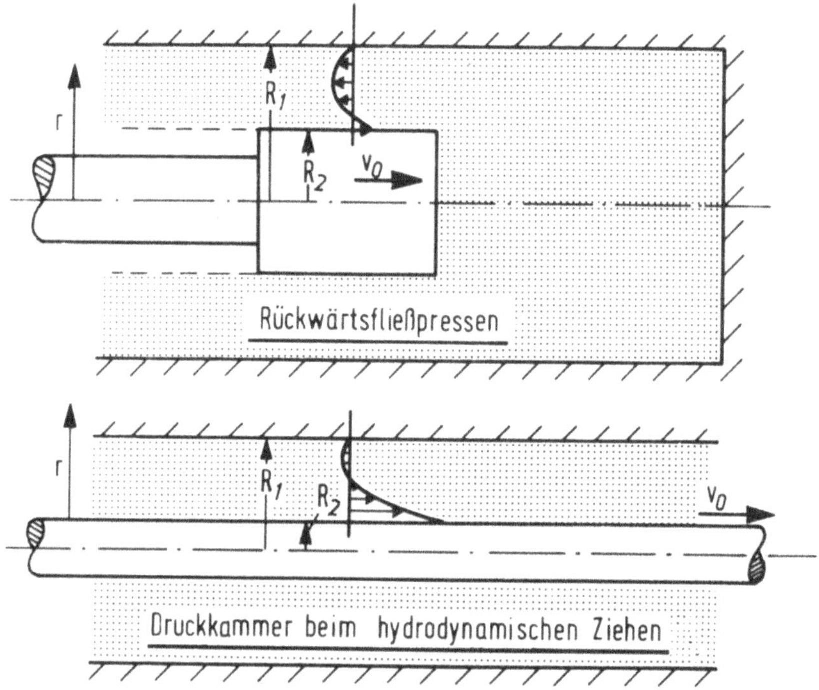

Bild 4: Geschwindigkeitsverteilung beim Rückwärtsfließpressen und beim hydrodynamischen Ziehen

2.3.2.1. Analyse mit der Annahme von Newton-Verhalten der Seife

Das Schrankenverfahren stellt eine recht einfache Methode dar, das Fließverhalten beim Rückwärtsfließpressen zu beschreiben. Bei der hier angesetzten "Oberen-Schranke-Methode" wird unter der Annahme eines beliebigen Geschwindigkeitsfeldes die Gesamtleistung berechnet und unter Variation dieses Geschwindigkeitsfeldes minimiert. Voraussetzung dazu ist die Tatsache, daß die berechnete Gesamtleistung stets größer ist als die des wahren Prozesses. Die Leistungsberechnung des in Bild 5 vorgeschlagenen Geschwindigkeitsfeldes mit den Variablen h und φ geht von folgenden Annahmen aus:

a) die Seife ist ein Newton-Fluid,
b) die Fließzone wird in drei Gebiete geteilt,
c) Haftung der Seife an den Werkzeugwänden,
d) keine Geschwindigkeitssprünge zwischen den drei Regionen,
e) ebene Formänderung in den Regionen II und III.

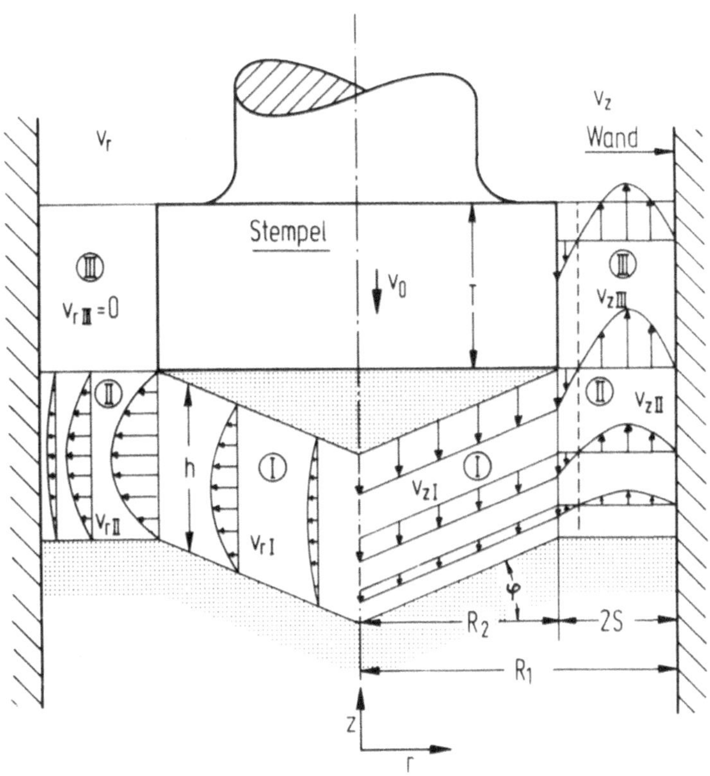

Bild 5: Angenommenes Geschwindigkeitsfeld beim Rückwärtsfließpressen

Unter Berücksichtigung der Geschwindigkeitsrandbedingungen, die an den Wänden von Zylinder und Stempel herrschen und die durch die starren Seifenzonen unter dem Stempel und am Werkzeugboden vorgegeben sind, können bei Beachtung der Inkompressibilitätsbedingung die Komponenten des Geschwindigkeitsfeldes in den Regionen I, II und III ermittelt werden. In den Gleichungen gehen die Geschwindigkeit des Stempels v_0 und die Geometrie des Werkzeuges mit R_1, R_2, S und T ein. Die freien Parameter zur Variation des Geschwindigkeitsfeldes sind $\alpha = \tan\varphi$ und h. Für das in Zylinderkoordinaten r, ϑ, z berechnete Geschwindigkeitsfeld gilt

Region I

$$v_r = -\frac{3v_0}{h^3}\left[\alpha^2 r^3 + (\alpha h - 2\alpha z)\, r^2 + (z^2 - hz) r\right] \tag{18}$$

$$v_z = \frac{3v_0}{h^3}\left[\frac{2}{3}\, z^3 - (3\alpha r + h) z^2 + (4\alpha^2 h^2 + 3\alpha r h) z - \frac{5}{3}\,(\alpha r)^3 - 2h(\alpha r)^2\right]$$

Region II

$$v_r = -\frac{3v_0}{h^3}\,(z^2 - hz)(R-S)(Ar^3 + Cr^2 + Dr + E) \tag{19}$$

$$v_z = \frac{3v_0}{h^3}\,(R-S)\left(\frac{1}{3}z^3 - \frac{h}{2}\, z^2\right)(3Ar^2 + 2Cr + D)$$

Region III

$$v_r = 0 \tag{20}$$

$$v_z = -\frac{R-S}{2}\, v_0\, (3Ar^2 + 2Cr + D).$$

Die Größen A bis E sind so definiert

$$A = \frac{1}{4S^3} \cdot \frac{R_1}{R_2} \qquad D = -\frac{1}{2S} - A(R-S)^2 - 2BR \tag{21}$$

$$B = \frac{1}{4S^2} - A\,(R_1 + R_2) \qquad E = (R+S)\left[\frac{1}{2S} + A(R-S)^2 + B(R-S)\right]$$

$$C = B - A\, R_1$$

mit $R = (R_1 + R_2)/2$.

Die Gesamtleistung berechnet sich nun in allgemeiner Form

$$P = \int_V \dot{\varepsilon}_{ij}\, s_{ij}\, dV. \tag{22}$$

Dies ist ein Volumenintegral über der Formänderungsgeschwindigkeit $\dot{\varepsilon}_{ij}$ und dem Deviator des Spannungstensors s_{ij}. Für eine Newtonsche Flüssigkeit gilt

$$s_{ij} = 2\eta\dot{\varepsilon}_{ij} \tag{23}$$

so daß für den Leistungsausdruck folgt

$$P = 2\eta \int_V \dot{\varepsilon}_{ij}\dot{\varepsilon}_{ij}\, dV. \tag{24}$$

Berechnet man nun die Formänderungsgeschwindigkeit gemäß

$$\dot{\varepsilon}_{ij} = \frac{1}{2}\left(\frac{\partial v_x}{y} + \frac{\partial v_y}{x}\right), \tag{25}$$

so ergibt sich die Gesamtleistung zu

$$P = \eta v_o^2\, (G_I + G_{II} + G_{III}) = \eta v_o^2 G. \tag{26}$$

wobei $G = G_I + G_{II} + G_{III}$ Geometriefunktionen darstellen.

Die Ergebnisse für ein Berechnungsbeispiel sind in Bild 6 wiedergegeben. Für $\varphi = 0$ wird das Minimum der Funktion G gesucht, das der Minimalleistung entspricht. Eine Rechneranalyse findet für h ungefähr einen Wert von 2,7 mm. Während G_{III} unabhängig von h ist und G_I monoton mit h abfällt, zeigt der Faktor G_{II} ein Minimum.

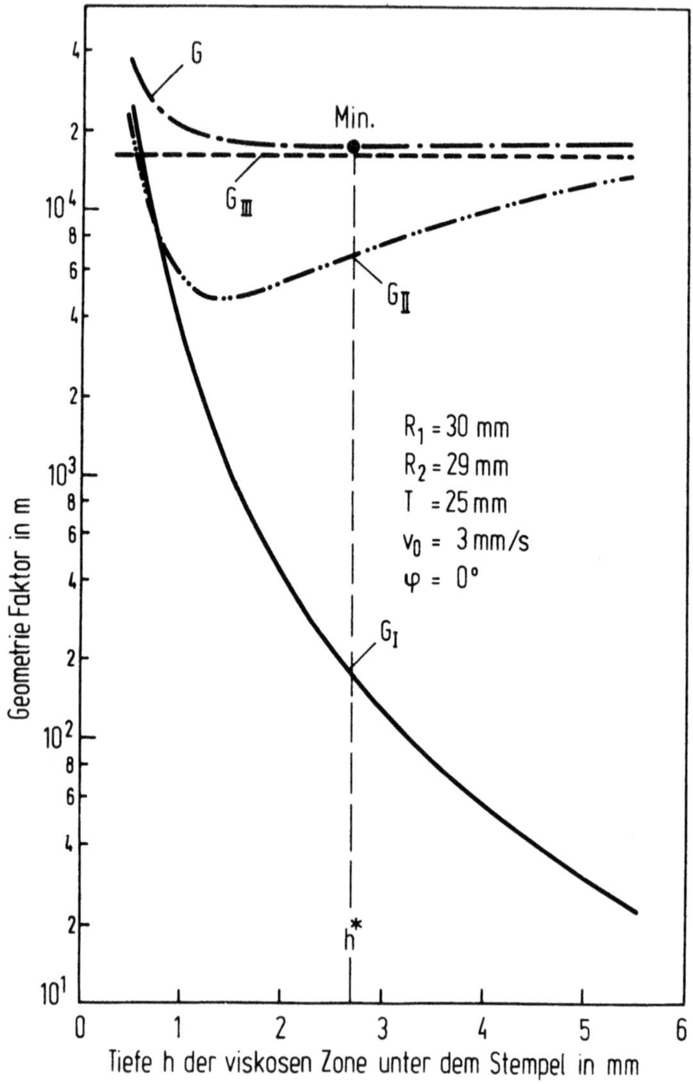

Bild 6: Minimale Tiefe der viskosen Zone (Obere-Schranke-Lösung)

Offensichtlich ist die Bestimmung von G_{III} allein ausreichend, die Gesamtleistung genügend genau zu bestimmen. Bild 7 zeigt für Geometrieverhältnisse $R_2/R_1 > 0,95$, wie sie in den Versuchen vorkommen, daß der Fehler immerhin kleiner als 7 % bleibt. Deshalb soll das rheologische Verhalten der Seife in einem neuen Ansatz besser berücksichtigt werden, wobei jedoch nur Region III betrachtet wird.

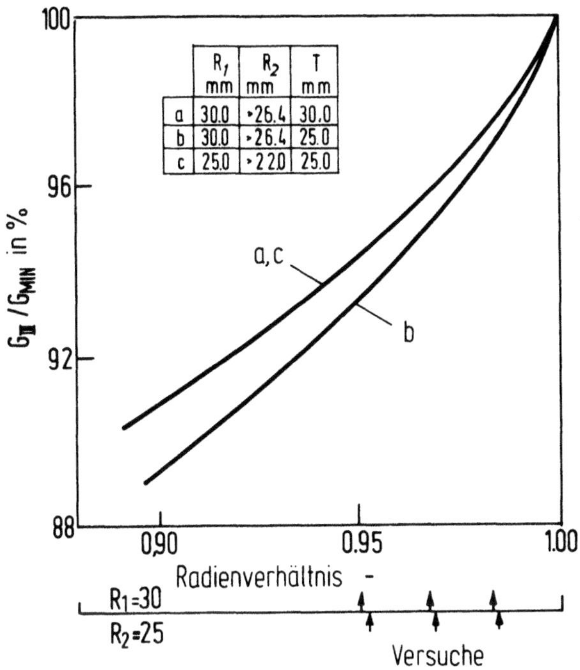

Bild 7: Einfluß des Radienverhältnisses R_2/R_1 auf den Geometrie-Faktor G_{III}/G_{Min}

2.3.2.2 Analyse mit der Annahme von Bingham-Verhalten der Seife

Wie in Bild 8 angedeutet, soll für diese Überlegungen nur das Fließen im Spalt analysiert werden. Unter der Bedingung einer Newtonschen Flüssigkeit würde die Lösung von den Navier- Stokes- Gleichungen abgeleitet werden. Wird das Medium jedoch als Bingham-Körper idealisiert, so führen die Gleichgewichts- und Kontinuitätsbedingung, die Randbedingungen und das Fließgesetz nach

Bingham zu einem linearen Ausdruck der Kraft F als Funktion der Stempelgeschwindigkeit v_o

$$F = a_1 + a_2 \, v_o \tag{27}$$

mit

$$a_1 = \bar{a} \, \tau_o \left[\frac{2(R_1^3+R_2^3)-2(a^3+b^3)}{R_1^2+R_2^2} - \frac{3(R_1^2-R_2^2)[(R_1+R_2)-(a+b)]}{(R_1^2+R_2^2)\ln \frac{R_1}{R_2}} \right] \tag{28}$$

$$a_2 = 3 \, \bar{a} \, \eta_B \quad , \tag{29}$$

wobei für

$$\bar{a} = \frac{2}{3} \frac{\pi h}{\ln \frac{R_1}{R_2} - \frac{R_1^2-R_2^2}{R_1^2+R_2^2}} \tag{30}$$

gilt. Die Werte a und b hängen von der τ-Verteilung, Bild 8, ab.

Durch Messung der Kraft F bei verschiedenen Geschwindigkeiten v_0 kann die lineare Funktion von Gl. (27) gefunden werden. Mit den Koeffizienten a_1 und a_2 können die Bingham-Viskosität η_B aus Gleichung (29) und die Fließspannung τ_0 aus Gleichung (28) bestimmt werden.

3. Versuche

3.1 Rheometer

Wie schon im theoretischen Teil dieses Berichtes erwähnt, werden zwei Arten von Rheometern eingesetzt, nämlich ein "Extrusiograph"+ und eine Rückwärtsfließpreßapparatur.

Der Extrusiograph, Bild 9, wird von vier voneinander unabhängigen Wicklungen temperaturgeregelt. Die Durchmesser der Kapillaren betragen 2, 4 und 6 mm. Die Drehzahl der Schraube liegt zwischen 3 und 60 min^{-1}. Die Meßwerte wie Temperatur, Druckabfall und Durchfluß werden erfaßt und gespeichert und können auf einen graphischen Bildschirm zur Bearbeitung abgerufen werden.

+Firma Brabender oHG, Duisburg, Deutschland

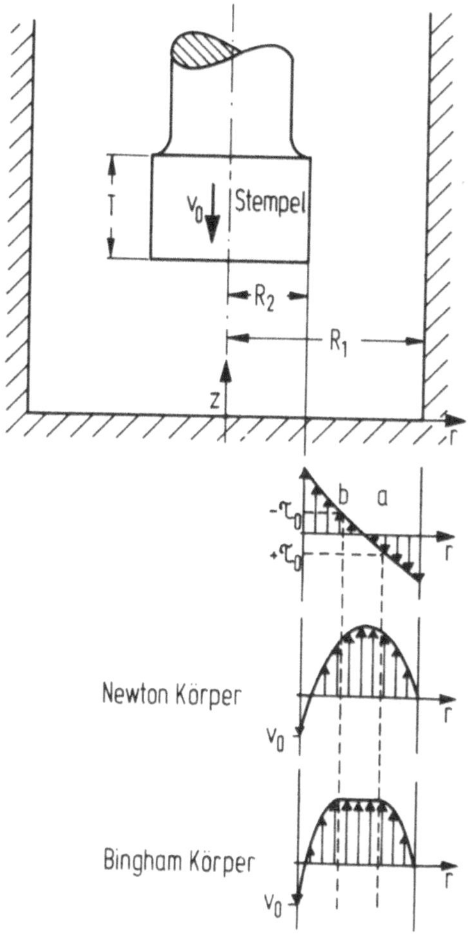

Bild 8: Modell für den Bingham-Pfropfenfluß im Spalt

Die Rückwärtsfließpreßvorrichtung ist in Bild 10 schematisch dargestellt. Über ein Führungssystem ist die Selbstzentrierung so gut wie möglich garantiert. Die Meßwerte Temperatur, Kraft und Weg bzw. Geschwindigkeit werden ebenfalls erfaßt und über einen X-Y-Schreiber ausgewertet.

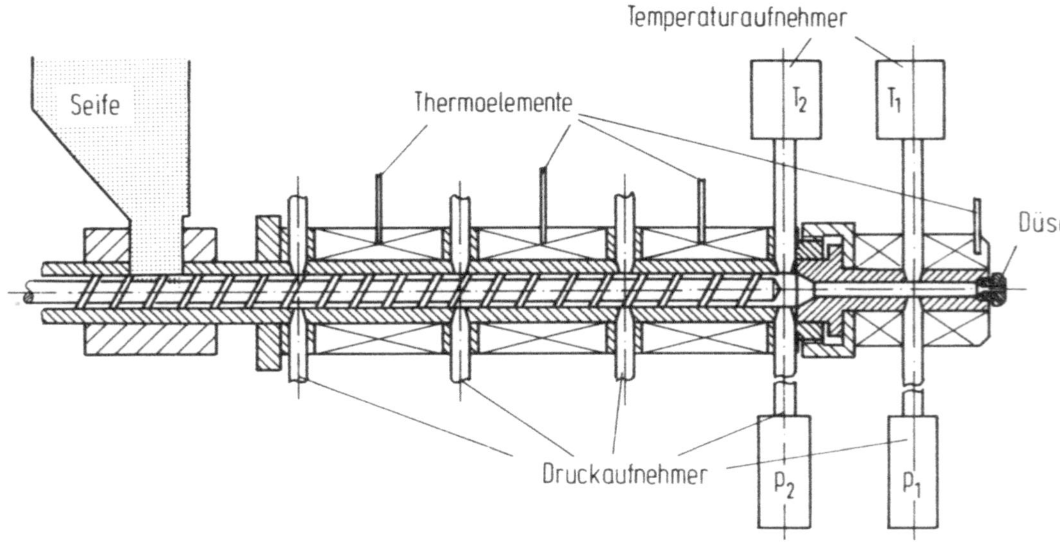

Bild 9: Extrusiograph (schematisch)

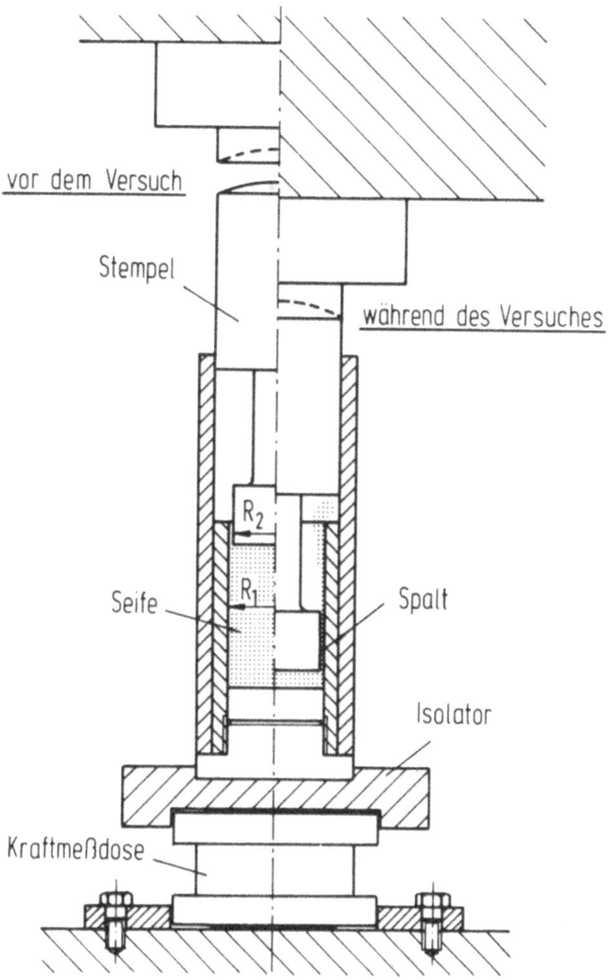

Bild 10: Rückwärtsfließpreßvorrichtung (schematisch)

3.2 Versuchsseifen

Drei typische Trockenziehseifen, wie sie die Industrie benutzt, wurden untersucht. An Hand von Tafel 1, in der die chemische Zusammensetzung der Seife wiedergegeben ist, erkennt man, daß es sich um ein Na-Stearat, ein Na-Ca-Stearat und um ein Ca-Stearat handelt. Bei unterschiedlichem Volumenanteil wurden die Feststoffe Kalk (CaO), Borax ($CaCO_3$) oder Soda ($NaCO_3$) als Schmiermittelträger benutzt.

Name der Seife	Seife		Feststoff		
	Na-Stearat %	Ca-Stearat %	Na_2CO_3 %	$CaCO_3$ %	CaO %
Na-Stearat	73	-	27	-	-
Na-Ca-Stearat	22	30	-	30	18
Ca-Stearat	-	41	-	18	41

Tafel 1: Chemische Zusammensetzung der Seifen

3.3 Ergebnisse

3.3.1 Extrusiograph

Die für die Übertragung auf das Drahtziehen wichtigsten Fließkurven wurden mit der 2 mm Düse gewonnen, da Düsen mit 4 und 6 mm Durchmesser ein zu geringes Schergeschwindigkeitsgefälle erlauben. Bild 11 zeigt Fließkurven für Na-Stearat, die bei Temperaturen zwischen 80 und 140 ^{o}C aufgenommen wurden.

Das Schergeschwindigkeitsgefälle liegt bei maximal $750 s^{-1}$. Zu jeder Temperatur wurden zwei mathematische Ansätze versucht. Dabei beschreiben die ausgezogenen Linien die Seife als Bingham-Körper, während das Potenz-Gesetz des Ostwald-Körpers durch die gestrichelten Linien wiedergegeben ist. Dem Bild kann entnommen werden, daß das Ostwald-Gesetz die Meßwerte zunächst gut beschreibt, daß aber mit steigendem Schergeschwindigkeitsgefälle der Bingham-Ansatz zu einer besseren Übereinstimmung zwischen Meßwerten und Theorie führt. Dies wird noch deutlich, wenn man berücksichtigt, daß bei hohen Schergeschwindigkeitsgefälle die tatsächliche Temperatur leicht ansteigt, so daß die dargestellte Nominaltemperatur nicht immer eingehalten werden konnte. Darüber hinaus ist zu sehen, daß bei hohen Temperaturen wie auch bei großen Schergeschwindigkeitsgefällen der Newton-Ansatz einer Proportionalität zwischen Schubspannung und Schergeschwindigkeitsgefälle eine gute Näherung darstellt.

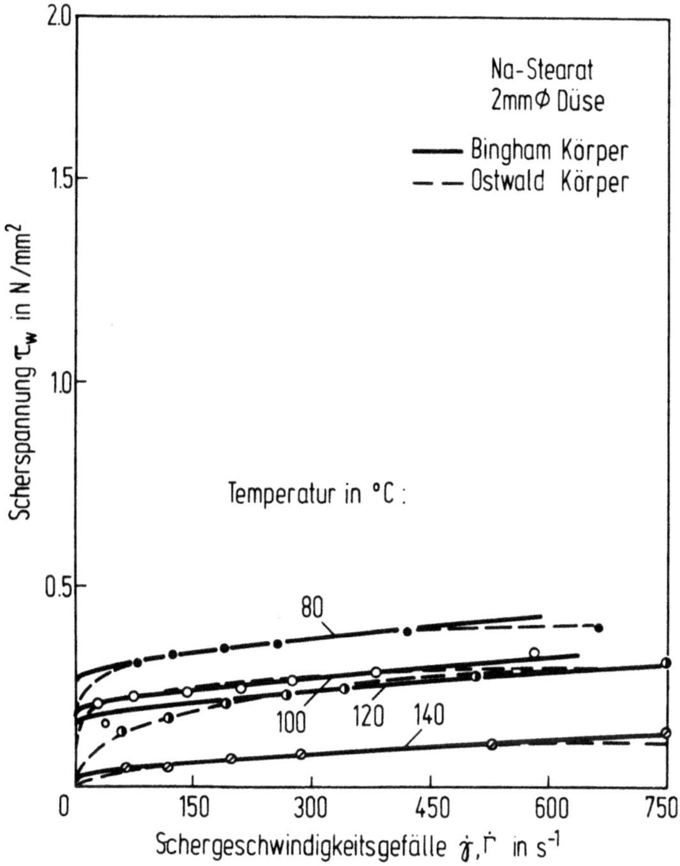

Bild 11: Fließkurven (Extrusiograph)

Während die Scherspannung an der Wand beim Na-Stearat den Wert 0,4 N/mm² nicht übersteigt, wird er, wie Bild 12 zeigt, beim Ca-Stearat bis zu 1,4 N/mm² groß.

Die maximale Temperatur beträgt jedoch nur 110 °C, da die Fließkurve keinen einheitlichen Verlauf mehr hat und nicht mehr durch einen Ostwald- bzw. Bingham-Ansatz beschrieben werden kann. Alle anderen Meßwerte werden von beiden Theorien gut beschrieben. Ein Vergleich zur Na-Stearat-Seife läßt erkennen, daß die Bingham-Fließspannung sich beim Ca-Stearat in der gleichen Größenordnung befindet. Der Anstieg der Fließkurven ist jedoch ungleich größer, was bedeutet, daß die Bingham-Viskosität Werte von über 3000 Pas erreicht, während sie beim Na-Stearat bei ungefähr 200 Pas liegt.

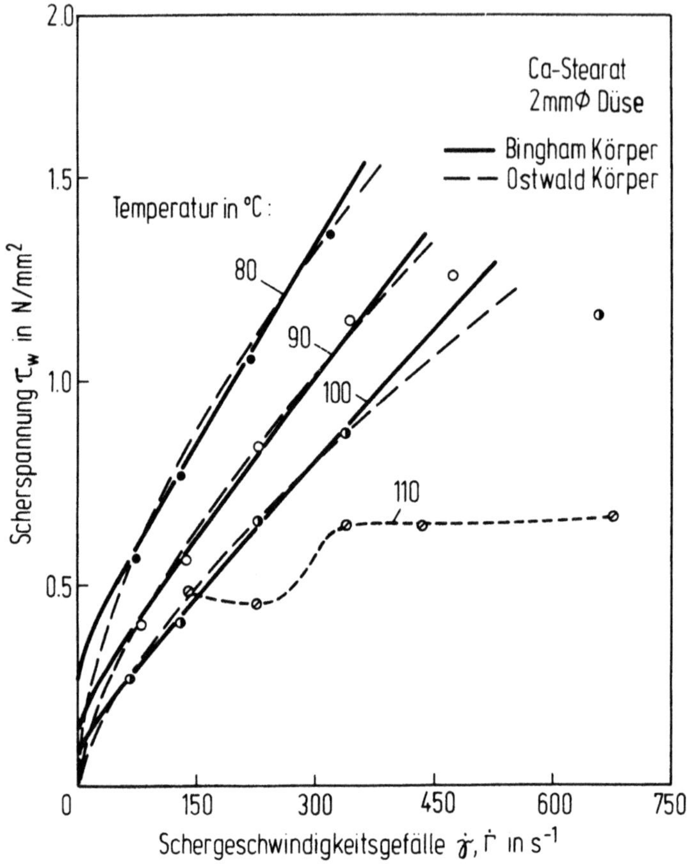

Bild 12: Fließkurven (Extrusiograph)

In Bild 13 sind die Fließkurven für das Na-Ca-Stearat dargestellt. Für die Bingham-Fließspannung wurden wiederum Werte zwischen 0,1 und 0,3 N/mm² ermittelt. Die Bingham-Viskosität liegt in der Größenordnung, wie sie beim Na-Stearat gefunden wurde. Sie steigt jedoch bei niedrigen Temperaturen auf sehr hohe Werte an.

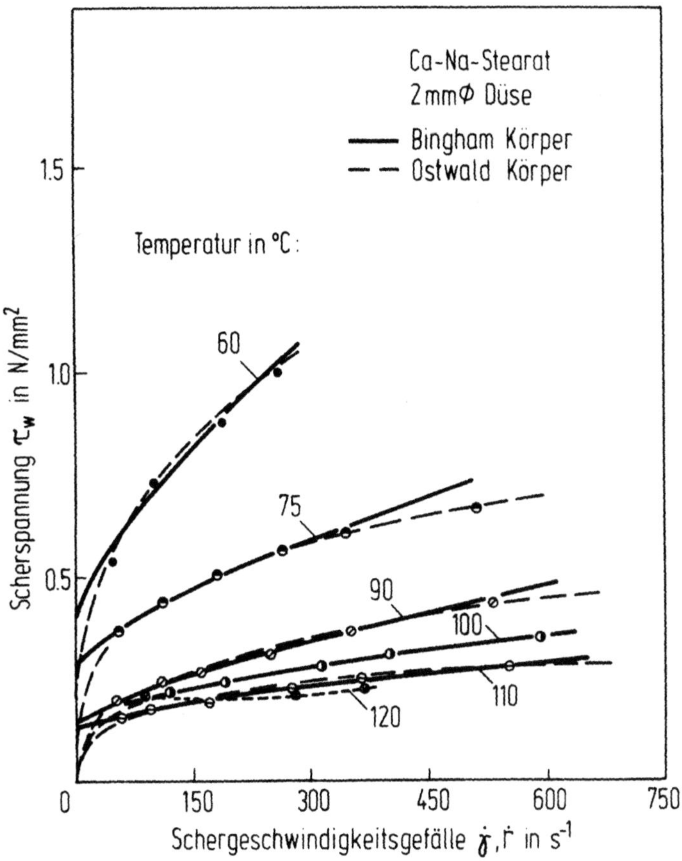

Bild 13: Fließkurven (Extrusiograph)

3.3.2 Rückwärtsfließpreßapparatur

Die Annahme der Bingham-Fließregel führt beim Rückwärtsfließpressen zu einer Fließkurve, die, wie bereits beschrieben, als lineare Funktion der gemessenen Stempelkraft F über der Stempelgeschwindigkeit v_0 gezeichnet werden kann. Als Beispiel sind in Bild 14 für Na-Stearat bei 100°C die Fließkurven dargestellt. Während der Radius des Behälters bei allen Versuchen konstant R_1 = 30 mm ist, wird der Stempelradius und damit der Parameter R_2/R_1 bei drei verschiedenen Kurven verändert.

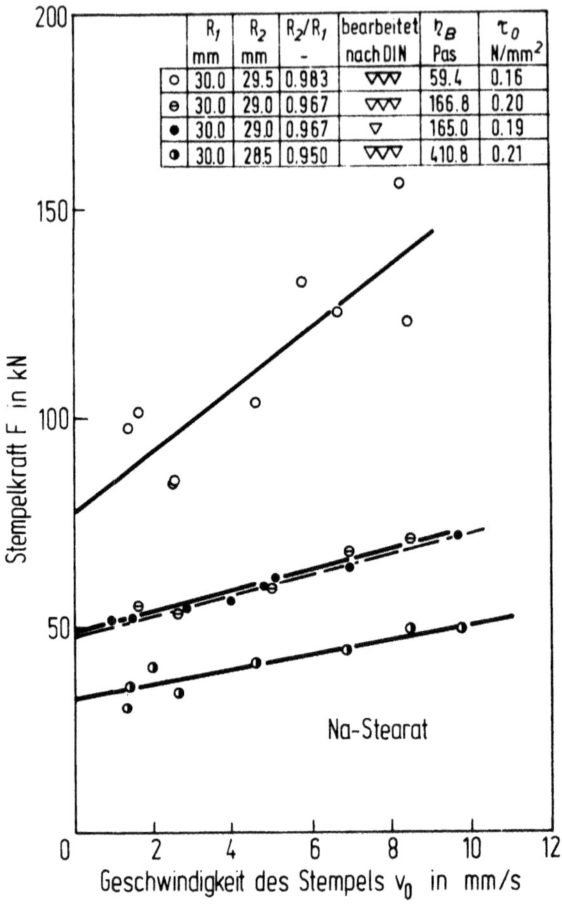

Bild 14: Fließkurven (Rückwärtsfließpressen)

Entsprechend der vorgestellten Theorie führt der Achsabschnitt der Ordinate zur Bingham-Fließspannung τ_0, für die Werte zwischen 0,16 und 0,21 N/mm² ausgerechnet werden können, was offensichtlich in guter Übereinstimmung mit den Ergebnissen des Extrusiographen steht. Für die Bingham-Viskosität, die aus der Steigung der Kurven berechnet wird, wurden dagegen sehr unterschiedliche Werte gefunden. Bei einem schmalen Spalt von 0,5 mm beträgt die Viskosität nur ca. 60 Pas, während sie bei einem Spalt von 1,5 mm auf über 400 Pas ansteigt.

Im Vergleich zu den Ergebnissen von Bild 14 wurden in Bild 15 Meßpunkte eingetragen, die mit einem Behälter gewonnen wurden, dessen Radius 25 mm betrug.

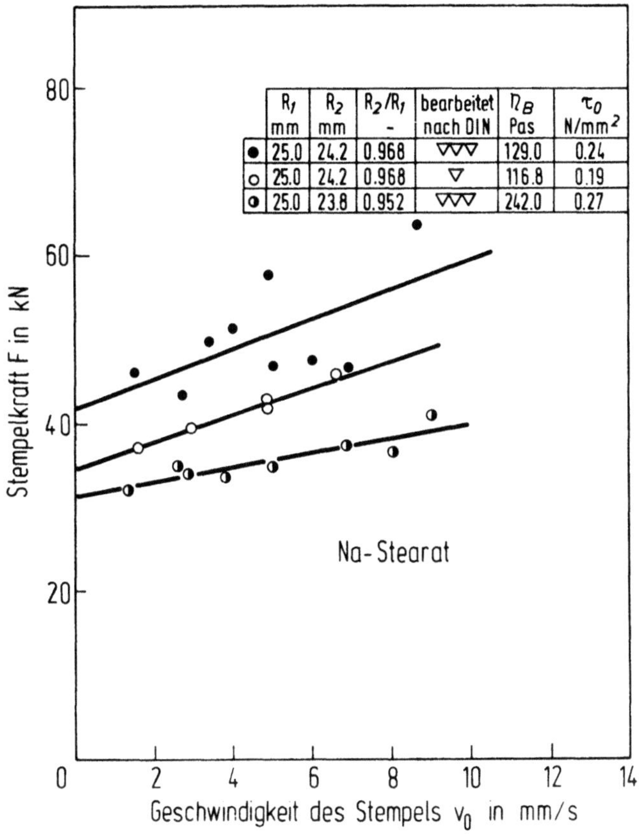

Bild 15: Fließkurven (Rückwärtsfließpressen)

Unterschiedliche R_2/R_1-Verhältnisse führen wiederum zu großen Unterschieden der Bingham-Viskosität. Es scheint, daß die Zentrierung des Stempels von ausschlaggebender Bedeutung ist, da eine unterschiedliche Bearbeitungsgüte der Werkzeuge zur gleichen Viskosität führt, was auch in Bild 14 zu sehen ist. Daneben spielt die Temperaturführung beim Rückwärtsfließpressen eine große Rolle. In einer besonderen Versuchsreihe konnte gezeigt werden, daß abhängig von der Stempelgeschwindigkeit und dem R_2/R_1-Verhältnis die Versuchstemperatur um bis zu 10 ^{o}C von der Nominaltemperatur abweichen kann.

3.3.3 Bingham-Kennwerte

Die Kennwerte für das Fließgesetz nach Bingham, die Fließspannung τ_0 und die Viskosität η_B, sind in Bild 16 gegen die Temperatur aufgetragen. Dabei wurde jeder Punkt aus den Fließkurven von Bild 11 bis Bild 15 errechnet.

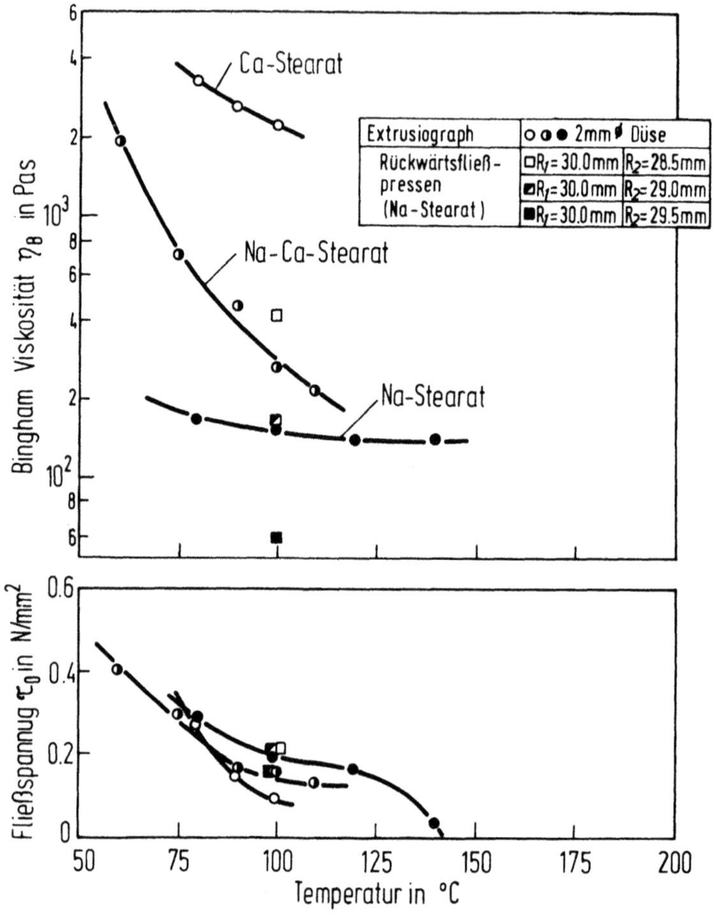

Bild 16: Kennwerte der Seifen bei Bingham Fluß

Für alle drei Seifenarten liegt die Bingham-Fließspannung im Bereich zwischen 0,1 und 0,3 N/mm^2. Sie fällt mit steigender Temperatur genau wie die Bingham-Viskosität im oberen Teilbild, wobei der Abfall bei Ca- und bei Na-Ca-Stearat schon erheblich ist.

Ein Vergleich zwischen den Versuchsmethoden Rückwärtsfließpressen und Extrudieren kann für das Na-Stearat bei 100 ^{0}C durchgeführt werden. Obwohl die Ergebnisse des Rückwärtsfließpressens erheblich streuen, kann doch gesagt werden, daß sie in der Größenordnung mit denen des Extrudierens übereinstimmen.

4. Schlußbemerkung

Zur Ermittlung der rheologischen Kennzahlen von Trockenziehseife kann die Fließkurve, d.h. die Abhängigkeit der Scherspannung vom Schergeschwindigkeitsgefälle, mit Rheometern wie dem Extrusiographen oder einer Rückwärtsfließpreßapparatur ermittelt werden. Drei rheologische Modellkörper, nämlich der Newton-, der Ostwald- und der Bingham-Körper, sind in der Lage, die Seife in Abhängigkeit von der Temperatur genügend genau zu beschreiben. Im allgemeinen zeigt die Seife bei kleinem Geschwindigkeitsgefälle ein Ostwald-Verhalten, so daß dort das Potenz-Gesetz gültig ist. Bei höherem Schergeschwindigkeitsgefälle stellt der lineare Bingham-Ansatz eine gute Näherung dar. Bei höheren Temperaturen sowie bei sehr hohem Schergeschwindigkeitsgefälle erhält das Newton-Gesetz seine Berechtigung, da die kleine Bingham-Fließspannung dann vernachlässigt werden kann.

In Übereinstimmung mit der Erfahrung der Drahtindustrie unterscheidet sich das Ca-Stearat sehr stark von Na-Stearat. Während die Viskosität des Ca-Stearates bei ca. 2000 Pas liegt, beträgt die Viskosität beim Na-Stearat nur etwa 200 Pas. Die Bingham-Fließspannung liegt bei beiden Seifen bei etwa 0,2 N/mm^2. Dieser Wert ist zwar sehr klein, aber offensichtlich groß genug, um den beim hydrodynamischen Ziehen beobachteten Propf-FLuß zu verursachen. Die Viskosität des untersuchten Na-Ca-Stearates liegt zwischen den Werten, die für die reinen Stearate gefunden wurden.

Zahlreiche Experimente zeigten, daß die Handhabung des Extrusiographen bei weitem einfacher und schneller ist als die der Rückwärtsfließpreßapparatur. Außerdem sind die Ergebnisse besser reproduzierbar.

5. Schrifttum

1) PAWELSKI, O.: Erhöhung des Umformvermögens beim Ziehen. In: Mitt. aus dem Institut für Bildsame Formgebung der TH Aachen 1976

2) TEGEL, L.: Untersuchung der Anwendbarkeit der Doppelziehsteine zur Verbesserung der Schmierung beim Ziehen von Stahldraht. Dr.-Ing. Diss. Aachen 1978

3) REINER, M.: Rheologie in elementarer Darstellung. München: Carl Hanser Verlag 1968

4) DARBY, R.: Viscoelastic Fluids. New York, Basel: Marcel Dekker 1976

5) RABINOWITSCH, B.: Über die Viskosität und Elastizität von Solen. Z. physikal. Chem. Abt. A. 145 (1929) 1, S. 1-26

Dem Bundesminister für Forschung und Technologie sei für die Förderung dieses Projektes gedankt.

Untersuchung der Möglichkeiten für eine technologische Schmierstoffprüfung für Verfahren der Kaltmassivumformung

von

Prof. Dr.-Ing. K. Lange

Dipl.-Ing. T. Gräbener

Institut für Umformtechnik

Universität Stuttgart

Förderkennzeichen: 01 ZT 408S

Laufzeit: 1.9.1978 - 31.8.1980

Zusammenfassung

Im Rahmen des bearbeiteten Forschungsvorhabens "Untersuchung der Möglichkeiten für eine technologische Schmierstoffprüfung für Verfahren der Kaltmassivumformung" wurden zunächst die höchsten Werte der relativen mittleren Flächenpressung, der Relativgeschwindigkeit und der Oberflächenvergrößerung von sechs Kaltmassivumformverfahren zu einem Anforderungsprofil an Schmierstoffprüfverfahren zusammengefaßt.
Aus den bisher bekannten Prüfverfahren für Schmierstoffe der Kaltmassivumformung wurden zwei Stauch- und vier Ziehverfahren für die nähere Untersuchung der Belastungsverhältnisse in der Umformzone ausgewählt. Der entsprechenden Literatur wurden die höchsten Werte der drei Beanspruchungsgrößen entnommen. Das damit erstellte Leistungsprofil wurde mit dem Anforderungsprofil der Kaltmassivumformverfahren verglichen.
Aus dieser Gegenüberstellung zeigte sich, daß die hohen Belastungen, die bei manchen Umformverfahren (z.B. NRFP) auftreten, von keinem Reibzahlermittlungsverfahren simuliert werden können. Das Ergebnis der Schmierstoffprüfung kann daher nur mit Einschränkungen auf die Praxis übertragen werden.
Mit den Verfahren Streifenziehen und Ringstauchen wurden Schmierstoffprüfversuche durchgeführt und die Abhängigkeit der Flächenpressung von der Reibzahl und der Geschwindigkeitseinfluß ermittelt. Während beim Streifenziehen die Flächenpressung mit zunehmender Reibzahl abnimmt, erfolgt beim Ringstauchen ein Anstieg, wobei die Absolutwerte höher liegen. Die Gründe für dieses abweichende Verhalten werden in den unterschiedlichen Reibverhältnissen, die bei Mischreibung auftreten können, vermutet. Ausgehend von dieser Betrachtung empfiehlt sich das Ringstauchen als (qualitatives) Schmierstoffprüfverfahren für das Kaltfließpressen, während das Streifenziehen zur Prüfung von Schmierstoffen für das Abstreckgleitziehen oder Verjüngen geeignet scheint.
Eine Variation der Ziehgeschwindigkeit beim Streifenziehen zeigte, daß eine mittlere Geschwindigkeit bei weitem die niedrigsten Reibzahlen liefert. Dies wird auf ein optimales Zusammenwirken reibungsvermindernder Einflüsse wie Temperatur (Viskosität), Schmierstofftransport u.a. zurückgeführt.

Gliederung

0 Einleitung

Im Gegensatz zur Lagerreibung, bei der sich beide Reibpartner im elastischen Zustand befinden, wird die Reibung bei den Verfahren der Umformtechnik dadurch gekennzeichnet, daß einer der Reibpartner - das Werkstück - plastisch verformt wird, während der andere - das Werkzeug - elastisch bleibt.

Die Auswirkungen der Reibung beim Umformen sind sehr vielfältig. Die dem Werkstück zugeführte Energie wird z.T. für die Formänderung verbraucht, während ein beträchtlicher Anteil zur Überwindung der Reibung benötigt wird. Je größer die Reibzonen,also die Gebiete, in denen das Werkstück unter hohem Druck am Werkzeug gleitet, im Verhältnis zum Werkstückvolumen sind, desto größer sind die Reibverluste. Um den Einsatz von Maschinen erhöhter Preßkraft zu vermeiden, sollten diese Reibungsverluste minimiert werden.

Ebenfalls von großer Bedeutung ist die meist unerwünschte Temperaturentwicklung in der Wirkfuge zwischen Werkstück und Werkzeug. Abgesehen von möglichen Veränderungen im Kristallgefüge zersetzen sich einige Schmierstoffe unter Wärmeeinwirkung. Andere Schmierstoffe wie z.B. Seifen werden erst bei Druck - und Temperatureinwirkung schmierwirksam.

Tritt an den Reibflächen direkter metallischer Kontakt der Werkstück - bzw. Werkzeugoberflächen auf, ist die Gefahr von Werkstoffübertrag des weicheren auf den härteren Werkstoff gegeben. In den meisten Fällen ist dann Nacharbeit oder ein neues Werkzeug erforderlich. Es sollte deshalb zwischen diesen Aufwendungen und den Kosten für höherwertige,teurere Schmierstoffe ein Optimum gefunden werden.

Der Stofffluß bei der Umformung ist in starkem Maße von den herrschenden Reibbedingungen abhängig. So bilden sich die Geometrien zweier Stauchkörper bei unterschiedlicher Reibung zwischen Stauchbahn und Stirnflächen vollkommen verschieden aus. Es treten also bei nahezu allen Verfahren der Kaltmassivumformung verstärkt Probleme hinsichtlich der Oberflächenausbildung, der Vermeidung von Kaltverschweißungen, erhöhter Preßkraft usw. auf, denen man mit der Auswahl eines geeigneten Schmierstoffes zu begegnen versucht.

Zur Schmierstoffprüfung in Modellversuchen stehen mehrere Verfahren zur Verfügung, die die tatsächlichen Umformbedingungen wie Normalspannung, Relativgeschwindigkeit, Oberflächenvergrösserung, Temperaturentwicklung usw. unterschiedlich genau wiedergeben. Eine theoretische und experimentelle Untersuchung an Umform- und Schmierstoffprüfverfahren soll die vorhandenen Unterschiede aufzeigen und mögliche Lösungswege zur Verbesserung und Erweiterung der Aussagefähigkeit von Reibzahlermittlungsverfahren vorschlagen.

1 Aufgabenstellung

Reibung und Verschleiß werden bei den Verfahren der Kaltmassivumformung hauptsächlich durch die drei Parameter Flächenpressung in der Reibfuge, Relativgeschwindigkeit von Werkstück zu Werkzeug und Oberflächenvergrößerung des Rohteils zum fertigen Preßteil bestimmt.
Im Sinne einer repräsentativen Schmierstoffprüfung für die Umformtechnik erhebt sich daher die Forderung einer möglichst genauen Nachbildung dieser Reibverhältnisse bei den Schmierstoffprüfverfahren.
Das Ziel dieser Arbeit war demnach eine Gegenüberstellung der Reibbedingungen bei ausgewählten Kaltmassivumformverfahren und bei Schmierstoffprüfverfahren. Aus diesem Vergleich von Anforderung der Umformtechnik und Leistung der Schmierstoffprüfung sollte, soweit möglich, eine Zuordnung hergeleitet werden, die eine gezielte Schmierstoffprüfung ermöglicht.

1.1 Stand der Erkenntnisse

In den letzten Jahren hat die Bedeutung der Schmierstoffprüfung nicht nur für die Gleit- und Wälzlagertechnik sowie im Motoren- und Getriebebau sondern auch für die Verfahren der Umformtechnik stark zugenommen.

Der Anwender von Schmierstoffen kann eine Auswahl der geeigneten Produkte durch Schmierstoffprüfung im Modellversuch oder am Umformteil im Produktionsprozeß selbst durchführen. Während die zweite Möglichkeit meist mit hohen Kosten und Risiken für

Maschine und Werkzeug verbunden ist und meist nur in Pilotserien nach einer Vorauswahl angewandt wird, gewährleisten möglichst universelle Modellversuche mindestens qualitative Aussagen zur Eignung verschiedener Schmierstoffe.
Die Reibbedingungen bei den Prüfverfahren sind sehr unterschiedlich und geben meist nicht die tatsächlichen Verhältnisse bei Kaltmassivumformverfahren wieder. Die so ermittelten Reibzahlen sind daher nicht unbedingt auf die Praxis der Umformtechnik übertragbar.

Zunächst gibt es eine Reihe von Stauchverfahren, die z.T. mit kegeligen Werkzeugflächen [1], gleichzeitiger Parallelverschiebung zur Stauchbahn [2,3] oder, wie beim Ringstauchen [4], mit charakteristischer Probenform arbeiten. Weiterhin spielt die große Gruppe der Streifen- und Stabziehverfahren [5,6,7,8,9] eine bedeutende Rolle. Zusätzlich besteht die Möglichkeit der Reibzahlermittlung durch Messung der Umformkraft und Rückrechnung aus bekannten Kraftformeln. Dies setzt aber die genaue Kenntnis der Fließkurve voraus und stellt im übrigen streng genommen nur die Berechnung eines Korrekturfaktors dar. Problematisch bleibt bei allen Modellverfahren die Simulation der Reibverhältnisse in der Umformzone beim Fließpressen, Stauchen, Abstrecken usw. und damit auch die Übertragbarkeit der gewonnenen Ergebnisse.

1.2 Zielsetzung der Arbeit

1.2.1 Theoretische Untersuchungen

Ziel dieser Arbeit war es, zunächst eine Beanspruchungsanalyse im tribologischen System bei sechs verschiedenen Kaltmassivumformverfahren und sechs Schmierstoffprüfverfahren vorzunehmen. Die wichtigsten zu ermittelnden Größen waren dabei die auftretende Flächenpressung, die Relativgeschwindigkeit und die Oberflächenvergrößerung des umgeformten Teiles.
Bei den sechs Kaltmassivumformverfahren handelt es sich um Verjüngen, Abstreckgleitziehen, Stauchen, Voll-Vorwärts-Fließpressen (VVFP), Hohl-Vorwärts-Fließpressen (HVFP) und Napf-Rückwärts-Fließpressen (NRFP). Die Berechnung der drei o.a.

Parameter erfolgte nach der elementaren Plastizitätstheorie, wobei für die Verfahrensgrenzen (φ, α etc.) jeweils maximale, aber doch praxisnahe Werte angenommen wurden [10].
Aus einer Zusammenstellung der berechneten Maximalwerte entsteht ein sogenanntes Anforderungsprofil der Umformverfahren.
Für die sechs Schmierstoffprüfverfahren (Ringstauchen, Stauchen mit Relativbewegung, 4 Streifenziehverfahren) wurden die entsprechenden Maximalwerte der Beanspruchung in der Reibzone der Literatur entnommen und ebenfalls zusammengefaßt zu Leistungsprofilen. Aus einem Vergleich von Anforderungs- und Leistungsprofil soll hervorgehen, welches Schmierstoffprüfverfahren für welches Umformverfahren zufriedenstellend oder gar nicht geeignet ist bzw. bei welchem der drei Parameter übereinstimmende Werte vorliegen.

1.2.2 Experimentelle Untersuchungen

Ausgehend von den vorgenannten theoretischen Betrachtungen werden zwei Schmierstoffprüfverfahren für experimentelle Untersuchungen ausgewählt. Es handelt sich dabei einerseits um das Streifenziehen nach Schlosser [8] und andererseits um das Ringstauchen nach Burgdorf [4].
Anhand dieser Verfahren sollten im Versuch die erreichbaren Maximalwerte der Flächenpressung sowie deren Abhängigkeit von der Reibzahl ermittelt werden. Weiterhin war beim Streifenziehen vorgesehen, durch mehrmalige Variation der Ziehgeschwindigkeit deren Einfluß auf die Reibzahl zu ermitteln.
Werden bei den Versuchen Werkstoffpaarungen eingesetzt, die eine ausgeprägte Adhäsionsverschleißneigung zueinander haben, in unserem Fall die Kombination Werkstückwerkstoff 1.4301 - Werkzeugwerkstoff 1.2601, so treten auf der Werkzeugoberfläche sehr leicht Anfresser auf. Zur Einschränkung bzw. Vermeidung dieser Verschleißerscheinungen wurde der Einsatz von Keramikhartstoffen als Werkzeugwerkstoff ins Auge gefaßt. Das Reibungs- und Verschleißverhalten dieser Alternativ-Werkstoffe für die Umformtechnik sollte so untersucht werden.
Anschließend an die Ergebnisauswertung ist ein Vergleich der mit beiden Verfahren ermittelten Reibzahlen und Normalspannungs-

abhängigkeiten vorgesehen. Der Einfluß der Oberflächenvergrößerung wurde nicht untersucht, da mit beiden Verfahren keine besonders hohen Werte erzielt wurden und sich dieser Parameter nicht losgelöst von erhöhter Flächenpressung, insbesondere beim Ringstauchen, betrachten läßt.

2 Beanspruchungsanalyse im tribologischen System

Nach DIN 50320 [11] läßt sich ein tribologisches System wie im Bild 1 dargestellt beschreiben. Die Struktur des Tribosystems

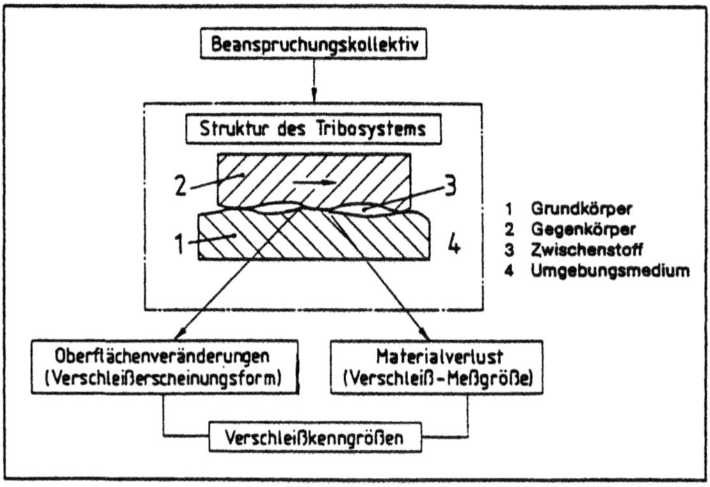

Bild 1: Darstellung eines Tribosystems [11].

wird durch Grund- und Gegenkörper, also Werkzeug und Werkstück, durch den Zwischenstoff, in unserem Falle den Schmierstoff und das Umgebungsmedium, i.a. die Atmosphäre, gekennzeichnet. Das Beanspruchungskollektiv beinhaltet die von außen einwirkenden Belastungen wie z.B. Spannungen, Temperatur, Gleitgeschwindigkeit, Oberflächenvergrößerung u.a.

Die in dieser Arbeit an den Umform- und Schmierstoffprüfverfahren vorgenommene Beanspruchungsanalyse beschränkt sich auf die Ermittlung der Maximalwerte von Flächenpressung, Relativgeschwindigkeit und Oberflächenvergrößerung.

2.1 Maximalbelastung bei Kaltmassivumformverfahren

Wie bisher dargelegt wurde, ist die Reibung bei den Verfahren der Umformtechnik von sehr vielen Faktoren abhängig.
Am Beispiel von 6 Kaltmassivumformverfahren werden die erreichbaren Maximalwerte der Normalspannung, der Relativgeschwindigkeit und der Oberflächenvergrößerung untersucht und zu einem Anforderungsprofil an die Schmierstoffe zusammengefaßt. Bei der Berechnung der Werte wurde von der elementaren Plastizitätstheorie [10] ausgegangen, da diese für die gewünschten Anhaltspunkte ausreichend genau erscheint. Zur besseren Vergleichbarkeit wurden die 3 Größen auf k_{f0}, v_{Wz} und A_0 bezogen. Als Modellwerkstoff diente der Einsatzstahl Ck 15 mit folgenden Kennwerten:

$$R_m = 420 \text{ N/mm}^2$$

$$k_{f0} = 323 \text{ N/mm}^2$$

$$n = 0.2233$$

Die entsprechende Fließkurve wurde im Stauchversuch aufgenommen und wird durch

$$k_f(\varphi) = 724 \cdot \varphi^{0.2233} \text{N/mm}^2$$

beschrieben.

Das Verjüngen ist eine Sonderform des Voll-Vorwärts-Fließpressens, wobei hier keine Abstützung des Rohteils vor der Umformzone besteht. Es entfällt demnach der Wandreibkraftanteil. Bedingt durch die Verfahrensgrenzen Ausknicken und Aufstauchen kann bei vorverfestigtem Material ein maximaler Umformgrad von $\varphi = 0{,}4$, bei weichgeglühtem Ck 15 von $\varphi = 0{,}25$ erreicht werden. Beim optimalen Matrizenöffnungswinkel $2\alpha = 20°$ ergeben sich die in Tabelle 1 angegebenen Werte. Aus diesen nicht sehr hohen Belastungen ergeben sich keine besonderen Anforderungen an den Schmierstoff.

Beim Abstreckgleitziehen liegen an Dorn und Matrize zwei Reibzonen mit unterschiedlicher Belastung vor. Während an der Matrizenschulter möglichst niedrige Reibzahlen erwünscht sind, wird die Reibung am Dorn meist größer eingestellt, um über die dort herrschende Reibkraft die Hülsenwand beim Ziehen zu unterstützen und den Boden zu entlasten. Das Verfahren des Abstreckgleitziehens wird häufig in Mehrfachzügen z.T. ohne Zwischenschmierung angewandt, weswegen der Schmierstoff eine sehr gute Trennfähigkeit und Haftung auf der evtl. künstlich aufgerauhten Werkstückoberfläche besitzen muß.

	Verjüngen	Abstreckgleitziehen	Stauchen	HVFP	VVFP	NRFP
$\bar{p}_{max}/k_{f0}$	1,9	2,1	5,9	5,5	6,4	9
v_R/v_{Wz}	1,5	2,3	2,4	5	5,7	6,3
A_1/A_0	1,2	2,2	4,5	4	2	11

Tabelle 1: Maximalwerte bei sechs Umformverfahren.

Die in Tabelle 1 aufgeführten Maximalwerte der drei untersuchten Größen liegen nicht viel höher als beim Verjüngen, die größte Normalspannung tritt am Dorn auf.

Die Normalspannungsverteilung beim Stauchen ist aus Bild 2 ersichtlich [12]. Der Maximalwert tritt demnach in Probenmitte auf,

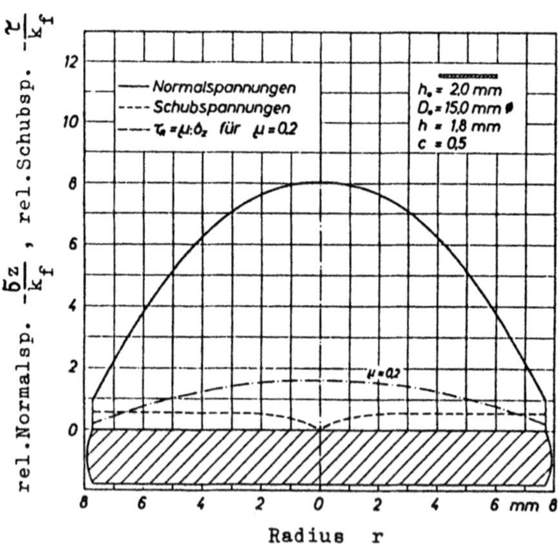

Bild 2: Spannungs- und Reibzahlverteilung beim Stauchen [12].

während am Rande gerade die Fließspannung erreicht wird. In Tabelle 1 sind die Höchstwerte der 3 Größen aufgeführt, wobei die Flächenpressung vom Stauchverhältnis l_0/d_0, μ und φ abhängt und die Relativgeschwindigkeit, ebenfalls von l_0/d_0 abhängig, am Probenrand am größten ist. Die Oberflächenvergrößerung gilt für den Fall des homogenen, reibungsfreien Stauchens. Der Schmierstoff beim Stauchen muß im Bereich hoher Spannungen hoch belastbar sein und gleichzeitig bei kleinen Drücken relativ hohe Geschwindigkeiten durch gute Scherstabilität ausgleichen.
Vergleichbar mit dem Abstreckgleitziehen liegen auch beim Hohl-Vorwärts-Fließpressen (HVFP) unterschiedliche Reibzonen mit dadurch erhöhten Reibkräften im plastischen Bereich vor. Zur Minimierung der Gesamtumformkraft muß deshalb möglichst ein optimaler Öffnungswinkel der Matrize eingehalten werden.

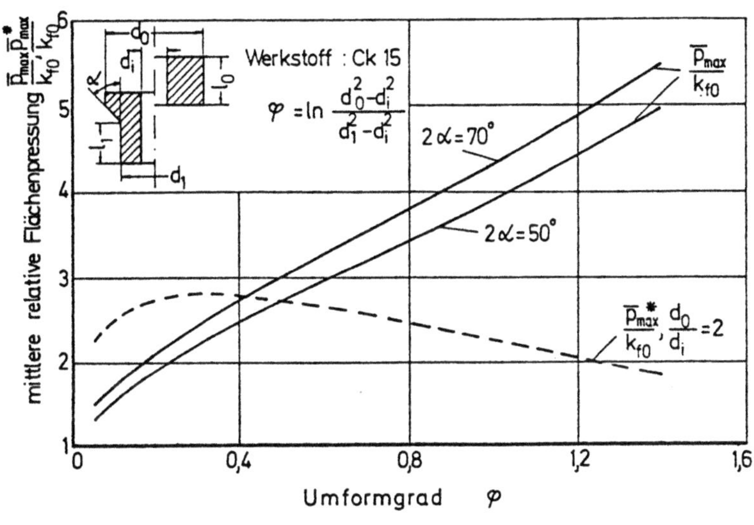

Bild 3: Flächenpressung $\bar{p}_{max}/k_{fo}$ auf die Schulter und $\bar{p}^*_{max}/k_{fo}$ auf den Dorn beim Hohl-Vorwärts-Fließpressen.

Wie Bild 3 zeigt, ist die Flächenpressung am Dorn sehr viel weniger vom Umformgrad abhängig als an der Schulter, wo sie einen Maximalwert von $\bar{p}_{max}/k_{f0} = 5.5$ erreichen kann. Dort ist auch die Relativgeschwindigkeit mit $v_R/v_{Wz} = 5$ bei $\varphi = 1.4$ sehr hoch. Die Oberflächenvergrößerung erreicht auf der Innenseite des Werkstücks den Wert 4 und liegt damit erheblich höher als auf der Außenseite.

Beim Voll-Vorwärts-Fließpressen (VVFP) liegen die Reibzonen an der Schulter und der Matrizenwand vor.

Die mittlere Flächenpressung an der Schulter - es können örtlich weit höhere Spannungsspitzen auftreten - ist stark vom Winkel 2α abhängig und erreicht, wie Tabelle 1 wiedergibt, bei einem üblichen Winkel von $2\alpha = 90°$ und $\varphi = 1{,}4$ $\bar{p}_{max}/k_{f0} = 6{,}4$. Gleichzeitig mit diesem hohen Druck treten große Relativgeschwindigkeiten und mittlere Oberflächenvergrößerungen auf.

Das Kaltmassivumformverfahren mit den schwierigsten Reibverhältnissen ist das Napf-Rückwärts-Fließpressen (NRFP).

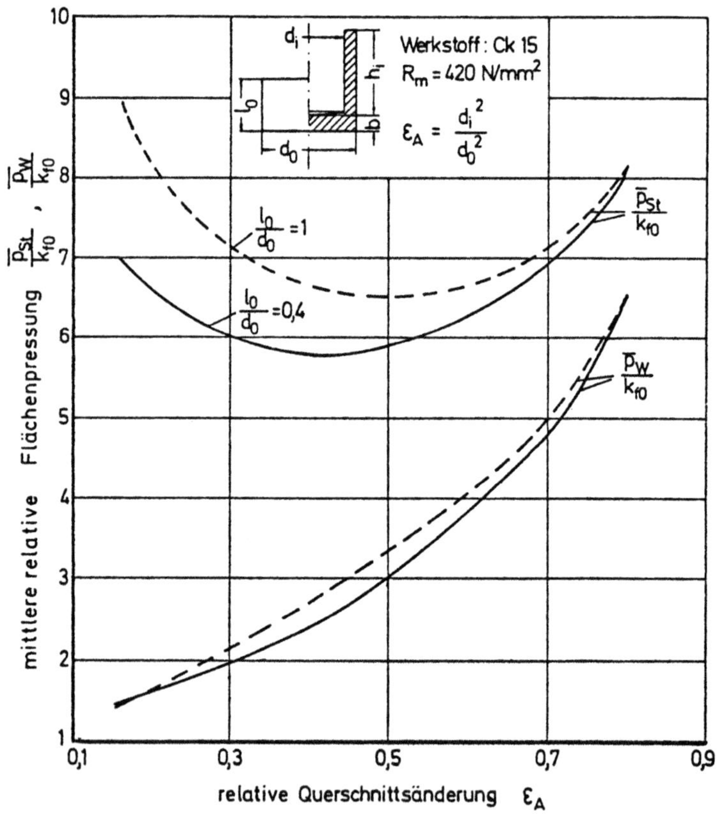

Bild 4: Flächenpressung $\bar{p}_{st}/k_{f0}$ auf den Stempel und $\bar{p}_w/k_{f0}$ auf die Preßbüchsenwand beim Napf-Rückwärts-Fließpressen.

Reibflächen liegen an Stempel, Boden und Matrizenwand vor. Durch das Eintauchen des Stempels in das Rohteil und den zunächst seitwärts und anschließend entgegengesetzt nach hinten abfliessenden Werkstoff ergeben sich im Werkzeug sehr unterschiedlich belastete Bereiche. Wie Bild 4 zeigt, treten die höchsten Flächenpressungen bis zu $\overline{p}_{max}/k_{f0} = 9$, abhängig vom l_0/d_0 - Verhältnis, am Stempel auf. Dies ist besonders zu berücksichtigen, da bei Mehrstufenpressen der Werkstückwerkstoff in Form von bereits phosphatierten und evtl. beseiften Drahtringen vorliegt, die dann in einer Scherstufe in die direkt zugeführten Rohteile zertrennt werden. Die blanke Scherfläche muß nun vor dem Eintauchen des Stempels mit einem geeigneten Schmierstoff versehen werden, was bei Ölen z.B. sehr oft zu Problemen mit der Oberflächenhaftung und der Druckbeständigkeit führt.
Die Flächenpressungen an Wand und Boden sind bei mittleren bis größeren Querschnittsänderungen in etwa um den Faktor ε_A kleiner als am Stempel. Die Relativgeschwindigkeit zwischen dem Stempel und dem entgegengesetzt fließenden Werkstoff ist mit $v_R/v_{Wz} = 6.3$ sehr hoch. Noch problematischer für den Schmierstoff ist allerdings die Oberflächenvergrößerung von der Stirnfläche des Rohteils zur Napfinnenfläche bis zu $A_1/A_0 = 11$ für den Ck 15. Sie kann örtlich im Bereich der Napfinnenwand noch sehr viel höhere Werte annehmen, was an die Haftung und Trennfähigkeit des Schmierstoffs natürlich höchste Anforderungen stellt.

Die für die sechs Kaltmassivumformverfahren theoretisch ermittelten Maximalwerte der drei Größen Flächenpressung, Relativgeschwindigkeit und Oberflächenvergrößerung sollten von einem für die Umformtechnik geeigneten Schmierstoffprüfverfahren nachgebildet werden können. Sie werden daher zu sogenannten Anforderungsprofilen zusammengefaßt, um später Vergleiche mit den Leistungsprofilen zu ermöglichen.
Aus dem Bild 5 ergeben sich demnach ähnliche Anforderungen beim Verjüngen und Abstreckgleitziehen sowie Voll- und Hohl-Vorwärts-Fließpressen. Beim Stauchen fällt die relativ hohe Spannung auf, während beim Napf - Rückwärts - Fließpressen eine besonders große Oberflächenvergrößerung auftritt.

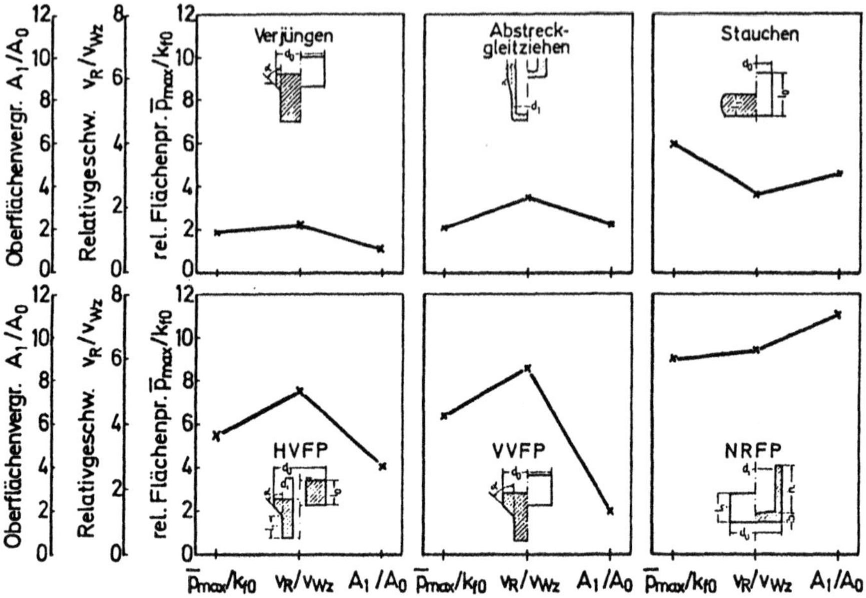

Bild 5: Anforderungsprofile der sechs Umformverfahren.

2.2 Maximalwerte bei Schmierstoffprüfverfahren

Ausgehend von den oben ermittelten sehr unterschiedlichen Beanspruchungen in der Reibzone der Kaltmassivumformverfahren kann als wichtigste Forderung, die an ein Schmierstoffprüfverfahren gestellt werden muß, die möglichst getreue Nachbildung der tatsächlich bei den Umformverfahren vorkommenden Verhältnisse festgehalten werden. Dies ist strenggenommen nur am Umformteil in der Fertigung möglich, was aber aus Kosten- und Zeitgründen nicht wünschenswert ist. Es wird vielmehr angestrebt, die Reibverhältnisse bei bestimmten technologischen Prüfverfahren denen bei den Umformverfahren anzupassen. Inwieweit bestehende Schmierstoffprüfungen diese Forderung erfüllen, soll eine Analyse von 2 Stauch- und 4 Ziehverfahren hinsichtlich der erreichbaren Flächenpressungen, Relativgeschwindigkeiten und Oberflächenvergrößerungen zeigen.

Das Ringstauchen [4] ist das am weitesten verbreitete Reibzahlermittlungsverfahren. Ein ringförmiger Körper festgelegter

Abmessungen wird auf bis zu 50% seiner Höhe gestaucht. Die dabei entstehende Geometrie stellt sich abhängig von der Reibung ein. Besonders der Innendurchmesser d_i reagiert sehr empfindlich auf die Änderung der Reibverhältnisse. Aus dem rechnerischen Ansatz eines Geschwindigkeitsfeldes ergibt sich für verschiedene Reibzahlen das in Bild 6 dargestellte Nomogramm, aus welchem sich nach Messung des Innendurchmessers und der Stauchhöhe die Reibzahl ablesen läßt.

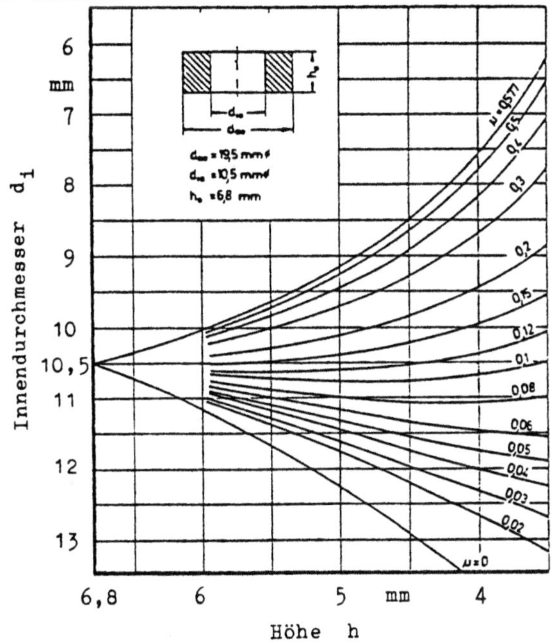

Bild 6: Nomogramm zur Ermittlung der Reibzahl [13].

Die maximal erreichbaren Werte sind in Tabelle 2 angegeben, wobei der höchste Druck im Bereich der Fließscheide auftritt. Der besondere Vorteil des Ringstauchens ist die leichte Durchführbarkeit auch bei erhöhten Temperaturen.

Beim Stauchversuch nach Nittel [2] wird der Probekörper über einen Keil gleichzeitig gestaucht und parallel zur Stauchbahn verschoben. Aus gemessener Normal- und Reibkraft wird die Reibzahl bestimmt. Die mit dieser Vorrichtung erreichbaren Maximalwerte von $\bar{p}_{max}/k_{f0} = 3$, $v_R/v_{Wz} = 2$ und $A_1/A_0 = 2,5$ liegen nicht sehr hoch. Die Spannungs- und Geschwindigkeitsverteilung auf der Stirnfläche der Probe sind nicht genau bekannt.

	Ringst. (Burgdorf)	Stauchen (Nittel)	Stabzug (Pawelski)	Streifenzug (Schlosser)	Streifenzug (Wiegand)	Streifenzug (Kawai)
	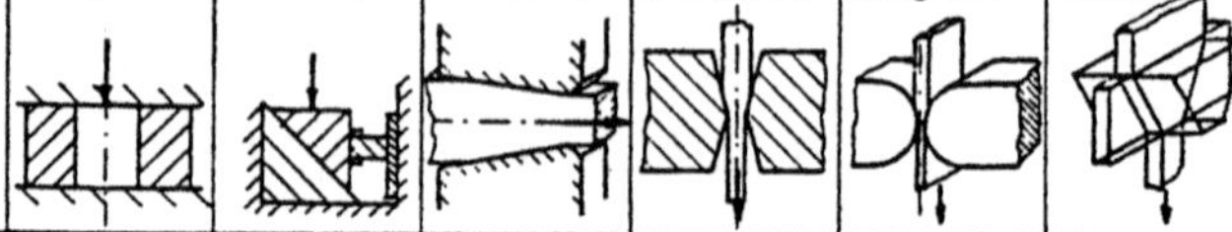					
$\bar{p}_{max}/k_{f0}$	5,7	3	4,5	4,8	4,9	2,3
v_R/v_{Wz}	2	2	1	1	1	1
A_1/A_0	2	2,5	1,5	2,2	-	1,7

Tabelle 2: Maximalwerte bei sechs Schmierstoffprüfverfahren.

Der Stabzug nach Pawelski [5] gehört zu den 4 untersuchten Ziehverfahren. Ein rechteckiger Stab wird, wie in Bild 7 angedeutet, durch 2 Ziehbacken gezogen und in seiner Dicke verringert.

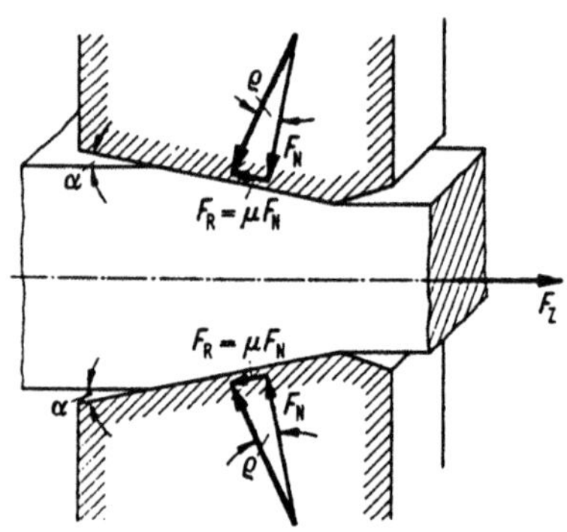

Bild 7: Prinzipskizze des Stabzuges [5].

Aus Quer- und Ziehkraft sowie dem Öffnungswinkel α ergibt sich die Reibzahl μ. Pawelski stellte fest, daß die Reibzahl mit steigender Normalspannung zunimmt. Tabelle 2 zeigt als Maximalwert für den Druck

$$\bar{p}_{max}/k_{f0} = 4,5.$$

Die Relativgeschwindigkeit wird bei allen Ziehverfahren höchstens so groß wie die Ziehgeschwindigkeit, also $v_R/v_{Wz} \leq 1$. Die Oberflächenvergrößerung wird bei $\varphi_{max} = 0{,}4$ zu $A_1/A_0 = 1{,}5$. Die von Pawelski entwickelte Vorrichtung ist zur Schmierstoffuntersuchung hinsichtlich Verfahren mittlerer Belastungen geeignet. Die Relativgeschwindigkeit kann über die Ziehgeschwindigkeit eingestellt werden. Die bei Ziehverfahren erreichbaren Oberflächenvergrößerungen sind im Vergleich zu den Kaltmassivumformverfahren zu gering.

Mit der Streifenzieheinrichtung nach Schlosser [8], die im Prinzip ähnlich wie die Vorrichtung von Pawelski aufgebaut ist, wurden umfangreiche Untersuchungen bei verschiedenen Werkstoffpaarungen insbesondere mit nichtrostendem austenitischem Stahlblech durchgeführt. Die maximale mittlere Flächenpressung $\bar{p}_{max}/k_{f0}$ erreicht, wie in Tabelle 2 angegeben, den Wert 4,8. Die Oberflächenvergrößerung betrug $A_1/A_0 = 2{,}2$. Dieses Verfahren hat sich besonders zur Schmierstoffprüfung für das Abstreckgleitziehen bewährt [14].

Beim Streifenziehen nach Wiegand [7] werden die zylindrischen Ziehbacken mit definierter Kraft in einen Blechstreifen eingedrückt, welcher dann zwischen diesen hindurchgezogen und in seiner Dicke verringert wird. Als maximale Flächenpressung wurde $\bar{p}_{max}/k_{f0} = 4{,}9$ berechnet. Die absolute Ziehgeschwindigkeit lag mit 2,5 mm/sec im Vergleich zu den Umformverfahren zu niedrig. Über die Oberflächenvergrößerung kann aus der Literatur heraus keine Angabe gemacht werden. Die mit diesem Verfahren erzielten Ergebnisse zeigten Übereinstimmungen mit beim Tiefziehen gemessenen Kräften.

Das Streifenzuggerät nach Kawai u.a. [9] ist dem Keilzug von Reihle [6] zur Schmierstoffuntersuchung für das Tiefziehen sehr ähnlich. Wie Bild 8 zeigt, entstehen beim Durchziehen des Blechstreifens 2 Bereiche unterschiedlicher Reibverhältnisse. Auf die umformenden Ziehbacken wirkt ein maximaler Druck von $\bar{p}_{max}/k_{f0} = 2{,}3$, wobei hier schon Kaltverschweißungen auftreten. Die Flächenpressung auf die Bremskörper erreicht etwa den Wert von k_f. Hier stellte Kawai fest, daß die Reibzahl von einem Flächendruck $< k_f$ unabhängig ist, solange kein Werkstoffüber-

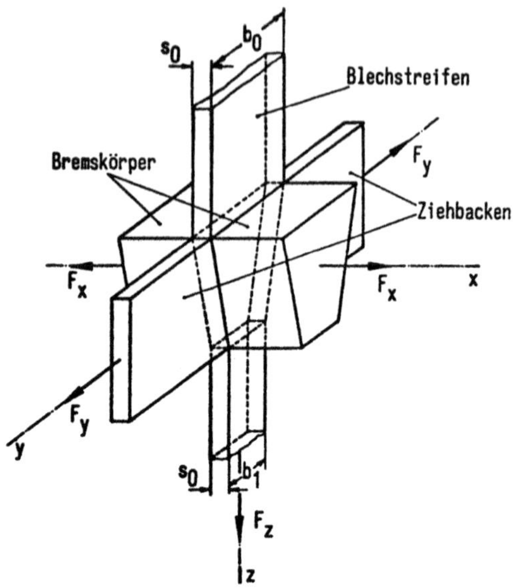

Bild 8: Prinzipskizze des Streifenziehens [9].

trag auftritt. Auch die Werte für die absolute Ziehgeschwindigkeit v_{Wz} = 0,6 mm/sec und die Oberflächenvergrößerungen von A_1/A_0 = 1,4 liegen im Vergleich zu den üblichen Kaltmassivumformverfahren zu niedrig. Die Verhältnisse dort können mit diesem Gerät nicht simuliert werden.

Ähnlich den Anforderungsprofilen der Umformverfahren werden nun die in Tabelle 2 enthaltenen Maximalwerte zu den in Bild 9 dargestellten Leistungsprofilen der sechs Schmierstoffprüfverfahren zusammengefaßt. Insgesamt die höchsten Werte treten demnach beim Ringstauchen auf obwohl auch hier Relativgeschwindigkeit und Oberflächenvergrößerung nicht sehr hoch sind. Der Stauchversuch nach Nittel ist auch hinsichtlich des Normaldruckes unbefriedigend. Den Ziehverfahren ist gemeinsam, daß die bisher herangezogene Relativgeschwindigkeit v_R/v_{Wz} maximal den Wert 1 annehmen kann. Die Oberflächenvergrößerungen sind sehr gering (bei Wiegand nicht bekannt) und erreichen mit A_1/A_0 = 2,5 den höchsten Zahlenwert.

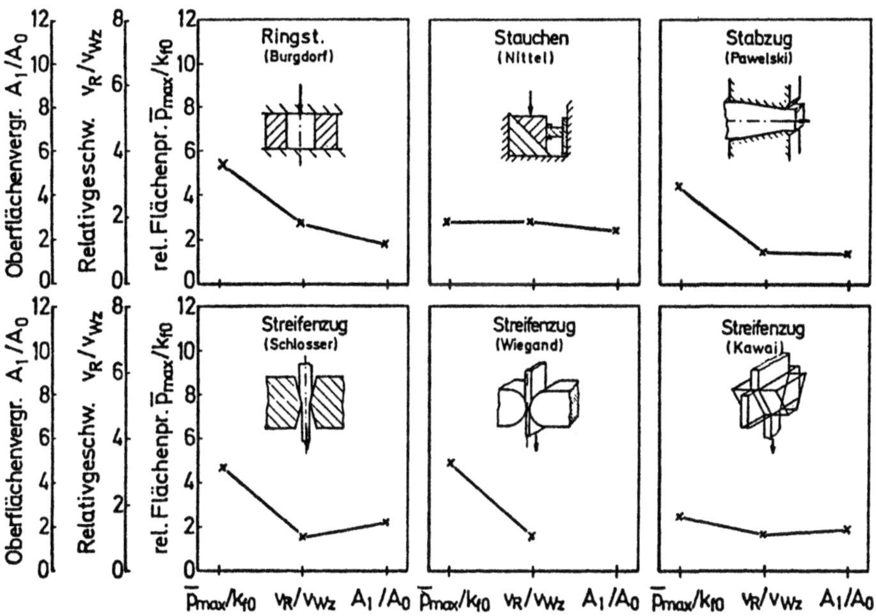

Bild 9: Leistungsprofile der sechs Schmierstoffprüfverfahren.

2.3 Vergleich von Anforderungs- und Leistungsprofil

Bei einer direkten Gegenüberstellung der in Bild 5 und 9 enthaltenen Diagramme entsteht eine sehr unübersichtliche Vielzahl von Darstellungen. Daher sollen die Vergleiche nicht für jedes einzelne Verfahren, sondern für jeden der Belastungsparameter, also Flächenpressung, Relativgeschwindigkeit und Oberflächenvergrößerung durchgeführt werden.

In Bild 10 sind die Maximalwerte der relativen mittleren Flächenpressung für die Umformverfahren einerseits und die Schmierstoffprüfverfahren andererseits aufgetragen.
Demnach gibt es unter den sechs betrachteten Kaltmassivumformverfahren nur zwei, das Verjüngen und das Abstreckgleitziehen, bei denen die Normaldrücke in der Umformzone so gering sind, daß sie von allen Schmierstoffprüfverfahren nachvollzogen werden können. Schon die Verhältnisse beim HVFP, Stauchen und VVFP sind nur noch in etwa durch die Streifenzugverfahren von Pawelski, Schlosser und Wiegand sowie den Ringstauchversuch simulierbar.

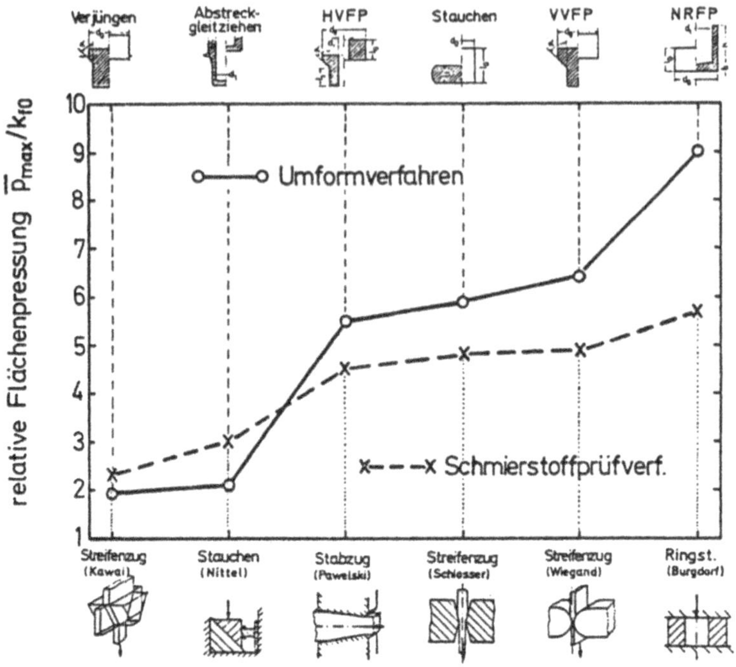

Bild 10: Vergleich der maximalen, relativen Flächenpressungen.

Das NRFP liefert sehr hohe Werte, die bisher noch nicht durch ein Reibzahlermittlungsverfahren nachgebildet werden konnten.

Kaum von Bedeutung bei der Entscheidung für ein bestimmtes Ziehverfahren ist die Relativgeschwindigkeit. Aus Bild 11 ist ersichtlich, daß nur die Verhältnisse beim Verjüngen durch die beiden Prüfverfahren des Stauchens in etwa simulierbar sind. Bei den Ziehverfahren kann grundsätzlich der Wert 1 nicht überschritten werden. Ausschlaggebend ist die, im Idealfall einstell- und meßbare, absolute Ziehgeschwindigkeit, so daß reale Verhältnisse bei den Umformverfahren in guter Näherung nachgebildet werden können.

Der Vergleich der höchsten Werte für die Oberflächenvergrößerung in Bild 12 zeigt, daß auch hier eine sehr große Diskrepanz zwischen der Kurve für die Umformverfahren und derjenigen für die Schmierstoffverfahren besteht. Eine Erhöhung des Umformgra-

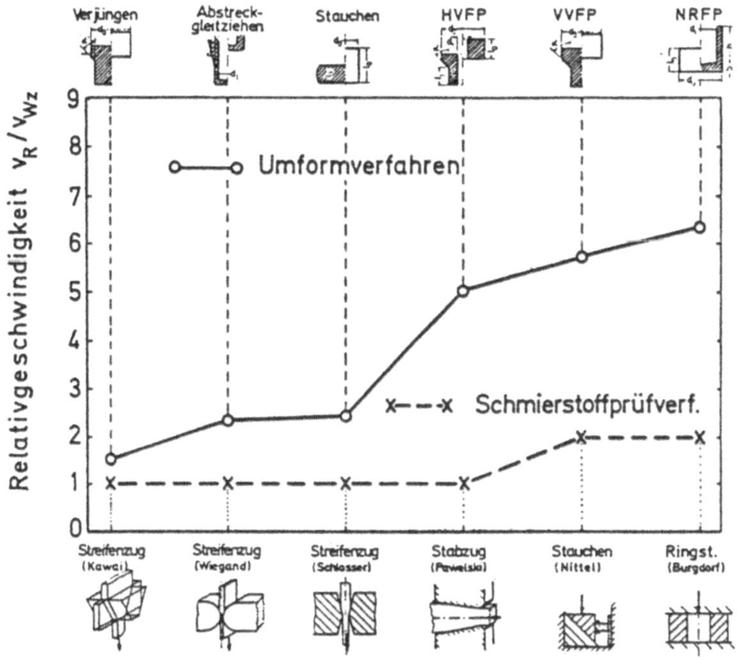

Bild 11: Vergleich der maximalen Relativgeschwindigkeiten.

des könnte bei den bestehenden Verfahren diese Differenz sicher verkleinern. Dies ist aber beim Streifenziehen aufgrund der Verfahrensgrenze Reißen und bei den Stauchverfahren durch den für eine sinnvolle Rechnung maximal zulässigen Stauchgrad beschränkt.

Zusammenfassend läßt sich aus dem Vergleich der theoretisch ermittelten Maximalwerte der Beanspruchungsparameter der Schluß ziehen, daß bisher kein Schmierstoffprüfverfahren existiert, welches die Reibverhältnisse bei konkreten Kaltmassivumformverfahren, besonders beim Fließpressen (HVFP, VVFP, NRFP) im Bereich höherer Umformung nachbilden kann. Insbesondere können hohe Oberflächenvergrößerungen und für die Umformverfahren repräsentative Spannungs- und Geschwindigkeitsverteilungen nicht simuliert werden.
Die Aussagefähigkeit der Ziehverfahren ist im Hinblick auf Ver-

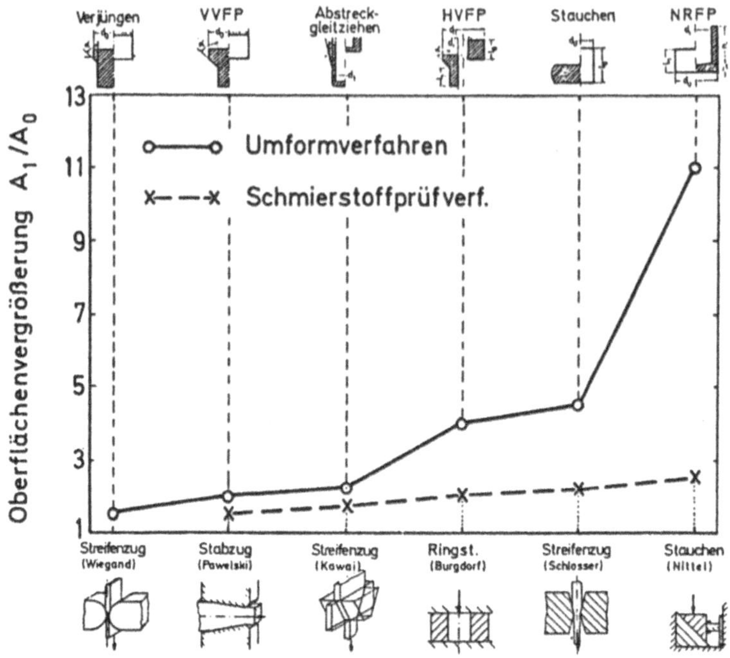

Bild 12: Vergleich der maximalen Oberflächenvergrößerungen.

jüngen und Abstreckgleitziehen zumeist befriedigend. Höhere Umformgrade und damit höhere Flächenpressungen und Oberflächenvergrößerungen allerdings lassen sich aufgrund der Verfahrensgrenzen nicht erreichen.

3 Schmierstoffprüfungen

Ausgehend von den theoretischen Überlegungen war nun im weiteren Arbeitsablauf die experimentelle Ermittlung der Einflüsse der Beanspruchungsparameter auf die Reibung vorgesehen. Die Abhängigkeit der Reibzahl von der Oberflächenvergrößerung wurde nicht untersucht, da die erreichbaren Werte zu gering sind und andererseits sich dieser Einfluß aufgrund der sich gleichzeitig ändernden Flächenpressung nicht isolieren läßt. Ausgehend von dem vorgenannten Vergleich wurden die beiden Schmierstoffprüfverfahren Streifenziehen nach Schlosser [8] und Ringstauchen nach Burgdorf [4] ausgewählt.

3.1 Versuchsprogramm

Um auch Vergleiche zwischen dem Streifenziehen nach Schlosser und dem Ringstauchen nach Burgdorf zu ermöglichen, wurden beide Versuchsreihen unter gleichen Bedingungen durchgeführt. Die Umformwerkzeuge bestanden aus dem Werkzeugstahl 1.2601 mit TiC-Beschichtung und den Hartstoffen Zirkonoxid, Mischkeramik (ZrO_2+Zusätze) und Aluminiumoxid. Bei den Werkstückwerkstoffen handelt es sich um den Kohlenstoffstahl C 45 und den austenitischen Werkstoff 1.4301 (X 5 CrNi 18 9).

Schmierstoff	Hersteller	Bezeichnung
S1	Klüber Lubric.	Presspate SEM 95
S2	Klüber Lubric.	Pressoil S 82
S3 ×	Klüber Lubric.	Pressoil S 32
S4	Klüber Lubric.	Pressoil S 77
S5 ×	Stuart Oil	Drawsol 3003
S6 ×	Stuart Oil	Drawsol 1622
S7 ×	Acheson	Dag V 657 (MoS_2)
S8	Acheson	Dag V 711 (Zinkst.)
S9 ×	Acheson	Gleitdag (Teflon)

Tabelle 3: Beim Streifenziehen und Ringstauchen (x) eingesetzte Schmierstoffe.

In Tabelle 3 sind die neun verwendeten Schmierstoffe, davon drei Festschmierstoffe, aufgeführt. Die Flüssigschmierstoffe wurden durch Bestreichen mit einem Pinsel aufgetragen, während der Festschmierstoff MoS_2 und Zinkstearat in Lösungsmitteln vorlagen und durch Tauchen und anschließendes Trocknen aufgebracht wurden. Der kunstharzgebundene Teflon-Schmierfilm wurde beim Streifenziehen aus der Dose aufgesprüht, beim Ringstauchen durch Tauchen aufgetragen.

3.1.1 Streifenzugversuch nach Schlosser [8]

Für die durchzuführenden Versuche wurde das in Bild 13 dargestellte Streifenziehgerät nach Schlosser eingesetzt, wobei lediglich die Form der Ziehbacken wie in Bild 14 für die Stahl- und die Keramikwerkzeuge geändert wurde.

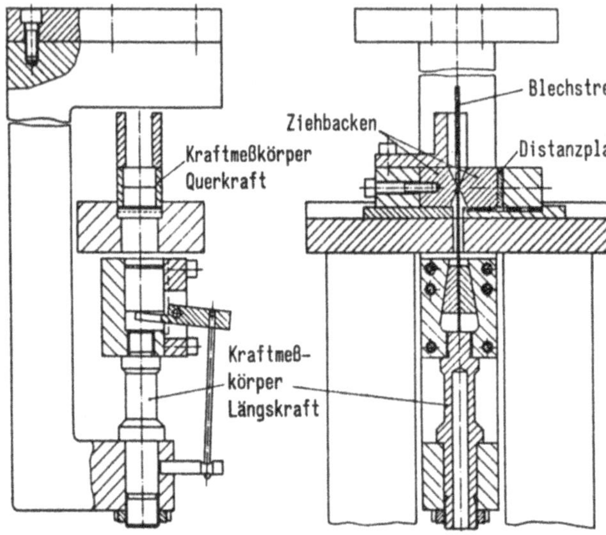

Bild 13: Aufbau der Streifenziehvorrichtung.

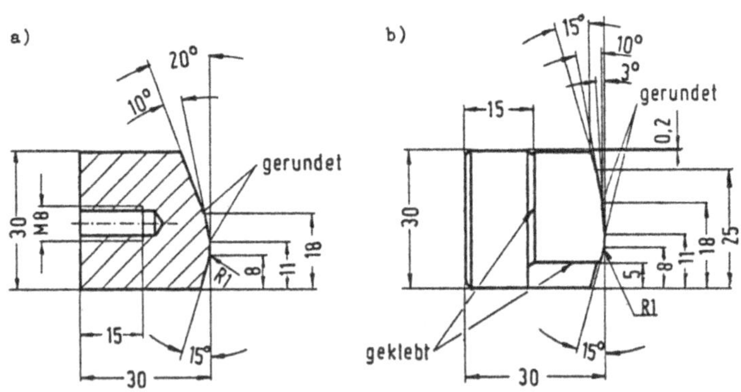

Bild 14: Verwendete Ziehbackenprofile (a. 1,2601, TiC-besch., b. Oxidkeramikeinsätze).

Durch Auswechseln der Distanzscheiben konnte der Abstand der Ziehbacken variiert und damit der Umformgrad gezielt verändert werden. Zur Versuchsauswertung wurde aus den gemessenen Längs- und Querkräften F_Z und F_Q über das Kräftegleichgewicht beim Umformen ein Beiwert f_r definiert, der den Reibungszustand charakterisiert aber nicht mit μ gleichgesetzt werden kann.

$$f_r = \frac{\frac{F_Z}{2 \cdot F_Q} - \tan\alpha}{1 + \frac{F_Z}{2 \cdot F_Q} \cdot \tan\alpha} \qquad (1)$$

Der Winkel α ergibt sich aus Verbindung des Ein- und Auslaufpunkts der Probe zur Senkrechten und ist daher von der Ziehbackengeometrie und der Blechdickenabnahme abhängig. Ebenfalls aus dem Kräftegleichgewicht wurde der mittlere Werkzeugdruck p_m definiert [8]:

$$p_m = \frac{2 \cdot F_Q \cdot \sin\alpha \cdot \cos\alpha + F_Z \cdot \sin^2\alpha}{b \cdot \Delta s} \qquad (2)$$

Die Blechstreifen mit den Ausgangsmaßen von 15 mm Breite und 3,02 mm Dicke wurden bis zum maximal möglichen Umformgrad, also bis zum Eintreten eines Versagens (Reißen, Werkstoffübertrag) gezogen.

Zur Untersuchung des Einflusses der Relativgeschwindigkeit zwischen Werkstück und Werkzeug wurden bei dem TiC-beschichteten Werkzeug drei verschiedene Ziehgeschwindigkeiten mit v_1 = 16,5 m/s, v_2 = 50 m/s und v_3 = 75 m/s eingestellt, bei den Zirkonoxidbacken dagegen lediglich die Ziehgeschwindigkeiten v_1 und v_2, wobei es sich jeweils um die Bewegungsgeschwindigkeit der Maschine in Ziehrichtung handelt, die betragsmäßig der Relativgeschwindigkeit am Werkzeugauslauf entspricht.

3.1.2 Ringstauchen nach Burgdorf [4]

Das Ringstauchen ist bis heute das in der Praxis am weitesten verbreitete Reibzahlermittlungsverfahren. Es wurde von Burgdorf zur Einsatzreife entwickelt und von einer Vielzahl von Autoren zur Schmierstoffprüfung benutzt.

Wie bereits in Abschn.2.2 erwähnt, läßt die beim Stauchen eines kreisringförmigen Probekörpers entstehende Geometrie direkte Rückschlüsse auf den Reibzustand, also die Reibzahl zwischen Werkstück und Werkzeug, zu.Für den realen Versuch bedeutet das, daß die Messung der Höhe und des Innendurchmessers eines gestauchten Ringkörpers genügen, um aus dem Nomogramm in Bild 6 die entsprechende Reibzahl μ zu entnehmen.
Die Flächenpressung wurde aus der gemessenen maximalen Umformkraft und der Querschnittsfläche der verformten Ringstauchprobe bestimmt. Die Messung der Stauchkraft wurde mit Hilfe einer Kraftmeßdose vorgenommen.

Für diese Versuchsreihe kamen die vier Werkzeugwerkstoffe (Werkzeugstahl 1.2601 mit TiC-Beschichtung, ZrO_2, Mischkeramik, Al_2O_3) in Kombination mit den Flüssigschmierstoffen S3, S5, S6 und den Festschmierstoffen S7 und S9 zum Einsatz. Zur Variation des Umformgrades wurden die kreiszylindrischen Ringteile auf fünf verschiedene Höhen gestaucht.

3.2 Versuchsergebnisse beim Streifenziehen

In den folgenden Abschnitten sollen die wichtigsten Ergebnisse aus den Schmierstoffprüfversuchen für die verschiedenen untersuchten Parameter dargestellt werden. Wichtigstes Augenmerk wird dabei auf die Abhängigkeit der Reibung von der Normalspannung, Relativgeschwindigkeit und Werkzeug-Schmierstoffkombination des Tribosystems gelegt. Ebenso werden die Temperaturentwicklung in der Umformzone und die Verschleißerscheinungen bei oxidkeramischen Werkzeugen besprochen.

3.2.1 Einfluß der Flächenpressung

In Bild 15 ist die Abhängigkeit der nach Gl. (2) berechneten, mittleren Flächenpressung von der aus Gl.(1) ermittelten Reibzahl f_r für verschiedene Schmierstoffe dargestellt. Es zeigt sich eine deutliche Tendenz zunehmender Reibzahl mit abnehmender Normalspannung. Dies gilt, von einigen Ausnahmen abgesehen, für nahezu alle untersuchten Kombinationen der Reibpartner und Zwischenstoffe. Eine Besonderheit tritt beim Schmierstoff S7,

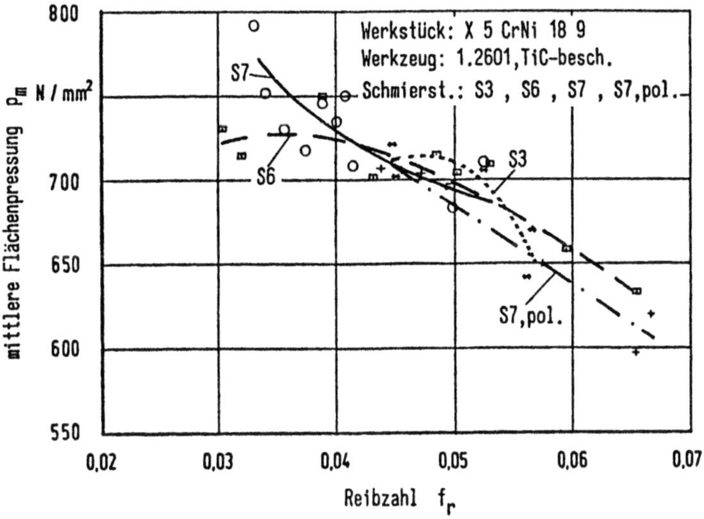

Bild 15: Abhängigkeit der mittleren Flächenpressung p_m von der Reibzahl f_r beim Streifenziehen.

dem Molybdändisulfid (MoS_2), auf. Dieser Schmierstoff zählt zu den Festschmierstoffen mit lamellarer Schichtgitterstruktur, wobei einzelne Ebenen der MoS_2-Kristalle aufgrund schwacher gegenseitiger Bindungen leicht aufeinander gleiten können. Die Rohteile werden nun durch Trommeln oder Polieren mechanisch behandelt, um diese Lamellen in etwa parallel zur Werkstückoberfläche auszurichten und fester dort zu verankern. Die dadurch angestrebte, verminderte Reibung konnte in den Versuchen nicht bestätigt werden, bei denen der Streifen vor dem Ziehen mit einem Lappen unter leichtem Druck poliert wurde. Bei geringerer Normalspannung traten erheblich höhere Reibzahlen auf. Gründe für dieses abweichende Verhalten können in der fehlenden Schmierstoffträgerschicht oder in zu großem Partikelabrieb beim Polieren liegen.

Die Absolutwerte der Flächenpressung lagen bei dem C 45 mit 500 bis 700 N/mm² um etwa 100 N/mm² niedriger als beim austenitischen X 5 CrNi 18 9. Ein Vergleich der Reibzahlen bei diesen zwei Werkstückstoffen ergab folgendes. Für S6, den einzigen Flüssigschmierstoff, mit dem sich der austenitische Streifen umformen ließ, lag die Reibzahl beim C 45 niedriger als beim X 5 CrNi 18 9. Bei den Festschmierstoffen MoS_2 (S7) und Teflon (S9) war dagegen die Reibzahl beim Kohlenstoffstahl höher. Diese Unterschiede dürften in der verschiedenen Oberflächenbeschaffenheit der beiden Blechwerkstoffe begründet sein. Während der Werkstoff 1.4301 als kaltgewalztes Band mit einer Rauhtiefe von R_t = 1,5 µm vorlag, handelte es sich bei dem Kohlenstoffstahl um warmgewalztes Material, dessen Oberfläche vor dem Ziehversuch durch leichtes Anwalzen auf R_t = 7 µm gebracht wurde. Diese relativ rauhe Oberfläche begünstigt die Schmiertaschenbildung beim Einsatz von Flüssigschmierstoff, während die sehr glatte Oberfläche des X 5 CrNi 18 9 - Bleches die Schmierwirkung der Festschmierstoffe verbessert.

Wie Bild 16 zeigt, wurde auch bei Verwendung der Zirkonoxid - Ziehbacken beim austenitischen Werkstoff höhere Reibung und höherer Druck bei Flüssigschmierung (mit S4, S5 und S6) beobachtet.

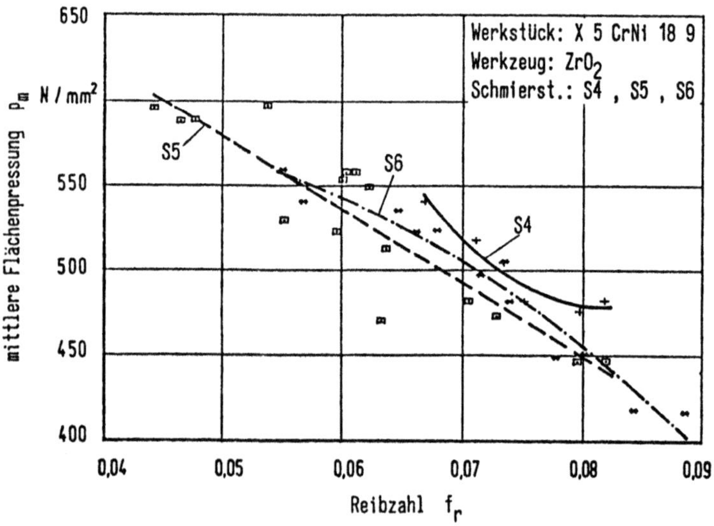

Bild 16: Abhängigkeit der mittleren Flächenpressung p_m von der Reibzahl f_r beim Streifenziehen.

Die Festschmierstoffe liefern auch hier beim C45 die höheren Reibzahlen. Die geringeren Absolutwerte der Flächenpressung beim Einsatz der ZrO_2 - Werkzeuge sind auf deren kleineren Öffnungswinkel zurückzuführen, der zur Vermeidung scharfkantiger Übergänge im Umformbereich diente.

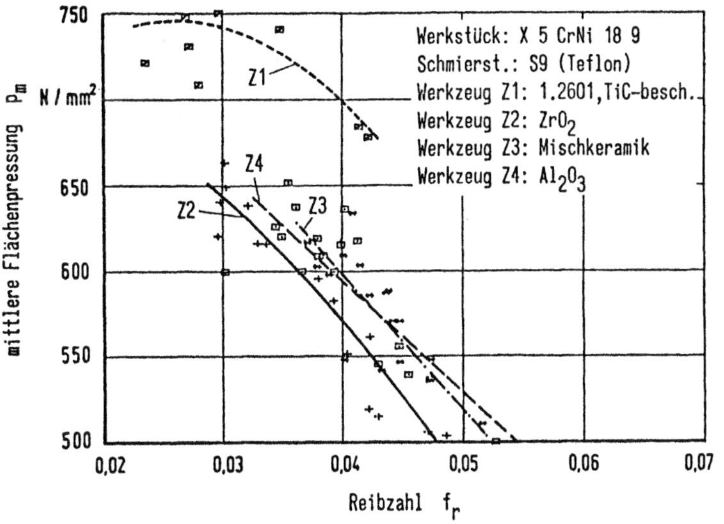

Bild 17: Abhängigkeit der mittleren Flächenpressung p_m von der Reibzahl f_r beim Streifenziehen.

In Bild 17 sind die vier Verläufe von Normalspannung über Reibzahl für den TiC-beschichteten Stahl (Z1) und die drei Oxidkeramikwerkstoffe (Z2, Z3 und Z4) beim Teflon-Schmierstoff S9 aufgetragen. Die sehr steilen Kurvenverläufe der drei Keramikwerkstoffe liegen so nahe beieinander, daß ein Werkstoffeinfluß auf diesen Schmierfilm fast nicht feststellbar ist. Geringfügig verminderte Reibung ergab sich bei ZrO_2.

Zusammenfassend kann hier festgehalten werden, daß zunächst die Absolutwerte von Flächenpressung und Reibzahl beim Streifenziehen sehr niedrig sind und vermutlich in dieser Form nicht auf reale Fließpressverfahren übertragen werden können.

Weiterhin konnte festgestellt werden, daß die Werkstückoberfläche einen beträchtlichen Einfluß auf die Reibung besitzt. Während die relativ rauhe Oberfläche des C 45 beim Einsatz von

Flüssigschmierstoffen niedrigere Reibzahlen lieferte, so war die Reibung bei Verwendung von Festschmierstoffen auf der glatteren Oberfläche des austenitischen Streifens geringer. Die Tendenz abnehmender Flächenpressung mit zunehmender Reibung wurde beim Streifenziehen schon von einigen anderen Autoren beobachtet [6,7,8], wohingegen besonders für Öl- und Fettschmierstoffe teils genau gegenteilige Ergebnisse ermittelt wurden [5]. Eine nähere Betrachtung der Reibverhältnisse wird beim Vergleich mit dem Ringstauchen in Kapitel 4 folgen.

3.2.2 Erreichbare Blechdickenabnahme

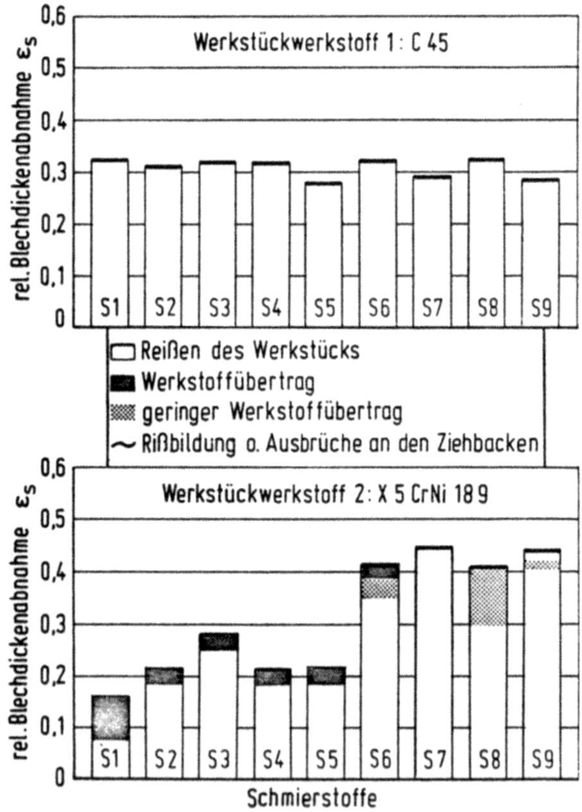

Bild 18: Erreichbare Blechdickenabnahmen beim Streifenziehen mit TiC-beschichteten Werkzeugen.

Wie erwähnt, ist der mit dem Streifenziehgerät erreichbare Umformgrad begrenzt. So ist die Ziehkraft entsprechend der Bruchfestigkeit des bereits umgeformten Streifenteils beschränkt. Werden die Kräfte in der Umformzone zu hoch, so wird die Verfahrensgrenze Reißen erreicht. Ein weiteres Versagenskriterium beim Ziehen des austenitischen Blechstreifens ist (adhäsiver) Werkstoffübertrag auf die Ziehbacken, der aber z.T. nur in kleinen Spuren am Rand der Umformzone auftritt. Werkstoffübertrag wurde auch bei den oxidkeramischen Ziehbacken beobachtet,wo ebenso bei zu hoher Belastung in der Umformzone Risse oder sogar Ausbrüche ganzer Werkstoffbereiche an den Ziehbacken festgestellt wurden.

In Bild 18 sind die erreichten Blechdickenabnahmen für alle Werkstoff-Schmierstoff Kombinationen bei den TiC - beschichteten Werkzeugen angegeben. Demnach gewährleisten alle 9 Schmierstoffe beim Ziehen des C 45 - Blechstreifens eine Dickenabnahme von $\varepsilon_s \approx 0,3$. Die Verfahrensgrenze wird vom Schmierstoff praktisch nicht beeinflußt. Anders sind die Verhältnisse beim austenitischen Werkstückstoff. Mit den Flüssigschmierstoffen S1 bis S6 treten an den Ziehbacken starke Anfresser auf, die bei S1 bis S3 und S5 zum Abbrechen der Versuchsreihen zwangen. Bei S4 und S6 riß gleichzeitig der Streifen hinter der Umformzone ab. Gute Ergebnisse wurden mit MoS_2 (S7) und Teflon (S9) erzielt.
Bei der durch adhäsiven Verschleiß besonders gefährdeten Werkstoffpaarung 1.4301 - 1.2601 mit TiC - Beschichtung zeigte sich, daß die Flüssigschmierstoffe keine vollständige Trennung der beiden Reibpartner gewährleisteten.

Kommt nun Zirkonoxid als Werkzeugwerkstoff zum Einsatz, so ergeben sich die in Bild 19 enthaltenen maximalen Dickenabnahmen. Für den Kohlenstoffstahl sind hier große Differenzen bei den einzelnen Schmierstoffen zu beobachten, wobei besonders das frühe Versagen bei MoS_2 auffällt.
Bei der Umformung des austenitischen X 5 CrNi 18 9 wurden, entgegen den Erwartungen, bei vielen Schmierstoffen mehr oder minder starke Verschleißerscheinungen beobachtet. Sie erlaubten bei den Flüssigschmierstoffen S1 - S4 nur geringe Umformung.

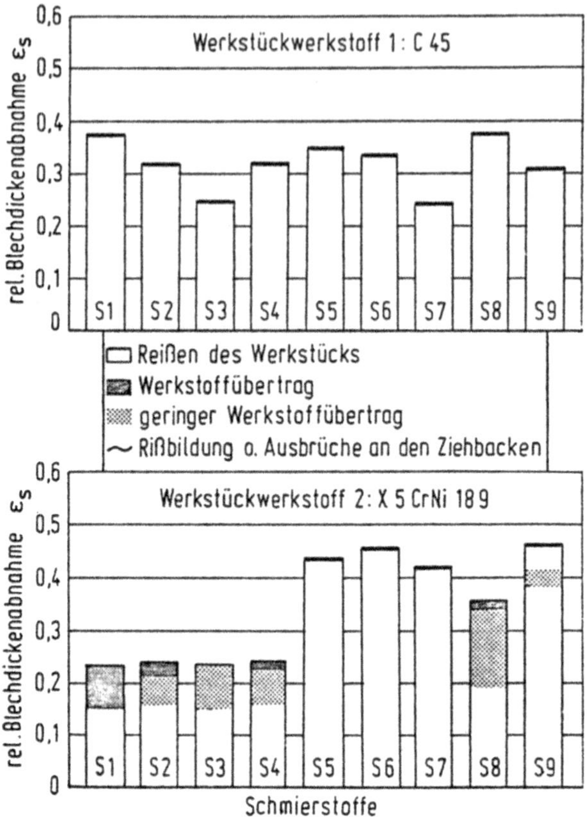

Bild 19: Erreichbare Blechdickenabnahmen beim Streifenziehen mit Zirkonoxid-Werkzeugen.

Sehr gute Ergebnisse hingegen liefern hier die Flüssigschmierstoffe S5 und S6 sowie MoS_2 (S7) und Teflon (S9).
Auf die Verschleißerscheinungen auf den Oxidkeramikoberflächen wird im Abschnitt 3.2.5 näher eingegangen.
Zusammen mit der Mischkeramik als Werkzeugwerkstoff und dem X 5 CrNi 18 9 stellten sich die Schmierstoffe S3, S7 und S9 als besonders günstig heraus, während bei Aluminiumoxid die Schmierstoffe S1, S2, S5 und S9 gut geeignet waren.

3.2.3 Geschwindigkeitseinfluß

Die bisher besprochenen Ergebnisse wurden bei einer Ziehgeschwindigkeit von 16,5 mm/sec ermittelt. Zur Untersuchung des Geschwindigkeitseinflusses wurden Versuche bei v_2 = 50 mm/sec und v_3 = 75 mm/sec durchgeführt. Für nahezu alle Schmierstoffe zeigte sich die in Bild 20 wiedergegebene Tendenz einer minimalen Reibzahl bei der mittleren Geschwindigkeit von 50 mm/sec und eines steilen Abfallens auf Werte unter f_r = 0,01 mit zunehmender Umformung.

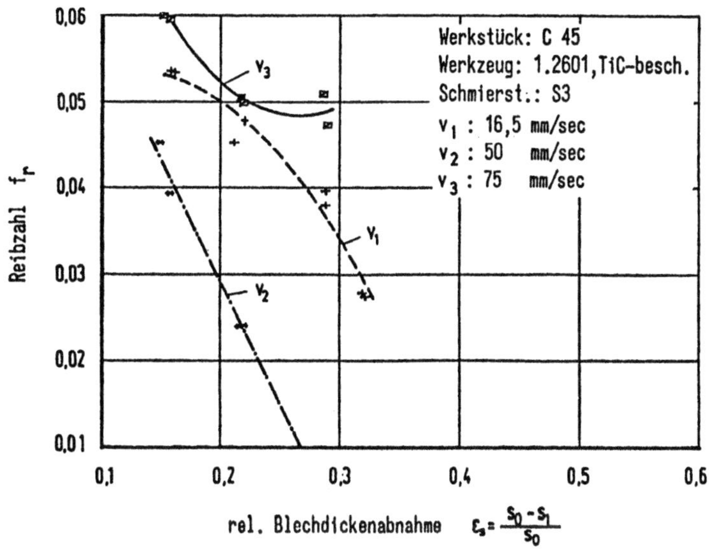

Bild 20: Geschwindigkeitseinfluß auf die Reibzahl f_r beim Streifenziehen.

Die stärkste Reibung tritt bei der Geschwindigkeit v_3 auf, während die niedrigste Ziehgeschwindigkeit teils ähnliche, teils mittlere Reibzahlen liefert.

Diese Ergebnisse kommen durch Zusammenwirken mehrerer Einflußfaktoren zustande. Wie bereits von anderen Autoren berichtet, nimmt die Reibzahl bei Geschwindigkeitszunahme ab [14,15], was zum einen auf verbesserte Schmierstoffzufuhr in die Umformzone und damit höhere Schmierfilmdicke, zum anderen auf die aus der Erwärmung resultierende Abnahme der Viskosität und eventuelle

Beeinflussung der chemischen Grenzflächenreaktionen zurückzuführen ist. Weiterhin kann davon ausgegangen werden, daß der Blechstreifen bei einer höheren Geschwindigkeit ein besseres Einlaufverhalten in die Ziehbacken aufweist, d.h. seine Dicke ist unmittelbar vor Werkzeugberührung schon etwas verringert, und es bildet sich ein Schmierkeil aus, der ebenfalls den Schmierstofftransport fördert.
Es gibt also eine optimale Gleitgeschwindigkeit und daraus resultierende Temperatur in der Reibfuge, bei der die geringste Reibung auftritt.

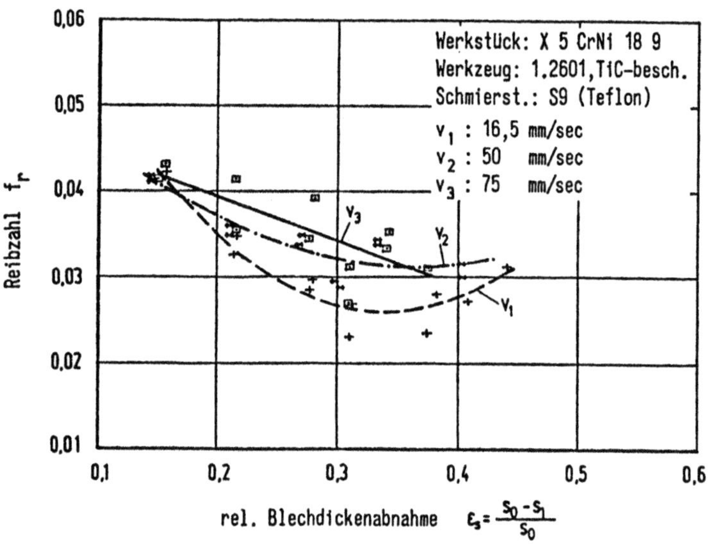

Bild 21: Geschwindigkeitseinfluß auf die Reibzahl f_r beim Streifenziehen.

Eine Ausnahme bildet der Teflon - Schmierfilm, bei dem, wie Bild 21 zeigt, die Reibung mit steigender Ziehgeschwindigkeit zunimmt, aber insgesamt nur in einem relativ schmalen Bereich von f_r = 0,025 bis 0,045 für alle Geschwindigkeiten liegt. Der Teflon - Schmierstoff reagiert also unempfindlich auf eine Geschwindigkeitsveränderung.
Ähnliche Untersuchungen zum Geschwindigkeitseinfluß auf die Reibung wurden auch mit den Zirkonoxidwerkzeugen durchgeführt,

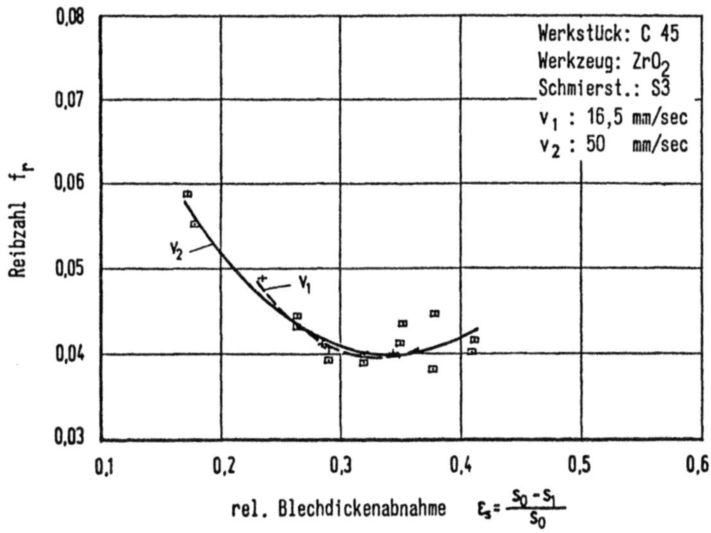

Bild 22: Geschwindigkeitseinfluß auf die Reibzahl f_r beim Streifenziehen.

wobei allerdings wegen zu hoher Temperaturentwicklung maximal v_2 = 50 mm/sec realisiert werden konnte. Wie aus Bild 22 hervorgeht, kann bei den oxidkeramischen Werkstoffen und den meisten Schmierstoffen praktisch kein Geschwindigkeitseinfluß auf die Reibzahl festgestellt werden. Die Einflüsse von Temperatur und Ziehbackengeometrie spielen also hier eine erhebliche Rolle.

Wie aus diesen Ergebnissen zu entnehmen ist, werden die Reibverhältnisse bei Umformbedingungen, die denen des Streifenziehens entsprechen, ganz erheblich von der Relativgeschwindigkeit des am Werkzeug entlanggleitenden Materials beeinflußt. Die rein geometrischen Parameter wie Ziehbackenwinkel und Oberflächenbeschaffenheit sind für einen ausreichenden Schmierstofftransport in die Umformzone verantwortlich. Sie bestimmen in starkem Maße die Reibbedingungen und damit das Auftreten von Grenz-, Misch- und hydrodynamischen Schmieranteilen, die alle verschieden auf Geschwindigkeitsänderung reagieren. Somit sollte bei weiteren Betrachtungen die Veränderung der Oberflächenstruktur durch die Umformung, also z.B. eventuelle Riefen, Rauheitstäler für Schmierstofftransport, Traganteil des Oberflächenprofils etc., mitbeachtet werden. Zudem muß für die verwendeten Schmierstoffe die Reaktion schmierwirksamer Komponenten auf eine Temperaturerhöhung untersucht werden.

3.2.4 Temperaturentwicklung in der Reibfuge

Wie bereits mehrfach erwähnt, spielt die Temperatur in der Umformzone hinsichtlich ihrer Beeinflussung der Schmierstoffeigenschaften eine entscheidende Rolle. So wird einerseits i.a. bei Temperaturerhöhung die Viskosität des Schmierstoffes herabgesetzt - die Reibung nimmt also ab -, andererseits wird dadurch die Gefahr des Abreißens des Schmierfilms erhöht - es kann zu direktem metallischem Kontakt kommen. Eventuelle EP-Zusätze werden erst durch höhere Temperaturen aktiviert, ebenso wie solche Schmierstoffe, deren Wirksamkeit auf Grenzschichtreaktionen beruht. Steigt die Temperatur auf zu hohe Werte, können sich Schmierstoffe zersetzen ("cracken") und völlig ihre Schmiereigenschaften verlieren. Bei den Streifenziehversuchen erfolgte die Messung der Temperatur etwa 3 sec. nach dem Ziehen.

Leichte Temperaturunterschiede wurden zwischen den beiden Werkstückwerkstoffen C 45 und X 5 CrNi 18 9 festgestellt, die auf das stärkere Verfestigungsverhalten des austenitischen Stahles und evtl. auf die glattere Oberfläche mit innigerem Kontakt zum Werkzeug zurückzuführen sind.
Bei den TiC-beschichteten Ziehwerkzeugen wurden, ausgehend von den relativ gleichmäßigen Blechdickenabnahmen beim C 45, Temperaturen um etwa 78°C gemessen. Durch die sehr verschiedenen möglichen Umformgrade beim 1.4301 kann hier kein Mittelwert angegeben werden. Die höchsten Absolutwerte traten bei den Festschmierstoffen auf.
Der Einsatz von Zirkonoxid als Werkzeugwerkstoff verändert die Temperaturverhältnisse in der Reibzone stark. Aufgrund des gegenüber Stahl etwa um ein Viertel niedrigeren Wärmeleitkoeffizienten kann die durch die Umformung und die Reibung erzeugte Temperatur nur begrenzt an das Werkzeug abgegeben werden. Die verbleibende Wärme heizt also das Werkstück, den Schmierstoff und die Oberflächenbereiche der Ziehbacken auf. So konnten auf den Streifen nach dem Ziehen erheblich höhere Temperaturen als

im vorstehenden Fall gemessen werden. Sie stieg, ausgehend von Raumtemperatur, bei großer Umformung auf bis zu 176°C beim Flüssigschmierstoff S6. Auch beim Kohlenstoffstahl liegt die Durchschnittstemperatur um mehr als 40% höher als im Fall der TiC-beschichteten Backen.
Bei den beiden anderen Keramikwerkstoffen, der Mischkeramik auf ZrO_2 - Basis und dem Al_2O_3, treten widerum etwas niedrigere Temperaturen auf.

Diese hohen Werte, die bei direkter Messung in der Reibfuge noch höher sein dürften, sind höchstwahrscheinlich für das Auftreten von Werkstoffübertrag bei der Umformung mitverantwortlich. Durch "Thermoschock" wird die Oberfläche der Keramik in bestimmten Bereichen zerstört, da aufgrund der fehlenden Wärmeleitung hohe thermische Spannungen entstehen.

3.2.5 Verschleißerscheinungen bei oxidkeramischen Werkstoffen

Die in den Versuchen gemessenen, sehr hohen Temperaturen sind hauptsächlich für das Verschleißverhalten der Oxidkeramik-Ziehbacken verantwortlich, das ausschließlich beim Ziehen des austenitischen Blechstreifens auftrat.
Anhand zweier REM-Aufnahmen soll der Mechanismus des Werkstoffübertrages auf die Ziehbackenoberfläche verdeutlicht werden.

Die geschliffene, noch unbelastete Oberfläche einer Oxidkeramik wird durch relativ ungleichmäßig verteilte, mehr oder weniger große Löcher charakterisiert, deren Vorhandensein einerseits auf die natürliche Porosität dieser Materialien und andererseits auf bei der Schleifbearbeitung herausgerissene Körner oder ganze Werkstoffbereiche zurückzuführen ist. Wird eine solche Oberfläche nun durch Druck, Temperatur und Reibschubspannung belastet, so ergibt sich für die Mischkeramik eine Struktur wie in Bild 23.
Gegenüber dem Ausgangszustand haben sich die vorhandenen Poren etwas vergrößert, ihre Ränder scheinen nach innen herunter gedrückt zu sein.

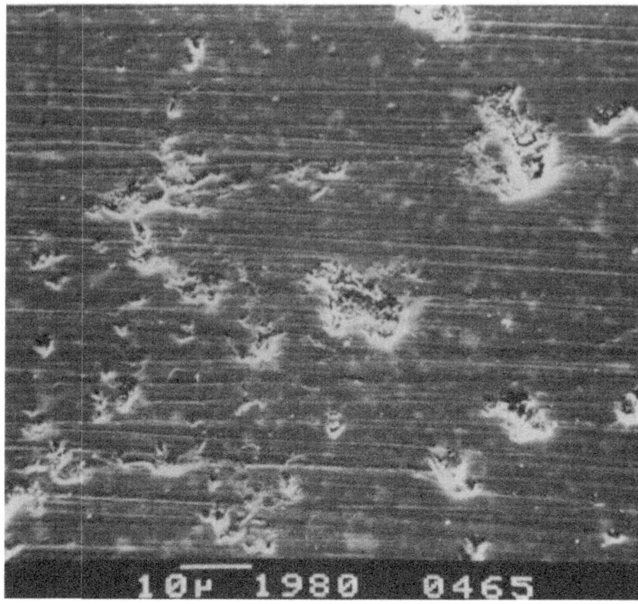

Bild 23: REM-Aufnahme einer belasteten Mischkeramik-Oberfläche.

Bild 24: REM-Aufnahme einer verschlissenen Mischkeramik-Oberfläche mit Werkstoffübertrag.

Bei weiterer Erhöhung der Belastung, wodurch Flächenpressung, Reibschubspannung und Temperatur steigen, lösen sich z.T. ganze Oberflächenbereiche ab und es bleibt eine sehr rauhe, zerklüftete Struktur zurück. Bild 24 zeigt eine derart verschlissene Mischkeramikoberfläche. Links im Bild ist ein stark aufgerauhter Bereich zu erkennen, in dem sich dann das in der Bildmitte befindliche Material anlagert. Rechts daneben ist noch eine Region intakter Oberfläche mit Schleifriefen zu sehen. In einem Schliff senkrecht zur Werkzeugoberfläche konnte keine feste chemische Bindung von Metall und Keramik festgestellt werden. So muß also davon ausgegangen werden, daß der aufgetretene Verschleißmechanismus nicht durch Adhäsion der beiden Reibpartner sondern vielmehr durch Oberflächenzerrüttung und anschließende Furchung der Rauheitsspitzen der Keramik hervorgerufen wird.

Maßnahmen zur Verhinderung des Werkstoffübertrages sollten demnach vornehmlich die thermischen Belastungen verringern, um ein Zerstören der Keramikoberfläche und damit die Bildung von Kratern und aufgerauhten Oberflächenbereichen zu vermeiden. Zu diesem Zweck sollte sichergestellt sein, daß der auftretende Temperaturgradient, also die Differenz zwischen Ruhe- und Arbeitstemperatur, nicht zu hoch wird. Wie gesehen, reicht schon eine Differenz von etwa 100°C aus, um die spröde Keramik durch Thermoschock zu zerstören. Ein künstliches Erwärmen des Umformwerkzeuges auf die zu erwartende Arbeitstemperatur dürfte hier Abhilfe schaffen. Leider liegen aus der Industrie von keramikherstellenden und -verbrauchenden Firmen hierzu keine Angaben vor.

3.3 Versuchsergebnisse beim Ringstauchen

Die Ringstauchversuche wurden hauptsächlich unter dem Aspekt durchgeführt, die Abhängigkeit der Reibzahl von der Flächenpressung bei einem Stauchverfahren zu untersuchen. Es handelt sich sowohl bei der Reibzahl als auch bei der Normalspannung um Mittelwerte, deren Maxima und Minima erheblich davon abweichen können. So hat der Spannungsverlauf im Bereich der Fließscheide seinen Höchstwert, während die Reibzahl mit zunehmender Relativgeschwindigkeit zum Rand der Probe hin ansteigt. Die mittlere

Flächenpressung errechnet sich also als Quotient aus der gemessenen Stauchkraft und der Probenstirnfläche am Ende des Stauchvorganges.

3.3.1 Einfluß der Flächenpressung

Die Bilder 25 und 26 zeigen die Abhängigkeit der Flächenpressung von der Reibzahl bei den TiC-beschichteten Stauchbahnen, den beiden Werkstückstoffen und drei Flüssigschmierstoffen S3, S5 und S6.

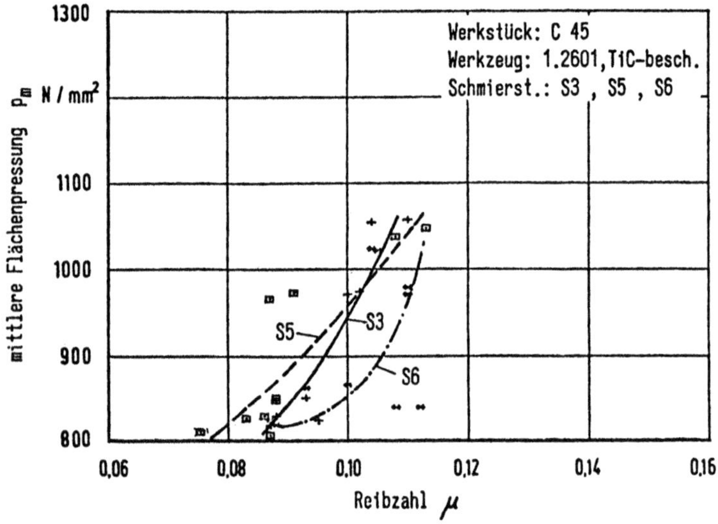

Bild 25: Abhängigkeit der mittleren Flächenpressung p_m von der Reibzahl μ beim Ringstauchen.

Auffallend ist zunächst, daß sowohl Normalspannung als auch Reibzahl mit Werten bis p_m = 1200 N/mm² und μ = 0,18 sehr viel höher liegen als beim Streifenziehen. Zudem ist hier die eindeutige Tendenz erkennbar, daß die Flächenpressung mit steigender Reibzahl stark zunimmt. Erwartungsgemäß liegt die mittlere Normalspannung bei den austenitischen Stauchproben höher. Weiterhin ist hier ein sehr viel größerer Reibzahlbereich mit etwa μ = 0,1 bis 0,18 gegenüber μ = 0,08 bis 0,11 beim C 45 zu beobachten.

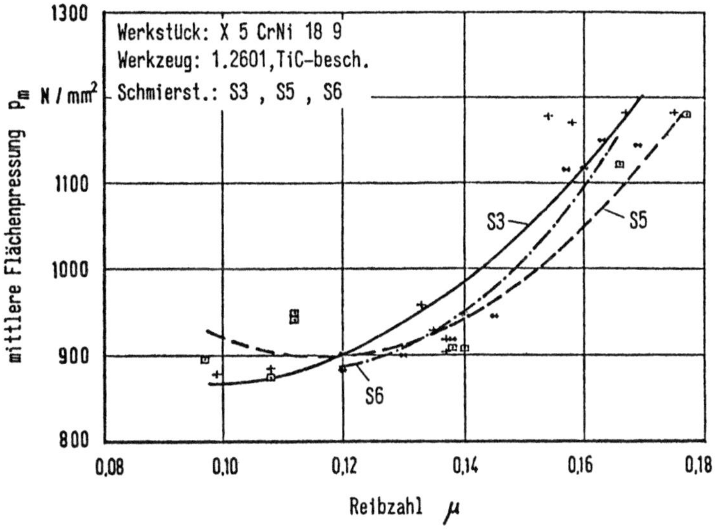

Bild 26: Abhängigkeit der mittleren Flächenpressung p_m von der Reibzahl μ beim Ringstauchen.

Auch bei den Festschmierstoffen S7 und S9 ist dieselbe Abhängigkeit bei steigendem Druck zu beobachten. Bemerkenswert ist, daß der Teflon - Schmierfilm, der beim Streifenziehen sehr niedrige Reibzahlen ergab, beim Ringstauchen die höchste Reibung erzeugt. Obwohl dies auf mangelnde Druckbeständigkeit zurückgeführt werden kann, konnte an den Stauchbahnen keinerlei Werkstoffübertrag festgestellt werden.

Wie sich nun der Einsatz oxidkeramischer Werkstoffe beim Ringstauchen auswirkt, ist am besten aus der Gegenüberstellung einzelner Kurvenverläufe herauszulesen.
In Bild 27 sind die Reibzahlen der Flächenpressung über der Reibzahl μ beim Schmierstoff S6 und drei verschiedenen Stauchbahneinsätzen aufgetragen. Während die erreichten Drücke nahezu gleich groß sind, tritt minimale Reibung beim Zirkonoxid auf. Die Mischkeramikeinsätze konnten die Reibzahl gegenüber den TiC-beschichteten Ziehbacken nicht verringern.

Bei annähernd gleicher, spezifischer Werkzeugbelastung ermöglichen also die oxidkeramischen Werkstoffe, besonders das Zirkonoxid, beim Ringstauchen verminderte Reibung und damit

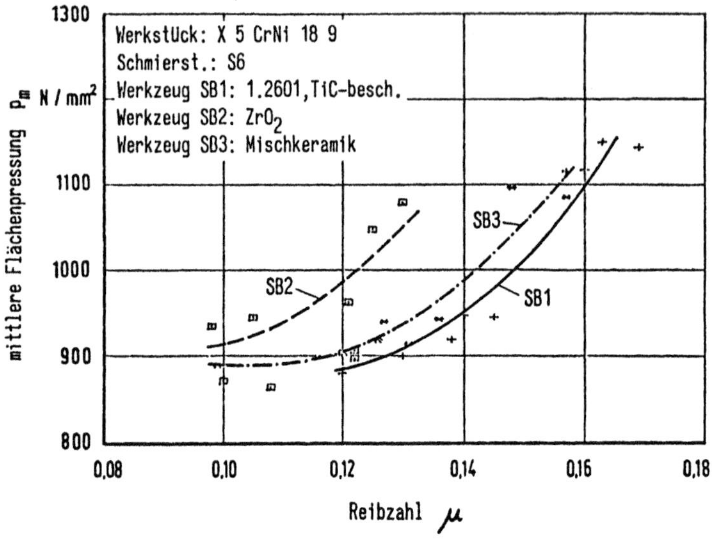

Bild 27: Abhängigkeit der mittleren Flächenpressung p_m von der Reibzahl μ beim Ringstauchen.

geringere Stauchkräfte. Der Einsatz dieser Werkstoffe ist aber begrenzt, da ihre Bruchspannung relativ schnell erreicht wird. So wurden die Al_2O_3- und z.T. die Mischkeramikeinsätze schon bei kleinen Umformgraden durch Bruch zerstört.

Aufgrund der höheren erreichbaren Flächenpressungen beim Ringstauchen scheint dieses Verfahren zur Schmierstoffprüfung hinsichtlich des Fließpressens besser geeignet als das Streifenziehen. Nachteilig sind allerdings die ungleichmäßigen Verteilungen der Normalspannung auf der Stirnfläche (mit einem ausgeprägten Maximum im Bereich der Fließscheide) und der Relativgeschwindigkeit (mit einer Haftzone und zunehmenden Werten zu den Probenrändern hin). So bringt schon die Mittelwertbildung der Flächenpressung als Quotient aus Stauchkraft und Stirnfläche einen gewissen Unsicherheitsfaktor mit sich. Die Möglichkeit qualitativer Schmierstoffprüfung mit dem Ringstauchen bleibt davon unberührt.

4 Gegenüberstellung der Schmierstoffprüfverfahren Streifenziehen und Ringstauchen

Die in den Versuchen ermittelte Abhängigkeit der Flächenpressung von der Reibzahl sowie deren Absolutwerte weichen bei den beiden Prüfverfahren z.T. erheblich voneinander ab. Obwohl versucht wurde, die Tribosystem - Struktur (Reibpartner, Zwischenstoff, Umgebungsmedium) bei beiden Versuchsreihen konstant zu halten, können doch die äußeren Belastungen bei den Verfahren nicht übereinstimmen. So besteht beim Ringstauchen in der Wirkfuge eine vollkommen andere Spannungsverteilung als beim Streifenziehen. Zudem sind die Geschwindigkeitsverhältnisse und die Erwärmung durch die Umformung sehr verschieden. Obwohl damit die Voraussetzungen für einen direkten Vergleich nicht günstig sind, soll doch eine kurze Gegenüberstellung der mit beiden Verfahren ermittelten Ergebnisse versucht werden.

Im Bild 28 sind die Streubereiche der Reibzahlen μ und f_r aus Versuchsreihen mit den Flüssigschmierstoffen S3 und S6 dargestellt. Es ergaben sich Unterschiede (μ - f_r) bis zu 0,12 beim jeweils höchstens Umformgrad für X 5 CrNi 18 9 und etwa 0,075 für C 45. Die unterschiedlichen Reib- und Schmierverhältnisse wirken sich bei den Verfahren also sehr stark aus. Weiterhin können systematische Fehler sowohl bei der Versuchsdurchführung

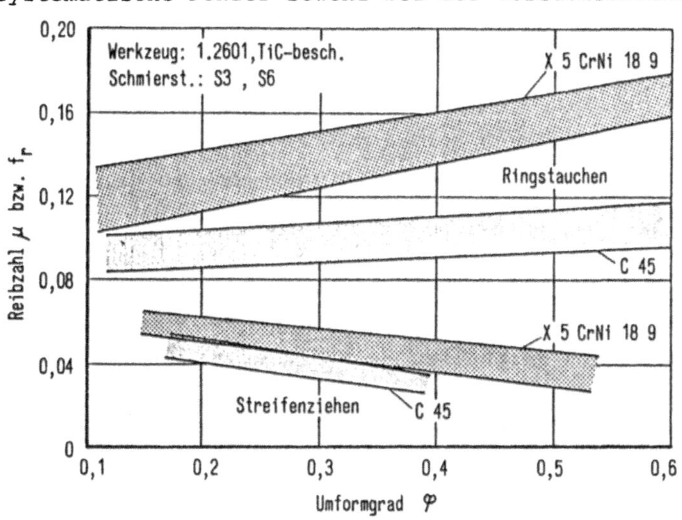

Bild 28: Gegenüberstellung der Streubereiche der Reibzahlen f_r beim Streifenziehen und μ beim Ringstauchen.

mit dem Streifenziehgerät als auch bei den Ringstauchversuchen (Ausbauchen der Probe) nicht ausgeschlossen werden.

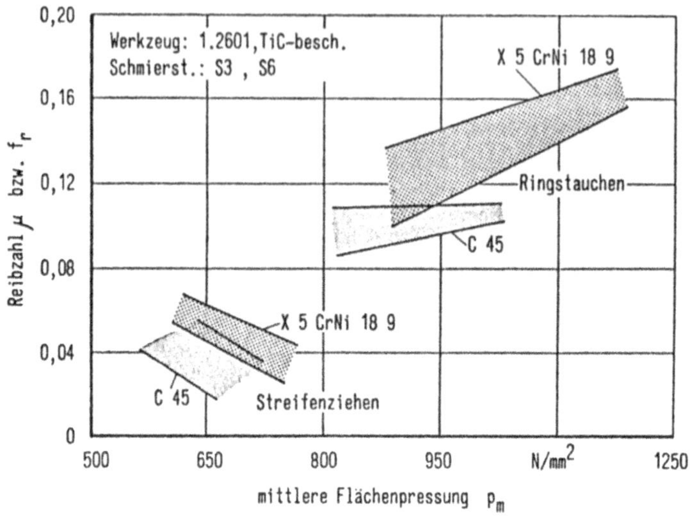

Bild 29: Gegenüberstellung der Streubereiche der mittleren Flächenpressung beim Streifenziehen und Ringstauchen.

Die in Bild 29 grob angedeuteten Verläufe der Reibzahlen über der Flächenpressung für beide Verfahren zeigen wiederum den fallenden Verlauf beim Streifenziehen und den Anstieg von µ beim Ringstauchen. Werden die in Bild 28 angegebenen Reibzahlunterschiede berücksichtigt und die Kurven angenähert, so ergibt sich für die beiden Werkstoffe folgender Verlauf.

Mit steigender Flächenpressung nimmt zunächst die Reibzahl ab, durchläuft ein Minimum, um bei höheren Drücken wieder anzusteigen. Eine ähnliche Abhängigkeit wurde von Wiegand und Kloos [7] sowie von Schlosser[8] gefunden. Dies deutet darauf hin, daß die bei der vorliegenden Mischreibung bestehenden hydrodynamischen und Grenzschmieranteile in Abhängigkeit vom Flächendruck auftreten, so daß die Reibzahl nicht die gleiche Abhängigkeit von der Normalspannung aufweist. Wird davon ausgegangen, daß beim Fließpressen ähnliche Verteilungen der Grenzreibungsgebiete vorliegen, so erweist sich das Ringstauchen als für dieses Verfahren besser geeignetes Prüfverfahren.

Schrifttum:

[1] Siebel, E., Pomp,A.: Die Ermittlung der Formänderungsfestigkeiten von Metallen durch den Stauchversuch. K.-W.-Inst.f.Eisenforschung 9 (1927), S. 157 - 171.

[2] Nittel, J.: Neues Verfahren zur Schmierstoffprüfung in der Umformtechnik durch Reibwertmessung. Fertigungstechnik und Betrieb 18 (1968), S. 301-304.

[3] Dobrucki, W., Odrzywolek,E.: Analyse des méthodes de détermination par refoulage du coefficient de frottement-entre l'outilage et le métal en état d'écoulement plastique-basées sur le procédé de refoulage.J.Mech. Working Technol. 4 (1980), S.155-176.

[4] Burgdorf, M.:Über die Ermittlung des Reibwertes für Verfahren der Massivumformung durch den Ringstauchversuch. Industrie-Anzeiger 89 (1967) 39,S. 15-20.

[5] Pawelski, O.: Ein neues Gerät zum Messen des Reibungsbeiwertes bei plastischen Formänderungen. Stahl und Eisen 84 (1964), S. 1233-1243.

[6] Reihle, M.: Verhalten des Gleitreibungskoeffizienten von Tiefziehblechen bei hohen Flächenpressungen. Dr.-Ing.-Diss., TH Stuttgart 1959.

[7] Wiegand, H., Kloos, K.H.: Der Reibungs- und Schmierungsvorgang in der Kaltformgebung und Möglichkeiten seiner Messung. Werkstatt und Betrieb 93 (1960) 4,S. 181-187.

[8] Schlosser, D.: Werkzeugwerkstoffe zum Umformen von nichtrostendem austenitischem Stahlblech. Bänder Bleche Rohre 17 (1976), S. 97-102.

[9] Kawai, N., Nakamura, T., Iwata, M.: The Frictional Mechanism on the Surface of Metal Being Plastically Deformed by Drawing. Transactions of the ASME, Journal of Engineering for Industry 99 (1977), S. 242-249.

[10] Lange, K.: Lehrbuch der Umformtechnik. Massivumformung. Band 2. Berlin/Heidelberg: Springer 1974.

[11] DIN 50 320: Verschleiß-Begriffe-Systemanalyse von Verschleißvorgängen-Gliederung des Verschleißgebietes. Beuth, Dez. 1979.

[12] Burgdorf, M.: Untersuchungen über das Stauchen und Zapfenpressen. Berichte aus dem Institut für Umformtechnik, Universität Stuttgart Nr.5. Essen: Girardet 1966.

[13] Geiger, R.: Untersuchung von Schmierstoffen zum Warmumformen von Stahl. Industrie-Anzeiger 92 (1970) 30, S. 623-629.

[14] Kerspe, J.H.: Abstreckgleitziehen von nichtrostenden austenitischen Stählen. Berichte aus dem Institut für Umformtechnik, Universität Stuttgart, Nr.53. Berlin/Heidelberg/New York: Springer 1980.

[15] Male, A.T., Cockcroft, M.G.: A method for the determination of the coefficient of friction of metals under conditions of bulk plastic deformation.J. Inst. Metals 93 (1964/65),S.38-46.

Untersuchung der Reibungsverhältnisse beim Tiefziehen

von
Prof. Dr.-Ing. E. Doege
Dr.-Ing. K.-P. Witthüser
Dipl.-Ing. R. Grahnert

Institut für Umformtechnik und Umformmaschinen
Universität Hannover
Welfengarten 1A

Förderkennzeichen: 01ZT 438 S
Laufzeit: 1.10.1978-31.3.1981

Zusammenfassung

Für eine optimale Auslegung von Werkzeugen für das Tiefziehen von großen Teilen oder dünnen Blechen ist eine genaue Kenntnis der Reibungsverhältnisse zwischen Blech und Werkzeug von großer Bedeutung. Bei solchen Teilen haben die Reibungskräfte einen großen Anteil an der gesamten Tiefziehkraft.

Es wurden deshalb Modellverfahren entwickelt und erprobt, um die Reibungsverhältnisse, insbesondere die Reibungskräfte, experimentell und rechnerisch bestimmen zu können. Außerdem sind Schmierstoffe sowie Blech- und Werkzeugwerkstoffe im Hinblick auf die Reibungsverhältnisse untersucht worden.

Eine besondere Bedeutung kommt dabei dem Streifenziehversuch mit Umlenkung zu. Die Untersuchungsergebnisse weisen darauf hin, daß beim Tiefziehen mit pastösen und flüssigen Schmierstoffen vor allem die Reibung an der Ziehkantenrundung von ausschlaggebender Bedeutung ist. Es zeigte sich, daß die Reibungsverhältnisse in diesem Bereich in zufriedenstellender Weise durch den Streifenziehversuch mit Umlenkung nachgebildet werden können.

Durch Messungen konnte festgestellt werden, daß sich bei pastösen und flüssigen Schmierstoffen und ziehringseitiger Schmierung zwischen Ziehring und Blech ein hydrostatischer Druck aufbaut. Die Reibungskraft wird daher in diesem Bereich nur geringfügig vom Schmierstoff und den Reibpartnern (Werkzeug- und Werkstückwerkstoff) beeinflußt.

Inhalt

1. Einleitung

Um bestehende Verfahren in der Umformtechnik zu optimieren, ist es notwendig, den Umformvorgang zu analysieren und von den ihn beeinflussenden Größen auszugehen.

Beim Tiefziehen ist das Ziehergebnis von vier übergeordneten Einflußgrößen 1. Ordnung, Bild 1 [1], abhängig.

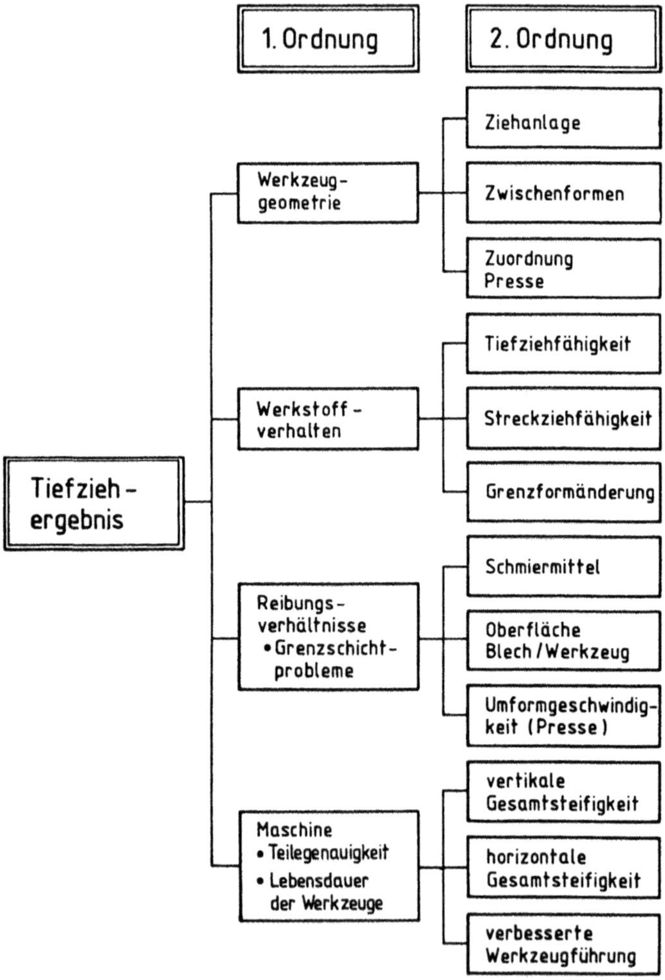

Bild 1: Einflußgrößen auf das Tiefziehergebnis [1]

Der Werkzeuggeometrie kommt hierbei die größte Bedeutung zu, da die richtige Gestaltung der Werkzeuge Voraussetzung für eine Optimierung des Tiefziehvorganges ist. Es wird noch viel Arbeit notwendig sein, um zu diesem Themenkreis zuverlässige Aussagen machen zu können.

Die Prüfung des Werkstoffverhaltens hat seit längerem einen recht hohen Stand erreicht. Mit Hilfe des Verfestigungsexponenten n (n-Wert) und des Anisotropiewertes R (R-Wert) kann die Streckzieh- und Tiefziehbarkeit von Blechen zufriedenstellend beurteilt werden.

Der Einflußgröße Reibung kommt ein besonderer Stellenwert zu, da beim Tiefziehen großer Teile oder dünner Bleche die Reibungsverhältnisse einen entscheidenden Einfluß auf das Ziehergebnis und den Versagensfall haben. Der nachfolgende Beitrag versucht den Stand der Kenntnisse zu bessern. Insbesondere ist es erstmals gelungen, die einzelnen Anteile an der Gesamtkraft meßtechnisch zu erfassen. Dies führte zu gänzlich neuen Erkenntnissen bezüglich der Reibungsverhältnisse.

Der Einfluß der Maschine ist durch die Erfassung der horizontalen und vertikalen Steifigkeit meßbar [2]. Langfristig gesehen wird es darauf ankommen, das System Maschine/Werkzeug sinnvoll aufeinander abzustimmen. Speziell im Bereich der Blechumformung arbeiten die Führungen von Werkzeug und Presse häufig gegeneinander; entsprechender Verschleiß ist die Folge.

2. Reibungszonen beim Tiefziehen

In Bild 2 sind die beim Tiefziehen auftretenden Reibungszonen dargestellt. Die Reibung im Bereich der Ziehkantenrundung sowie im Ziehteilflansch (Umformzone) und der Stempelkantenrundung (Zone der Krafteinleitung) übernimmt eine grundsätzlich unterschiedliche Funktion.

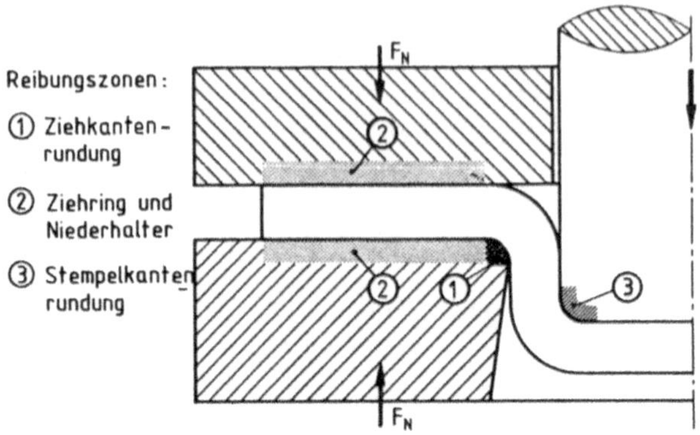

Bild 2: Reibungszonen beim Tiefziehen

An der Ziehkantenrundung und im Flanschbereich sollte eine möglichst geringe Reibkraft, d.h. eine kleine Reibzahl μ vorliegen, um das Nachfließen des Blechmaterials während des Ziehvorganges zu erleichtern.

Im Bereich der Stempelkantenrundung dagegen muß die Reibkraft hoch sein, wenn eine große Kraft in den Flanschbereich übertragen werden soll.

Eine kleine Kraft im Flanschbereich (Umformzone) bei großer übertragbarer Stempelkraft (Bodenbereich) führt zu dem bestmöglichen Ziehergebnis.

Um bei guter Schmierung im Ziehteilflansch dennoch hohe Reibkräfte an der Stempelkantenrundung übertragen zu können, sind unterschiedliche Vorgehensweisen möglich:

- Der Schmierstoff wird stempelseitig nur am Außenrand des Zuschnittes dünn aufgetragen, während die Platine ziehringseitig vollständig mit Schmierstoff bedeckt ist [3]. Ein dünner stempelseitiger Schmierfilm führt bei flüssigen und pastösen Schmierstoffen weitestgehend zur Unterdrückung der Falten 1. Art und zudem zu einer vergleichsweise großen übertragbaren Stempelkraft.

- Bei gleichmäßiger Schmierung der Rondenseiten kann durch eine unterschiedliche Oberflächenstruktur des Werkzeuges das gewünschte Reibungsverhalten eingestellt werden (glatte Ziehring- und Niederhalteroberfläche, angerauhte Stempelkante) [4].

Die maximal übertragbare Stempelkraft läßt sich durch die Bodenreißkraft F_{BR}

$$F_{BR} = a_R \cdot \pi \cdot d_o \cdot s_o \cdot R_m$$

ermitteln [7].

a_R = Reißfaktor
d_o = Stempeldurchmesser
s_o = Ausgangsblechdicke
R_m = Zugfestigkeit des Bleches

Der Reißfaktor a_R ist hierbei als Verhältnis der gemessenen Bodenreißkraft (Stempelkraft) zur fiktiven übertragbaren Kraft im unverminderten Zargenquerschnitt definiert [7]:

$$a_R = \frac{F_{BR}(\text{gemessen})}{\pi \cdot d_o \cdot s_o \cdot R_m(\text{berechnet})}$$

Dieser Wert liegt bei Tiefziehblech zwischen 1,2 und 1,5. Er ist im wesentlichen abhängig von der Geometrie des Ziehteilbodens (Verhältnis der Stempelkantenrundung r_s zum Stempeldurchmesser d_o), dem Anisotropiewert R und den Reibungsverhältnissen. Auch für rechteckige und quadratische Ziehteile kann die übertragbare Stempelkraft in guter Näherung errechnet werden.

Um die im Flanschbereich entstehende Gesamtkraft berechnen zu können, ist es notwendig, die Reibungskräfte auch in diesem Gebiet quantitativ erfassen zu können. Dabei ist es sinnvoll, von der bekannten Beziehung für die auftretende maximale Tiefziehkraft nach E. Siebel und W. Panknin [5] auszugehen:

$$F_{ges} = F_{id} + F_{rb} + F_{RN} + F_{RZ} \qquad [5] \quad (1)$$

F_{id}: Kraft für den verlustlosen Umformvorgang
F_{rb}: Rückbiegekraft zur Biegung des Bleches um die Ziehkantenrundung
F_{RN}: Reibkraft zwischen Ziehring und Niederhalter
F_{RZ}: Reibkraft an der Ziehkantenrundung

Max. Ziehkraft F_{ges} [n. E. Siebel und W. Panknin] :

$$F_{ges} = F_{id} + F_{rb} + F_{RN} + F_{RZ}$$

im einzelnen

$$F_{id} = A_o \cdot k_{fm} \cdot \ln\beta$$

$$F_{rb} = A_o \cdot k_{f1} \cdot \frac{s_o}{4 \cdot r_M}$$

$$F_{RN} = \frac{\pi}{2} \cdot \mu \cdot d_o^2 \cdot \frac{\beta_o^2 - 1}{\beta} \cdot p_N$$

$$F_{RZ} = (e^{\mu \cdot \frac{\pi}{2}} - 1) \cdot (F_{id} + F_{RN})$$

Spez. Niederhalterpressung p_N [n. E. Siebel]

$$p_N = 0{,}0025\left[(\beta_o - 1)^2 + 0{,}5 \cdot \frac{d_o}{s_o \cdot 100}\right] R_m$$

A_o : Zargenquerschnittsfläche
k_{fm} : Mittlere Formänderungsfestigkeit zwischen Rondenrand und Ziehkante
k_{f1} : Formänderungsfestigkeit an der Ziehkantenrundung
β : Ziehverhältnis im jeweiligen Ziehstadium
β_o : Ziehverhältnis
s_o : Ausgangsblechdicke

r_M : Ziehkantenradius
μ : Reibzahl
d_o : Stempeldurchmesser
p_N : Spezifischer Niederhalterdruck
R_m : Zugfestigkeit

Bild 3: Formeln zur Berechnung der max. Ziehkraft F_{ges} [5]

Die Gesamtkraft F_{ges} ist in der Vergangenheit nach diesen Beziehungen berechnet worden. Eine getrennte experimentelle Bestimmung der Einzelkräfte an der Gesamtziehkraft [6] führte zu den nachfolgend beschriebenen neuen Erkenntnissen bezüglich der Reibungsverhältnisse im Flanschbereich.

3. Verfahren und Versuchseinrichtungen zur Ermittlung der Reibungsverhältnisse beim Tiefziehen

3.1 Die Untersuchung der Reibungsverhältnisse an der Ziehkante

Um die Reibungsverhältnisse an der Ziehkante für sich allein experimentell bestimmen zu können, ist es erforderlich, geeignete Modellverfahren zur Verfügung zu haben. Aus diesem Grunde wurde der Streifenziehversuch mit Umlenkung entwickelt.

Bei diesem Versuch wird ein Streifen des zu untersuchenden Tiefziehbleches bei einer Umlenkung von 90° über einen Formkopf gezogen, vergl. Bild 4 [6].

Dadurch werden die Verhältnisse an der Ziehkantenrundung nachgebildet (Radius des Formkopfes ≙ Radius der Ziehkantenrundung).

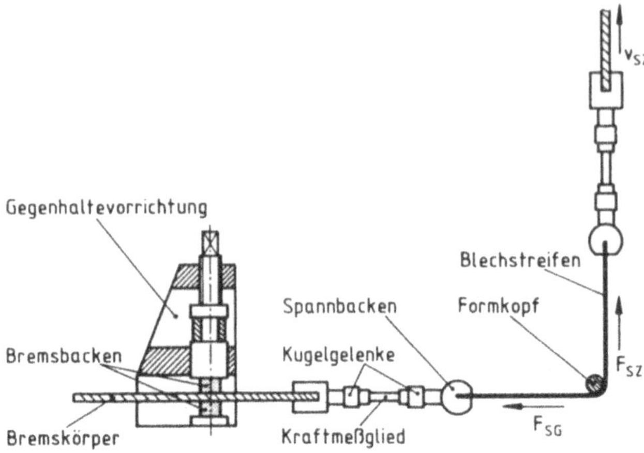

F_{SZ} : **Streifenziehkraft**
F_{SG} : **Streifengegenkraft**

Formkopf

1) Drehbar gelagert : F_{SB} : **Streifenbiegungsanteil**

$$F_{SB} = F_{SZ} - F_{SG}$$

2) Feststehend " : F_{SR} : **Streifenreibungsanteil**

$$F_{SR} = F_{SZ} - F_{SB} - F_{SG}$$

⟶ μ_{SZ} : **Streifenreibzahl**

$$\mu_{SZ} = \frac{2}{\pi} \ln \frac{F_{SZ} - F_{SB}}{F_{SG}}$$

Bild 4: Streifenziehversuch mit Umlenkung (schematisch) [6]

Es sind zwei Versuche erforderlich:

1. Der Blechstreifen wird um einen drehbaren Formkopf gezogen. Dabei tritt keine Relativbewegung zwischen Blech und Formkopf auf, so daß ausschließlich der Streifenbiegungsanteil F_{SB} ermittelt, s. Bild 4, und auf einen praxisüblichen Tiefziehversuch umgerechnet wird. Dieser Anteil wird durch die Streifenbreite b dividiert und das

Ergebnis mit dem Umfang U des Ziehringinnendurchmessers multipliziert. Bei der Bestimmung der Rückbiegekraft F_{rb} ist jedoch nur die Hälfte des Streifenbiegungsanteils ($F_{SB}/2$) zu berücksichtigen, weil sich beim Einlaufen in die Ziehkantenrundung der Blechwerkstoff im bildsamen Zustand befindet und keine wesentlichen Biegemomente aufnehmen kann. Beim Tiefziehversuch liegt somit nur eine Rückbiegung des Werkstoffes beim Auslaufen aus der Ziehkantenrundung vor, während beim Streifenziehversuch eine Hin- und Rückbiegung stattfindet.
Die Rückbiegekraft F_{rb} ergibt sich damit zu:

$$F_{rb} = \frac{F_{SB} \cdot U}{2 \cdot b} \qquad [6] \quad (2)$$

2. Man zieht den Streifen über einen feststehenden Formkopf, wobei die Streifenziehkraft F_{SZ} und die Streifengegenkraft F_{SG} ermittelt werden.

Unter Zugrundlegung der Eythelweinschen Gleichung kann aus beiden Versuchen die Streifenreibzahl μ_{SZ} berechnet werden, vergl. Bild 4.

3.2 Die Untersuchung der Reibungsverhältnisse im Flansch bei nur ziehringseitiger Schmierung

Während der Streifenziehversuch mit Umlenkung für die Untersuchung der Reibungsverhältnisse an der Ziehkante geeignet ist, können die Reibungsverhältnisse im Flansch durch den Tiefziehversuch selbst erfaßt werden. Dazu erfolgten Versuche mit rotationssymmetrischen Tiefziehwerkzeugen mit einem Stempeldurchmesser von d_0 = 200 mm.

Um die Verhältnisse nicht unnötig zu komplizieren, wurde lediglich ziehringseitig geschmiert. Auf der Seite des Niederhalters lag "Naturschmierung" vor (Walzölrückstände bzw. Korrosionsschutzöl).

Während des Ziehvorganges verdickt sich das Blech am Außenrand des Zuschnittes aufgrund der hier maximalen tangentialen Druckspannungen. Der relativ starre Niederhalter stellt sich entsprechend dieser Verdickung ein und liegt somit nur am Außenrand des Zuschnittes mit hoher Flächenpressung auf. Das bedeutet, daß ein Spalt zwischen Blech und Werkzeug gegeben ist.

Aufgrund der ebenfalls hohen Flächenpressung an der Ziehkante werden flüssige oder pastöse Schmierstoffe in diesem Spalt eingeschlossen. Dadurch entsteht während des Tiefziehens ein hydrostatischer Druck im Flanschbereich, dessen Messung Grundlage für die Untersuchung der Reibungsverhältnisse und zur Ermittlung der Reibkraft in diesem Bereich ist.
Wie aus Bild 5 zu ersehen ist, ergibt sich die Reibkraft F_{RN} zu:

$$F_{RN} = F_{RN1} + F_{RN2} + F_{RN3} + F_{RN4}$$

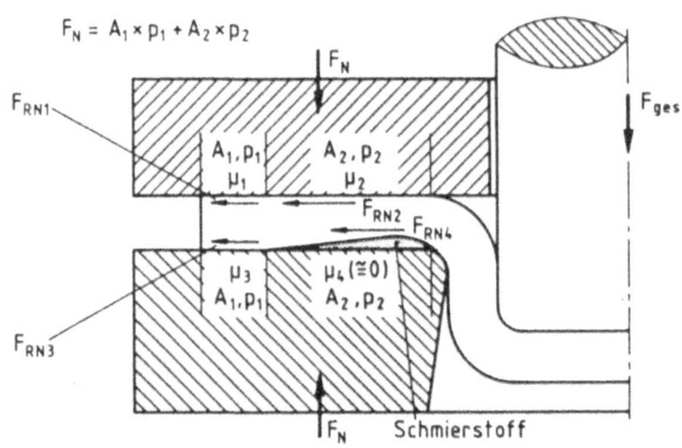

Bild 5: Druck- und Reibungsverhältnisse in der Reibungszone zwischen Ziehring und Niederhalter

Mit Hilfe piezoelektrischer Miniaturmeßgeber ist es möglich, diesen hydrostatischen Schmierstoffdruck zwischen Ziehring und Blech zu messen, Bild 6. Die Anordnung der Druckmeßbohrungen im Ziehring zeigt Bild 7.

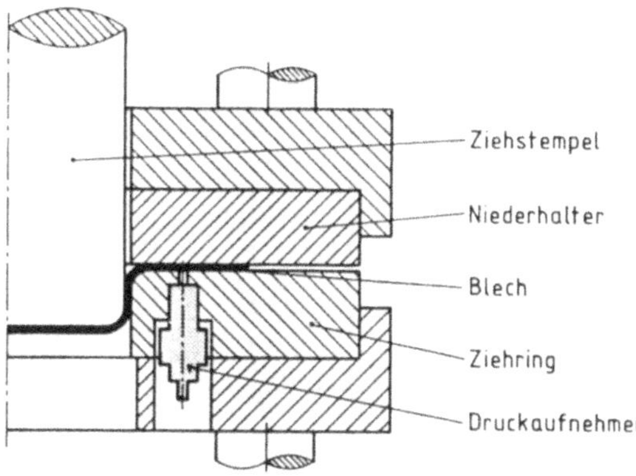

Bild 6: Tiefziehwerkzeug mit Druckaufnehmer

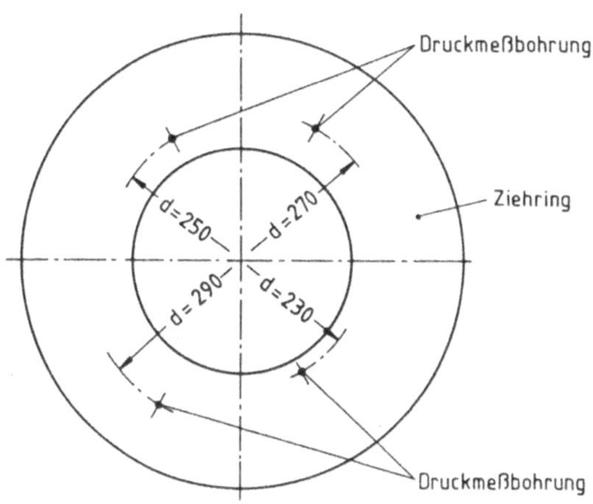

Bild 7: Anordnung der Druckmeßbohrungen im Ziehring

3.2.1 Messung des hydrostatischen Schmierstoffdruckes

Die Schmierstoffdruckmessung im Tiefziehversuch mit einem Stempeldurchmesser d_0 = 200 mm ergab einen Druckverlauf in Abhängigkeit vom Ziehkraft-Wegverlauf wie er in Bild 8 dargestellt wird.

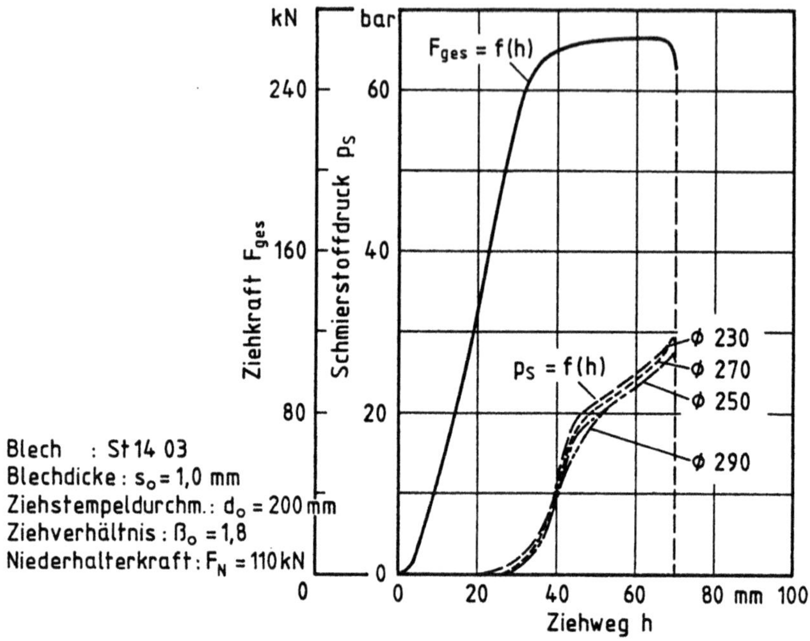

Bild 8: Ziehkraft- und Schmierstoffdruckverlauf

Es stellte sich heraus, daß erst nach einem Ziehweg von etwa 20 mm ein Druckaufbau stattfindet, was darauf zurückzuführen ist, daß der Flansch zur Verdickung des Bleches am Außenrand erst etwas eingezogen werden muß, um den Schmierstoff zwischen Ziehringkante und Rondenrand einzuschließen.

Danach steigt wie in Bild 8 dargestellt, der hydrostatische Schmierstoffdruck bis auf einen Wert, der dem eingestellten spezifischen Niederhalterdruck entspricht, an.
Hieraus ist zu schließen, daß kontinuierlich Schmierstoff über den Rondenrand aus dem Druckraum zwischen Blech und Ziehring herausfließt, sobald der mittlere Schmierstoffdruck die spezifische Niederhalterkraft erreicht und hierbei auch am Außenrand des Flansches eine Trennung zwischen Blech und Ziehring stattfindet.

3.2.2 Die Ermittlung der Reibungskraft zwischen Ziehring und Niederhalter

Wird der Tiefziehversuch nach Erreichen des Ziehkraftmaximums abgebrochen, so kann man entsprechend Bild 5 bei Betrachtung des Flansches "niederhalterseitig" die unterschiedlichen Wirkzonen des Niederhalters erkennen. Diese unterschiedlichen Flächen können betragsmäßig gut bestimmt werden. Sie werden in Bild 5 mit A_1 (A_1 = "Hauptdruckfläche" des Niederhalters) und A_2 (A_2 = "Schmierstoffdruckfläche") bezeichnet.

Die Niederhalterkraft F_N verteilt sich auf die Fläche A_1 und A_2 entsprechend der Gleichung

$$F_N = A_1 \cdot p_1 + A_2 \cdot p_2,$$

worin mit p_1 und p_2 die entsprechenden spezifischen Flächenpressungen bezeichnet sind.

Da der hydrostatische Schmierstoffdruck p_2 der spezifischen Flächenpressung p entspricht und A_1 und A_2 bekannt sind, kann die auf die Hauptdruckfläche wirkende spezifische Flächenpressung p_1 ermittelt werden.

Nachdem die wirkenden spezifischen Flächenpressungen bekannt sind, können mit Hilfe der Ergebnisse des Streifenziehversuchs mit Umlenkung bei verschiedenen Flächenpressungen die entsprechenden Reibzahlen, vergl. Bild 5, ermittelt werden. In Bild 9 sind diese Streifenreibzahlen für die einzelnen Schmierstoffe und für den ungeschmierten Zustand in Abhängigkeit von der spezifischen Flächenpressung dargestellt.

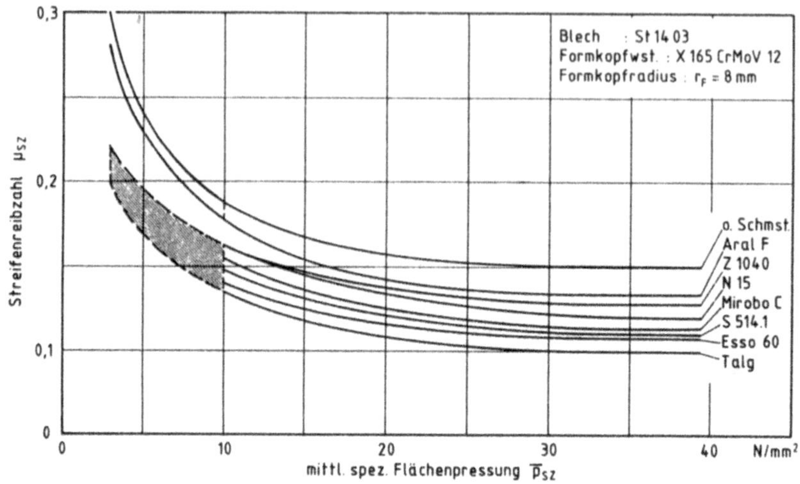

Bild 9: Streifenreibzahl μ_{SZ} in Abhängigkeit von der mittleren spezifischen Flächenpressung p_{SZ} bei unterschiedlichen Schmierstoffen

Da die spezifischen Flächenpressungen p_1 und p_2 einen Unterschied aufweisen, liegen bei ziehringseitiger Schmierung insgesamt vier unterschiedliche Reibzahlen vor, Bild 5.

Aus der Aufteilung der Niederhalterkraft nach der Gleichung $F_N = A_1 \cdot p_1 + A_2 \cdot p_2$ ergeben sich die Teilkräfte F_{N1} $(=A_1 \cdot p_1)$ und F_{N2} $(= A_2 \cdot p_2)$, Da die Reibzahlen bekannt sind, können nach dem Coulombschen Gesetz die im einzelnen auftretenden Reibungskräfte bestimmt werden. Aufsummiert ergibt sich die Reibungskraft F_{RN} zwischen Ziehring und Niederhalter zu:

$$F_{RN} = F_{N1} \cdot \mu_1 + F_{N1} \cdot \mu_3 + F_{N2} \cdot \mu_2 + F_{N2} \cdot \mu_4 .$$

Erreicht der Schmierstoffdruck während des Tiefziehvorgangs den spezifischen Niederhalterdruck p_N - der Zeitpunkt hängt von der Schmierstoffviskosität ab - entweicht der Schmierstoff am Außenrand der Ronde, vergl. Bild 10. Dann existieren nur noch zwei Reibungszonen, wobei die Beträge für die Reibzahlen aus dem Streifenziehversuch bekannt sind.
Daraus folgt, daß die Reibkräfte F_{RN} für alle Phasen des Tiefziehvorgangs vorliegen.

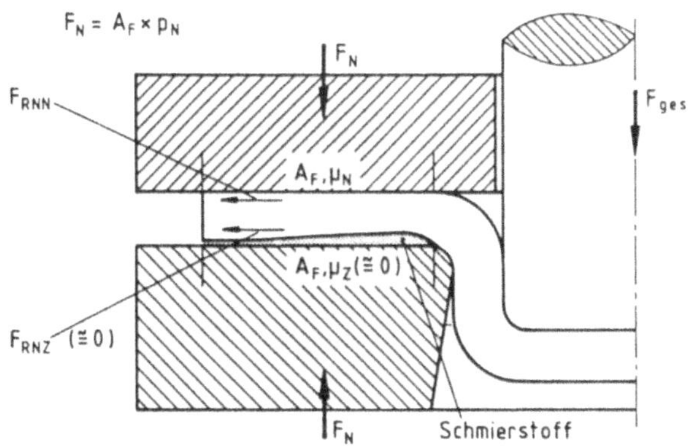

Bild 10: Druck- und Reibungsverhältnisse in der Reibungszone zwischen Ziehring und Niederhalter (mittlerer Schmierstoffdruck p_S = spez. Niederhalterdruck p_N)

Betrachtet man die Reibkraft F_{RZ} an der Ziehkante, siehe Bild 3, so sieht man, daß zu deren Berechnung zusätzlich die ideelle Ziehkraft F_{id} bekannt sein muß.

3.3 Das Zwei-Bleche-Verfahren

Zur Bestimmung der idellen Umformkraft F_{id} wurde das am Institut für Umformtechnik und Umformmaschinen (IFUM) an der Universität Hannover entwickelte 2-Bleche-Verfahren [6] angewandt. Dabei wird durch den Tiefziehversuch sowohl mit einem Blech als auch mit zwei übereinandergelegten Blechen die maximale Ziehkraft F_{1ges} und F_{2ges} bei sonst gleichen Verhältnissen gemessen und die ideelle Umformkraft nach folgendem Gleichungssystem ermittelt:

$$F_{1ges} = F_{id} + F_{RN} \quad \text{(1 Blech)} \qquad (3)$$

F_{RN} beinhaltet die jeweils gleichbleibenden Reibungsanteile

$$F_{2ges} = 2F_{id} + F_{RN} \quad \text{(2 Bleche)} \qquad (4)$$

Hieraus folgt die ideelle Umformkraft zu

$$F_{id} = F_{2ges} - F_{2ges}. \qquad (5)$$

Das Zwei-Bleche-Verfahren basiert also auf der Überlegung, daß sich beim Tiefziehen von zwei übereinandergelegten Blechen gegenüber einem Blech im Bereich Ziehring/Niederhalter die ideelle Umformkraft F_{id} verdoppelt und die Reibungskraft F_{RN} konstant bleibt.

Vermittels der beschriebenen Modellversuche
- dem Streifenziehversuch mit Umlenkung
- der Messung des hydrostatischen Schmierstoffdruckes
- dem Zwei-Bleche-Verfahren

ist somit die Bestimmung aller Einzelkräfte - basierend auf meßtechnischen Werten - entsprechend der Ziehkraftgleichung nach [5] möglich.

4. Die besondere Bedeutung des Streifenziehversuchs mit Umlenkung zur Beurteilung der Reibungsverhältnisse beim Tiefziehen

Aus den dargestellten Ergebnissen ist zu schließen, daß die Streifenreibzahlen eine eindeutige Bewertungsmöglichkeit der unterschiedlichen Reibungszustände ergeben. Bei einem Vergleich der errechneten Kräfte mit den maximalen gemessenen Ziehkräften konnten keine Widersprüche festgestellt werden, Bild 11. Die ideelle Umformkraft F_{id} und die Rückbiegekraft F_{rb} wurden als Umformanteil F_U zusammengefaßt.

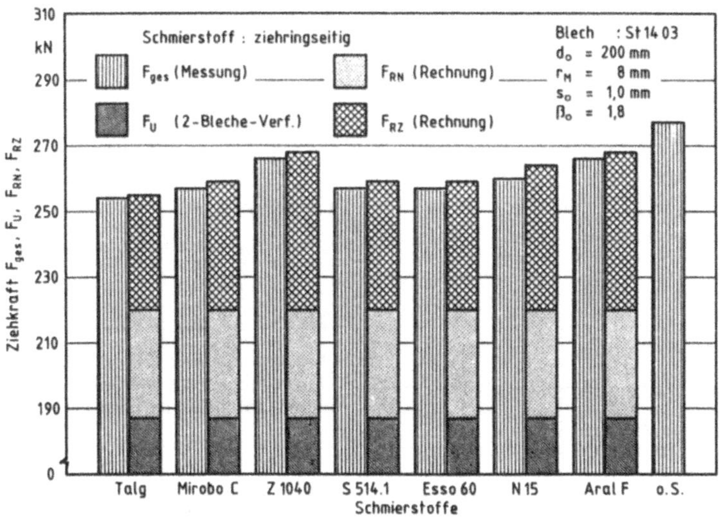

Bild 11: Vergleich der max. Ziehkraft F_{ges} mit der Summe der Ziehkraftanteile bei verschiedenen Schmierstoffen und ziehringseitiger Schmierung

Die Übereinstimmung dieser Kräfte ist recht gut, obwohl mit Hilfe des Streifenziehversuchs mit Umlenkung ursprünglich nur Verhältnisse an der Ziehkantenrundung betrachtet wurden. Daraus ergibt sich die Schlußfolgerung bei ziehringseitiger Schmierung, daß die komplexen Reibungsverhältnisse in der Reibungszone zwischen Ziehring und Niederhalter von Schmierstoff zu Schmierstoff nur geringfügig schwanken und die Ziehkraft in erster Linie durch die unterschiedlichen Reibungs- und Schmierungszustände an der Ziehkantenrundung beeinflußt wird.

Dem Praktiker steht mit dem Streifenziehversuch mit Umlenkung somit ein einfach durchzuführendes Prüfverfahren zur Verfügung, welches es erlaubt, in kurzer Zeit Aufschlüsse über vorliegende bzw. zu erwartende Reibungsverhältnisse zu erhalten.

5. Der Einfluß der Ziehgeschwindigkeit v_{SZ} auf die Reibungsverhältnisse

Auf die Reibungsverhältnisse beim Tiefziehen hat auch die Ziehgeschwindigkeit einen Einfluß. Diese Abhängigkeit läßt sich ebenfalls mit dem Streifenziehversuch mit Umlenkung ermitteln, vergl. Bild 12.

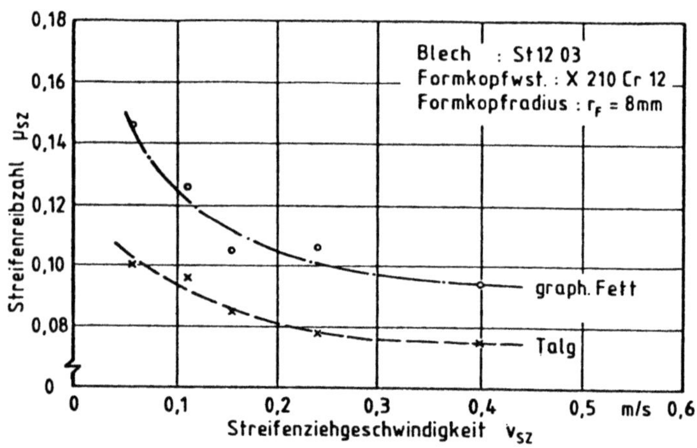

Bild 12: Streifenreibzahl μ_{SZ} in Abhängigkeit von der Streifenziehgeschwindigkeit v_{SZ} bei verschiedenen Schmierstoffen

Es fällt auf, daß mit zunehmender Ziehgeschwindigkeit unabhängig vom Schmierstoff eine Abnahme der Streifenreibzahl μ_{SZ} auftritt. Die Schmierstoffart legt lediglich die Höhe der Reibzahlbeträge fest. Hier wird deutlich, daß das Herausdrücken des Schmierstoffes aus dem Kontaktbereich geschwindigkeitsabhängig ist.

6. Der Einfluß verschiedener Werkzeug-/Werkstückwerkstoffe auf die Streifenreibzahl μ_{SZ} und auf die gesamte Ziehkraft F_{ges}

In Ergänzung zu den durchgeführten Untersuchungen wurde der Einfluß des Werkzeugwerkstoffes auf die Streifenreibzahl μ_{SZ} und auf die maximale Ziehkraft F_{ges} bei verschiedenen Schmierstoffen, Tabelle 1, ermittelt.

Schmierstoff (Hersteller)	Viskosität bei 22°C $[mm^2/s]$	Schmierstoffgrundsubstanz
Talg (N.N.)	n.b. (pastös fest)	Ölsäure
Mirobo C (Fa.Mirobo, Dortmund)	n.b. (pastös weich)	Mineralöl
Z 1040 (Fa.Dietz, Düsseldorf)	n.b. (zähflüssig)	H_2O
S 514.1 (Fa.Matthes, Solingen)	360	Mineralöl
Esso 60 (Fa.Esso AG, Hamburg)	240	Mineralöl
N 15 (Fa.Dietz,Düsseldorf)	195	Mineralöl
Aral F (Fa. Aral AG,Bochum)	5	Mineralöl

n.b. = nicht bestimmbar

Tabelle 1: Übersicht über die untersuchten Schmierstoffe

In Bild 13 und 14 sind die Streifenreibzahlen und die maximalen Ziehkräfte aus Tiefziehversuchen für verschiedene Schmierstoffe und Werkzeugwerkstoffe dargestellt.

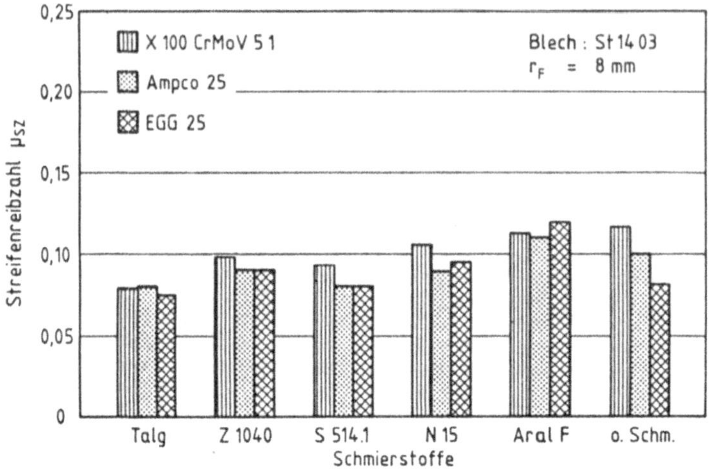

Bild 13: Streifenreibzahlen μ_{SZ} für verschiedene Schmierstoffe und Werkzeugwerkstoffe

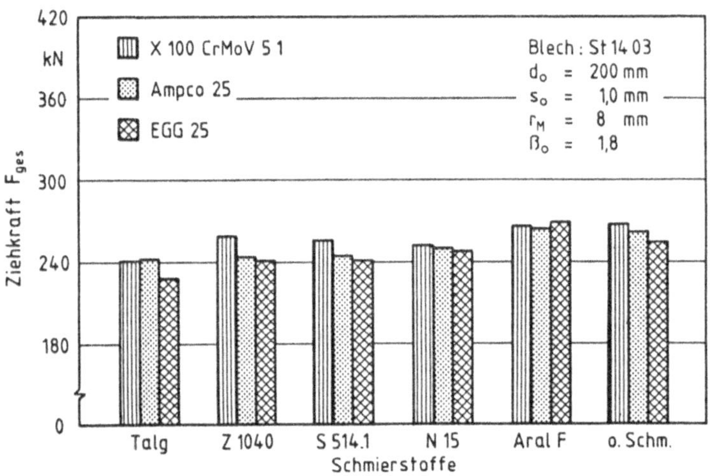

Bild 14: Maximale Ziehkraft F_{ges} bei verschiedenen Schmierstoffen und Werkzeugwerkstoffen bei ziehringseitiger Schmierung

Aus diesen Untersuchungen ist eine geeignete Kombination von Schmierstoff und Werkzeugwerkstoff hinsichtlich minimaler Ziehkraft F_{ges} bzw. Streifenreibzahl μ_{SZ} zu erkennen. Außerdem ist festzustellen, daß die Abweichungen der Gesamtziehkraft F_{ges} - abhängig vom Schmierstoff und Werkzeugwerkstoff - durch die Streifenreibzahl μ_{SZ} aus den entsprechenden Streifenziehversuchen mit Umlenkung gut wiedergegeben werden, vergl. Bild 13 und 14.

7. Die Auswirkung verschiedener Gußwerkzeugwerkstoffe auf die maximale Ziehkraft

Als Werkzeugwerkstoff für Tiefziehwerkzeuge kommen in der Massengüterindustrie hauptsächlich Grauguß, Stahlguß und unlegierte Werkzeugstähle zur Anwendung.
Für Tiefziehwerkzeuge in Großpreßwerken werden heute vorwiegend Gußwerkstoffe, z.B. der Qualität GG 25, eingesetzt. Für die Versuche wurde daher dieser Werkstoff zugrundegelegt, wobei verschiedene Erschmelzungsarten (GG 25, VSGG 25 und EGG 25) berücksichtigt wurden.

In Bild 17 ist der Einfluß des Werkzeugwerkstoffes Grauguß auf die maximale Ziehkraft F_{ges} bei verschiedenen Schmierstoffen dargestellt.

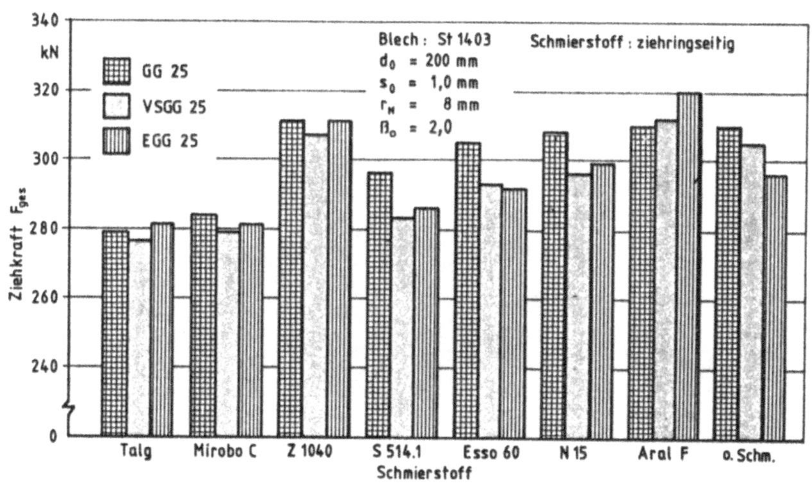

Bild 17: Maximale Ziehkraft F_{ges} bei verschiedenen Schmierstoffen und Graugußwerkzeugwerkstoffen bei ziehringseitiger Schmierung

Der Vergleich der ohne Schmierung durchgeführten Tiefziehversuche mit den Werkzeugwerkstoffen Grauguß und Stahl zeigt, daß bei Graugußwerkzeugen relativ geringere Ziehkräfte auftreten. Dieser Effekt ist dadurch zu erklären, daß die im Gußgefüge, Bild 18, enthaltenen Graphiteinschlüsse positive Auswirkungen auf die Reibungsverhältnisse haben.

Beim Einsatz einiger Schmierstoffe, z.B. Z 1040, ist die Ziehkraft größer als im ungeschmierten Zustand. Man konnte in diesen Fällen das Auftreten eines geschlossenen Schmierfilms unterstellen, so daß die günstige Schmierwirkung des Graphits aufgehoben wird.

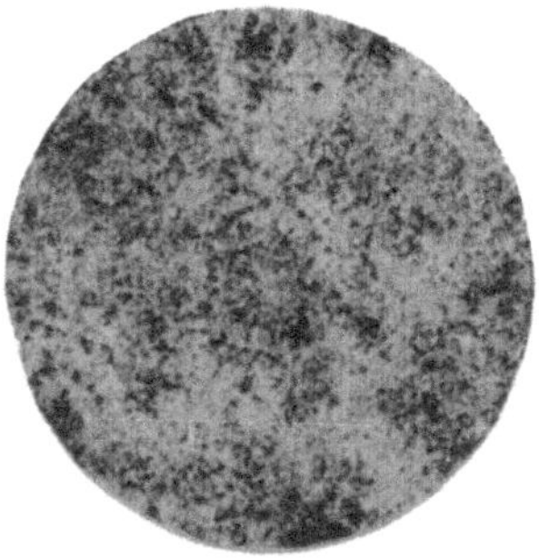

Vergütungsgefüge: Kaltarbeitsstahl X 100 CrMoV 5 1

Graugußgefüge: E GG 25

Bild 18: Schliffbilder eingesetzter Werkzeugwerkstoffe (Vergrößerung: 200)

Schrifttum

[1] Doege, E.: Wichtige Einflußgrößen beim Tiefziehen.
wt.-z. ind. 66 (1976) 11,
S. 615-619

[2] Doege, E., G. Bockel, D. Mohagheghi und M. Hanisch:
Einfluß der Umformmaschine auf die Werkstückgenauigkeit.
HFF-Bericht*) 6 (1980), S. 25/1-25/13

[3] v. Finckenstein, E. u. K.-J. Lawrenz: Beschichtete
Bleche umformen.
Bänder Bleche Rohre 19 (1978) 3. S. 130-136

[4] Möller, J. und F. Fischer: Über den Einfluß von
Ultraschall auf die Reibung beim Tiefziehen.
Bänder Bleche Rohre 16 (1975) 11, S. 457-460

[5] Siebel, E. und W. Panknin: Ziehverfahren der
Blechverarbeitung.
Z. Metallkunde 47 (1956) 4, S. 207-212

[6] Witthüser, K.-P.: Untersuchung von Prüfverfahren zur
Beurteilung der Reibungsverhältnisse beim Tiefziehen.
Dr.-Ing. Diss., Universität Hannover, 1980

[7] Doege, E.: Untersuchung über die maximal übertragbare
Stempelkraft beim Tiefziehen rotationssymmetrischer
Teile.
Berlin, Techn. Uni, Dr.-Ing. Diss., 1963

*) HFF-Bericht: Hannoversches Forschungsinstitut für Fertigungsfragen e.V. (HFF), Welfengarten 1A, 3000 Hannover 1

Schmiermittelprüfung für die Halbwarmumformung

von

Dr. rer. nat. F. Dausinger
Dr. rer. nat. R. Conrad

Robert Bosch GmbH
Abt.FSM
Stuttgart

Förderkennzeichen: 01 ZT 458 S
Laufzeit: 1.8.1978 - 31.12.1980

Zusammenfassung

Die Halbwarmumformung (Fließpressen von Stahl zwischen 750 K und 1000 K) ist ein neuer Verfahrenszweig der Umformtechnik. Bei vergleichbar niedrigem Energie- und Werkstoffeinsatz wie bei der Kaltumformung ermöglicht sie das Fließpressen von Werkstücken mit hoher Festigkeit, die bisher nur durch Warmumformung, Schmieden oder Spanen hergestellt werden konnten. Die Anwendung der Halbwarmumformung wird durch tribologische Probleme erheblich behindert. Angesichts der technischen und wirtschaftlichen Vorzüge dieses Verfahrens erschien es wichtig, eine Prüfmethode zu entwickeln, die es erlaubt, für betriebliche Anwendungsfälle geeignete Schmiermittel auszuwählen. In der vorliegenden Arbeit werden Oberflächenuntersuchungen an Preßteilen mittels Rasterelektronenmikroskopie zur Beurteilung herangezogen. Untersucht wurden drei Verfahren mit unterschiedlicher Oberflächenvergrößerung (Durchschieben, Voll - Vorwärts - Fließpressen, Napf - Rückwärts - Fließpressen) und drei typische Schmiermittelsysteme: Graphit/Wasser, Salz/Wasser und Graphit/Zylinderoel.

Inhaltsverzeichnis

1. Einleitung

Die Halbwarmumformung ist ein neuer Zweig der Verfahrenstechnik. Man bezeichnet damit das Fließpressen von Stahl zwischen 750 K und 1000 K, d.h. in einem Temperaturbereich, der zwischen dem Kaltumformen und dem konventionellen Warmumformen liegt. Die Stärken dieses neuen Verfahrens resultieren daraus, daß es die wesentlichen Vorteile der beiden anderen Methoden vereinigt, deren Hauptnachteile jedoch vermeidet. Mit der Halbwarmumformung lassen sich jetzt auch Werkstoffe mit sehr hohen Festigkeitswerten fließpressen. Diese Stähle konnten bisher nur mit den weniger wirtschaftlichen Verfahren, wie z.B. Schmieden und Spanen, bearbeitet werden. Dabei benötigt die Halbwarmumformung einen Energie- und Werkstoffeinsatz, der genauso niedrig ist wie der bei der Kaltumformung.

Die Werkzeugstandmengen bei der Halbwarmumformung erreichen jedoch bis heute nur etwa ein Drittel der Mengen, die bei der Kaltumformung üblich sind. Dies zeigt die Dringlichkeit von Entwicklungen zur Verringerung des Verschleißes. Deshalb wird im Rahmen dieses Projektes versucht, eine neue Prüfmethode über Verschleißvorgänge und Schmiermittel zu entwickeln. Mit Hilfe der Rasterelektronenmikroskopie werden dazu zunächst Werkstückoberflächen nach dem Umformen begutachtet. Danach wird eine Schadensanalyse betrieben, um Verschleißschäden aufzuklären und gleichzeitig die Erkenntnisse über die Verschleißmechanismen zu vervollständigen. Mit der so angestrebten Untersuchungsmethode ist es dann möglich, Verschleiß frühzeitig zu erkennen und Vergleiche zwischen benutzten Schmiermittelsystemen zu ziehen.

2. Rasterelektronenmikroskopie, eine Methode zur Analyse von tribologischen Problemen

Die Tribologie hat in den letzten Jahren wichtige Beiträge zur Lösung von Verschleißproblemen geleistet. Wichtig dabei war die sowohl von der Forschung als auch von der technischen Praxis betonte Erkenntnis, daß der Verschleiß nicht als eine "Werkstoffeigenschaft", sondern nur als eine Eigenschaft eines "tribologischen Systems" aufgefaßt werden kann [1].

Der Verschleiß ist nach DIN 50320 [2] definiert als Vorgang eines fortschreitenden Materialverlustes aus der Oberfläche eines festen Körpers infolge tribologischer Beanspruchung. Czichos [3] versteht unter tribologischer Beanspruchung die Beanspruchung der Oberfläche eines festen Körpers durch Kontakt und Relativbewegung eines festen, flüssigen oder gasförmigen Gegenkörpers.

Bei der Halbwarmumformung besteht damit das tribologische System im wesentlichen aus dem Werkzeug als Grundkörper, dem ständig erneuerten Werkstück sowie dem Schmiermittel. Die tribologische Beanspruchung ist durch die folgenden Beanspruchungsgrößen gekennzeichnet:

- die Vergrößerung der Werkstückoberfläche
- die Kontaktspannung in der Wirkfuge
- die Temperaturen (Werkzeug, Werkstück und Schmiermittel)
- die Relativbewegung zwischen Werkzeug und Werkstück

Diese Beanspruchungsgrößen sind einer Messung nur äußerst schwer zugänglich [4,5]. Vor allem die Bestimmung der Kontaktspannungen und der Temperaturen in der Wirkfuge sind, wenn überhaupt, jedenfalls im Betrieb unmöglich. Damit fehlt eine wichtige Voraussetzung für eine hinreichend genaue Simulation des betrieblichen tribologischen Systems durch Modellversuche. Dazu kommt, daß es mit den bekannten Simulationsverfahren nicht möglich ist, die in der Praxis auftretenden Werte der Beanspruchungsgrößen zu erreichen [6]. Andererseits ist in der Umformtechnik eine direkte Messung des Verschleißes durch Bestimmung der Standmengen äußerst zeit- und kostenaufwendig. Ideal wäre daher eine Methode, die eine Verschleiß- und Schmiermittelbeurteilung schon nach wenigen Versuchspressungen erlaubte.

Im folgenden wird untersucht, in wieweit sich die Rasterelektronenmikroskopie als eine solche Methode eignet. Dabei werden in Verbindung mit einer energiedispersiven Mikroanalyse und einer Auger - Analyse Werkstückoberflächen beurteilt. Danach wird versucht, schon nach einer kleinen Anzahl von Pressungen Kriterien zu entwickeln, wie Verschleiß verringert werden kann. Bestätigen sich diese Kriterien in einer anschließenden Serienfertigung, dann stellt die angestrebte Untersuchungsmethode zweifelsohne ein Verfahren zur Früherkennung von Verschleißschäden dar.

3. Versuchsdurchführung

3.1 Bau von Modellwerkzeugen

3.1.1 Auswahl der Modellwerkzeuge

Die Modellwerkzeuge dienen zur Herstellung der Werkstücke unter kontrollierter tribologischer Beanspruchung. Zur Variation der Werkstückoberfläche wurden drei Verfahrensvarianten gewählt (siehe Bild 1 - 3):

- Durchdrück - Werkzeug
- Voll - Vorwärts - Fließpress - Werkzeug
- Napf - Rückwärts - Fließpress - Werkzeug

3.1.2 Durchdrück - Werkzeug

Bei der in Bild 1 dargestellten Werkzeugversion werden die Proben praktisch nicht umgeformt, folglich verändert sich auch ihre Oberfläche nur unwesentlich (ca. 2 % beim Überwinden des Einlegespiels). Die Proben werden lediglich unter Druck in einer Matrize (Pos. 1) verschoben, wobei sie von einem Preßstempel (Pos. 2) auf einen hydraulisch abgefederten Gegenstempel (Pos. 3) gedrückt werden. Die axiale Druckspannung auf die Stirnflächen der Probe ist über die Gegenkraft einstellbar und wird durch die Belastbarkeitsgrenze der Stempel auf ca. 2000 N/mm^2 begrenzt. Bei dieser Spannung, die weit über der Fließspannung des Werkstücks liegt (Größenordnung: 500 N/mm^2), verhält sich das Werkstück beinahe hydrostatisch; in der Wirkfuge (Probenmantelfläche) werden somit Normalspannungen bis zu annähernd 2000 N/mm^2 erreicht.

Unter dieser Belastung können die Proben bis zu 60 mm in der Matrize verschoben werden. Dabei bewegt der Gegenstempel einen Hydraulikkolben (Pos. 4), der seinerseits das Hydrauliköl über ein einstellbares Überströmventil in einen Vorratsbehälter preßt. Dadurch wird erreicht, daß die Gegenkraft über den Preßweg ausreichend konstant bleibt. Bei Entlastung des Gegenstempels wird das Öl mit

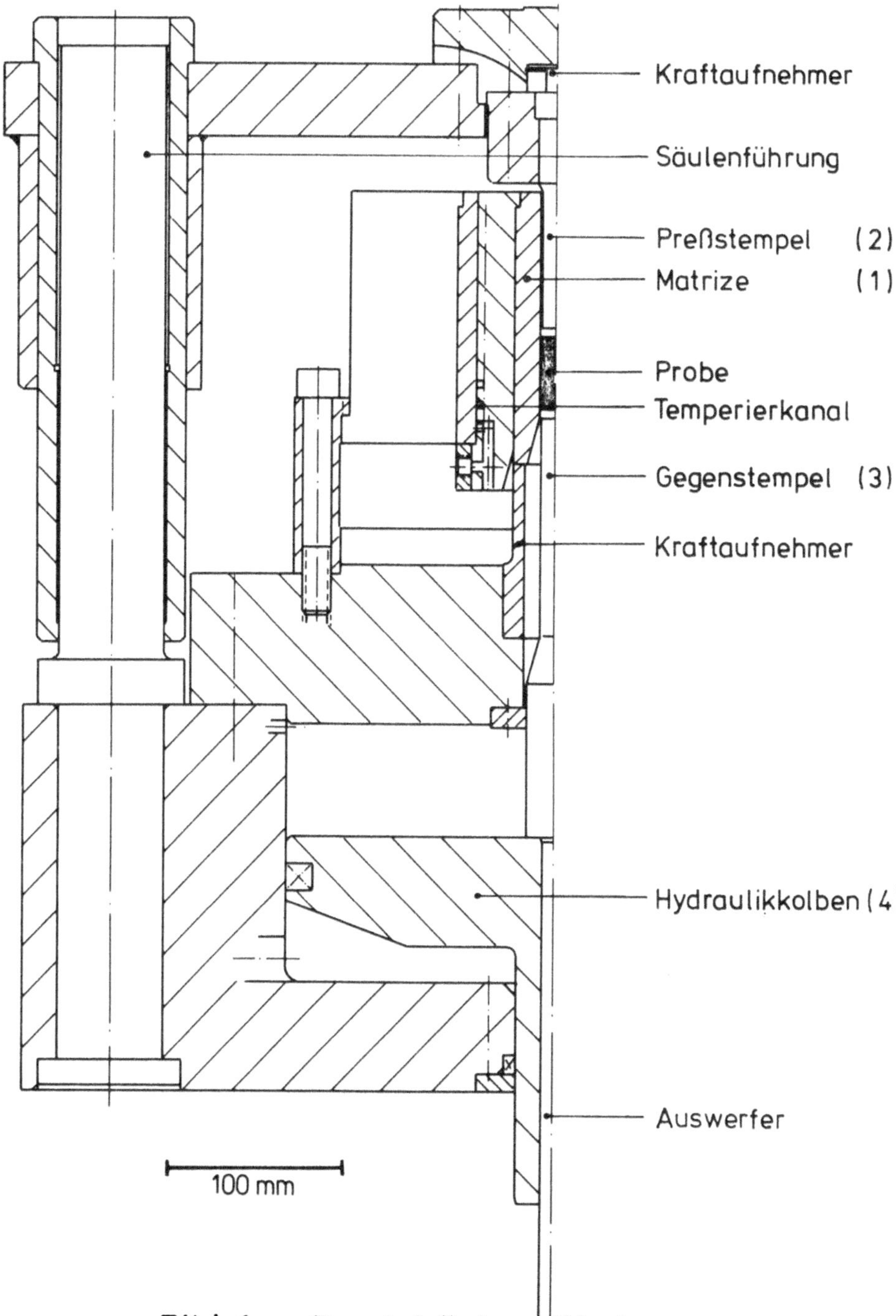

Bild 1: **Durchdrück - Werkzeug**

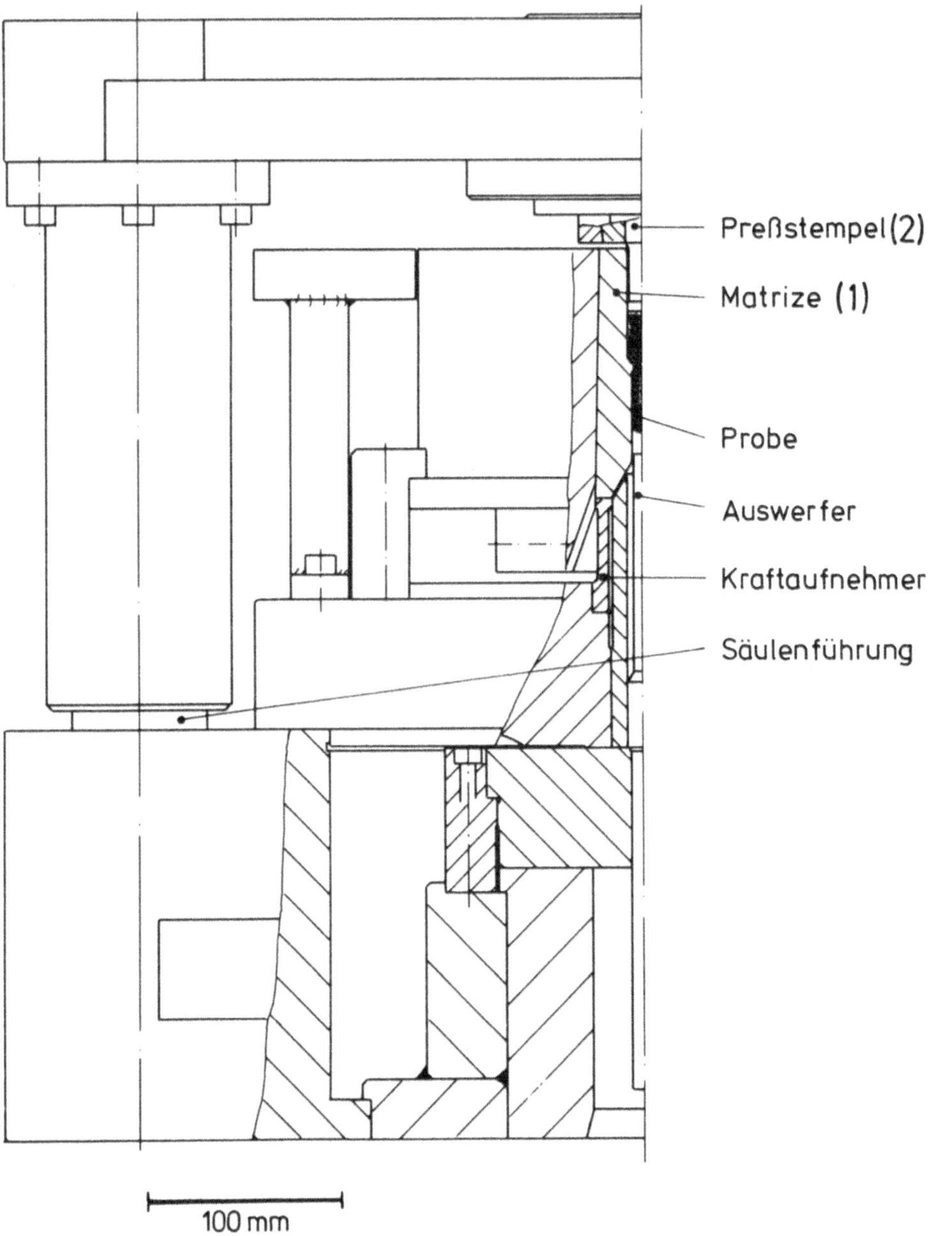

Bild 2 : Voll - Vorwärts - Fließpress - Werkzeug

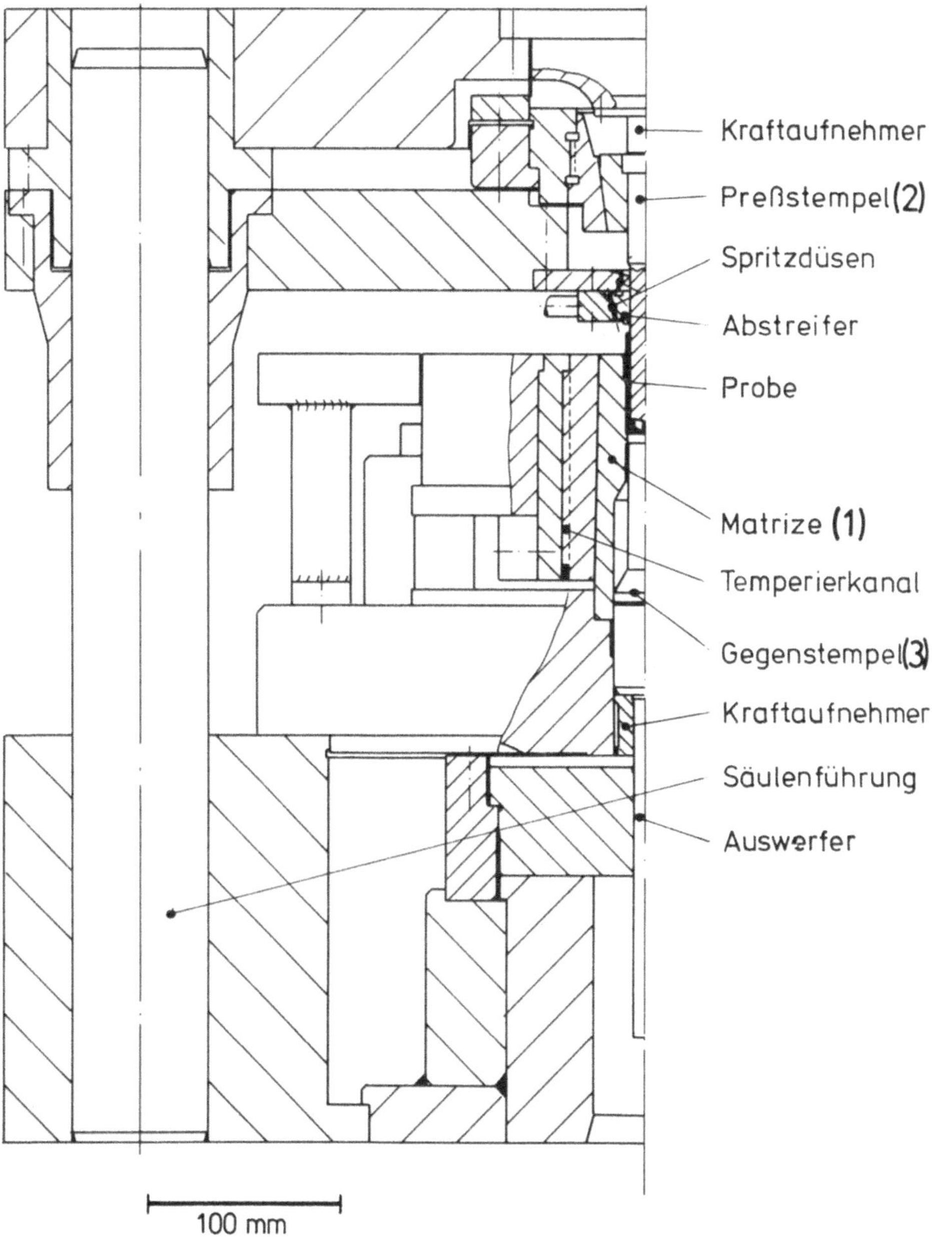

Bild 3 : Napf - Rückwärts - Fließpress - Werkzeug

Preßluft in den Hydraulikzylinder zurückgedrückt.

3.1.3 Voll - Vorwärts - Fließpress - Werkzeug

Bei dieser Werkzeugversion wird zur Realisierung einer mittleren Oberflächenvergrößerung eine andere Matrize eingesetzt und die hydraulische Feder gegen einen starren Aufbau ausgetauscht. Nach den in Bild 4 dargestellten Fließpressteilen wird auf der Mantelfläche eine Oberflächenzunahme um 70 % erreicht.

3.1.4 Napf - Rückwärts - Fließpress - Werkzeug

Durch Napffließpressen läßt sich die stärkste Oberflächenzunahme erreichen. Nach Bild 4 ergibt sich für die Napfinnenwand rein rechnerisch eine Oberflächenvergrößerung von 780 %, wobei eine gleichmäßige Formänderung angenommen ist. Bei der zu erwartenden inhomogenen Formänderung kann dieser Wert jedoch lokal noch weit überschritten werden. Durch Zurücknahme der Pressenzustellung - der Stempel taucht weniger weit ins Werkstück ein - kann die Formänderung auf einfache Weise vermindert werden. Das für die Herstellung der Napfteile konzipierte Werkzeug ist in Bild 3 dargestellt. Es verwendet weitgehend den Aufbau der beiden anderen Versionen, ergänzt durch eine Abstreiferplatte.

3.1.5 Eignung der Werkzeuge für Standmengenversuche

Bei den Standmengenversuchen werden unter den gleichen Bedingungen alle drei genannten Verfahren mehrmals durchfahren. Dadurch entstanden in Zusammenhang mit den vorher durchgeführten Kurzversuchen innerhalb dieses Projektes 35 Versuchsserien mit einigen zehntausend Teilen. Insbesondere für die Standmengenversuche war es daher erstrebenswert, daß die Teile erstens mechanisiert eingelegt und zweitens einer weiteren Verwendung zugeführt werden konnten. Durch extreme Säulenabstände und Freiräume zwischen Werkzeugober- und Unterteil wurde

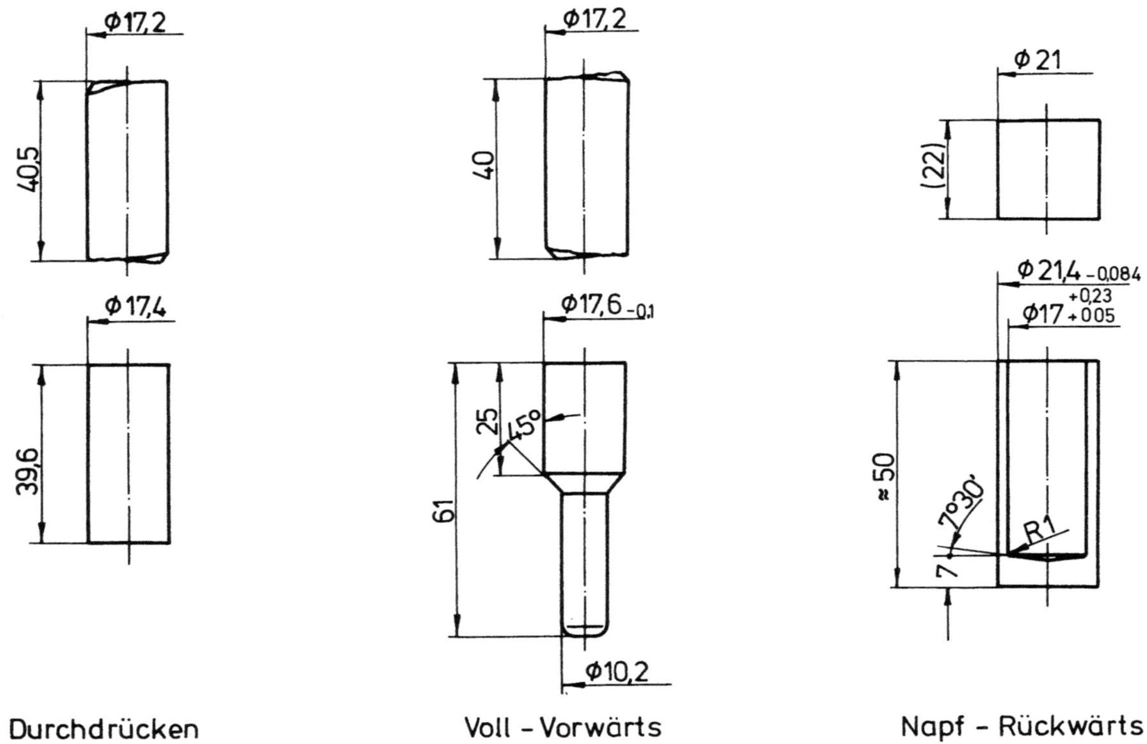

Bild 4 : Fließpressteile

Platz für Greiferschienen gelassen. Diese Schienen liegen auf der Matrize sowie Auflagewinkeln und werden von einem direkt von der Presse angetriebenem Stufenvorschubgerät bewegt. Wegen der Weiterverwendbarkeit wurde für das Voll - Vorwärts - Fließpressen ein Serienteil ausgewählt. Die Durchdrück - Teile können ihrerseits als Rohteile für das Voll - Vorwärts - Fließpressen verwendet werden.

In Bild 5 ist schematisch der Aufbau der Anlage zur Halbwarmumformung gezeigt. Mit ihr wurden im Rahmen die-

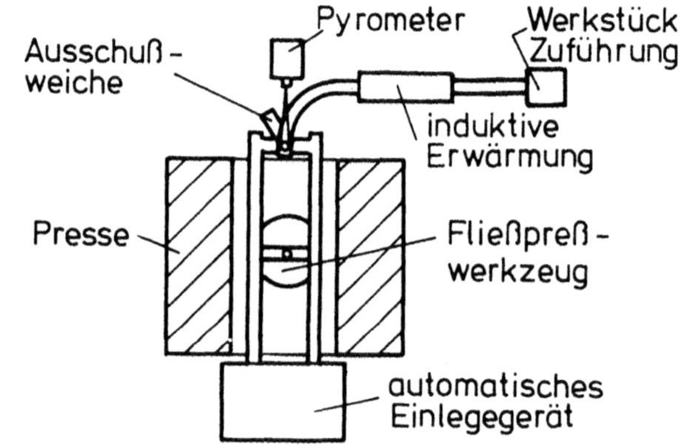

BILD 5: Anlage zur Halbwarmumformung (schematisch)

ses Projektes ca. 20000 Teile gefertigt, von denen mehr als 15000 Teile in der Einspritzdüsenfertigung weiterverwendet werden können.

3.2 Meßgrößen

3.2.1 Kontaktspannungen in der Wirkfuge

Zur genauen Beschreibung des vorliegenden tribologischen Systems ist eine Erfassung und Kontrolle der unter 2. genannten Beanspruchungsgrößen erforderlich. Bei den

Untersuchungen der Kontaktspannungen in der Wirkfuge ist zu beachten, daß keine ortsunabhängigen Spannungen vorliegen. Zur genauen Beschreibung sind daher lokale Messungen möglichst an verschiedenen Werkzeugpartien erforderlich. Sie sind nur mit Meßstiften möglich [7], jedoch mit großem Aufwand und mit störendem Einfluß auf die Produktion verbunden. Sie kommen daher für eine Serienfertigung nicht in Frage.

Versuche, die Normalspannungen über die radiale Dehnung des Matrizenverbandes zu erfassen, waren nicht durchführbar. Die Dehnmeßstreifen auf dem Schrumpfring ergaben keine reproduzierbaren Meßergebnisse. Erfasst wurden schließlich die folgenden Größen:

- Preßkraft mit Kraftaufnehmer über Preßstempel
- Axialkraft auf die Matrize mit Kraftaufnehmer unter der Matrize
  bzw.
- Auflagekraft des Werkstücks beim Napffließpressen sowie
- Hydraulikdruck beim Durchdrück - Verfahren

3.2.2 Temperaturen in der Wirkfuge

Die Temperaturen des vorliegenden tribologischen Systems werden in der Kontaktfläche Werkzeug - Werkstück auf vielfache Weise beeinflußt:

- die Verschweißung nimmt mit steigender Temperatur zu
- die Schmiermitteleigenschaften sind stark temperaturabhängig
- der Werkzeugverschleiß steigt bei thermischer Belastung
- die Reibschubspannung wird durch die Temperaturabhängigkeit der Fließspannung der Werkstückstoffe beeinflußt.

Zusätzlich führt die während des Pressens durch Umformen und Reibung enstehende Wärme zu einer unerwünschten Temperatursteigerung der Werkzeuge.Daher haben wir für die Versuchswerkzeuge ein kombiniertes Heiz - und Kühlsystem gewählt. Das Werkzeug wurde mit Temperierkanälen versehen, durch die thermostatisiertes Wärmeträgeroel gepumpt wird. Bei der Auslegung der Temperierkanäle mußten folgende Gesichtspunkte beachtet werden:

- Die Kanäle sollten möglichst nahe am Arbeitsraum der Matrize liegen.
- Die Festigkeit des Matrizenverbandes soll möglichst wenig geschwächt werden.
- Der dem Verschleiß unterworfene Matrizenkern sollte möglichst einfach ausgeführt sein.

Unter diesen Gesichtspunkten entschieden wir uns für eine doppelte Wendel (Vor- und Rücklauf) im mittleren Ring des Schrumpfverbandes. An diesen Temperierkanal wird über den Anschlußring und flexible Schläuche ein Temperaturregler angeschlossen, wie er auch bei Spritzgußwerkzeugen eingesetzt wird.

Die eigentliche Temperaturmessung in der Wirkfuge ist genauso problematisch wie die Messung der Kontaktspannungen [8]. Da in der Wirkfugenpartie die größte tribologische Belastung zu erwarten ist, haben wir Thermoelemente bis auf 0.5 mm an diese Stellen herangeführt. Beim Durchdrück - und Voll - Vorwärts - Werkzeug sollten die Temperaturen in der Matrize, beim Napf - Rückwärts - Werkzeug dagegen im Preßstempel näherungsweise bestimmt werden. Auswertbare Meßergebnisse waren jedoch auf diese Weise nicht zu erhalten. Bei der endgültigen Werkzeugversion verzichteten wir daher auf diese Messungen und kontrollierten nur, daß die Matrizen und Stempel des Werkzeuges durch thermostatisiertes Wärmeträgeroel auf 150° C vorgewärmt waren. Zusätzlich war gewährleistet, daß die Werkstück - Rohteile in 60 s auf $700 \pm 20^{\circ}$ C erwärmt werden.

3.2.3 Relativbewegungen

Das Werkzeuggestell kann sowohl in eine hydraulische als auch in eine Kurbelpresse eingebaut werden. In beiden Fällen ist der Preßweg über die Stößelzustellung wählbar. Bei der hydraulischen Presse liegt die Preßgeschwindigkeit fest, sie ist während des Pressens weitgehend konstant. Bei den Kurbelpressen nimmt dagegen die Geschwindigkeit gemäß der Pressenkinematik von der Auftreffgeschwindigkeit bis auf Null ab. Die Auftreffgeschwindigkeit ist über die Hubzahl stufenlos einstellbar.

Bei unseren Pressungen war das Werkzeug in eine Kurbelpresse mit 6.3 MN Nennpreßkraft eingebaut, deren Umlaufgeschwindigkeit auf 20 Hübe pro min eingestellt war.

3.3 Schmiermittelsysteme

3.3.1 Schmiermittelarten

Es ist bekannt, daß im Temperaturbereich der Halbwarmumformung (750 K bis 1000 K) Schmiermittelsysteme, die beim Kaltfließpressen Verwendung finden, zu unbeständig werden. Wir haben daher auf Schmiermittel zurückgegriffen, die teilweise schon bei der Warmumformung verwendet wurden. In Tabelle 1 sind die von uns untersuchten Systeme angegeben:

Graphit/Zylinderöl	Graphit/Wasser	Salz/Wasser
Kolloidalgraphit in Zylinderöl (5 gew %) Bosch-eigene Rezeptur	10 vol % Graphit Basis: Delta 144 der Firma Acheson	5 gew % Salz (Alkali-phosphate und -borate) Basis WF 54 EJ der Firma Matthes

Tabelle 1: verwendete Schmiermittelsysteme

3.3.2 Auftragung der Schmiermittel

In den Kurzversuchsreihen gelangte beim Durchdrück- und Voll - Vorwärts - Werkzeug das Schmiermittel, das außerhalb des Werkzeuges vernebelt wird, über Schmierkanäle von unten in die Matrizenbohrung. Beim Napfwerkzeug wurde dagegen die Matrize von oben und außerdem noch der Napfstempel besprüht.

Bei den Standmengenversuchen mit dem Voll - Vorwärts - Werkzeug zeigte sich, daß das Schmiermittelsystem Graphit/Wasser schon nach 100 Pressungen die Schmierkanäle zusetzte. Nur mit dem Schmiermittel Graphit/Zylinderoel ließen sich mehr als 100 Pressungen durchführen, wenn auch unter Schwierigkeiten, da auf der Zuführseite mit einem Druck von mehr als 5 bar gearbeitet werden mußte. In den abschließenden Standmengenversuchen wurden mit einem Sprühsystem die Matrize und der Stempel durch drei Düsen von oben besprüht. Sprühdauer und -druck wurden durch das von der Presse angetriebene Einlegegerät gesteuert.

3.4 Verwendete Werkstoffe

Bei den Werkstück - Werkstoffen handelte es sich um Rohlinge aus 18 CrNi8 und 100 Cr 6, die jeweils aus feingeschälten Stangen abgesägt wurden. Beim Durchdrücken und Voll - Vorwärts - Fließpressen wurden nur Rohlinge aus 18 CrNi8, beim Napf - Rückwärts - Fließpressen dagegen nur aus 100 Cr 6 verwendet.

Alle formgebenden Werkzeuge wurden aus dem Schnellarbeitsstahl S6 - 5 - 2 angefertigt. Die Rockwell - Härte betrug wegen der größeren Zähigkeit nur 57 statt 63.

3.5 Beurteilung von Werkstückoberflächen

Im Rasterelektronenmikroskop werden zunächst Oberflächenerscheinungen auf den Werkstücken untersucht. Dabei werden topographische Effekte, wie z.B. Fresser und Auftragungen, begutachtet. Danach wird versucht, die aufgetretenen Schäden zu analysieren. Dabei ist zu berücksichtigen, daß die Elementanalysen im Rasterelektronenmikroskop nur ausreichen, um Schmiermittel zu analysieren, die bei der Kaltumformung angewendet werden (wie z.B. MoS_2). Für die bei der Halbwarmumfornung üblichen Schmiermittel muß dagegen noch eine Auger - Analyse hinzugezogen werden. Denn damit lassen sich auch die Elemente Sauerstoff (Oxidschicht) und Kohlenstoff (die Schmiermittel: Graphit/Wasser und Graphit/Zylinderoel) nachweisen, die mit einer energiedispersiven Mikroanalyse nicht zu bestimmen sind. Zusätzlich bietet die Auger - Analyse die Möglichkeit, die Randzonen der Werkstücke zu untersuchen. Damit ließe sich vielleicht klären, ob dieser Werkstückbereich z.B. aus FeO besteht [9,10], was eine wichtige Erkenntnis wäre, da dieses Eisenoxid wesentlich bessere duktile Eigenschaften hat als die höherwertigen Oxide Fe_2O_3 und Fe_3O_4.

Um einen Vergleich der drei verwendeten Schmiermittel zu erhalten, wurden die Oberflächen der Werkstücke jeweils nach der gleichen Anzahl von Pressungen untersucht.

4. Versuchsergebnisse

4.1 Oberflächenerscheinungen im Kurzversuch nach 50 Pressungen

4.1.1 Durchschieben

Beim Schmiermittel Graphit/Wasser ist eine deutliche 3-Schichtenstruktur zu erkennen (siehe Bild 6). Es folgen aufeinander: eine schwarze und eine graue Schicht sowie helle Aufbrüche, die durch Reinigen der Probe im Ultraschallbad entstanden. Energiedispersive Mikroanalyse und Auger-Analyse ergaben, daß die schwarze Schicht sich aus Schmiermittelresten, die graue Schicht sich aus einer Kombination von Schmiermittelresten und Eisenoxiden zusammensetzt (wobei über den prozentualen Anteil wegen der geringen Schichtdicke keine Aussage möglich ist) und die hellen Aufbrüche aus dem Stahlgrundmaterial bestehen. In Skizze 1 sind die Ergebnisse der zugehörigen Mikro-

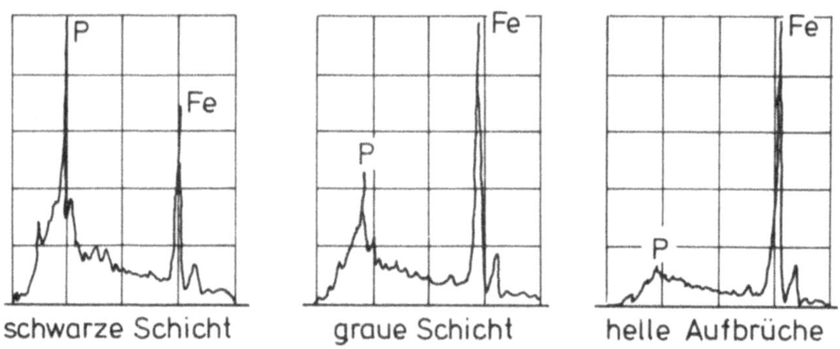

Skizze 1: energiedispersive Mikroanalyse z. Bild 6

analyse angegeben und in Bild 7 die Ergebnisse einer entsprechenden Auger - Analyse bzgl. der Elemente Fe, O und C. In Skizze 1 sieht man deutlich die unterschiedlichen P - Elementpeakhöhen in den drei Schichten. (Die Schmierschicht wurde in diesem Fall über den Phosphorgehalt untersucht, weil der Kohlenstoffgehalt der

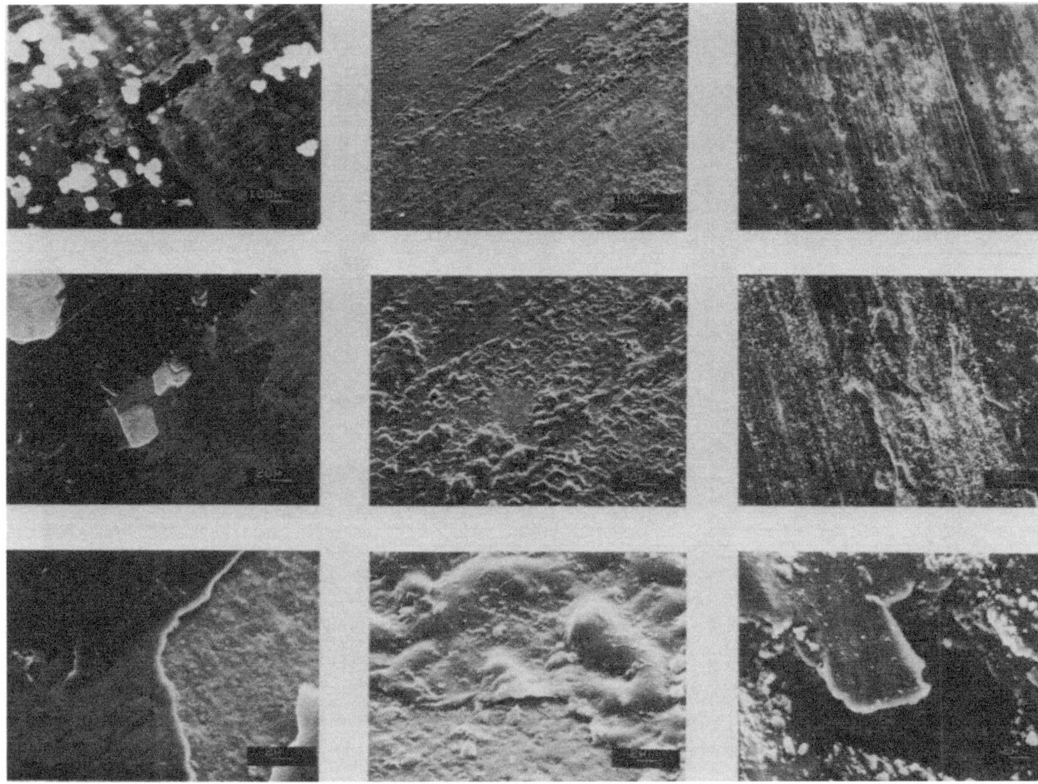

Graphit / Wasser | Salz / Wasser | Zylinderoel / Graphit

Bild 6 : Durchschieben

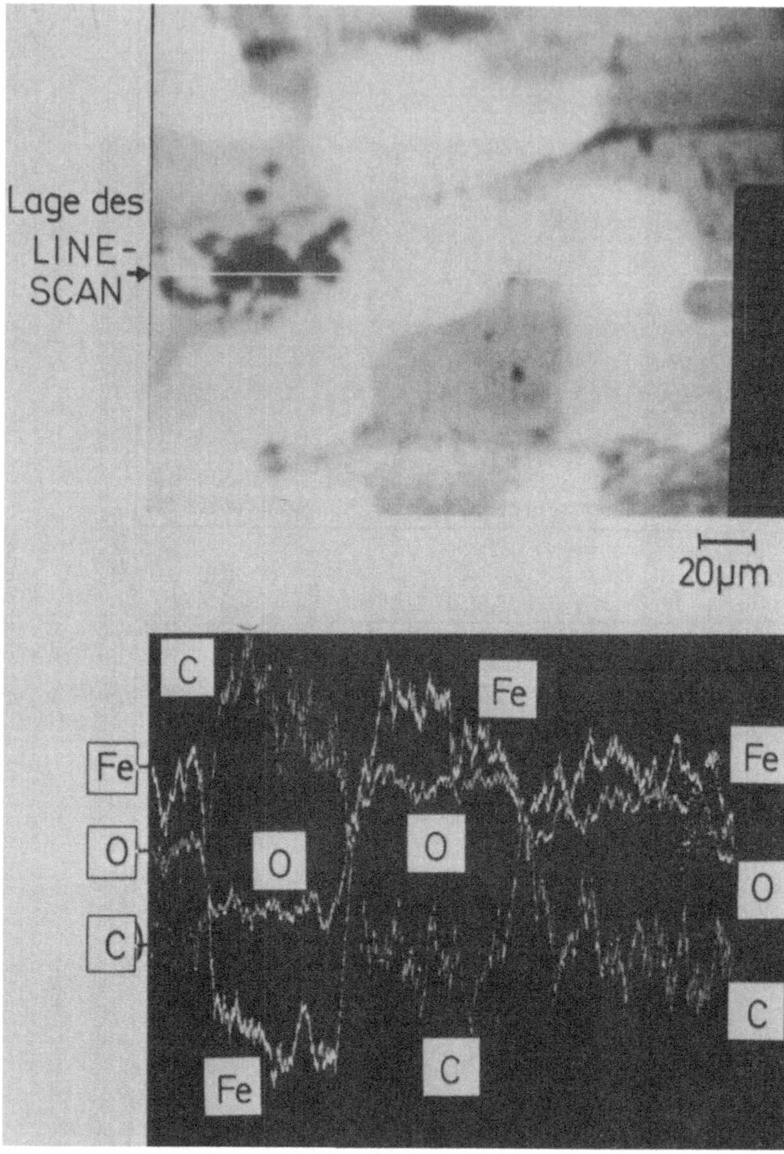

Bild 7: Augeranalyse zu Bild 6
bzgl. Fe, O und C (Durchsch.m.Graphit/Wasser)

Schicht mit einer energiedispersiven Mikroanalyse nicht zu bestimmen ist und die Emissionsspektralanalyse des Schmiermittels 10 % Phosphor ergab.) Die Auger - Analyse in Bild 7 bestätigt diese Ergebnisse. Man sieht, wie in der schwarzen Schicht der C - Gehalt stark ansteigt und beim Übergang zum hellen Aufbruch im gleichen Maße wieder abfällt. Dafür überwiegt in diesem Probenbereich der Fe - Gehalt, während die graue Schicht durch einen in etwa gleichen Fe - und O - Gehalt ausgezeichnet ist.

Auf der Werkstückoberfläche sind keinerlei Fresser zu erkennen, die auf Mikroverschweißungen zwischen Werkzeug und Werkstück hindeuten würden. Die Pressriefenstruktur ist nur schwach ausgeprägt. Beim Schmiermittel Graphit/Zylinderoel sind dagegen starke Fresser, eine ausgeprägte Pressriefenstruktur und größere Aufplatzungen auf der Werkstückoberfläche zu erkennen. Durch Analysen waren keinerlei Schmiermittelreste mehr nachzuweisen. Beim Schmiermittel Salz/Wasser ist deutlich zu erkennen, daß die Werkstückoberfläche nach dem Umformen von einer direkt aus dem flüssigen Zustand erstarrten Schmierschicht zugedeckt wurde. Dadurch wurden die umformspezifischen Erscheinungen verfälscht und eine Beurteilung der entstandenen Oberflächenformation ist nicht mehr möglich.

4.1.2 Voll - Vorwärts - Fließpressen

Beim Schmiermittel Graphit/Wasser ist eine deutliche Pressriefenstruktur zu beobachten. Diese Pressriefen nehmen in Tiefe und Ausdehnung kontinuierlich von Teil 1 bis Teil 10 von etwa 0 bis 90 % zu. Nach weiteren Pressungen bleibt die entstandene Struktur dann konstant. (Bei den dunklen Auftragungen handelt es sich um Schmiermittelreste, wie es durch Auger - Analysen bestätigt werden konnte. Sie wurden vermutlich nach dem Umformen auf die Werkstückoberfläche gespritzt.) Beim Schmiermittel Graphit/Zylinderoel sind neben diesen Pressriefen ausgesprochen schuppige Strukturen mit größeren Auftragungen zu erkennen. Sie erreichen eine Höhe

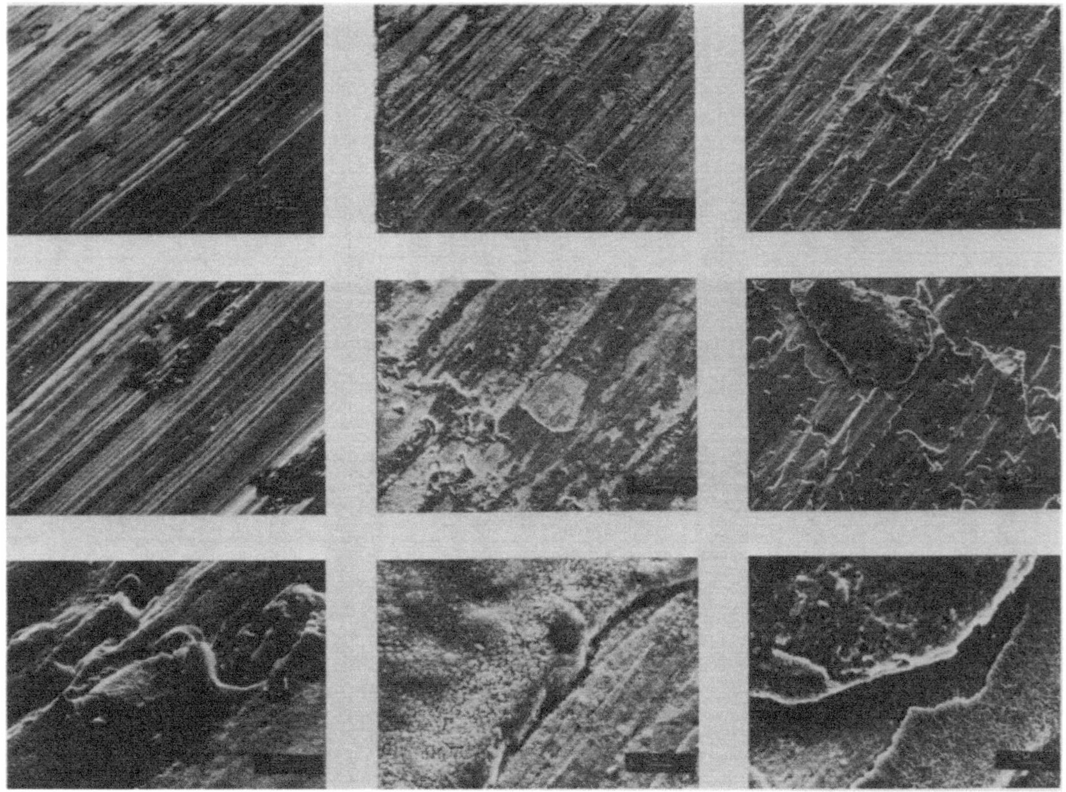

Graphit / Wasser | Salz / Wasser | Zylinderoel / Graphit

Bild 8: Voll - Vorwärts - Fließpressen

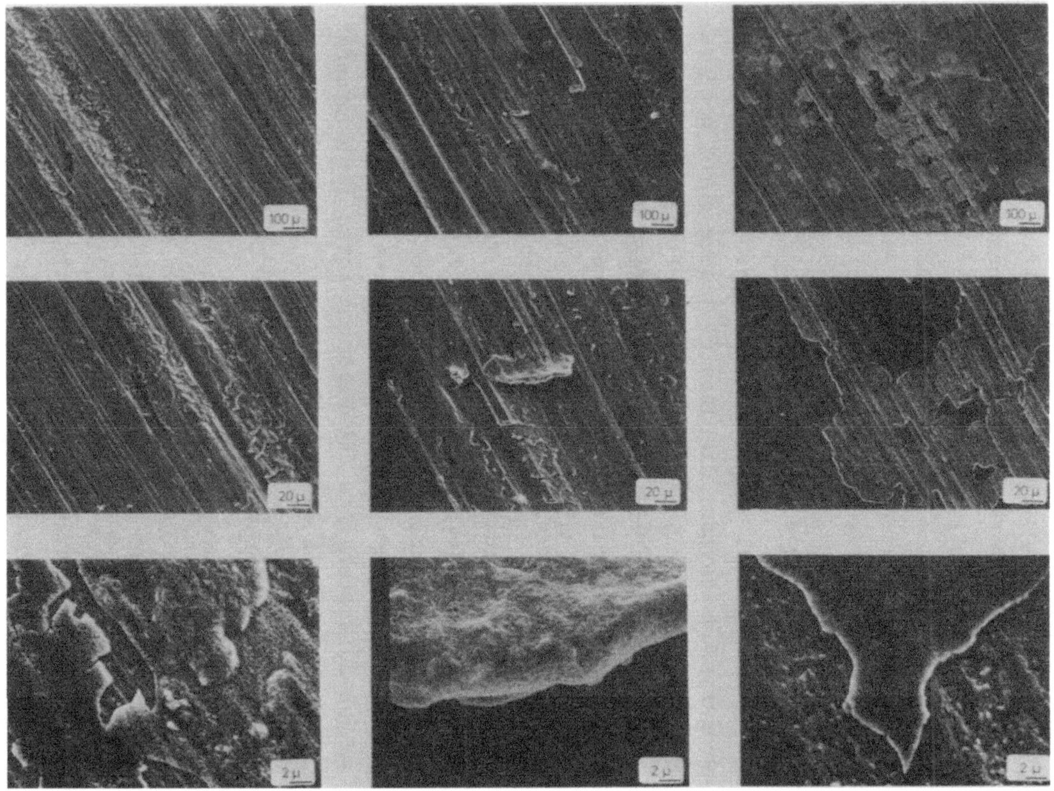

Graphit / Wasser — Salz/Wasser — Zylinderoel/Graphit

Bild 9: Napf-Rückwärts-Fließpressen

von ungefähr 30 μm und deuten darauf hin, daß es bei diesem Schmiermittel zu Mikroverschweißungen gekommen ist. Beim Schmiermittel Salz/Wasser ist dagegen die Werkstückoberfläche durch Pressriefen, Querriefen und Aufbrüche der obersten Schicht gekennzeichnet. Da der Abstand der Querriefen zueinander verschieden ist, kann man davon ausgehen, daß sie keine Schälriefen sind. Sie zeigen einen steten Wechsel zwischen Haft- und Gleitreibung an der Werkstückoberfläche.

4.1.3 Napf - Rückwärts - Fließpressen

Das Schmiermittel Graphit/Wasser ergibt eine deutliche Pressriefenstruktur, die diesmal jedoch durch größere Mikroverschweißungen und daraus resultierenden Ausreißern von Material innerhalb der Riefen gekennzeichnet ist. Das Schmiermittel Graphit/Zylinderoel ergibt dagegen eine relativ glatte Oberfläche, obwohl es größere Bereiche gibt, in denen die Oxidschichten abgeplatzt sind. Im Gegensatz zu Graphit/Wasser kommt es jedoch nicht zu den ausgeprägten örtlichen Fressern. Das Schmiermittel Salz/Wasser liefert auch eine relativ glatte Oberfläche, obwohl lokal starke Mikroverschweissungen auftreten. Die abgebildeten Aufroller enstanden beim Herausziehen des Stempels.

4.2 Randschichten der Werkstücke

Die in diesem Kapitel untersuchten Werkstücke befinden sich bis unmittelbar vor dem Umformen für etwa 60 s auf 973 K. Dabei sind die warmen Oberflächen einer ständigen Sauerstoffzufuhr ausgesetzt, so daß Oxidschichten am Werkstückrand entstehen. Bekanntlich [9,10] bilden sich bei diesen Temperaturen auf Eisenoberflächen bevorzugt FeO - Schichten. Diese sind auf Grund ihrer Duktilität besser geeignet, die Funktion einer Notschmierung zu übernehmen, als die höherwertigen Oxide Fe_3O_4 und Fe_2O_3, die sich bei Temperaturen $T < 833$ K bilden. Bild 10 zeigt eine typische Randschichtformation und in Bild 11 sind

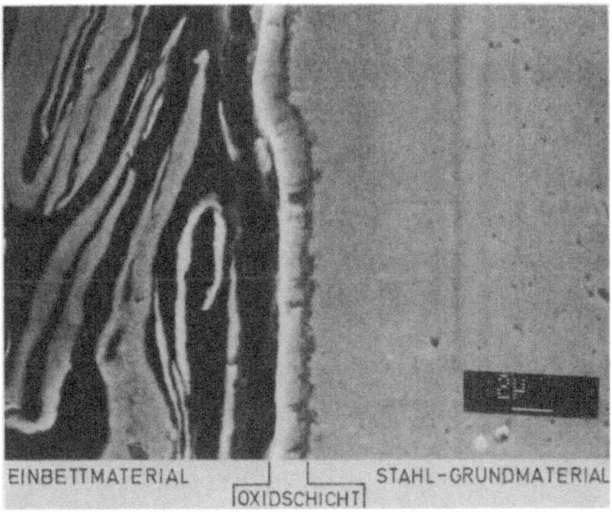

Bild 10: Randschicht eines Voll-Vorwärts-Fließpressteil

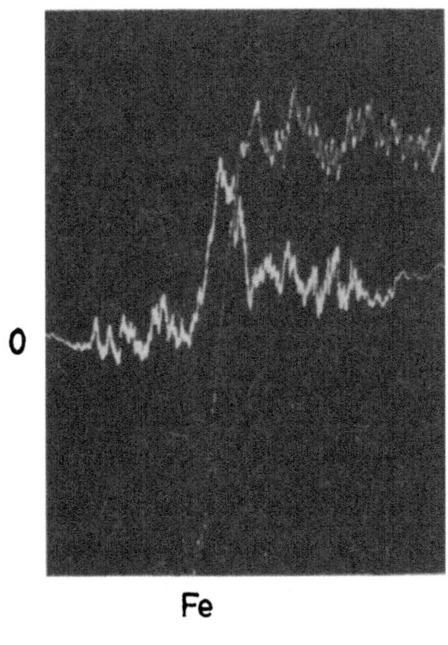

Bild 11: Eisen- und Sauerstoff-Verteilungen in Bild 10

die zugehörigen Fe - und O - Verteilungen über den Probenrand angegeben. Die Elementverteilung der Auger-Analyse deutet darauf hin, daß der Sauerstoffanteil in der Randzone variiert. Allerdings ist bei einer Gesamtschichtdicke von etwa 3 - 5 μm ein prozentualer Nachweis der höherwertigen Oxide mit dieser Analysenmethode nicht möglich. Mikrohärte-Messungen, die mit unterschiedlichen Auflagegewichten zwischen 20 und 200 g immer den für FeO typischen Wert von 350 HV ergaben, scheinen jedoch zusammen mit der Auger - Analyse die Annahme zu erhärten, daß der Hauptbestandteil der Randschichten aus FeO besteht.

Um zu beurteilen, ob eine Oxidschicht die Funktion einer Notschmierung übernehmen kann, ist es wichtig zu wissen, ob die Oxidschichten den Umformprozess mitmachen und in welcher Weise sie sich eventuell dabei verändern. Daher wurden bei einem nicht erwärmten und bei einem erwärmten Werkstück - Rohteil sowie bei einem umgeformten Werkstück die Randschichten näher analysiert. Bild 12 gibt an, an welchen drei Stellen die Schichtdicken untersucht wurden. Um zu gewährleisten, daß nur Oxidschichten analysiert wurden, die vor dem Umformprozeß entstanden sind, wurde der Werkzeugstempel im Werkstück belassen und damit eine Oxidation beim Abkühlen unterdrückt. Bild 13 zeigt Beispiele entsprechender Randschichtformationen. Es zeigte sich, daß bei den nur erwärmten Werkstück - Rohteilen die Oxidschichtdicken zwischen 3 μm und 7 μm liegen, bei den umgeformten Werkstücken dagegen zwischen 2.5 μm und 1 μm.

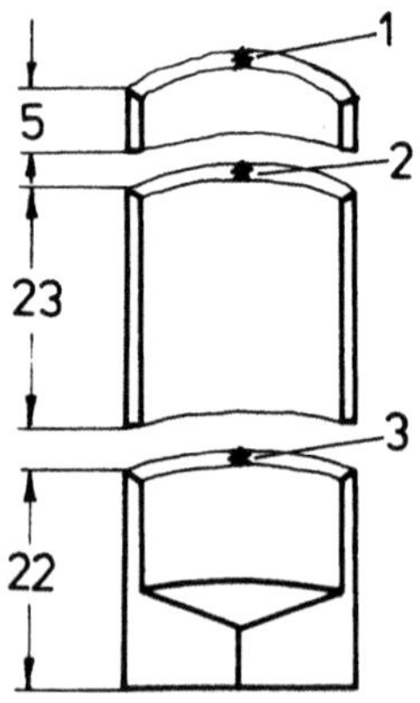

BILD 12 :

Analysenpunkte beim Napfen

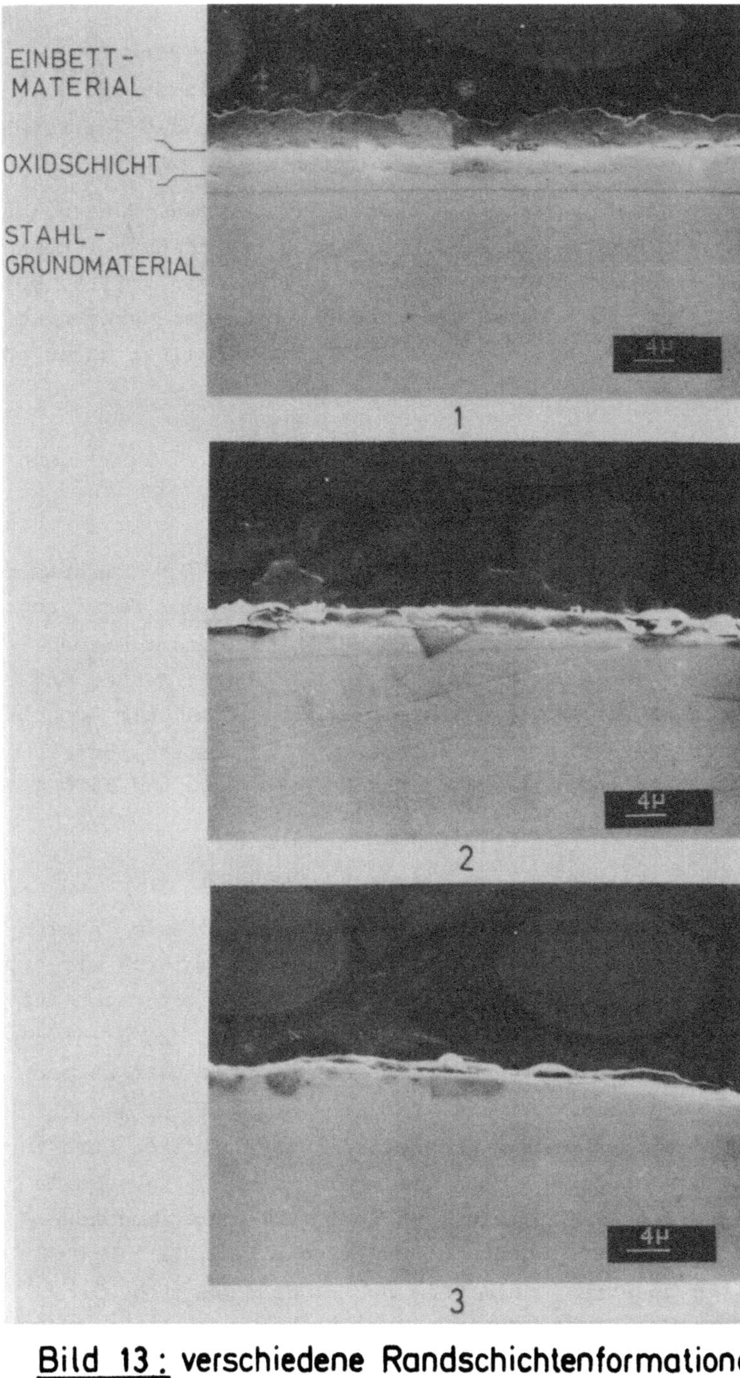

Bild 13: verschiedene Randschichtenformationen b. Napfen

Es scheint, daß Oxidschichten beim Napf - Rückwärts - Fließpressen Oberflächenvergrößerungen bis zu 780 % überstehen und nur in ihrer Dicke um etwa die Hälfte abnehmen.

Zusammen mit diesen Ergebnissen und den o.a. Untersuchungen (Auger - Analyse und Mikrohärtemessungen) kann also angenommen werden, daß es möglich sein sollte, bei der Halbwarmumformung die Randschichten der Werkstücke mit heranzuziehen, um auch im Notfall zu einer guten Lösung der Schmierprobleme zu gelangen.

4.3 Standmengenversuche

4.3.1 Versuchsdurchführung

Für die Standmengenversuche wurde das Voll - Vorwärts - Fließpreßteil verwendet, da es weiter in der Fertigung eingesetzt werden konnte. Es wurden insgesamt sieben Serien mit den Schmiermittelsystemen Graphit/Wasser und Graphit/Zylinderoel gefahren. Dabei wurden beim Graphit/Wasser Standmengen mit 50, 180 und 500 Teilen, beim Graphit/Zylinderoel dagegen mit 1400, 2600 und 7600 Teilen erreicht.

4.3.2 Graphit/Zylinderoel im Standmengenversuch

Nach den Kurzversuchen ergaben sich bei diesem Schmiermittelsystem ausgesprochen schuppige Strukturen mit größeren Auftragungen und vielen Mikroverschweißungen auf der Werkstückoberfläche. Eine Beobachtung, die nicht ganz in die Serienfertigung zu übernehmen ist. Zwar zeigen sich auch hier teilweise ähnliche Strukturen, jedoch treten sie in stetem Wechsel mit einer glatten Struktur auf, wie es Bild 14 bei Teile Nr. 50 und 51 aus Serie 32 zeigt. Diese Erscheinung spricht für einen abrupten Wechsel bei den Schubspannungen, die beim Teil Nr. 50 sehr groß waren, beim Teil Nr. 51 dagegen nicht mehr auftreten. Dieser Übergang von rauh zu glatt auf der Werkstückoberfläche ist nicht unbedingt mit zwei aufein-

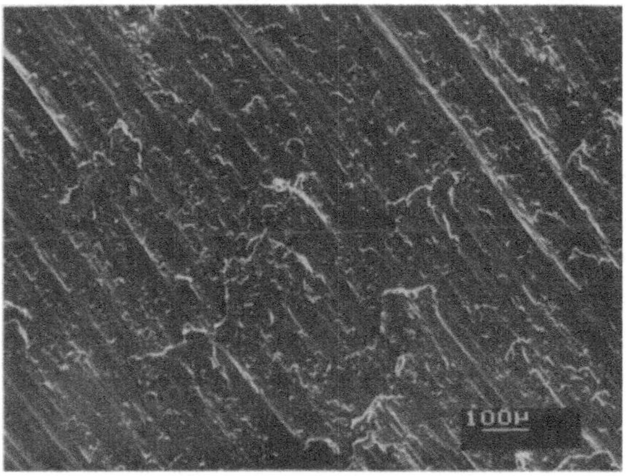

Teil - Nr. 50

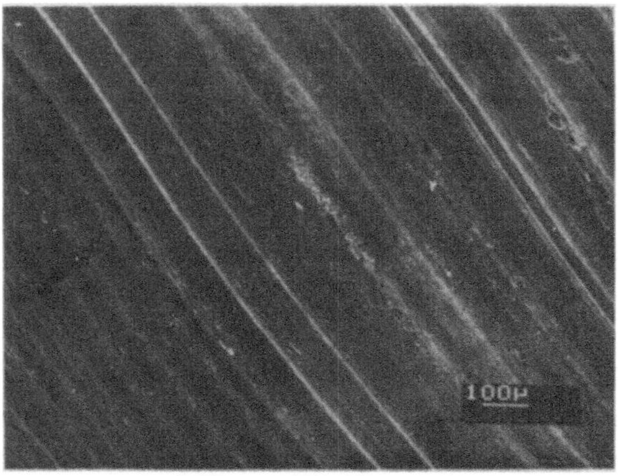

Teil - Nr. 51

Bild 14: Wechsel der Oberflächenformation rauh - glatt beim Graphit / Zylinderoel im Standzeitversuch

anderfolgenden Teilen verbunden, sondern kann sich auch über mehr als 30 Pressungen hinziehen.

Eine nicht gleichmäßige Auftragung des Schmiermittels (z.B. durch Verstopfung oder Ausfall einer Sprühdüse) führt zu starken örtlichen Mikroverschweißungen. Sie macht sich besonders im Rundheitsdiagramm und der zugehörigen Werkstückoberflächenformation bemerkbar. In Bild 15 ist gezeigt, daß die Abweichung von der Rundheit in dem Oberflächenbereich, in dem es zu Verschweißungen gekommen ist, um mehr als das 10-fache höher liegt als in den anderen Bereichen. Kann die Unstetigkeit in der Schmiermittelauftragung nicht behoben werden, setzen sich die Mikroverschweißungen zwischen Werkstück und Werkzeug durch Materialauftragungen auf die Matrize ständig fort. Wie in Bild 16 gezeigt, haben sich schon nach wenigen Umformvorgängen die beschriebenen Erscheinungen um mehr als 45 % vergrößert. Die Bilder 17 bis 22 zeigen die Entwicklungsstufen des Verschleißes an Werkstück und Werkzeug in Serie 33 bis zum endgültigen Ausfall des Werkzeuges nach 7651 Pressungen. Bild 17 gibt die verschiedenen Stadien am Werkstück im Bereich der Spritzkante der Matrize an, Bild 18 den Anfangszustand der Matrize in diesem Bereich und Bild 19 einen Ausschnitt aus dem entsprechenden Bereich des Werkzeuges nach 7651 Pressungen. Man erkennt deutlich, daß sich der Ausfall der Matrize in einer nicht mehr existierenden Spritzkante widerspiegelt. Analog dazu nehmen die Mikroverschweißungen und Fresser zwischen Werkzeug und Werkstück mit steigender Stückzahl zu (siehe dazu Bild 20 und 21). Ebenso steigen die Ausschläge im Rundheitsdiagramm um mehr als 100 % (siehe dazu Bild 22). Schliffuntersuchungen des in Bild 19 gezeigten Spritzkantenbereiches zeigen, daß es in diesem Teil des Werkzeuges zu Mikrorissen und mehreren Anlaßzonen von unterschiedlicher Tiefe (bis zu 40 μm) gekommen ist. Sie entstanden vermutlich durch die verschiedenen Stadien des Selbst-

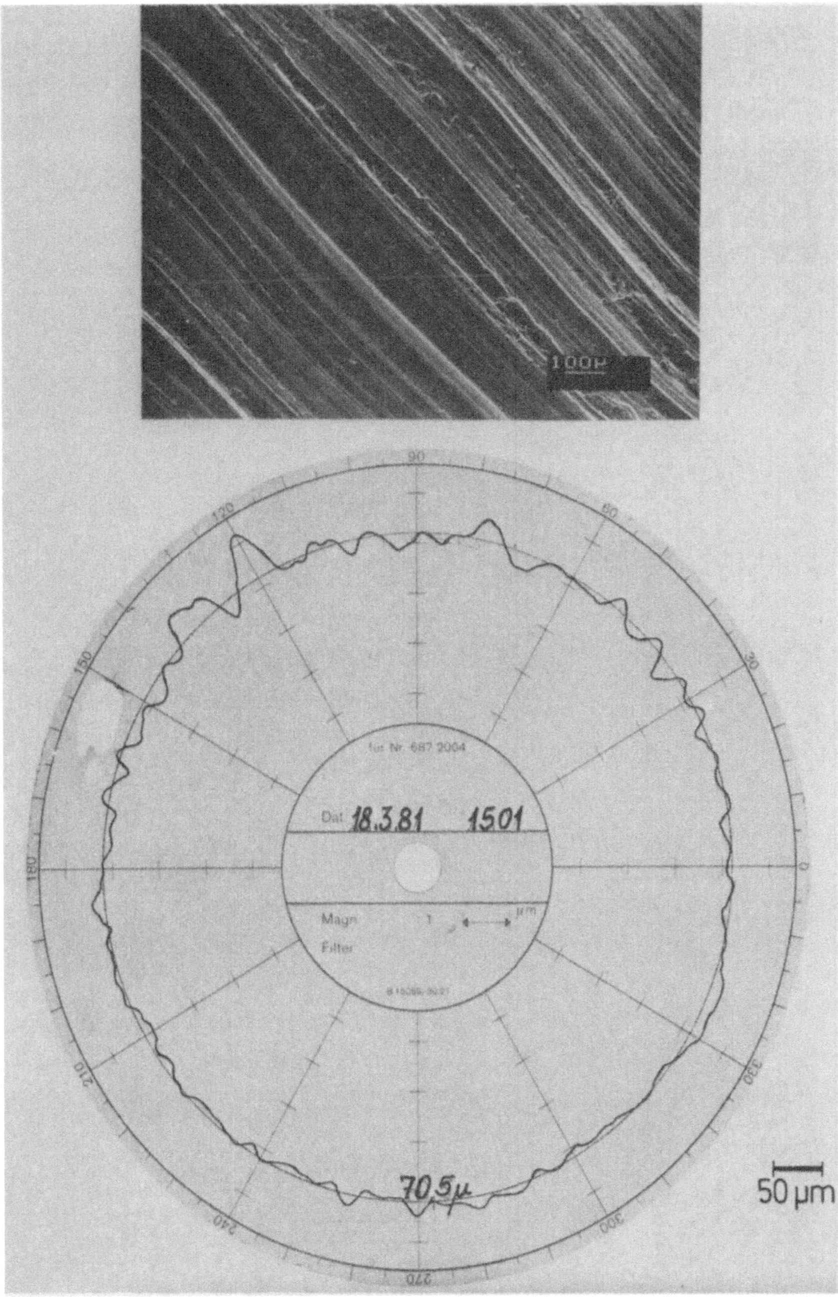

<u>Bild 15:</u> Oberflächenformation und Rundheitsmessung bei beginnender unstetiger Schmierung
(Teile-Nr. 1501 in Serie 32)

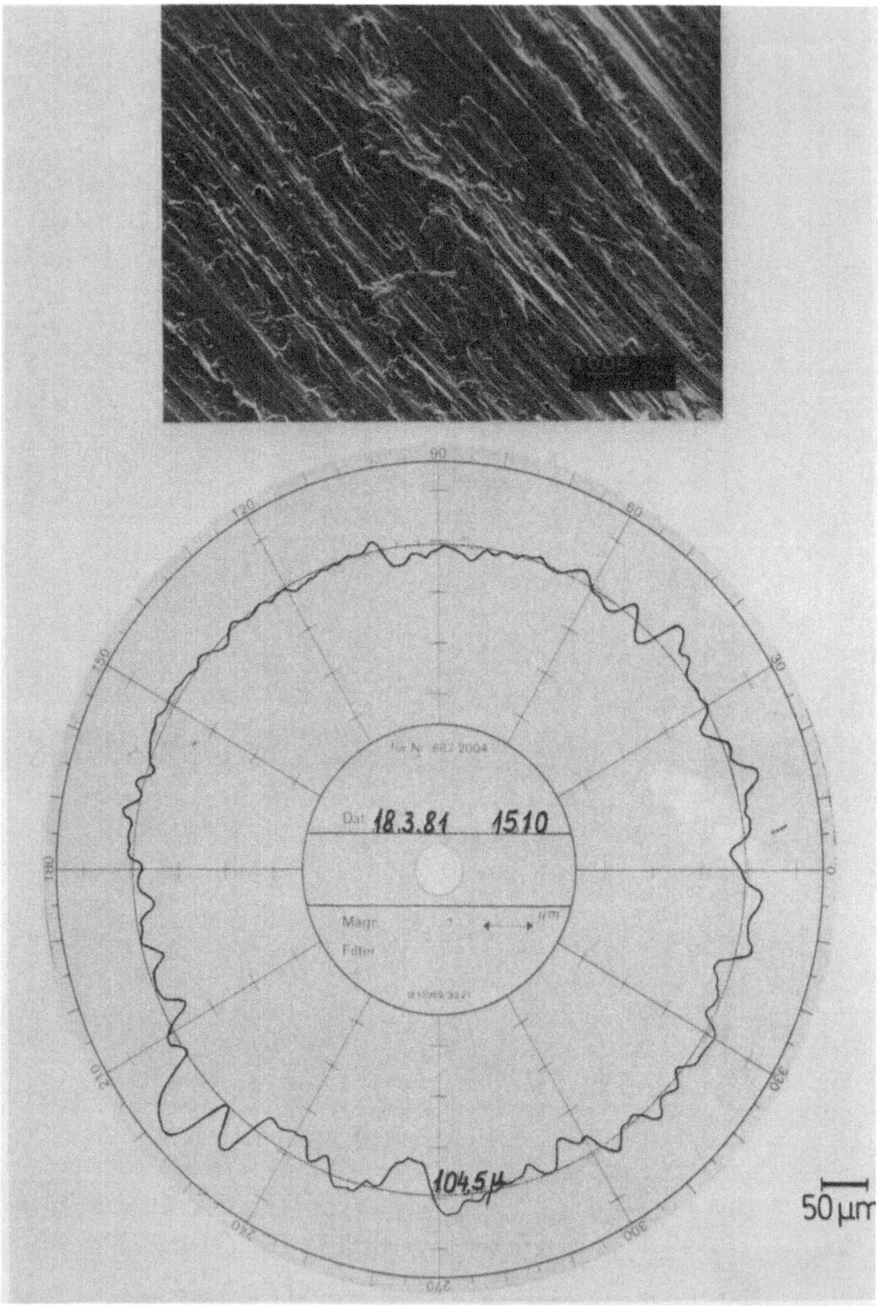

Bild 16: Oberflächenformation und Rundheitsmessung nach 10 Versuchen bei unstetiger Schmierung (Teile-Nr. 1510 in Serie 32)

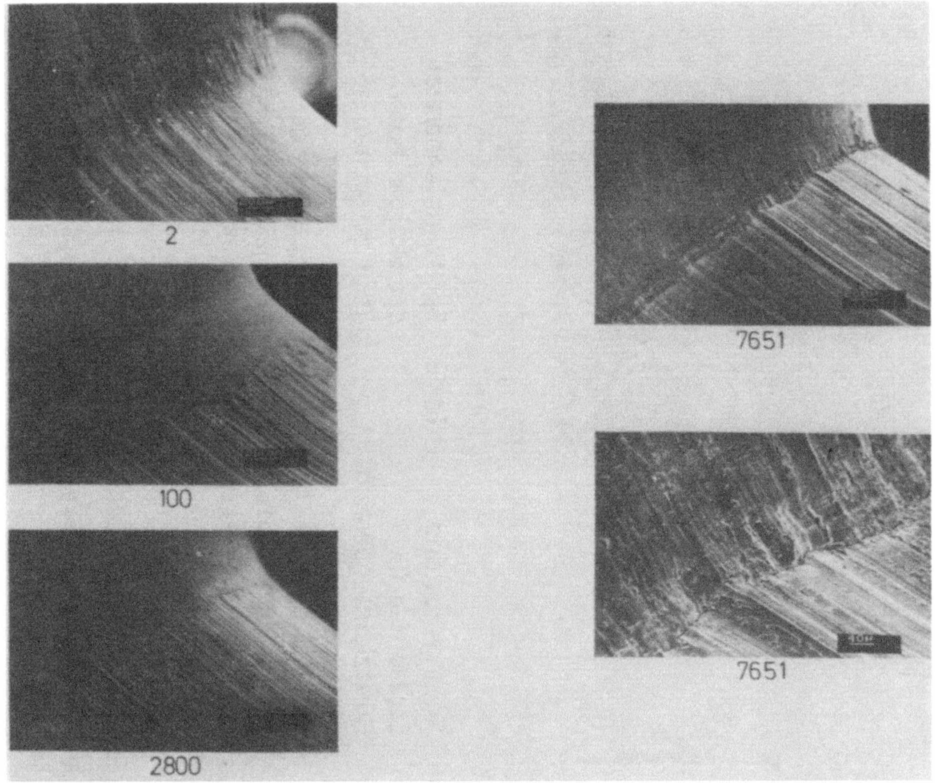

Bild 17 : Entwicklung des Verschleißes im Bereich der Spritzkante bei Serie 33

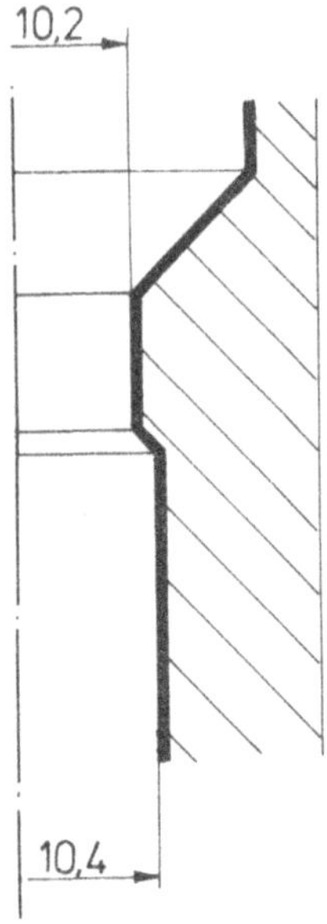

Bild 18: Anfangszustand der Matrize im Spritzkantenbereich

Bild 19: Spritzkantenbereich der Matrize nach 7651 Versuchen

reinigungseffektes (das ist der geschilderte Wechsel in der Oberflächenformation von rauh zu glatt) und die extrem starken Verschweißungen zwischen Werkstück und Werkzeug kurz vor Ausfall der Matrize.

4.3.3 Graphit/Wasser im Standmengenversuch

Die Kurzversuche ergaben, daß mit Graphit/Wasser beim Voll - Vorwärts - Fließpressen sehr glatte Werkstückoberflächen entstehen. Außerdem sind keinerlei Auftragungen oder Mikroverschweißungen zwischen Werkstück und Werkzeug zu beobachten. Diese Ergebnisse werden durch unsere Standmengenversuche bestätigt. Die Bilder 23 - 25 zeigen die Stufen der Verschleißentwicklung bei Serie 34, innerhalb der 500 Teile produziert werden konnten. Man

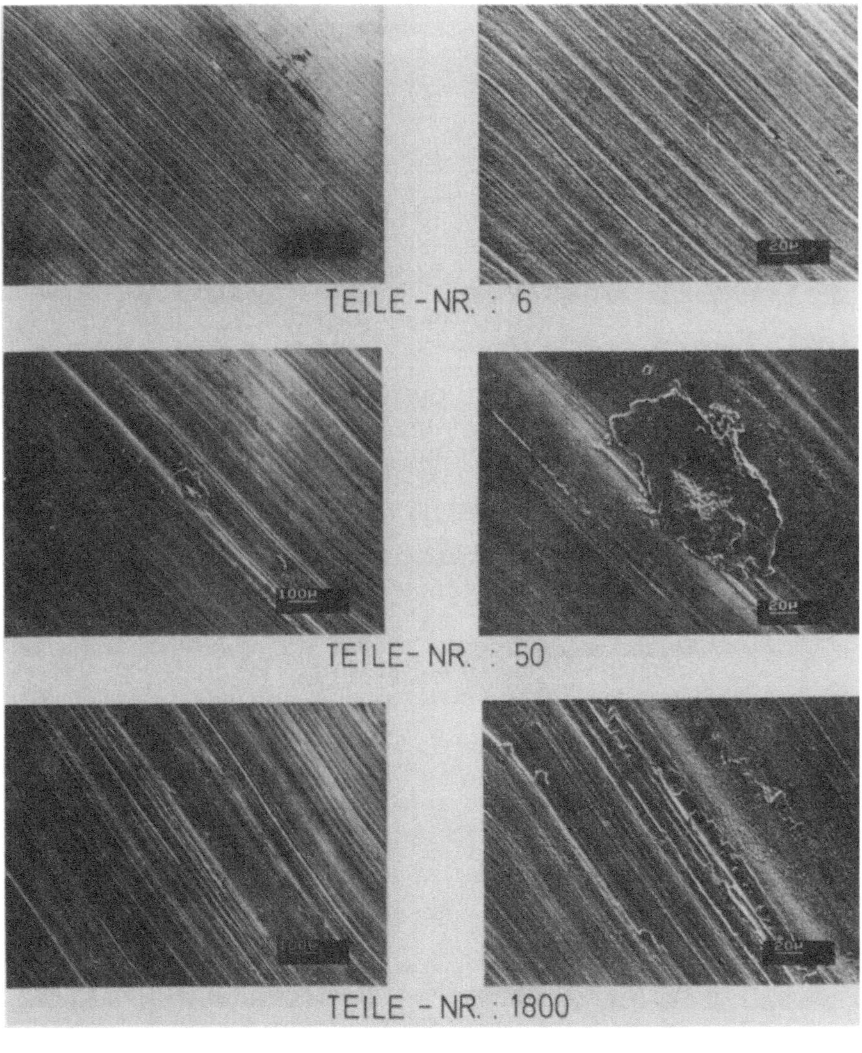

Bild 20 : Entwicklung des Verschleißes im Bereich des Schaftes bei Serie 33

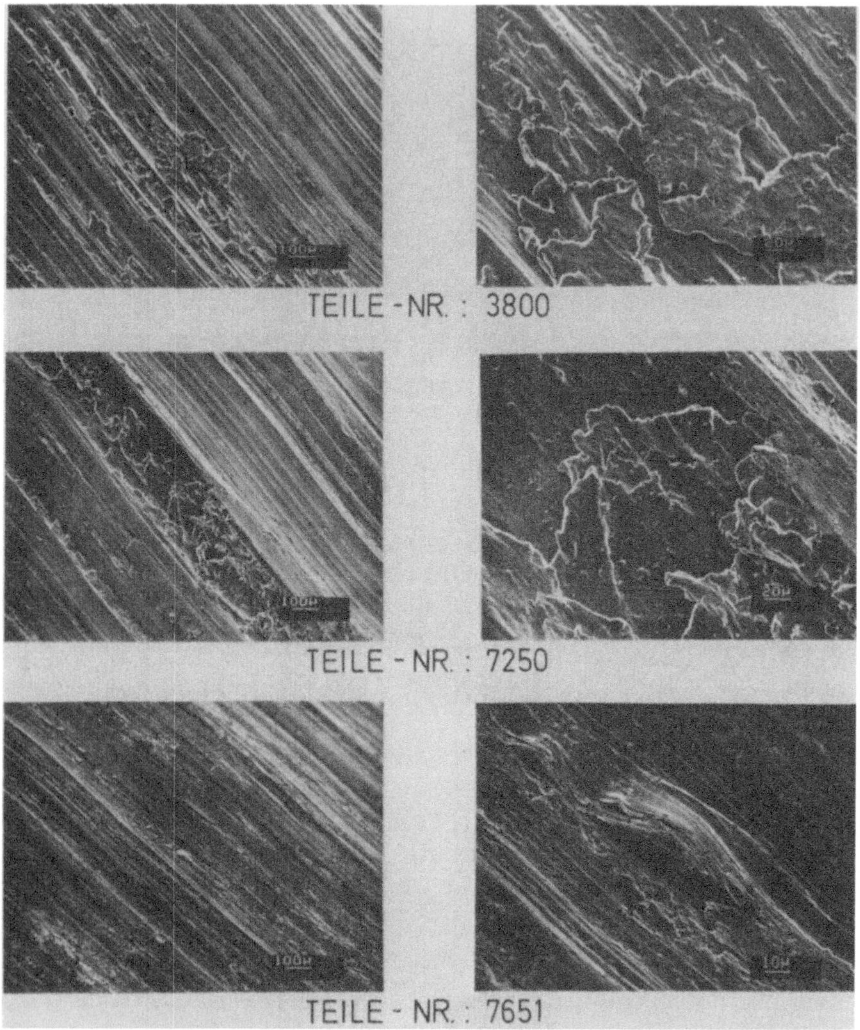

Bild 21: Entwicklung des Verschleißes im Bereich des Schaftes bei Serie 33

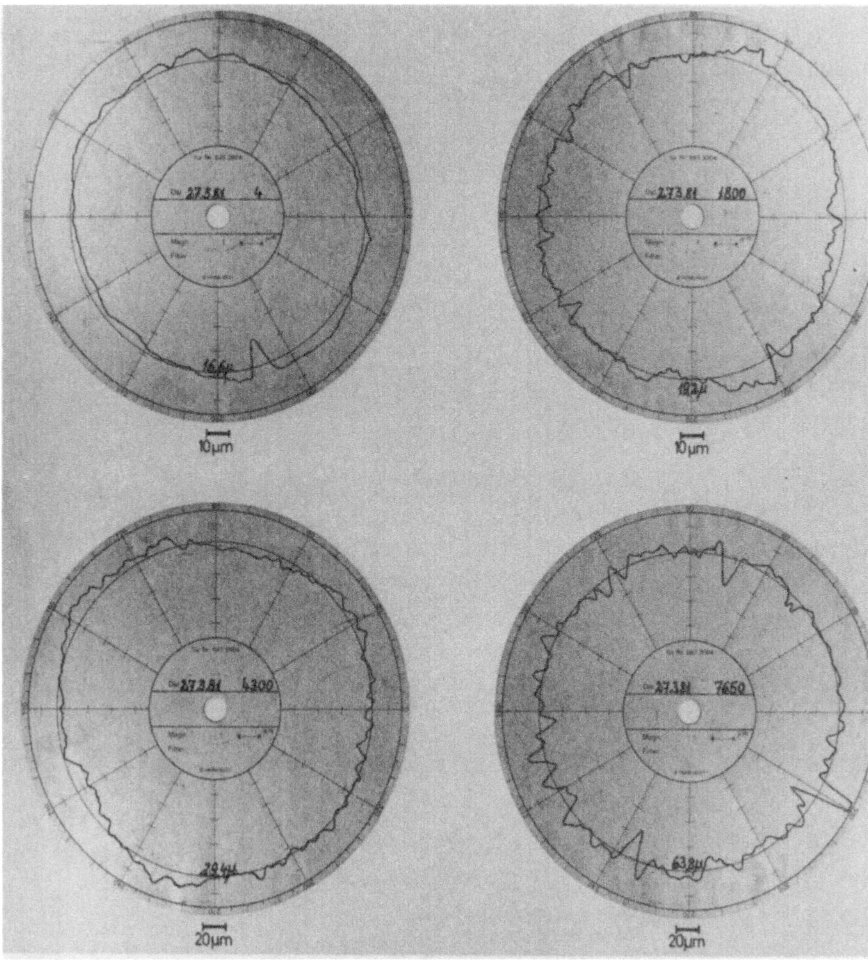

Bild 22: Entwicklung der Rundheitsbestimmungen bei verschiedenen Teilen der Serie 33

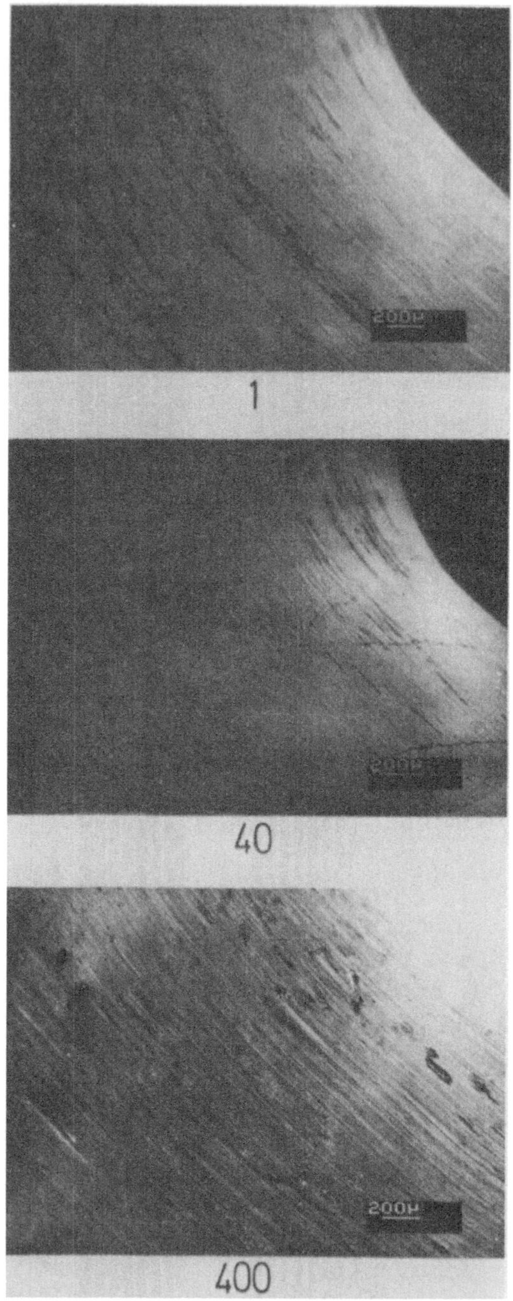

Bild 23: Entwicklung des Verschleiß im Bereich der Spritzkante bei Serie 34

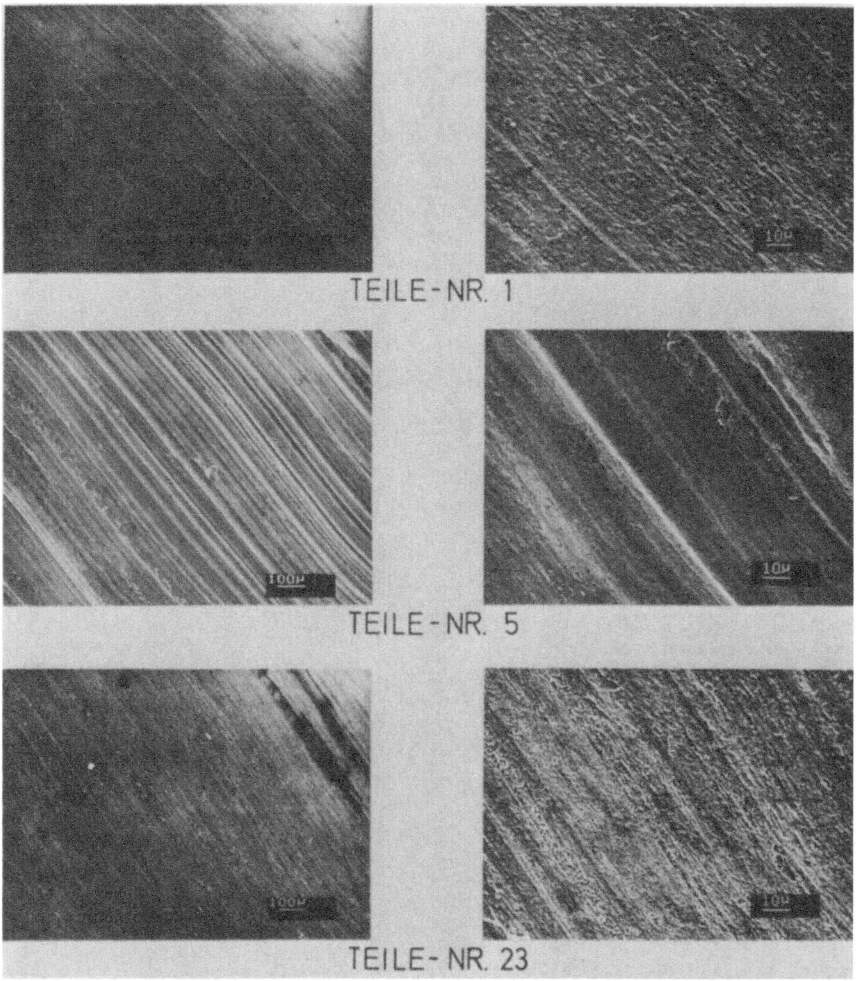

Bild 24 : Entwicklung des Verschleiß im Schaftbereich b. Serie 34

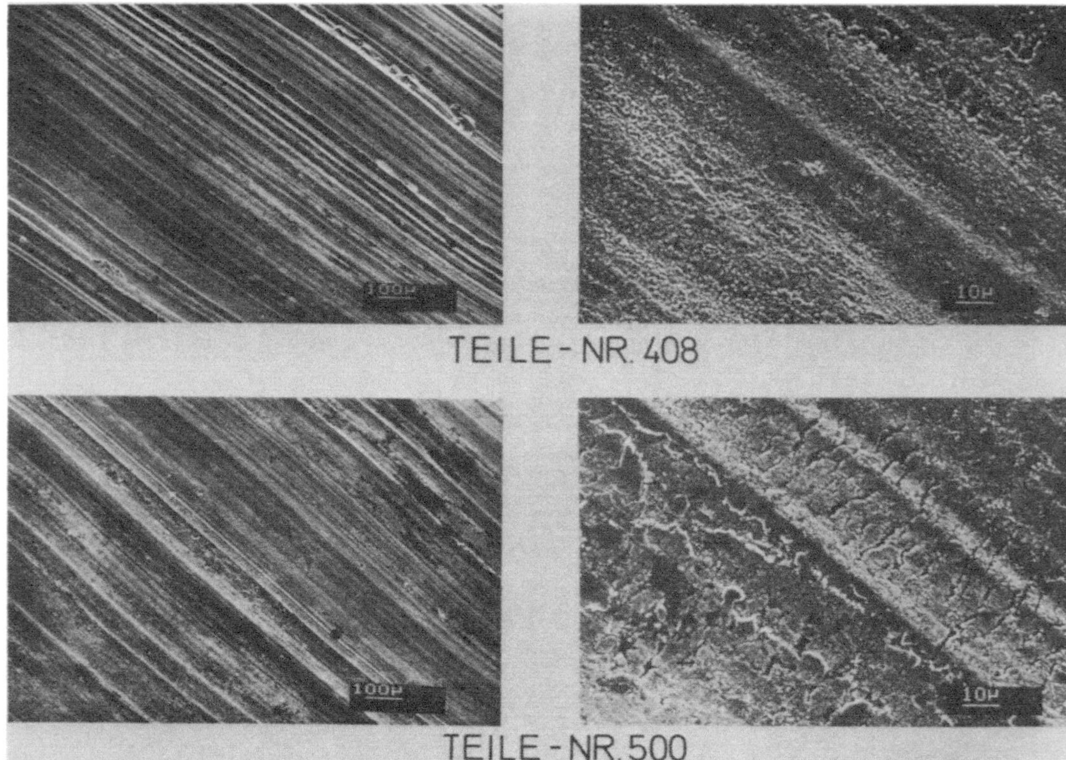

Bild 25: Entwicklung des Verschleiß im Schaftbereich bei Serie 34

erkennt deutlich die sehr glatten Werkstückoberflächen und die geringen Tiefen der Preßriefen. Auch nach 400 Pressungen sind im Bereich der Spritzkante noch keinerlei Einkerbungen im Werkstück zu beobachten, wie sie z.B. beim Graphit/Zylinderoel schon nach weniger als 100 Pressungen auftraten. Im Bereich des Schaftes zeigt sich, daß außerdem die Anfangsrauhigkeit sehr gering ist, siehe dazu besonders Teil 1 und Teil 5 in Bild 24. Erst nach mehr als 400 Pressungen entstehen leichte Mikroverschweißungen innerhalb der Preßriefen. Nach 500 Pressungen mußte dieser Standmengenversuch allerdings abgebrochen werden, da das Werkzeug wegen Überlastung ausfiel.

Bei der Durchführung der Standmengenversuche ergaben sich einige technische Schwierigkeiten, die auf den speziellen Eigenschaften des Schmiermittels Graphit/Wasser beruhen. Werden die heißen Werkstücke in die Matrize eingeführt, so entsteht in dem darunter liegenden abgeschlossenen Raum durch das verdampfende Wasser kurzfristig ein Überdruck. Dadurch können die Teile wieder aus der Matrize herausgedrückt werden, noch bevor der Stempel sie niederdrückt, wenn nicht durch spezielle Kanäle im Werkzeugbereich für eine Entlüftung gesorgt wird. Außerdem müssen das Aufsprühen des Schmiermittels und die Werkstückeinführung unmittelbar aufeinander erfolgen, da sonst die Teile durch das angetrocknete Graphit nicht mehr in die Matrize gleiten können.

5. Diskussion

5.1 Aussagen nach den Kurzversuchen

In der Umformtechnik zeichnet sich ein gutes Schmiermittel dadurch aus, daß es den metallischen Kontakt zwischen Werkstück und Werkzeug so gut wie möglich verhindert und damit den Verschleiß so klein wie eben möglich hält. Betrachtet man unter diesem Gesichtspunkt die drei vorliegenden Schmiermittelsysteme, so ergeben sich für die drei Umformprozesse die folgenden Aussagen:

- Beim Durchschieben erhält man mit Graphit/Wasser relativ glatte Oberflächen. Dagegen werden beim Graphit/Zylinderoel Anzeichen für starke Schubspannungen in den Oberflächen und Mikroverschweißungen zwischen Werkstück und Werkzeug beobachtet. Man erwartet daher mit Graphit/Wasser einen geringeren Verschleiß.
- Beim Voll - Vorwärts - Fließpressen wirkt aus den eben genannten Gründen Graphit/Wasser ebenfalls geeigneter als Graphit/Zylinderoel.
- Beim Napf - Rückwärts - Fließpressen scheint dagegen Graphit/Zylinderoel das bessere Schmiermittel zu sein. Zwar ergeben sich hierbei viele Auftragungen auf der Werkstückoberfläche, aber dafür keine größeren Mikroverschweißungen wie beim Graphit/Wasser, die bei einer größeren Anzahl von Pressungen einen starken Werkzeugverschleiß hervorrufen.
- Das Schmiermittelsystem Salz/Wasser ist mit unserer Untersuchungsmethode nicht zu beurteilen. Nach dem Umformen bildet sich auf der Werkstückoberfläche durch das auskristallisierende Salz eine Schicht, die die eigentlichen umformspezifischen Erscheinungen zudeckt.

Es fällt auf, daß sich beim Napf - Rückwärts - Fließpressen eine andere Bewertungsreihenfolge ergibt als bei den beiden anderen Umformverfahren. Der Grund dürfte sicher in Zusammenhang mit der sehr starken Oberflächenvergrößerung beim Napfen stehen.

5.2 Aussagen nach den Standmengenversuchen

Aus den Kurzversuchen wurde geschlossen, daß das Schmiermittel Graphit/Wasser für das Voll - Vorwärts - Fließpressen geeigneter sein sollte als Graphit/Zylinderoel. In den Standmengenversuchen wurden mit Graphit/Zylinderoel jedoch deutlich höhere Standmengen erreicht als mit Graphit/Wasser. Damit scheint sich ein Widerspruch zur Aussagefähigkeit der Kurzversuche zu ergeben.

Beim Graphit/Wasser ergaben sich unerwartet große experimentelle Schwierigkeiten (siehe 4.3.3). Dadurch war es nicht möglich, im Rahmen dieses Projektes mit diesem Schmiermittel einen einwandfreien Standmengenversuch mit mehr als 500 Pressungen zu fahren. Zur Schmiermittelbeurteilung muß man sich daher auf eine Untersuchung der Verschleißentwicklung auf Werkstück und Werkzeug beschränken.

Beim Graphit/Zylinderoel wurden schon nach weniger als 100 Pressungen auf dem Werkzeug Auftragungen bis zu 8 mm^2 Ausdehnung und 50 μm Höhe festgestellt. Die daraus resultierende und in den Kurzversuchen beobachtete Neigung zu Mikroverschweißungen wird in den Standmengenversuchen deutlich bestätigt. Beim Graphit/Wasser werden dagegen erst nach 500 Pressungen ganz leichte Mikroverschweißungen festgestellt. Daraus kann geschlossen werden, daß dieses Schmiermittel seine Hauptaufgabe, den metallischen Kontakt zwischen Werkzeug und Werkstück zu verhindern, besser erfüllt als das Schmiermittel Graphit/Zylinderoel. Insofern wird die nach den Kurzversuchen getroffene Aussage von den Standmengenversuchen doch bestätigt. Es hat sich jedoch auch gezeigt, daß neben der tribologischen Eignung ebenso andere technologische Eigenschaften bei der Schmiermittelauswahl maßgebend sein können.

5.3 Beurteilung von Werkstückoberflächen mittels Rasterelektronenmikroskopie

Im Rahmen dieses Projektes wurde gezeigt, daß die Rasterelektronenmikroskopie in einem bisher nicht gekannten Maße die Analyse von Verschleißvorgängen und das Verfolgen von tribologischen Zusammenhängen erlaubt. Mit ihr lassen sich in Verbindung mit Mikroanalyse und Auger - Analyse Schmiermittelverteilungen und Randschichten von Werkstücken untersuchen.

Typische Verschleißerscheinungen lassen sich schon nach wenigen Versuchspressungen im Rasterelektronenmikroskop beobachten und analysieren. Damit ist eine Früherkennung von Verschleiß möglich. Dies ist besonders für die Umformtechnik interessant, da hier das Werkstück als Informationsträger einer Oberflächenuntersuchung bequem zugänglich ist.

Das im Rahmen dieses Projektes angestrebte Ziel, Aussagekriterien zu entwickeln, mit denen man für den jeweils vorliegenden Anwendungsfall ein geeignetes Schmiermittel auswählen kann, ist also weitgehend erreicht. Damit erscheint die Rasterelektronenmikroskopie als geeignetes Hilfsmittel, um Schmiermittel für die Halbwarmumformung zu prüfen.

6. Zusammenfassung

Für den Bereich der Halbwarmumformung wurde eine Methode entwickelt, mit deren Hilfe sich Schmiermittelsysteme vergleichen lassen. Dabei wurde geprüft, ob sich die Oberflächenbeurteilung von umgeformten Werkstücken im Rasterelektronenmikroskop zu einer Schmiermittelbewertung heranziehen läßt. Dazu wurden Modellwerkzeuge konstruiert und gebaut, mit denen Proben unter kontrollierter tribologischer Beanspruchung hergestellt wurden. Durch eine Realisierung der Umformprozesse Durchschieben, Voll - Vorwärts - Fließpressen und Napf - Rückwärts - Fließpressen wurde die Oberflächenvergrößerung der Werkstücke in einem sehr weiten Rahmen variiert. Angewandt wurde die genannte Methode auf drei typische Schmiermittelsysteme: Graphit/Wasser, Salz/Wasser und Graphit/Zylinderoel.

Mit diesen Schmiermittelsystemen und den o.a. Verfahren wurden jeweils 50 Teile in mehreren Versuchsserien gepreßt. Außerdem wurden mit den graphithaltigen Schmiermitteln und dem Voll - Vorwärts - Fließpress - Werkzeug Standmengenversuche gefahren. Eine definierte Anzahl der gepreßten Proben wurde im Rasterelektronenmikroskop untersucht. Als Ergänzung dazu wurden Oberflächen- und Randschichtenanalysen mit Hilfe der Rasterelektronenmikroskopie sowie mit der Auger - Mikrosonde durchgeführt und Oberflächenprofile der Werkstücke aufgenommen.

Im Rasterelektronenmikroskop ließen sich typische Verschleißerscheinungen beobachten: Riefen, Auftragungen, schuppenförmige Strukturen, Mikroverschweißungen, Abplatzungen der Oxidschichten und ungleichförmige Schmiermittelschichten. Bezüglich dieser Erscheinungen zeigen die einzelnen Schmiermittelsysteme charakteristische, schon nach wenigen Versuchspressungen deutlich erkennbare Unterschiede. Die Rasterelektronenmikroskopie erscheint daher als eine geeignete Methode zur Früherkennung von Verschleiß und zur Auswahl von Schmiermitteln.

Schrifttum

[1] G.Heinke, Verschleiß - eine Systemeigenschaft, Auswirkungen auf die Verschleißprüfung, Z.f.Werkstofftechnik/J. of Mat. Technology, 5 (1975), 164 - 169

[2] DIN 50320, "Verschleiß", Begriffe, Systemanlysen von Verschleißvorgängen, Gliederung des Verschleißgebietes, Dezember 1979

[3] H.Czichos, Festkörperreibung - Teilgebiet der Tribologie Umschau, 4 (1971), 116

[4] G.Heinke, Betriebs- und Laborprüftechniken von reibungs- und verschleißbeanspruchten Bauteilen, Esslingen, Lehrgang 3613 (1978)

[5] Lange/Meyer/Noelkemper, Gesenkschmieden, 2. Auflage, Springer - Verlag, Berlin (1977)

[6] K.Lange und Th.Graebener, Untersuchungen der Möglichkeiten für eine technologische Schmierstoffprüfung für Verfahren der Kaltmassivumformung, siehe dieses Buch FE - Vorhaben 01 ZT 408 S, Universität Stuttgart

[7] F.Dohmann, Die Messungen der mechanischen Kontaktspannungen in der Wirkfuge Werkzeug - Werkstück beim Umformverfahren, Giradet - Verlag, Essen (1974)

[8] K.H.Friedrich, Beitrag zur Messung der Strangenoberflächentemperatur beim Strangpressen, Giradet-Verlag, Essen (1975)

[9] L.Schaeben und A.Rahmel, Das Zunderverhalten der Stähle und die Prüfung auf Zunderbeständigkeit, B + R, Technische Mitteilungen (1974)

[10] F.K.Peters und H.J.Engell, Die Haftfestigkeit von Zunderschichten auf Stahl, Archiv für das Eisenhüttewesen, 5 (1959), 275 - 282

A U T O R E N V E R Z E I C H N I S

Adam, Peter, Dr.-Ing.habil.
Motoren- und Turbinen-Union München GmbH, München

Broszeit, Erhard, Dr.-Ing.
Technische Hochschule Darmstadt, Institut für Werkstoffkunde

Bunk, Wolfgang, Prof. Dr.rer.nat.
Deutsche Forschungs- und Versuchsanstalt für Luft- und Raumfahrt e.V.,
Institut für Werkstoff-Forschung, Köln-Porz
RWTH Aachen, Lehrstuhl für angewandte Metallkunde

Conrad, Reinhard, Dr.rer.nat.
Robert Bosch GmbH, Stuttgart

Czichos, Horst, Prof. Dr.-Ing.
Bundesanstalt für Materialprüfung (BAM), Berlin

Dausinger, Friedrich, Dr.rer.nat.
Robert Bosch GmbH, Stuttgart

Doege, Eckart, Prof. Dr.-Ing.
Universität Hannover, Institut für Umformtechnik und Umformmaschinen

Dürr, Fritz, Dipl.-Ing.
Forschungsinstitut der Forschungsgesellschaft für Uhren- und Feingerätetechnik e.V., Stuttgart

Geyer, Manfred, Dipl.-Ing.
Deutsche Forschungs- und Versuchsanstalt für Luft- und Raumfahrt e.V.,
Institut für Werkstoff-Forschung, Köln-Porz

Gräbener, Thomas, Dipl.-Ing.
Universität Stuttgart, Institut für Umformtechnik

Grahnert, Rolf, Dipl.-Ing.
Universität Hannover, Institut für Umformtechnik und Umformmaschinen

Grewe, Hans, Dr.-Ing.
Fried. Krupp GmbH, Krupp Forschungsinstitut, Essen

Hansen, Jörn, Dr.rer.nat.
Deutsche Forschungs- und Versuchsanstalt für Luft- und Raumfahrt e.V.,
Institut für Werkstoff-Forschung, Köln-Porz

Heinke, Günter, Dr.-Ing.
Robert Bosch GmbH, Stuttgart

Heinz, Rudolf, Dr.-Ing.
Robert Bosch GmbH, Stuttgart

Hirouchi, Tetsuya, Dipl.-Ing.
Max-Planck-Institut für Eisenforschung GmbH, Düsseldorf

Jost, H. Peter, CBE, DSc, CEng, FIMechE
Präsident des "International Tribology Council"
Angel Lodge Laboratories, London, Großbritannien
Chairman K S Paul Products Limited

Kloos, Karl-Heinz, Prof. Dr.-Ing.
Technische Hochschule Darmstadt, Institut für Werkstoffkunde

König, Udo, Dr.rer.nat.
Fried. Krupp GmbH, Krupp Forschungsinstitut, Essen

Kretschmer, Ingo, Dipl.-Ing.
Verschleiß-Technik Dr.Ing. Hans Wahl GmbH & Co, Stuttgart-Ruit

Lange, Kurt, Prof. Dr.-Ing.
Universität Stuttgart, Institut für Umformtechnik

Paterok, Lienhard, Dipl.-Ing.
Bernex GmbH, Langenfeld

Pawelski, Oskar, Prof. Dr.-Ing.
Max-Planck-Institut für Eisenforschung GmbH, Düsseldorf

Wahl, Hans, Dr.-Ing.
Verschleiß-Technik Dr.Ing. Hans Wahl GmbH & Co, Stuttgart-Ruit

Wahl, Wolfgang, Dr.-Ing.
Verschleiß-Technik Dr.Ing. Hans Wahl GmbH & Co, Stuttgart-Ruit

Wilms, Hans-Friedrich, Dipl.-Ing.
Fried. Krupp GmbH, Krupp Forschungsinstitut, Essen

Witthüser, Klaus-Peter, Dr.-Ing.
Universität Hannover, Institut für Umformtechnik und Umformmaschinen

D E S K R I P T O R E N L I S T E

0	**Technische Funktion des Tribosystems**	
01	Bewegungsübertragung, Führung	1-5, 1-7
02	Bewegungshemmung	1-5
03	Kraft-/Energieübertragung	1-1, 1-5, 1-6, 1-7
04	Informationsübertragung	
05	Materialtransport	1-1
06	Abdichtung	
07	Materialbearbeitung	1-3, 1-4
08	Materialumformung	1-8, 1-9, 1-10, 1-11
1	**Anwendungsgebiet**	
11	System, Maschine, Gerät, Bauteil, Produkt	
111	Antriebstechnik	
111.1	Lagerungen, Führungen	1-1, 1-5, 1-6, 1-7
111.11	Gleitlager	1-1, 1-7
111.12	Wälzlager	
111.13	Spitzenlager	
111.14	Magnetlager	
111.15	Sonstige Lager	
111.2	Getriebe	
111.21	Zahnradgetriebe	
111.22	Reibgetriebe	
111.23	Kurvengetriebe	
111.24	Sonstige Getriebe	
111.3	Kupplungen	
111.4	Bremsen	
111.5	Dichtungen	
111.6	Dämpfungselemente	
112	Kraft- und Arbeitsmaschinen	
112.1	Kolbenmotoren	
112.2	Pumpen, Verdichter	
112.3	Strömungsmaschinen	1-6
112.4	Hydraulik, Pneumatik	
112.5	Werkzeugmaschinen	
112.6	Textilmaschinen	
112.7	Papiermaschinen	
112.8	Druckmaschinen	
113	Feinwerktechnik	
113.1	Meßgeräte, Meßinstrumente	1-7
113.2	Optische Geräte	
113.3	Phono- u. fernsehtechn. Geräte	1-7
113.4	Büromaschinen	1-7
113.5	Datenverarbeitungsmaschinen	
113.6	Sonstige:	
113.61	Großuhren	1-7
114	Fahrzeuge, Verkehr	
114.1	Straßenverkehr	
114.2	Spurgebundener Verkehr	
114.3	Schiffahrt	1-6
114.4	Luftfahrt	1-6
114.5	Raumfahrt, Raketentechnik	

115 Elektrotechnik

115.1 Elektromotoren, Generatoren
115.2 Elektrische Kontakte, Schalter
115.3 Elektroden

116 Grundstoffindustrien, Nahrungsmittel, Energie

116.1 Aufbereitungstechnik 1-2
116.2 Bergbau, Tiefbohren 1-1, 1-2
116.3 Bau, Steine, Erden 1-1, 1-2
116.4 Land- und Forsttechnik 1-2
116.5 Meerestechnik 1-2
116.6 Hüttenwesen 1-2, 1-8
116.7 Fördertechnik, Lagertechnik 1-2
116.8 Gas- und Wasserversorgung
116.9 Kraftwerkstechnik 1-2

12 Fertigungsverfahren, Werkzeuge

121 Urformen

121.1 Gießen 1-2
121.2 Extrudieren 1-2
121.3 Pulvermetallurgie, Sintern

122 Umformen

122.1 Schmieden, Stauchen, Pressen 1-9, 1-11
122.2 Walzen 1-2
122.3 Ziehen 1-8, 1-9
122.4 Tiefziehen 1-4, 1-10
122.5 Biegen, Drücken

123 Fügen, Beschichten

123.1 Schrauben, Nieten
123.2 Klemmen, Einpressen
123.3 Schweißen 1-2
123.4 Beschichten 1-2, 1-4, 1-6

124 Trennen

124.1 Schneiden, Sägen
124.2 Drehen, Fräsen, Feilen 1-3, 1-4
124.3 Schleifen, Läppen, Polieren
124.4 Flüssigkeitsstrahlschneiden

125 Oberflächenbearbeitung

(Strahlen, Bürsten, Glattwalzen) 1-6

2 Beanspruchungsfall

21 Kontakt-/Bewegungsart

211 Kontaktform
211.1 Konformer Kontakt 1-1, 1-5, 1-6, 1-10, 1-11
211.2 Kontraformer Kontakt 1-1, 1-7

212 Bewegungsart
212.1 Normal-, Haftkontakt 1-1, 1-6
212.2 Gleiten 1-1, 1-2, 1-4, 1-5, 1-6, 1-7, 1-8, 1-11
212.3 Wälzen
212.4 Stoßen, Prallen 1-1, 1-2, 1-5
212.5 Strömen
212.6 Strahlen

22 Betriebsparameter

221 Bewegungsanblauf
221.1 kontinuierlich 1-2, 1-4, 1-7, 1-8
221.2 intermittierend, repetierend 1-1, 1-2
221.3 oszillierend, schwingend 1-5, 1-6, 1-7

3 Systemstruktur

31 Stoffliche Eigenschaften und Stoffparameter

311 Stoffpaarung Grund-/Gegenkörper

311.01 fest/fest 1-1, 1-4, 1-6, 1-11
311.02 fest/fluid

311.11 Metall/Metall 1-1, 1-3, 1-5, 1-6, 1-8, 1-9, 1-10, 1-11
311.12 Metall/Mineral 1-1, 1-2, 1-6, 1-9
311.13 Metall/Polymer 1-5, 1-7
311.14 Metall/Verbundwerkstoff 1-3, 1-4, 1-5

311.21 Mineral/Mineral
311.22 Mineral/Polymer
311.23 Mineral/Verbundwerkstoff

311.31 Polymer/Polymer 1-7
311.32 Polymer/Verbundwerkstoff

311.41 Verbundwerkstoff/Verbundwerkstoff

312 Stoffart von Grundkörper, Gegenkörper, Zwischenstoff, Umgebungsmedium

312.1 Metalle und Halbmetalle
312.11 Stahl, Eisen (vgl.312.42) 1-1, 1-2, 1-3, 1-4, 1-5, 1-7, 1-8, 1-9, 1-10, 1-11
312.12 Leichtmetalle
312.121 Aluminium + Legierungen
312.122 Titan + Legierungen 1-6
312.13 Schwermetalle
312.131 Blei/Zinn + Legierungen
312.132 Kupfer + Legierungen
312.133 Nickel + Legierungen 1-6
312.134 Kobalt + Legierungen 1-2, 1-6
312.135 Edelmetalle

312.2 Minerale, keramische Stoffe
312.21 Minerale 1-1, 1-2, 1-6
312.22 Glas
312.23 Keramische Stoffe 1-6, 1-9
312.24 Gestein 1-1
312.25 Beton
312.26 Schlacke, Asche

312.3 Organische Stoffe, Polymere
312.31 Einfache organische Stoffe 1-11
312.32 Polymere 1-1, 1-5, 1-7
312.321 Kunststoffe, Gummi
312.322 Pharmazeut. Stoffe, Kosmetika
312.323 Biologische Stoffe
312.324 Papier, Pappe
312.325 Holz (vgl. 312.45)
312.326 Leder

312.4 Verbundwerkstoffe
312.41 Verstärkte Kunststoffe 1-5

312.42 Verstärkte Metalle 1-1
312.43 Cermets, Hartmetalle 1-1, 1-3, 1-4
312.44 Schichtwerkstoffe 1-3, 1-6
312.45 Holzwerkstoffe

312.5 Faser- u. Textilstoffe

312.6 Körnige u. stückige Stoffe

312.7 Flüssigkeiten, Fette, flüssige Mehrphasensysteme 1-10
312.71 Anorganische Flüssigkeiten
312.711 Wasser, Elektrolyt 1-11
312.712 Metallschmelze
312.713 Salzschmelze
312.714 Flüssiggas
312.715 Silikonöl
312.72 Organische Flüssigkeiten 1-5
312.721 Organische Lösungsmittel 1-5
312.722 Mineralöle 1-9, 1-11
312.723 Synthetische Öle 1-9
312.724 Tierische u. pflanzl. Öle
312.725 Polymerschmelzen
312.726 Polymerlösungen 1-9
312.73 Additive 1-11
312.74 Suspensionen 1-11
312.75 Emulsionen
312.76 Fette 1-5
312.77 Seifen 1-8
312.78 Pasten, Kolloide

312.8 Gase
312.81 Luft 1-1, 1-2, 1-3, 1-5, 1-6, 1-7, 1-9, 1-11
312.82 Wasserdampf 1-1
312.83 Sauerstoff 1-1
312.84 Inertgas
312.85 Vakuum
312.86 Weltraumatmosphäre
312.87 Sonstige:
312.871 CO, CO_2, N_2, CHs 1-1

312.9 Verunreinigungen/Schmutz

313 <u>Stoffparameter</u>

313.1 Werkstoffabhängige Einflüsse
313.11 Zusammensetzung 1-1, 1-2, 1-3, 1-4, 1-5, 1-6, 1-10
313.12 Gefüge, Struktur 1-1, 1-2, 1-3, 1-4, 1-6, 1-7, 1-9
313.13 Physikal.Stoffdaten 1-1, 1-2, 1-3, 1-4, 1-5, 1-8
313.14 Festigkeitsdaten/Härte 1-1, 1-2, 1-3, 1-4, 1-6, 1-9
313.15 Eigenspannungen 1-4, 1-6
313.16 Verträglichkeit 1-6

313.2 Oberflächenabhängige Einflüsse
313.21 Oberflächenenergie
313.22 Mech. Oberflächenbearbeitung 1-6
313.23 Wärmebehandlung/Härten 1-2
313.24 Elektropolieren
313.25 Oberflächenreinigung
313.26 Sonstige:
313.261 Verfestigen durch Strahlen 1-6
313.262 Verarbeitungsbedingungen/Härte 1-7

313.3 Adsorbierte Schichten

313.4 Oberflächenüberzüge
313.41 Anorganische Überzüge 1-3
313.411 Galvanische Überzüge, stromlose Abscheidungen 1-6
313.412 Schmelztauchüberzüge
313.413 Gasphasenabscheidg.PVD, CVD 1-1, 1-3, 1-4, 1-6
313.414 Schweiß-, Guß-, Spritz- u. Sinterüberzüge 1-1, 1-2, 1-6
313.415 Walzüberzüge (Plattierg.)
313.416 Glasuren, Emails
313.417 Thermochem.Schichten 1-6
313.418 Chemische Überzüge (Oxid-, Phosphatschichten)
313.42 Organische Überzüge
313.421 Farben, Anstriche
313.422 Gleitlacke

32 Formparameter

321 Konstruktive Gestaltung
321.1 Geometrie (Maßstab, Eingriff, Toleranzen) 1-5, 1-7, 1-11
321.2 Konstruktion 1-11
321.3 Einbau/Montagefehler
321.4 Thermische Ausdehnung

322 Oberflächenrauheit, Welligkeit 1-6, 1-7, 1-10

4 Tribologische Problemstellung

41 Oberflächen und Kontakte

411 Oberflächenzustand
411.1 Oberflächenmorphologie 1-4, 1-6, 1-11
411.2 Oberflächenstruktur, Randgefüge, Kontaktzone 1-1, 1-4, 1-6, 1-7, 1-10
411.3 Oberflächenzusammensetzung 1-2, 1-4
411.4 Grenzflächenreaktionen

412 Kontaktbildung
412.1 Kontaktmechanik, -elastizität 1-6
412.2 Wärmeübergang 1-9
412.3 Bindungscharakter, Adhäsion 1-9, 1-10

413 Meßgrößen
413.1 Grenzflächenspannung, -energie 1-4
413.2 Rauheit 1-6, 1-7, 1-10
413.3 Kontaktfläche 1-6
413.4 Kontaktkräfte, -spannungen 1-6, 1-9, 1-11
413.5 Kontaktpotential, elektr. Kontaktwiderstand

42 Reibung und Energieverlust

421 Reibungsarten (vgl. 441)
421.1 Ruhereibung
421.2 Gleitreibung 1-1, 1-4, 1-7, 1-8, 1-9, 1-10
421.3 Wälz- u. Rollreibung, Traktion 1-1
421.4 Bohrreibung
421.5 Reibschwingungen 1-6

422 Reibungs- u. Energieverlustmechanismen

443.11 Umlaufschmierung
443.12 Verlustschmierung 1-10
443.13 Dauerschmierung
443.2 Schmiersystem-Vorrichtung
443.3 Schmierstoffverbrauch
443.4 Schmierstoffpflege
443.5 Sonstige:
443.51 Werkzeugbesprühung 1-11

444 Schmierstoffeigenschaften 1-8, 1-9, 1-10
444.1 Physikal. Eigenschaften
444.2 Mech. Eigenschaften (E.-Modul, Schermodul, Kompressibilität)
444.3 Rheol. Eigenschaften (Viskosität, Pumpfähigkeit, Konsistenz) 1-8, 1-10
444.4 Grenzflächen-Eigenschaften
444.5 Thermische Eigenschaften
444.6 Phys.-chem. Eigenschaften (Struktur, Löslichkeit, Wasserabscheidevermögen, Emulgierbarkeit, Schaumneigung u.ä.)
444.7 Chem. Eigenschaften (Flammpunkt, Oxydationsstabilität, Scherstabilität, Korrosionseigenschaften, chemische Zusammensetzung u.ä.) 1-10
444.8 Elektr. u. optische Eigenschaften
444.9 Anwendungsbezogene Eigenschaften (Geruch, Toxizität, bakteriol. Stabilität, Schmiereigenschaften, Verschleißschutzverhalten) 1-9

5 Untersuchungsverfahren

51 Charakter der Untersuchung

511 Tribologische Untersuchung
511.1 Modellprüfung 1-2, 1-5, 1-6, 1-7, 1-8, 1-9
511.2 Praxisnahe Laborprüfung 1-1, 1-2, 1-3, 1-6, 1-8, 1-10, 1-11
511.3 Feldversuch, praktischer Einsatz 1-1, 1-2, 1-3

512 Nicht-tribologische Werkstoff- u. Schmierstoffuntersuchung 1-3, 1-4

52 Art des Untersuchungsverfahrens

521 Mech.phys. Prüfung (Rheometrie, Teilchenmessg., Schwingungsmessg) 1-1, 1-4, 1-6, 1-7, 1-8, 1-9
522 Phys.-chem. Prüfung 1-1, 1-2
523 Chemische Analyse (einschl. spektrosk. Verfahren, Mikroanal.) 1-1, 1-2, 1-3, 1-4, 1-11
524 Akustische Verfahren 1-6
525 Thermische Verfahren
526 Elektr. u. magn. Verfahren
527 Optische Verfahren (visuelle Prüfung, photogr., mikroskop. u. hologr. Verfahren, Ellipsom.) 1-3, 1-4, 1-6, 1-7, 1-10, 1-11

528 Durchstrahlungsverfahren (Röntgen-, Gamma-, Neutronen-Radiographie) 1-3, 1-4
529 Radioisotopverfahren

6 Interessenhintergrund

61 Entwicklungsgeschichte

611 Historische Entwicklung
612 Stand der Forschung 1-10
613 Neuentwicklung 1-2, 1-3, 1-4, 1-6, 1-11
614 Forschungsstelle, Forscher 1-8, 1-11

62 Terminologie, Dokumentation Normung

63 Ausbildung, Lehre

64 Systemgrundlagen

641 Werkstoffe 1-1, 1-2, 1-3, 1-4, 1-5, 1-7, 1-9, 1-11
642 Schmierstoffe 1-5, 1-9, 1-10
643 Berechnung 1-1
644 Konstruktion 1-1
645 Fabrikation
646 Qualitätskontrolle
647 Auswahl 1-2
648 Vergleich 1-9
649 Toleranz

65 Einbau, Montage

66 Betrieb

661 Funktion 1-1, 1-6
662 Kinematik
663 Einlauf, allg. Betriebseigenschaften 1-7, 1-10
664 Stabilität 1-1
665 Zuverlässigkeit, Sicherheit 1-1, 1-5, 1-6

67 Instandhaltung, Überwachung 1-1

68 Organisation

69 Ergebnisanalyse

691 Schadensanalyse
692 Übertragbarkeit 1-1, 1-9, 1-10
693 Wirtschaftlichkeit, Kosten 1-1, 1-4, 1-11
694 Volkswirtschaftliche Bedeutung 1-1, 1-4, 1-6

D E S K R I P T O R E N L I S T E (alphabetisch)

GPSR Compliance

The European Union's (EU) General Product Safety Regulation (GPSR) is a set of rules that requires consumer products to be safe and our obligations to ensure this.

If you have any concerns about our products, you can contact us on ProductSafety@springernature.com

In case Publisher is established outside the EU, the EU authorized representative is:

Springer Nature Customer Service Center GmbH
Europaplatz 3
69115 Heidelberg, Germany

Batch number: 10682269

Printed by Printforce, the Netherlands